DATE DUE FOR RETURN

This book may be recalled before the

SEEDS

Physiology of Development and Germination

Second Edition

SEEDS

Physiology of Development and Germination

Second Edition

J. Derek Bewley
Department of Botany
University of Guelph
Guelph, Ontario, Canada

and

Michael Black
Division of Life Sciences
King's College
University of London
London, England

Plenum Press • New York and London

Library of Congress Cataloging-in-Publication Data

Bewley, J. Derek, 1943-
Seeds : physiology of development and germination / J. Derek Bewley and Michael Black. -- 2nd ed.
p. cm.
Includes bibliographical references and index.
ISBN 0-306-44747-9. -- ISBN 0-306-44748-7 (pbk.)
1. Seeds--Development. 2. Germination. I. Black, Michael. II. Title.
QK661.B49 1994
582'.0467--dc20

94-19278
CIP

J. Derek Bewley
Department of Botany
University of Guelph
Guelph, Ontario N1G 2W1
Canada

Michael Black
Division of Life Sciences
King's College
University of London
London W8 7AH
England

ISBN 0-306-44747-9 (Hardbound)
ISBN 0-306-44748-7 (Paperback)

A Division of Plenum Publishing Corporation
233 Spring Street, New York, N.Y. 10013

Though I do not believe
that a plant will spring up
where no seed has been,
I have great faith in a seed.
Convince me that you have a seed there,
and I am prepared to expect wonders.

Henry D. Thoreau

Preface to the Second Edition

Since the publication of the first edition of this book there have been enormous advances in certain areas of the subject, particularly those which are amenable to study using the techniques and approaches of molecular biology and mutant technology. This is especially apparent with respect to seed development, effects of abscisic acid and gibberellin on gene expression and certain aspects of dormancy. Sections of the book dealing with these topics have been extensively revised, to incorporate new information and concepts. In one case—seed development—it has been necessary to allocate two chapters to the subject, to separate reserve synthesis and its regulation from the more developmental aspects of embryogenesis and seed maturation. As in the previous edition, we have been mindful throughout of placing our discussion and considerations firmly in the context of the seed, and where new knowledge at the molecular level is included we have attempted always to emphasize its special relevance to seed biology.

The reader will discern that some sections have been altered relatively little from their state in the previous edition. These are areas in which, at least according to our perceptions, advances have been more modest, especially as regards the development of new concepts and interpretations. But some of these are now opening up to the new approaches which are already having such substantial impact in the realm of seed physiology.

We are grateful to those who have helped us in the production of this book. Many researchers have contributed material or given us permission to use their published findings. Laurie Winn of the Department of Botany, University of Guelph, has done sterling work in converting our scrawl into a splendidly presented transcript. Our thanks also to Sherry Hall, at Guelph, for helping out when the pressure became too high. The wonders of computer technology have been effectively harnessed by Wenjin Yu of the same Department in the production of many of the illustrations. We express our sincere appreciation of their skills and patience.

J.D. Bewley
M. Black

Contents

Chapter 3

Development—Regulation and Maturation

Chapter 4

Cellular Events during Germination and Seedling Growth

Chapter 5

Dormancy and the Control of Germination

Chapter 6

Some Ecophysiological Aspects of Germination

Chapter 7

Mobilization of Stored Seed Reserves

Chapter 8

Control of the Mobilization of Stored Reserves

Chapter 9

Seeds and Germination: Some Agricultural and Industrial Aspects

Chapter 1

Seeds
Germination, Structure, and Composition

1.1. INTRODUCTION

The new plant formed by sexual reproduction starts as an embryo within the developing seed, which arises from the ovule. When mature, the seed is the means by which the new individual is dispersed, though frequently the ovary wall or even extrafloral organs remain in close association to form a more complex dispersal unit as in grasses and cereals. The seed, therefore, occupies a critical position in the life history of the higher plant. The success with which the new individual is established—the time, the place, and the vigor of the young seedling—is largely determined by the physiological and biochemical features of the seed. Of key importance to this success are the seed's responses to the environment and the food reserves it contains, which are available to sustain the young plant in the early stages of growth before it becomes an independent, autotrophic organism, able to use light energy. People also depend on these activities for almost all of their utilization of plants. Cultivation of most crop species depends on seed germination, though, of course, there are exceptions when propagation is carried out vegetatively. Moreover, seeds such as those of cereals and legumes are themselves major food sources whose importance lies in the storage reserves of protein, starch, and oil laid down during development and maturation.

The biological and economic importance of seeds is obvious. In this book we will give an account of processes involved in their development, in germination and its control, and in the utilization of seed reserves during the early stages of seedling growth.

1.2. SEED GERMINATION—SOME GENERAL FEATURES

In the scientific literature the term *germination* is often used loosely and sometimes incorrectly, and so it is important to clarify its meaning. *Germination begins with water uptake by the seed (imbibition) and ends with the start of elongation by the embryonic axis, usually the radicle.* It includes numerous events, e.g., protein hydration, subcellular structural changes, respiration, macromolecular syntheses, and cell elongation, none of which is itself unique to germination. But their combined effect is to transform a dehydrated, resting embryo with a barely detectable metabolism into one that has a vigorous metabolism culminating in growth. Germination *sensu stricto* therefore does not include seedling growth, which commences when germination finishes. Hence, it is incorrect, for example, to equate germination with seedling emergence from soil since germination will have ended sometime before the seedling is visible. Seed testers often refer to germination in this sense because their interests lie in monitoring the establishment of a vigorous plant of agronomic value. Although, as physiologists, we do not encourage such a definition of the term *germination*, we acknowledge its widespread use by seed technologists. Processes occurring in the nascent seedling, such as mobilization of the major storage reserves, are also not part of germination: they are postgermination events.

A seed in which none of the germination processes is taking place is said to be *quiescent*. Quiescent seeds are resting organs, generally having a low moisture content (5–15%) with metabolic activity almost at a standstill. A remarkable property of seeds is that they are able to survive in this state, often for many years, and subsequently resume a normal, high level of metabolism. For germination to occur, quiescent seeds generally need only to be hydrated under conditions that encourage metabolism, e.g., a suitable temperature and the presence of oxygen.

Components of the germination process, however, may occur in a seed that does not achieve radicle emergence. Even when conditions are apparently favorable for germination, so that imbibition, respiration, synthesis of nucleic acids and proteins, and a host of other metabolic events all proceed, culmination in cell elongation does not occur, for reasons that are still poorly understood; such a seed expresses *dormancy*. Seeds that are dispersed from the parent plant already containing a block to the completion of germination show *primary dormancy*. Sometimes, a block(s) to germination develops in hydrated, mature seeds when they experience certain environmental conditions, and such seeds show *induced* or *secondary* dormancy. Dormant seeds are converted into germinable seeds (i.e., dormancy is broken) by certain "priming" treatments such as a light stimulus or a period at low or alternating temperature which nullify the block to germination but which themselves are not needed for the duration of the germination process. The relationships between these properties of seeds are shown in Fig. 5.1.

1.3. MEASUREMENT OF GERMINATION

The extent to which germination has progressed can be determined roughly, say by measuring water uptake or respiration, but these measurements give us only a very broad indication of what stage of the germination process has been reached. No universally useful biochemical marker of the progress of germination has been found. The only stage of germination that we can time fairly precisely is its termination! Emergence of the axis (usually the radicle) from the seed normally enables us to recognize when germination has gone to completion, though in those cases where the axis may grow before it penetrates through the surrounding tissues, the completion of germination can be determined as the time when a sustained rise in fresh weight begins.

We are generally interested in following the germination behavior of large numbers of seeds, e.g., all the seeds produced by one plant or inflorescence, or all those collected in a soil sample, or all those subjected to a certain experimental treatment. The degree to which germination has been completed in a population is usually expressed as a percentage, normally determined at time intervals over the course of the germination period. Figures 1.1 and 1.2 show some examples of germination curves, about which some general points should be made. Germination curves are usually sigmoidal—a minority of the seeds in the population germinates early, then the germination percentage increases more or less rapidly, and finally the relatively few late germinators emerge. The curves are often positively skewed because a greater percentage germinates in the first half of the germination period than in the second (see the discussion on uniformity). But although the curves have the same general shape, important differences in behavior between populations are evident. For example, curve d in Fig. 1.1 flattens off when only a low percentage of the seeds has germinated, showing that this population has a low *germination capacity*, i.e., the proportion of seeds capable of completing germination is low. Assuming that these seeds are viable, the behavior of the population could be related to dormancy or to environmental conditions, such as temperature or light, which do not favor germination of most of the seeds.

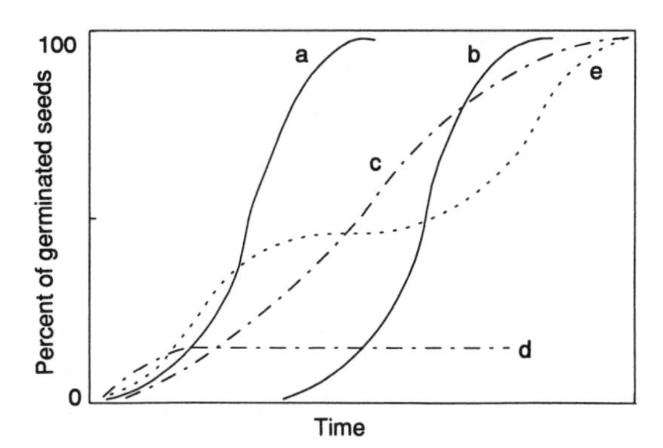

Figure 1.1. Generalized time courses of germination. See text for details.

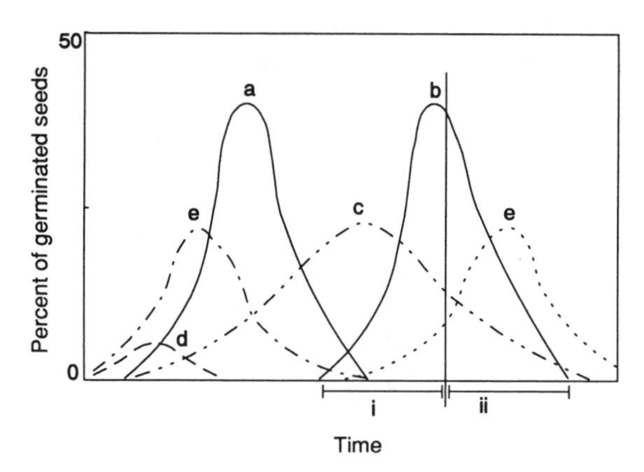

Figure 1.2. Distributions of germination with time, derived from Fig. 1.1. The vertical line on curve b indicates the midpoint of the germination period. Note that the curve is positively skewed, i.e., more seeds complete germination in the first half (i) of the period than in the second half (ii). See text for details.

The shape of the curves also depends on the *uniformity* of the population, i.e., the degree of simultaneity or synchrony of germination. It is especially clear that the seed population represented by curve e in Fig. 1.1 is not uniform. Here, a limited percentage of seeds succeeds in germinating fairly early, but the remainder begin to do so only after a delay. The population, therefore, seems to consists of two discrete groups: the quick and the slow germinators (see Fig. 1.2, curve e). This example also illustrates the point that populations with the same germination capacity (e.g., a and e in Fig. 1.1) can differ in other respects. The population depicted by curve c in Fig. 1.1 also lacks uniformity as individual seeds complete their germination in very different periods of time (see also Fig. 1.2, curve c). On the other hand, highly uniform behavior is displayed by seeds represented by curves a and b in Fig. 1.1, the steepness reflecting the fact that the majority of seeds complete germination over a relatively short time period (see also Fig. 1.2, curves a and b). As mentioned previously, it is important to note that distribution curves of the type shown in Fig. 1.2 are often positively skewed because more seeds germinate in the first than in the second half of the germination period, i.e., there is frequently not a normal temporal distribution of germination.

Although seed samples may be alike as far as both germination capacity and uniformity are concerned (e.g., a and b in Figs. 1.1 and 1.2), they may be very different in their *rate* of germination. The rate of germination can be defined as the reciprocal of the time taken for the process to be completed, starting from the time of sowing. This can be determined for an individual seed, but it is generally expressed for a population. Here, the mean time to complete germination (t) is equal to $\Sigma\,(t \cdot n)/\Sigma n$, where t is the time in days, starting from day 0, the day of sowing, and n is the number of seeds completing germination on day t. The mean germination rate (R), therefore, equals $\Sigma n/\Sigma(t \cdot n)$. A value sometimes used is the coefficient of the rate of germination (CRG), which equals $R \times 100$. Alternatively, a measure of the rate of germination can be based on the time required by an arbitrary percentage of seeds, generally 50%, to complete the germination process. At this point it is worthwhile returning briefly to the matter of uniformity. A seed population which is highly uniform is one in which individual germination rates are close to the mean rate of germination for the population as a whole. Uniformity, therefore, can be expressed as the variance of individual times around the mean time. If we assume a normal distribution of the time to complete germination (often this is not strictly the case, however—see the previous discussion), then the coefficient of uniformity of germination (CUG) is given by: $\mathrm{CUG} = \Sigma n/\Sigma[(\bar{t} - t)^2 \cdot n]$, and the higher the value, the greater is the uniformity.

It should now be clear that the populations represented by curves a and b in Fig. 1.1 have very different germination rates and that the parallelism of the curves indicates only that they have similar uniformity. Of course, populations with similar germination rates may differ markedly in uniformity or in other respects, such as germination capacity.

The behavior of a seed population with respect to germination, therefore, has several quantitative aspects that must be considered, and quantification obviously should not be limited to one parameter, say maximum germination percentage (germination capacity) or germination rate. Several attempts have been made to incorporate two parameters, rate and capacity, in a mathematical expression, so that population behavior can be described by a single statistic, but some of these are not entirely satisfactory, for several reasons. Mathematical treatments which take account of both the rates of germination and total germination, and which may be adequate for dealing with germination data in some circumstances, include use of the normal distribution function and of polynomial regression methods of curve fitting. Probit analysis can also be of great value in handling germination data. It is beyond the scope of this book to deal further with these mathematical methods, but details can be found in works listed at the end of this chapter.

1.4. SEED STRUCTURE

The seed develops from the fertilized ovule; this process is discussed in more detail in Chapter 2. At some stage during its development the angiosperm seed is usually comprised of (1) the embryo, the result of the fertilization of the egg cell in the embryo sac by one of the male pollen tube nuclei; (2) the

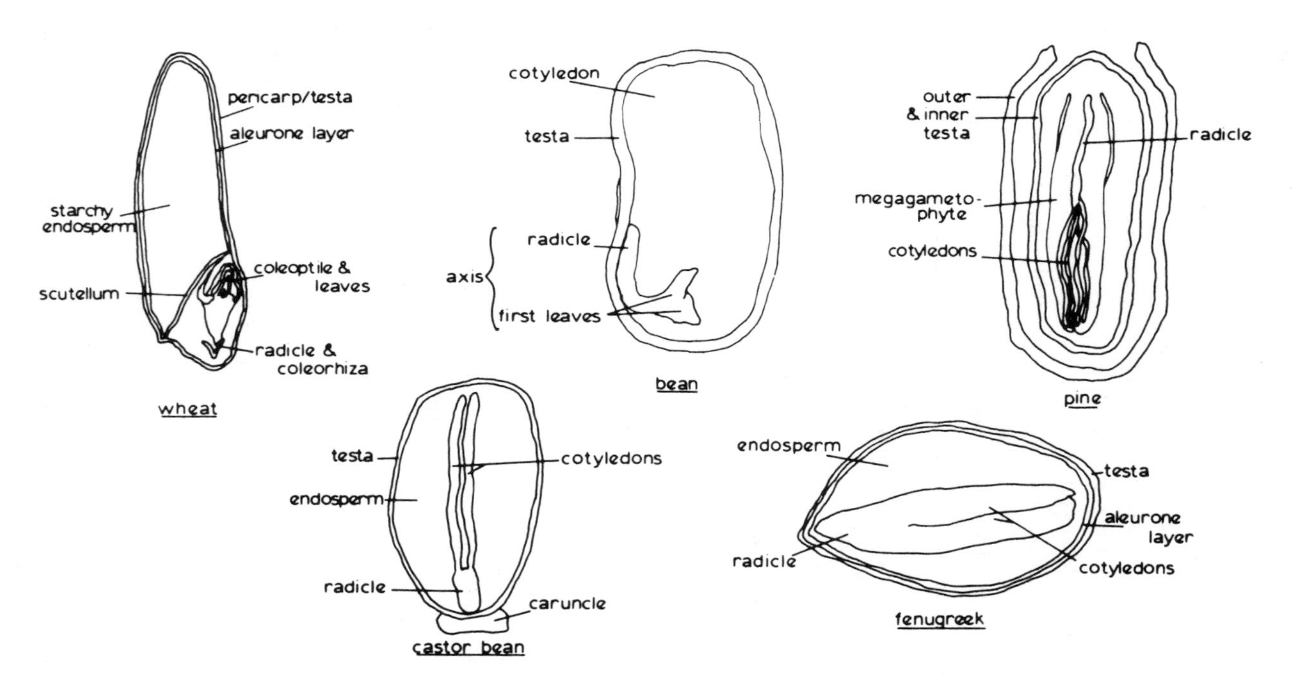

Figure 1.3. The location of tissues in monocot, dicot, and gymnosperm seeds. Not drawn to scale.

endosperm, which arises from the fusion of two polar nuclei in the embryo sac with the other pollen tube nucleus; (3) the perisperm, a development of the nucellus; and (4) the testa or seed coat, formed from one or both of the integuments around the ovule. Although all mature seeds contain an embryo (sometimes poorly developed), and many are surrounded by a distinguishable seed coat, the extent to which the endosperm or perisperm persists varies between species. Sometimes the testa exists in a rudimentary form only, the prominent outermost structure being the pericarp or fruit coat derived from the ovary wall; in these cases, the dispersal unit is not a seed, but a fruit (Table 1.1). In gymnosperm seeds there is no fusion of the male and polar nuclei leading to the formation of a triploid endosperm; in these the storage tissue in the mature seed (which is functionally similar to the true endosperm) is haploid and is the modified megagametophyte (Figs. 1.3 and 2.1).

More rarely, seeds are produced by nonsexual processes, such as by apomixis (i.e., from diploid cells in the ovule); such seeds are often indistinguishable from those of the same species resulting from sexual reproduction. The dandelion, and some other composites, have come to rely exclusively, or almost exclusively, on apomictic reproduction.

Let us consider briefly each of the seed components.

1.4.1. Embryo

The embryo is comprised of the embryonic axis and one or more cotyledons. The axis incorporates the embryonic root (radicle), the hypocotyl to which the cotyledons are attached, and the shoot apex with the first true leaves (plumule). These parts are usually easy to discern in the dicot embryo (Figs. 1.3 and 2.2), but in monocots—particularly the Gramineae—identifying these parts is considerably more difficult. Here, the single cotyledon is much reduced and modified to form the scutellum (Figs. 1.3 and 7.1), the basal sheath of the cotyledon is elongated to form a coleoptile covering the first leaves, and in some

Table 1.1. Is the Dispersal Unit a Seed or a Fruit? Some Examples

Seed	Fruit (and type)
Legumes (e.g., peas, beans)	Cereals (caryopsis)
Cotton	Lettuce, sunflower, and other Compositae (cypsela)
Rapeseed	Ash and elm (samara)
Castor bean	Hazel and oak (nut)
Tomato	Buttercup, anemone, and avens (a collection of achenes)
Squashes (e.g., cucumber, marrow)	
Coffee bean	

species (e.g., maize) the hypocotyl is modified to form a mesocotyl. The coleorhiza is regarded as the base of the hypocotyl sheathing the radicle.

The shapes of embryos and their sizes in relation to other structures within the seed are variable. In monocot and dicot species with a well-developed endosperm in the mature seed, the embryo occupies less of the seed than in nonendospermic seeds (Fig. 1.3). Cotyledons of endospermic seeds are often thin and flattened since they do not store much in the way of reserves (e.g., castor bean); in nonendospermic seeds, such as many of the legumes, the cotyledons are the site of reserve storage and account for almost all of the seed mass (Fig. 1.3). Cotyledons of nonendospermic, epigeal (Fig. 4.32) species (such as some members of the squash family) which are borne above the ground after germination and become photosynthetic are relatively not as large, nor do they contain as much stored reserves as the subterranean, hypogeal type. The cotyledons are absent from seeds of many parasitic species; in contrast, the embryos of many coniferous species contain several cotyledons (polycotyledonous) (Figs. 1.3 and 2.1).

Polyembryony, i.e., more than one embryo in a seed, occurs in some species, e.g., *Poa alpina*, *Citrus*, and *Opuntia* spp. This can arise because of cleavage of the fertilized egg cell to form several zygote initials, development of one or more synergids (accessory cells in the embryo sac), the existence of several embryo sacs per nucellus, and the various forms of apomixis and adventitious embryony (from diploid cells of the nucellus). In *Linum usitatissimum* and other species some of the embryos formed by polyembryony are haploid.

Not all seeds contain mature embryos when liberated from the parent plant. Orchid seeds (Fig. 1.4) contain minute and poorly formed embryos and no endosperm. The final developmental stages of the embryo in other species occur after the seed has been dispersed, e.g., *Fraxinus* (ash) species and *Heracleum sphondylium*.

1.4.2. Nonembryonic Storage Tissues

In most species the perisperm, derived entirely from the maternal nucellar tissue of the ovule, fails to develop and is quickly absorbed as the embryo becomes established. In a few species, of which coffee and *Yucca* are the most noted examples, the perisperm is the major store of the seed's food reserves. In these seeds the endosperm is absent, although in others it may be developed to a greater (*Acorus* spp.) or lesser (*Piper* spp.) extent than the perisperm. In beet seeds both the perisperm and cotyledons of the embryo contain substantial reserves—there is no endosperm.

Seeds can be categorized as endospermic or nonendospermic in relation to the presence or absence in the mature seed of a well-formed endosperm. Even

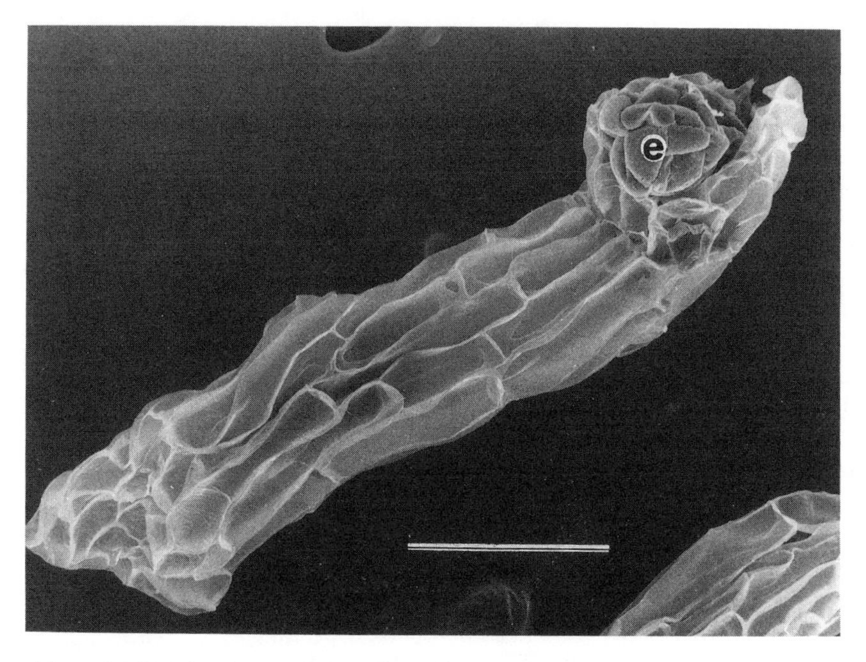

Figure 1.4. Scanning electron micrograph of a seed of the orchid *Spiranthes magnicamporum*. Note the diminutive and undifferentiated embryo (*e*). Bar = 100 μm. Courtesy of D. Herr.

though an endosperm is present, some seeds are not generally regarded as being endospermic because this tissue is broken down and is only a remnant of that formed during development (e.g., soybean and peanut), or it may be only one to a few cell layers thick (e.g., lettuce). In these cases, other structures, usually the cotyledons, are the principal storage organs. Some endosperms are relatively massive and are the major source of stored reserves within the mature seed, e.g., in cereals, castor bean, date palm, and the endospermic legumes such as fenugreek, carob, and honey locust. In the cereals and some endospermic legumes (e.g., fenugreek) the majority of cells in the endosperm are nonliving at maturity, the cytoplasmic contents having been occluded by the stored reserves during development. But on the outside of the endosperm remains a living tissue, the aleurone layer, which does not store many reserves, but rather may be responsible ultimately for the release of enzymes for their mobilization (Figs. 1.3 and 7.6). Endosperms with a high water retention capacity may have a dual role: to provide reserves for the germinated embryo, and to regulate the water balance of the embryo during germination (e.g., in fenugreek seeds). An unusual endosperm is that of the coconut, in that part of it remains acellular and liquid.

1.4.3. Seed Coat (Testa)

Variability in the anatomy of the testa is considerable, and it has been used taxonomically to distinguish between different genera and species. Hence, a discussion of the range of seed coat structures is well beyond the scope of all but specialist monographs. The testa is of considerable importance to the seed because it is often the only protective barrier between the embryo and the external environment (in some species the fruit coat, and even the endosperm, support or provide a substitute for this role). The protective nature of the seed coat can be ascribed to the presence of an outer and inner cuticle, often impregnated with waxes and fats, and one or more layers of thick-walled, protective cells (Fig. 5.5). Layers of crystal-containing cells (calcium oxalate or carbonate, silica) occur in the seed coats of many species; these may play a protective role also, discouraging insect predation, for example. Coats may contain mucilaginous cells which burst upon contact with water, providing a water-retaining barrier around the seeds. Such barriers may also restrict oxygen uptake, as will the presence of phenolics, and there are other structural features in some coats that restrict exchange of gases between the embryo and environment. Some coats, e.g., in many legumes, are largely impermeable to water and consequently can restrict the metabolism and growth of inner tissues.

The coloring and texture of seed coats are distinguishing features of many seeds but sometimes cannot be used taxonomically because they may change as a result of environmental and genetic influences during development (such polymorphism in seeds is discussed in Section 5.2). Upon detachment from the parent plant, the seed coat bears a scar, called the hilum, marking the point at which it was joined to the funiculus. At one end of the hilum of many seed coats can be seen a small hole, the micropyle. Rarely, hairs or wings develop on the testa to aid in seed dispersal (e.g., in willow, lily, *Epilobium* spp.); more usually the dispersal structures are a modification of the enclosing fruit coat. Outgrowths of the hilum region may be seen: the strophiole, which restricts water movement into and out of some seeds, and the aril, which often contains chemicals; in some species these attract animals important in dispersal of the seeds. In castor bean the aril is associated with the micropyle and is called the caruncle. Arils are variable in shape, forming knobs, bands, ridges, or cupules, and are often brightly colored. The aril of the nutmeg is used as a source of the spice mace; the seed coat contains different chemicals and is ground up for the spice named after the seed itself.

1.5. SEED STORAGE RESERVES

About 70% of all food for human consumption comes directly from seeds (mostly those of cereals and legumes), and a large proportion of the remainder is derived from animals that are fed on seeds. It is not surprising, therefore, that there is a wealth of literature concerned with the chemical, structural, and nutritional composition of seeds. Unfortunately, we will be able to do it only the briefest of justice. Most of our knowledge of the chemical composition of seeds is for cultivated species since they make up such a large share of our food source and also provide a great many raw materials for industry. Information on seeds of wild species and wild progenitors of our cultivated crops is relatively scarce. But with increasing interest in new food sources and in improved genetic diversity within domesticated lines, the seeds of wild plants are now receiving more attention.

In addition to the normal chemical constituents found in all plant tissues, seeds contain extra amounts of substances stored as a source of food reserves to support early seedling growth. These are principally carbohydrates, fats and oils, and proteins. Seeds contain other minor, but nevertheless important reserves (e.g., phytin); of these, several are recognized as being nutritionally undesirable or even toxic (e.g., alkaloids, lectins, proteinase inhibitors, phytin, and raffinose oligosaccharides).

The chemical composition of seeds is determined ultimately by genetic factors and hence varies widely among species and their varieties and cultivars. Some modifications of composition may result from agronomic practices (e.g., nitrogen fertilizer application, planting dates) or may be imposed by environmental conditions prevalent during seed development and maturation; but such changes are usually relatively minor. Through crossing and selection, plant breeders have been able to manipulate the composition of many seed crops to improve their usefulness and yield. Modern cultivars of many cereals and legumes store significantly higher quantities of food material than either earlier cultivars or their wild progenitors. Even so, some nutritional deficiencies remain to be rectified; e.g., the composition of the storage proteins of legumes and cereals is such that they do not provide all of the amino acids required by simple-stomached (monogastric) animals such as humans, pigs, and poultry (see Sections 1.5.3 and 2.3.4). Some indication of the variability in the composition of food reserves in seeds is to be found in Table 1.2: the important storage tissue within each seed is noted.

As indicated in Table 1.2, the major food reserves may be deposited within the embryo—usually the cotyledons, although in the Brazil nut they occur within the radicle/hypocotyl—or within extra-embryonic tissues such as the endosperm, the megagametophyte (in gymnosperms), or, rarely, the perisperm (e.g., coffee, *Yucca*). In many seeds the stored reserves may occur within both embryonic and extraembryonic tissues, but in different proportions, e.g., in maize (Table 1.3). Different reserves may even be located within different storage tissues; in fenugreek seeds, for example, the endosperm is the exclusive source of carbohydrate (as cell-wall galactomannan), but the cotyledons contain the protein and fats. The reserves may be distributed unevenly within any one

Table 1.2. The Food Reserves of Some Important Crop Species[a]

	Average percent composition			
	Protein	Oil	Carbohydrate[b]	Major storage organ
Cereals				
Barley	12	3[c]	76	Endosperm
Dent corn (maize)	10	5	80	Endosperm
Oats	13	8	66	Endosperm
Rye	12	2	76	Endosperm
Wheat	12	2	75	Endosperm
Legumes				
Broad bean	23	1	56	Cotyledons
Garden pea	25	6	52	Cotyledons
Peanut	31	48	12	Cotyledons
Soybean	37	17	26	Cotyledons
Other				
Castor bean	18	64	Negligible	Endosperm
Oil palm	9	49	28	Endosperm
Pine	35	48	6	Megagametophyte
Rape	21	48	19	Cotyledons

[a]After Crocker and Barton (1957) and Winton and Winton (1932).
[b]Mainly starch.
[c]In cereals, oils are stored within the scutellum, an embryonic tissue.

storage tissue; for example, in the maize grain, there are protein-rich regions (horny endosperm) and starch-rich regions (floury endosperm) within the endosperm. Chemical differences can also exist within a species in relation to the distribution of any particular reserve: in rapeseed, the oil in the cotyledons and hypocotyl contains different proportions of erucic and palmitic acids.

We shall now discuss each of the major reserves in more detail and briefly elaborate on the chemistry of some of the minor reserves. Chapter 2 deals with the deposition of stored reserves, and therein, of necessity, will be found some further discussion of their chemical composition and localization.

1.5.1. Carbohydrates

Carbohydrates are the major storage reserve of most seeds cultivated as a food source (Table 1.2). Starch is the carbohydrate most commonly found in seeds, although "hemicelluloses," "amyloids," and the raffinose series oligosaccharides may be present and may sometimes be the major carbohydrate reserve. Other carbohydrates that occur in non-storage forms are cellulose, pectins, and mucilages.

Starch is stored in seeds in two related forms, amylose and amylopectin; both are polymers of glucose. Amylose is a straight-chain polymer some 300–

Table 1.3. Percent Composition of Stored Reserves in Different Parts of a Maize (cv. Iowa 939) Kernel[a]

Reserve	Whole grain	Endosperm (starchy and aleurone layer)	Embryo (including scutellum)
Starch	74	88	9
Oil	4	< 1	31
Protein	8	7	19

[a]After Earle *et al.* (1946).

400 glucose units in length; adjoining glucose molecules are connected by $\alpha(1\rightarrow4)$ glucosidic linkages (Fig. 1.5A). Amylopectin is much larger (10^2–10^3 times); it consists of many amylose chains linked via $\alpha(1\rightarrow6)$ bonds to produce a multiple-branched molecule (Fig. 1.5B). Starch is laid down in discrete subcellular bodies called starch grains. Most grains are composed of about 50–75% amylopectin and 20–25% amylose. Certain mutants of cereals have a higher- or lower-than-average content of amylose (Table 2.1), and starch grains of waxy mutants of maize are devoid of this polymer. In wrinkled peas, amylose accounts for two-thirds or more of the starch compared with about one-third in smooth peas and other legumes.

Starch grains may have an appearance characteristic for the individual species; examples are predominantly spherical grains for barley, angular for maize, or elliptical for runner bean. Grain shape is to a large extent determined by amylose content: the higher the amount of amylose, the rounder the grain. Grain size is very variable, from 2 to 100 μm in diameter, even in the same seed. In the rye endosperm, for example, there are large oval starch grains which are up to 40 μm in diameter; there are also numerous smaller grains less than 10 μm in diameter embedded in a fine network of cytoplasmic protein (Fig. 1.6A). In barley, the spherical starch grains are separable into two groups, large ones and small ones. Although the latter account for about 90% of the total number of grains, they comprise only 10% of the total starch by weight.

Hemicelluloses are the major form of stored carbohydrates in some seeds, particularly in certain endospermic legumes. Usually starch is absent from tissues where hemicelluloses are present in appreciable quantities. The endosperm or perisperm of the ivory nut, date, and coffee is extremely hard because of hemicelluloses laid down as very thick cell walls; there are no hemicellulose storage bodies. Many of the hemicelluloses are mannans, i.e., long-chain polymers of mannose [linked by $\beta(1\rightarrow4)$ bonds] with variable but small quantities of sugar present as side chains [e.g., galactose (linked by $\alpha(1\rightarrow6)$ bonds) in galactomannans: Fig. 1.5C]. The number of galactose side chains affects the consistency of the galactomannan; it varies from being very hard in almost pure mannans (coffee mannans contain 2% galactose) to being mucilaginous at much

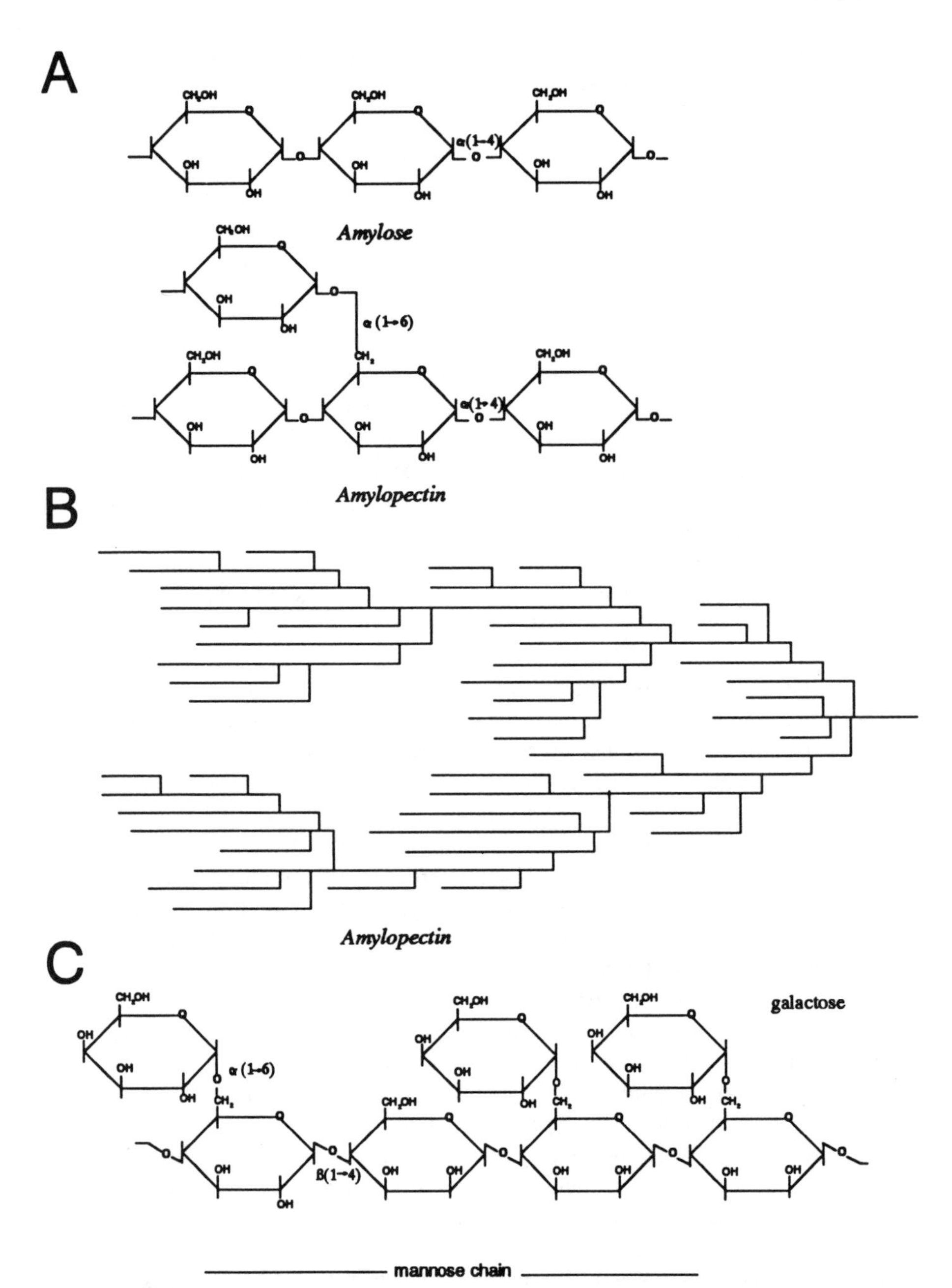

Figure 1.5. (A) The chemical composition of starch, and (B) a proposed structure of amylopectin. (C) The chemical structure of galactomannan. The chemical links between the molecules are noted.

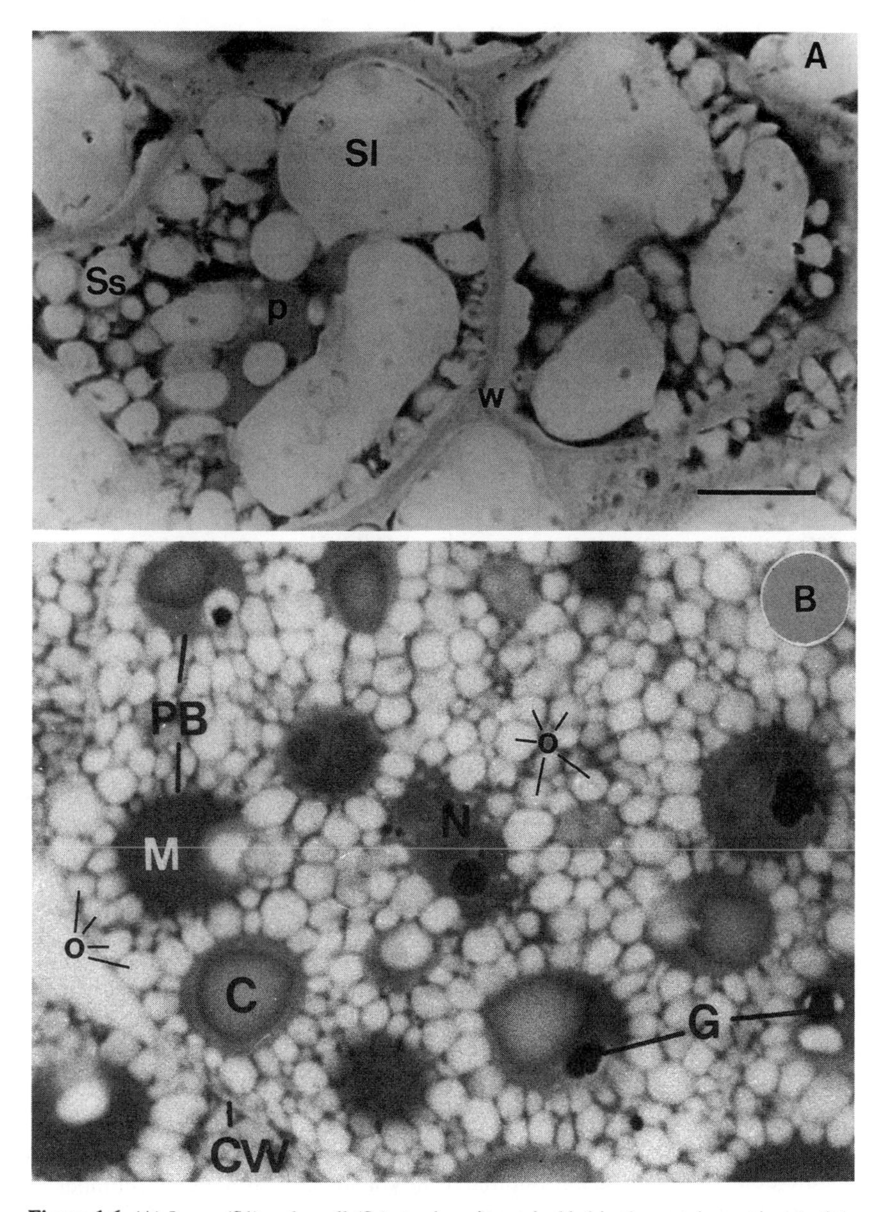

Figure 1.6. (A) Large (S1) and small (Ss) starch grains embedded in the protein matrix (p) of the starchy endosperm cells of rye. w, Cell wall. Bar = 10 μm. Courtesy of M. L. Parker (1981). (B) Endosperm of castor bean showing extensive packing with oil bodies (o). PB, protein body; C, crystalloid (protein); G, globoid (phytin); M, matrix (protein); N, nucleus; CW, cell wall. Courtesy of J. S. Greenwood.

higher galactose contents. Within the Leguminosae the mannose/galactose ratio may have a taxonomic significance as well. Glucomannans, in which a large proportion of the mannose residues is replaced by glucose, are restricted to the endosperm of certain monocots, specifically members of the Liliaceae and Iridaceae. They, too, are components of thickened cell walls. Other hemicelluloses include the xyloglucans, which, strictly, should be called galactoxyloglucans since they are substituted cellulose-type molecules, i.e., a linear $\beta(1\rightarrow4)$-linked glucose backbone with short xylose (some with a galactose attached) side chains. They are found in cell walls of certain dicot embryos (cotyledons) and endosperms, such as in members of the Caesalpinoideae, a subfamily of the Leguminosae, e.g., tamarind (*Tamarindus indica*). Xyloglucans are also called "amyloids" because they react positively to a stain for starch.

Free sugars are rarely the main storage carbohydrate, but in sugar maple, which lacks starch, they may account for nearly 11% of the dry weight of the mature seed. Disaccharides (sucrose) and oligosaccharides (raffinose-series oligosaccharides, Section 4.3.2) are commonly found as minor reserves in the embryo and reserve tissues. There is increasing evidence that they are an important source of sugars for respiration during germination and early seedling growth.

1.5.2. Fats and Oils (Neutral Lipids)

Chemically, these are triacylglycerols, most of which are oils, i.e., they are liquid above about 20°C; some seeds may also contain appreciable quantities of phospholipids, glycolipids, and sterols. Triacylglycerols (formerly called triglycerides) are insoluble in water but soluble in a variety of organic solvents including ether, chloroform, and benzene. They are esters of glycerol and fatty acids

$$\begin{array}{r} CH_2O \cdot OC \cdot R^1 \\ | \quad\quad\quad\quad \\ R^2 \cdot CO \cdot OCH \quad\quad\quad\quad \\ | \quad\quad\quad\quad \\ CH_2O \cdot OC \cdot R^3 \end{array}$$

The number of carbon atoms in the fatty acid chains, denoted as R^1, R^2, and R^3, may be identical but usually it is not.

Fatty acids are identified according to the number of carbon atoms and double bonds in their chain. Saturated fatty acids contain an even number of carbon atoms and no double bonds; e.g., palmitic acid (16:0—16 carbon atoms:no double bonds) is the most common saturated fatty acid in seed oils. But the predominant fatty acids in seeds are the unsaturated ones, and of these oleic (18:1Δ9—double bond in position 9 of the fatty acid chain) and linoleic (18:2Δ9,12) account for more than 60% by weight of all oils in oil seed crops.

Less common fatty acids are erucic acid (22:1Δ13), a component of some rapeseed oils (although it has now almost been bred out of most commercial cultivars, e.g., canola, because it is toxic) and of the oil of *Crambe abyssinica*, ricinoleic acid (12-hydroxy 18:1Δ9), the major component of castor (bean) oil, and petroselenic acid (18:1Δ6) which accounts for 85% of the total fatty acids of seeds of the Umbelliferae. The storage lipids of jojoba are unusual in that they are wax esters of long-chain fatty acids and alcohols, and they are liquid. They are an important source for the manufacture of high-pressure lubricants.

The fatty acid composition of the oils and fats of some important crop species is shown in Table 1.4. Both maize and sunflower oil are widely used as cooking oils and in margarine. Among their desirable properties for this purpose is their high linoleic acid content. Catalytic hydrogenation of seed oils changes their melting point and lowers their linolenic acid content. This triunsaturated fatty acid oxidizes readily during food storage and produces off-flavors. The degree of unsaturation left after hydrogenation determines whether a fat or oil is a solid at room temperature (required for margarine) or a liquid (cooking oils). Variable hydrogenation of peanut oil yields peanut butter of different consistencies. The advantage of using plant oils over animal fats in the diet is that because of their higher unsaturated fatty acid content (polyunsaturated) after hydrogenation they are purported to reduce atherosclerosis, a claim not wholly substantiated in medical tests. Fats such as lard (Table 1.4) from animal sources are high in oleic acid, and hydrogenation causes rapid saturation of the one double bond per molecule. Drying oils, such as those used in paints and lacquers (e.g., linseed oil from flax and rapeseed oil), contain a high linolenic acid content after processing. Upon exposure to air a free-radical polymerization reaction occurs, initiated by oxygen, which cross-links the oils to form a tough film.

The triacylglycerol reserves in seeds are laid down in discrete subcellular organelles–oil bodies (fat bodies or wax bodies, depending on the consistency of the stored product). They range in size from 0.2 to 6 μm in diameter, according to species. In high-triacylglycerol seeds the oil bodies occupy a substantial volume of the cell, as in the castor bean endosperm (Fig. 1.6B). The ontogeny of these bodies is discussed toward the end of Section 2.3.3.

1.5.3. Proteins

According to Osborne's classification, seed proteins can be divided into four classes on the basis of their solubility: (1) *albumins*—soluble in water and dilute buffers at neutral pHs; (2) *globulins*—soluble in salt solutions but insoluble in water; (3) *glutelins*—soluble in dilute acid or alkali solutions; (4) *prolamins*—soluble in aqueous alcohols (70–90%). This classification was proposed at about the turn of this century, and although it is far from ideal, a better

Table 1.4. The Major Fatty Acid Composition of Commercial Oils from Various Plant Sources[a]

Species	Palmitic (16:0)	Stearic (18:0)	Oleic (18:1)	Linoleic (18:2)	Linolenic (18:3)
Sunflower	6	4	26	64	0
Maize	12	2	24	61	< 1
Soybean	11	3	22	54	8
Rapeseed (canola)	5	2	55	25	12
Cotton	27	3	17	52	0
Peanut	12	2	50	31	0
Oil palm	49	4	36	10	0
Linseed (flax)	—	—	—	77	17
Animal fat (lard)	29	13	43	10	0.5

[a]After Weber (1980) and Miller (1931).

one has still to be devised. Solubilization of some proteins requires harsh extraction procedures [e.g., boiling in buffer containing the detergent sodium dodecyl sulfate (SDS)].

Most investigations on these various classes of proteins have been conducted on edible seeds of crop species. Hence, we will concentrate here on the seeds of cereals and legumes whose protein composition is important in both human and animal nutrition. The percentage of protein in cereals is debatably sufficient to satisfy the requirements of human beings, when adequate calories are obtained, but it is insufficient to support the growth of farm animals. Moreover, the amino acid composition of both cereal and legume proteins is not ideally suited for the diet of human beings or animals.

The approximate proportions of the main protein classes in cereals are shown in Table 1.5, along with the commonly used names of some proteins. The major storage protein in the endosperms of maize, barley, and sorghum is of the prolamin type, which is found only in monocots; in wheat it is glutelin, whereas in oats globulins predominate. Globulins are the storage proteins of cereal embryos, but as such they represent only a very small proportion of the total reserve protein of the whole grain. The amino acid composition of the total protein fraction present in cereal grains (Table 1.6) is strongly influenced, not surprisingly, by the nature of the major storage protein. Albumins and globulins are not seriously deficient in specific amino acids, and hence, from a nutritional point of view, oats are a good source of dietary protein, particularly for breeding stock. In normal barley and maize grains, however, where prolamins are high, lysine is seriously limiting; tryptophan is low and threonine levels are nutritionally inadequate. Prolamins are high in proline and glutamic acid/glutamine. Attempts have been made to modify the amino acid composition of the prolamin-rich cereals such as maize and barley, and in particular to increase their lysine

Table 1.5. The Approximate Percent Protein Composition of Some Cereals and the Common Names of Some Storage Proteins[a]

Cereal	Albumin	Globulin	Prolamin	Glutelin
Wheat	9	5	40 (gliadin)	46 (glutenin)
Maize	4	2	55 (zein)	39
Barley	13	12	52 (hordein)	23 (hordenin)
Oats	11	56	9 (avenin)	23
Rice	5	10	5 (oryzin)	80 (oryzenin)
Sorghum	6	10	46[b]	38

[a]Based on Payne and Rhodes (1982).
[b]This has been called kafirin, but because of the unsavory connotations of part of this word, a more suitable name is required (perhaps sorghin).

content (Section 2.3.4). High-lysine mutants of maize (*opaque-2*) and barley (*hiproly*) exist; these also contain less prolamin storage protein and more lysine-rich glutelin (Table 2.6), resulting in an altered overall amino acid composition (Table 1.6).

Characteristically, storage (holo-) proteins are oligomeric; that is, they are made up of two or more subunits that can be separated, after extraction, using mildly dissociating conditions. These subunits in turn may be made up of a number of polypeptide chains, which may vary slightly in amino acid composition between "homologous" subunits. This variation results in considerable heterogeneity of the native protein. For example, the prolamin fraction of maize (zein) consists of major subunits of 23 and 21 kDa molecular mass and a minor subunit of 13.5 kDa; separation of these subunits on the basis of their molecular charge (i.e., by isoelectric focusing) shows that together they are comprised of nearly 30 polypeptides. Similar complexity is shown by the wheat prolamin gliadin, which is separable into four major holoproteins (α, β, γ, and ω) made up of at least 46 discrete polypeptides. The wheat glutelins are even more complex, being polymeric molecules made up of 15 component proteins ranging from 11 to 133×10^2 kDa. Hence, most storage proteins should not be thought of as a single protein, but rather as a complex of individual proteins bound together by a combination of intermolecular disulfide groups, hydrogen bonding, ionic bonding, and hydrophobic bonding.

Some understanding has been reached about the evolutionary relationships among the prolamins and glutelins of cereals. In wheat and barley there are three types of prolamin: sulfur-rich, sulfur-poor and high-molecular-weight (HMW) (previously classed as a HMW glutenin). These possess similar amino acid domains (i.e., regions of the amino acid sequence). For example, both the sulfur-rich and the sulfur-poor contain similar proline-rich domains, so these two proteins probably share a common evolutionary origin. Three other conserved regions, A, B, and C, containing most of the cysteine, also are found but to

Table 1.6. Percentages of Amino Acids in Grain Protein of Oats, Normal and *Opaque-2* Maize, and Normal and *Hiproly* Barley[a]

	Maize		Barley		
Amino acid	Normal	*Opaque-2*	Normal	*Hiproly*	Oats
Alanine	10.0	7.2	3.8	4.3	5.0
Arginine	3.4	5.2	4.6	4.9	6.9
Aspartic acid	7.0	10.8	6.0	6.8	8.9
Cystine[b]	1.8	1.8	1.1	0.9	1.6
Glutamic acid	26.0	19.8	26.8	23.9	23.9
Glycine	3.0	4.7	3.6	3.9	4.9
Histidine	2.9	3.2	2.1	2.2	2.2
Isoleucine[b]	4.5	3.9	3.7	3.9	3.9
Leucine[b]	18.8	11.6	6.7	7.1	7.4
Lysine[b]	1.6	3.7	3.4	4.2	4.2
Methionine[b]	2.0	1.8	1.2	1.5	2.5
Phenylalanine[b]	6.5	4.9	5.9	5.9	5.3
Proline	8.6	8.6	12.6	11.3	4.7
Serine	5.6	4.8	4.3	4.4	4.2
Threonine[b]	3.5	3.7	3.4	3.6	3.3
Tryptophan[b]	0.3	0.7	—	—	—
Tyrosine[b]	5.3	3.9	2.8	2.8	3.1
Valine	5.4	5.3	4.8	5.3	5.3

[a]Based on Frey (1977).
[b]Essential amino acids which must be provided in the animal diet, since the animal itself cannot synthesize them.

different extents in the three protein types (Fig. 1.7A). These regions are flanked and interrupted by other, variable sequences which, because of their similarities in amino acid composition, are thought to have a common origin. The sulfur-rich and the HMW prolamins all have A, B and C, but in the sulfur-poor prolamins there occurs only a vestige of C. All three proteins may have originated from an ancestral protein subunit of about 90 amino acid residues containing A, B, and C (Fig. 1.7B). During evolution, the major changes in the sulfur-rich and HMW prolamins have been the insertion of repeated sequences among A, B, and C, and flanking regions. The repeated sequences have been amplified in the sulfur-poor prolamins, A and B have been lost, and only a part of C remains (Fig. 1.7A).

Similarities have been revealed between the above prolamins in wheat and barley and other proteins. β-Zein also has the A, B, and C regions but none of the repeat sequences: instead, a methionine-rich region is inserted between B and C (Fig. 1.7A). The proteinase/α-amylase inhibitor in barley consists almost exclusively of A, B, and C. And of the two subunits of the 2 S protein in castor bean (see later) one has the A sequence, and the other, the B and C regions. Finally, the 2 S albumin in sunflower possesses the B and C regions, though the cDNAs of this protein also have the base sequence coding for A; the reason for this discrepancy is unknown.

The number of genes encoding the different proteins varies considerably among the different prolamins. For example, there is a large family—more than 100—of wheat gliadin genes, about the same number for maize α-zein, while the other zeins (β, γ, δ) each have only one or two genes.

Legume seeds are the second most important protein sources, on a world basis, after cereals. Nutritionally, they are generally deficient in the sulfur-containing amino acids (cysteine and methionine), but unlike cereal grains, their lysine content is adequate (Table 1.7). The major storage proteins are globulins, which account for up to 70% of the total seed nitrogen. The globulins consist of two major families of proteins which differ in molecular mass and which sediment during ultracentrifugation with sedimentation coefficients (S values) of approximately 7 (average 7–8) and 11 (average 11–13). These are the 7 S (called the vicilin group) and 11 S (the legumin group) proteins; both are holoproteins composed of regularly assembled subunits. The holoproteins form a series of related polymers of the vicilin or legumin type, although each type may differ quantitatively and qualitatively in subunit composition.

The 11 S storage globulins occur as hexameric complexes of 320–400 kDa composed of six nonidentical subunits (52–65 kDa). Each subunit contains an acidic polypeptide (pI~6.5) of 33–42 kDa and a basic polypeptide (pI~9) of 19–23 kDa. The acidic and basic polypeptides are linked by a single disulfide bond (see Figs. 1.8 and 1.9 and accompanying discussion also). Compared with the 11 S proteins, the 7 S storage globulins have more varied structures. They are present in seeds as trimeric complexes of 145–190 kDa, composed of three nonidentical polypeptides of 48–83 kDa. In some legumes, e.g., pea, the polypeptides undergo a series of proteolytic cleavages, to range in size from 12 to 75 kDa (Table 1.8). In other species, including soybean and *Phaseolus vulgaris*, no proteolytic processing occurs. Most, if not all, 7 S proteins completely lack cysteine and thus the polypeptides are not linked by disulfide bonds.

Some examples of the complexities of the subunit polypeptide composition of legume storage proteins are given in Table 1.8. To highlight the storage proteins for just one species, soybean, which has been much studied, the major globulin that is present in the mature seed is the legumin glycinin, which is probably comprised of four acidic and four basic subunits making up the 12.3 S holoprotein of 330 kDa. The 7 S storage fraction, β- and γ-conglycinin, also contains a heterogenous group of proteins. β-Conglycinin, which is the predominant vicilin form, has an average mass of 160 kDa and is made up of three subunits: α and α′ of 57 kDa and β of 42 kDa. These may combine into six different isomeric forms made up of different subunit compositions (i.e., α′β, αβ, αα′β, αβ, αα, and α), each varying slightly in amino acid and carbohydrate composition (the carbohydrate is present usually as glucose, mannose, or glu-

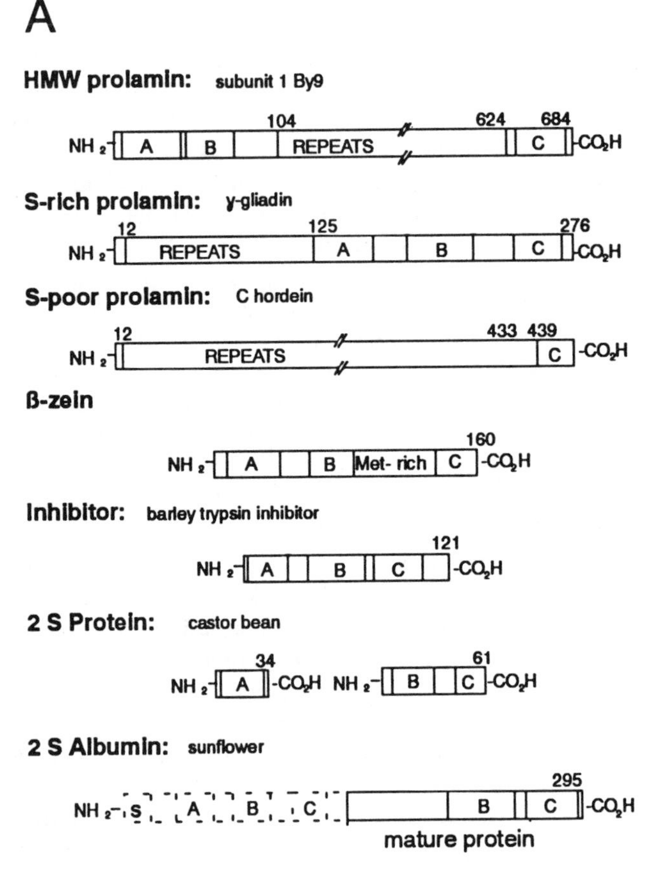

Figure 1.7. Relationships among cereal prolamins and other proteins. (A) Amino acid domains of several proteins. The different regions are explained in the text. Numbers refer to the amino acid position in the polypeptide chain. (B) Scheme showing the possible evolutionary changes occurring in the prolamins and related proteins. After Shewry and Tatham (1990).

cosamine, from 0.5 to 1.5%, because many subunits of 7 S storage proteins are glycosylated during their deposition; see Section 2.3.4). Soybean storage protein also contains a 2 S form, α-conglycinin, which is heterogeneous and includes several enzymes as well as trypsin inhibitor activity (Section 7.6.4). Storage proteins of other legumes show equal complexity.

To demonstrate how the polypeptide composition of the 7 S and 11 S storage proteins is discernible using polyacrylamide gel electrophoresis (PAGE),

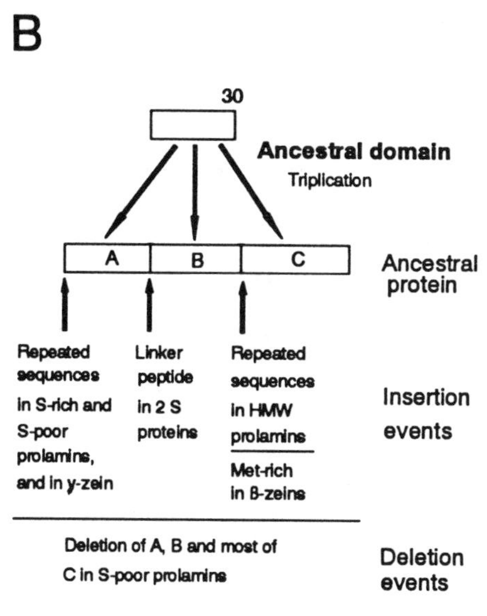

Figure 1.7. (*Continued*)

Table 1.7. Amino Acid Composition of the 11 and 7 S Globulin Storage Proteins of Soybean Seeds (Expressed as mole %)[a]

Amino acid	11 S	7 S
Alanine	6.2	3.7
Arginine	5.6	8.8
Aspartic acid	11.7	14.1
Cystine	0.6	0.3
Glutamic acid	21.4	20.5
Glycine	7.5	2.9
Histidine	1.7	1.7
Isoleucine	4.1	6.4
Lysine	3.9	7.0
Methionine	1.3	0.3
Phenylalanine	4.6	7.4
Proline	6.5	4.3
Serine	6.0	6.8
Threonine	3.8	2.8
Tryptophan	0.8	0.3
Tyrosine	2.7	3.6
Valine	5.2	5.1

[a]After Derbyshire *et al.* (1976).

Table 1.8. Subunit Composition of the Globulin Storage Proteins of Some Legumes[a]

Species	Approximate sedimentation coefficients (S)	Average mass (kDa)	Name of holoprotein	Subunits (kDa)
Pisum sativum	7–8	186	Vicilin	12, 14, 18, 24, 30, 50, 75
(garden pea)	12–13	360	Legumin	18, 20, 25, 27, 37, 40
Vicia faba	7	150	Vicilin	31, 33, 46, 56
(broad bean)	11–14	328	Legumin	20, 37
Phaseolus vulgaris	6.5–7.5	150	Glycoprotein I	43, 47, 53
(French bean)	11	340	Glycoprotein II	30, 32, 34
Medicago sativa	7	150	Alfin	14, 16, 20, 32, 38, 50
(alfalfa, lucerne)	11	360	Medicagin	59, 63, 64, 65, 67, 69
Glycine max	7–8	160	β-Conglycinin	42, 57
(soybean)	12	330	Glycinin	19, 37, 42

[a]Based on Miflin and Shewry (1981), Larkins (1981), and Krochko and Bewley (1988). See Casey *et al.* (1986) for a more complete account.

the vicilins and legumins from alfalfa (lucerne, *Medicago sativa*) are shown in Fig. 1.8, as an example. Several storage proteins are extracted in a low-salt-containing buffer, including the major 7 S storage protein (a vicilin called alfin, Table 1.8), some minor high-molecular-weight storage proteins (HMW), and a 2 S storage protein of low molecular-weight (LMW) (Fig. 1.8, lane A). This electrophoretic separation occurs on the basis of their size, and the subunits of alfin (α_1–α_6) range from 14 to 50 kDa (lane A). When this protein fraction is treated with β-mercaptoethanol, a reducing agent, which breaks disulfide bonds between adjacent polypeptide chains, the positions of the α subunits on the gel are unchanged (compare Fig. 1.8A and B). This shows that no interchain linkages have been broken, and hence the component polypeptides of the 7 S protein are not joined together by disulfide bonds. The 11 S storage protein, a legumin type called medicagin, is soluble only in high-salt-containing buffer. This separates by electrophoresis into a series of discrete bands ranging from 59 to 69 kDa (Fig. 1.8, lane C, Table 1.8). After treatment with β-mercaptoethanol, the 11 S protein runs on a gel as a group of subunits (A_{1-7}) of 40 to 49 kDa, and another group of 20 to 24 kDa (B_{1-3}) (lane D). This is because the interchain disulfide bonds which are present are broken by the reducing agent, and the individual component polypeptides of each subunit are released.

The A polypeptides are acidic, because their isoelectric points (pI) are about pH 6, and the B polypeptides are basic, with a pI of approximately 7.5. The proportion of acidic and basic amino acids in the polypeptides determines their pI. There is added complexity regarding the composition of the 11 S protein since individual polypeptides may vary slightly in their amino acid content, and hence exhibit minor variations in charge. There are, for example, some 30–40 charge isomers of the acidic polypeptides of alfalfa, having a pI ranging from 5.4 to 6.5.

The size of the 11 S holoprotein of alfalfa is about 360 kDa. The individual subunits of this protein are comprised of an acidic polypeptide, disulfide-bonded to a basic polypeptide. Since each subunit has a molecular mass of approximately 60 kDa, it is not surprising to find that the holoprotein is made up of six subunits.

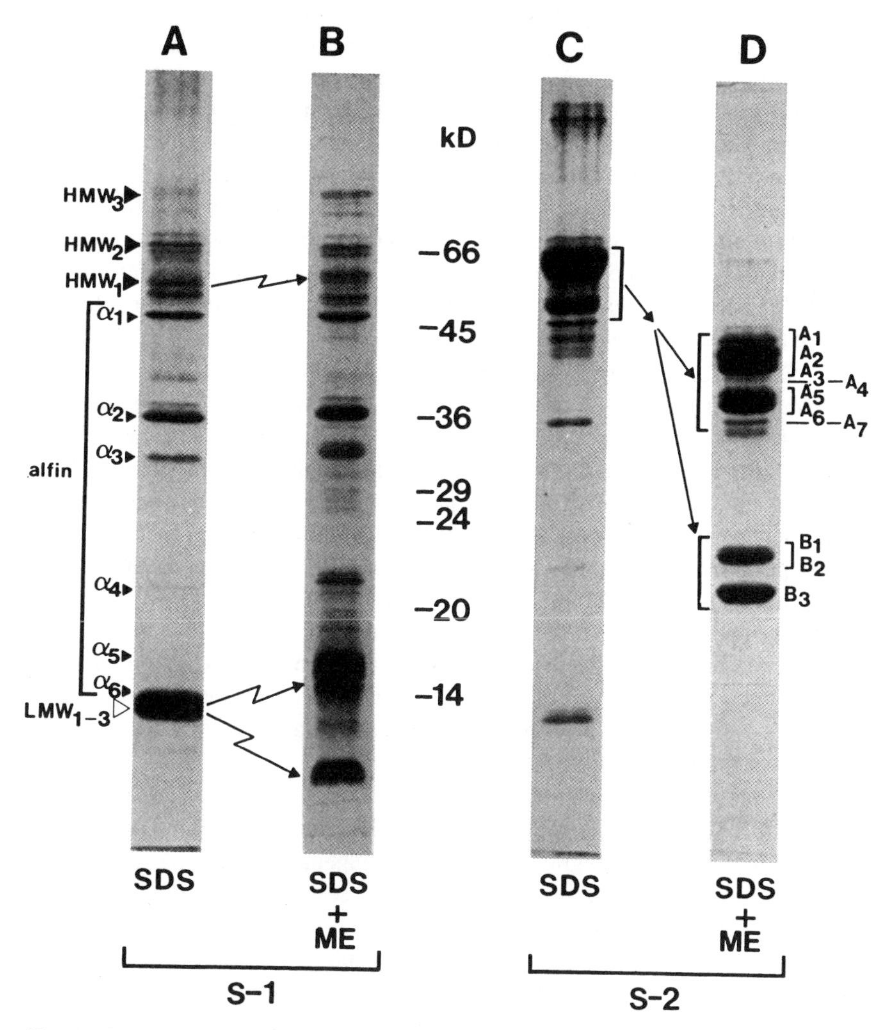

Figure 1.8. One–dimensional gel electrophoresis (SDS-PAGE) of low-salt-soluble vicilin and albumin (S-1), and high-salt-soluble legumin (S-2) storage proteins of alfalfa. Reduced (+ME, B and D) and nonreduced (A and C) samples are shown in adjacent lanes. The major proteins and molecular weight markers are indicated. After Krochko and Bewley (1988).

The composition of a typical 11 S legume protein and its relationship to its subunits and component polypeptides are detailed in Fig. 1.9.

Many storage proteins are encoded by a large number of genes (gene family), and slight variations in their codons result in similar small variations in the amino acid composition of the proteins they encode. Notwithstanding this, the genes (actually cDNAs) of all of the globulins of different species that have been sequenced show certain similarities within either the vicilin or legumin types. Certain base sequences, and hence amino acid sequences, are conserved in all of the legumins and in all of the vicilins regardless of the plant source. Therefore, these proteins are thought to have originated from two ancestral genes. Emphasizing the relationships is the fact that similar antigenic properties are shared by the legumins and by the vicilins of different seeds. During evolution, changes have occurred in these two ancestral genes, resulting in the many different types of globulins that we can now recognize. Some loss of genes has also taken place. For example, the vicilins do not occur in the families to which *Brassica* relatives (e.g., *Arabidopsis*, rapeseed) and sunflower relatives belong (Brassicaceae and Asteraceae, respectively).

Not all seed storage proteins have nutritionally desirable properties. Enzyme inhibitors may reduce the effectiveness of hydrolases, etc., in the digestive tracts of animals. Lectins, which are usually glycoproteins, have the capacity to bind to animal cell surfaces, sometimes causing agglutination (particularly of erythrocytes), e.g., concanavalin A in jack bean (*Canavalia ensiformis*) seeds; hence, lectins are sometimes called phytohemagglutinins. This property is probably superfluous as far as the seed is concerned, and many lectins are innocuous from a nutritional standpoint. Some lectins are highly toxic, however, including ricin D from castor bean (*Ricinus communis*) and abrin from *Abrus precatorius* (rosary pea), both of which are mixtures of nonagglutinating toxins and nontoxic agglutinins. Any toxic effect of lectins can usually be eliminated by proper heat treatment. For example, castor bean meal may be fed as a protein supplement to cattle after extraction of castor oil, and heating.

Mandelonitrile lyase (MDL) is a glycoprotein sequestered in the protein bodies of the endosperm of black cherry seeds (*Prunus serotina)* and is an enzyme involved in the degradation of the cyanogenic disaccharide amygdalin. It may be important in this and other cyanogenic seeds in producing HCN when tissues are damaged by pathogens. Many seeds contain proteins which may be part of their defense mechanisms against pests and predators; e.g., wild species of *Phaseolus vulgaris* contain the glycoprotein arcelin, which confers resistance against some brucid beetles, whereas seeds of domestic species of *P. vulgaris* appear to have such resistance conferred by an α-amylase inhibitor. Chitinase has been isolated from mature dry seeds of several monocots and dicots, an enzyme which increases resistance to fungal attack. Some of these enzymes may

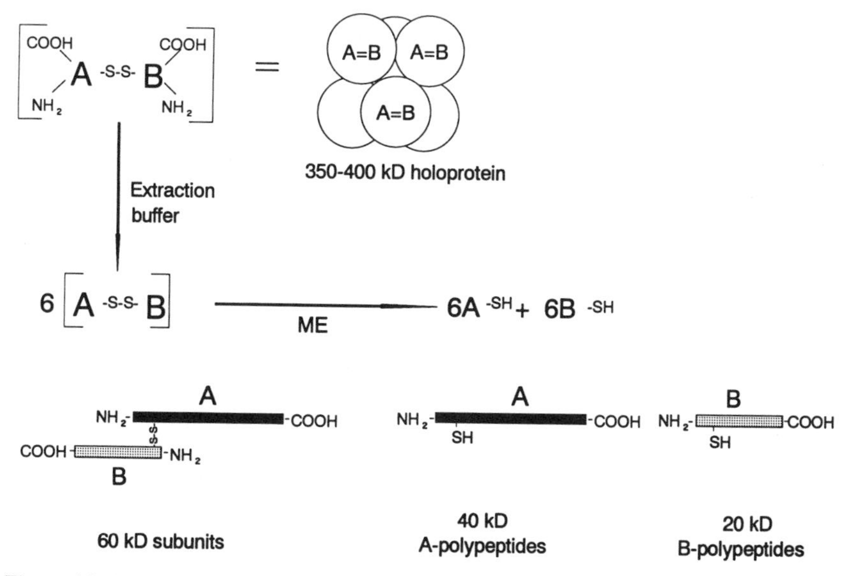

Figure 1.9. Components of an 11 S (legumin) storage protein as affected by a high-salt extraction buffer and the reducing agent β-mercaptoethanol (ME). The holoprotein is made up of six subunits, each containing one acidic (A) and one basic (B) polypeptide joined by a disulfide (-S-S-) bond. During extraction in high-salt buffer the protein is dissociated into its individual subunits because of the disruption of the noncovalent, electrostatic bonds holding them together. The subunits further dissociate into their component polypeptides in the presence of ME. After Krochko and Bewley (1988).

have the dual function of being storage proteins, as well as deterrents if the seed is the object of predation.

Seed storage proteins are usually deposited within special cellular organelles called protein bodies. These range in diameter from 0.1 to 25 μm and are surrounded, at least during development, by a single membrane. In mature, dry, storage tissues of some seeds, e.g., certain cereal grains (Section 2.3.4), the membrane is incomplete, or absent, leaving the protein dispersed in the cytoplasm. This is relatively uncommon, however. Some protein bodies are simple in that they consist of a protein matrix surrounded by a limiting membrane. Inclusions frequently occur, however, particularly crystalloids and globoids and, more rarely, druse (calcium oxalate) crystals. The crystalloids are insoluble (in water or buffers) proteinaceous inclusions embedded in the soluble protein matrix; e.g., in castor bean (Fig. 1.10A) the crystalloid is an insoluble 11 S protein, and the matrix is made up of 2 S and 7 S soluble proteins, including lectins. Globoids are noncrystalline, globular structures and are the most commonly occurring inclusion in protein bodies (Fig. 1.10A,B) although they vary in both size and number. In some species, the globoids are found in protein bodies of one region of a seed but not in another (e.g., the aleurone-layer protein bodies—aleurone grains—of cereals usually contain globoids, but protein bodies of the starchy endosperm never do). Globoids are the sites of deposition of phytin—the potassium, magnesium, and calcium salts of phytic acid (see Section 1.5.4); some globoids are surrounded by a soft-globoid region of unknown chemical composition. Barley aleurone-layer protein bodies also contain carbohydrate but in a unit distinct from the globoid, known as the protein–carbohydrate body (Fig. 1.10B). Various enzymes may occur within the protein body, and during reserve mobilization other enzymes may be added so that it eventually becomes an autolytic vacuole (Section 7.6.3).

Protein bodies may contain only one type of storage protein. In certain legumes, for example, some protein bodies contain only albumin, or vicilin, or legumin, although most seem to contain both vicilin and legumin. In maize, the small protein bodies in the starchy endosperm contain proportionately more zein than do the larger bodies; protein bodies in the aleurone layer of this and other cereals do not contain any of the major reserve proteins.

1.5.4. Phytin

Phytin is the insoluble mixed potassium, magnesium, and calcium salt of *myo*-inositol hexaphosphoric acid (phytic acid), and although present in relatively minor quantities compared with the aforementioned reserves, it is an important source to the seed of phosphate and mineral elements. Additionally, iron, manganese, copper, and, more rarely, sodium are to be found in some phytin sources (Table 1.9).

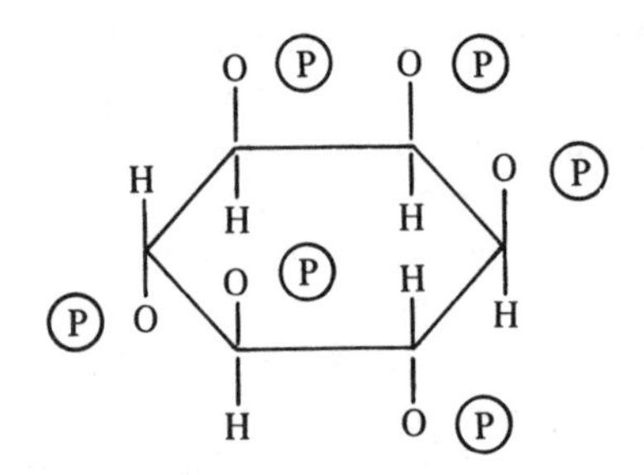

Phytic acid

The phytin is located exclusively within the globoid but, as mentioned in Section 1.5.3, not in all protein bodies of the seed, nor is the mineral element

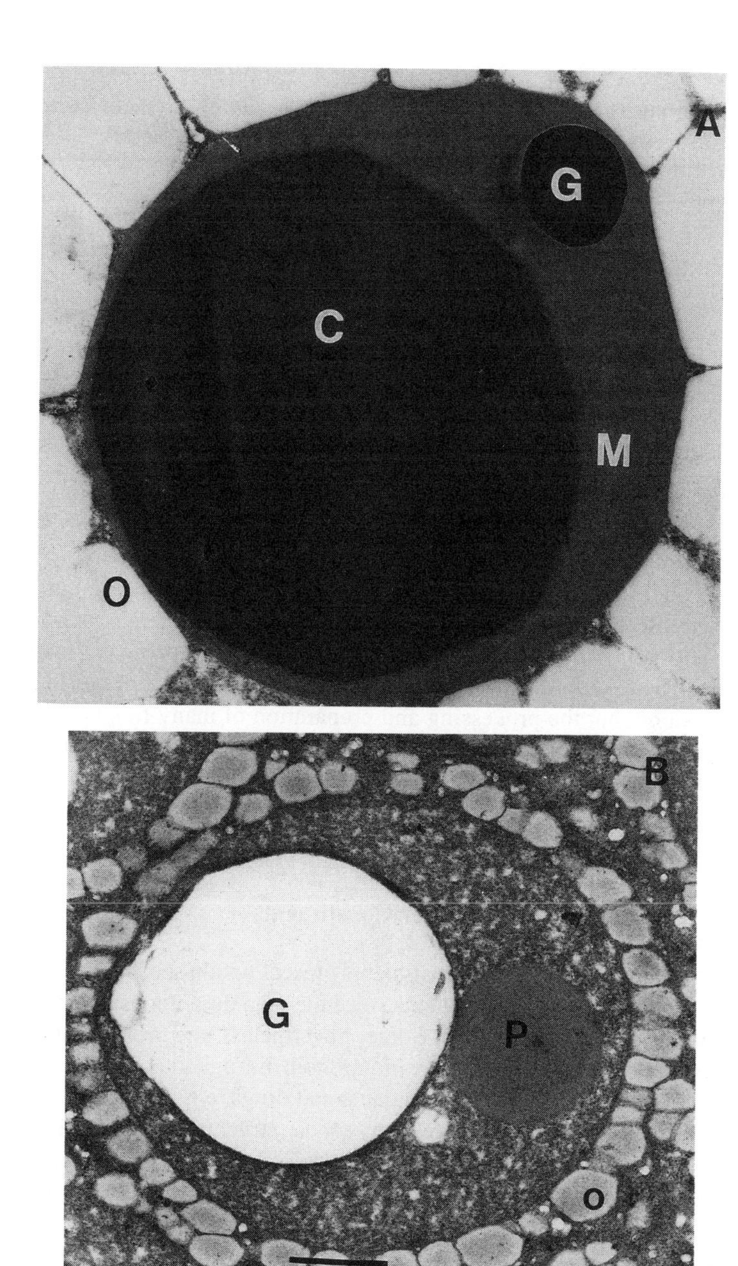

Figure 1.10. (A) Protein body of castor bean endosperm showing the phytin-containing globoid (G) and crystalloid protein (C) embedded in a protein matrix (M). The protein body is surrounded by oil bodies (O). Courtesy of J. S. Greenwood. (B) Aleurone layer cell of barley with aleurone grain (protein body) containing a globoid (G) and a protein-carbohydrate body (P) embedded in a protein matrix. The protein body is surround by oil bodies (o). Bar = 1 μm. From Jacobsen *et al.* (1971).

Table 1.9. The Content of the Main Inorganic Elements of Phytin in Various Species of Seeds, Expressed on a Percent Dry Weight Basis[a]

Species	Mg	Ca	K	P	Fe	Mn	Cu
Oat	0.4	0.19	1.1	0.96	0.035	0.008	0.005
Hazel	0.19	0.1	0.74	0.4			
Soybean	0.22	0.13	2.18	0.71			
Cotton	0.4	0.13	2.18	0.79	0.059	0.003	0.005
Barley	0.16	0.03	0.56	0.43			
Broad bean	0.11	0.05	1.13	0.51			
Sunflower	0.4	0.2	1	1.01			

[a]Taken from Weber and Neumann (1980).

composition of phytin the same in all cells; e.g., calcium is present in highest amounts in the globoids of the radicle and hypocotyl regions of the embryo, but little, if any, of this element is found in the globoids of the cotyledons.

Phytic acid and its conjugates are generally regarded as being nutritionally undesirable since they can bind essential dietary minerals (e.g., zinc, calcium, and iron), thus making them wholly or partially unavailable for absorption. The extent to which this is a problem in the Western world is not fully understood, but the processing and preparation of many foods from seed sources for human consumption may remove most of the phytin. In Third World countries, where food is often less refined, the problem may be more acute.

1.5.5. Other Constituents

There is a wealth of minor constituents present within seeds that cannot be regarded strictly as storage components since the seed does not use them during germination or subsequent growth. Some, nevertheless, are worth mentioning briefly. Certain alkaloids, which are nonprotein nitrogenous substances, are important commercial sources of stimulants and drugs, e.g., theobromine from the cacao bean, caffeine from coffee and cocoa, strychnine and brucine from *Strychnos nux-vomica* (present as 2.5% of seed dry weight), and morphine from certain types of poppy. In nature, such compounds can prevent insects and animals from using a seed for food. Phytosterols such as sitosterols and stimasterols are present in some seeds, e.g., in soybean. The latter is important pharmaceutically because it can be converted to the animal steroid hormone, progesterone. Certain nonprotein amino acids may be found in considerable quantities in some seeds; e.g., canavanine accounts for 8.3% of the dry weight of *Dioclea megacarpa* seeds, hydroxytryptophan accounts for 14% of the dry weight of *Griffonia*, and dihydroxyphenylalanine (*L*-dopa) makes up 6% of dry

Mucuna seeds. These amino acids are probably mobilized by the seed as a nitrogen source after germination. Glucosides are bitter-tasting components of some seeds, e.g., amygdalin from almonds, peaches, and plums and esculin from horse chestnut; some, such as saponin from tung seeds, may be deadly to humans and animals.

Phenolic compounds such as coumarin and chlorogenic acid, and their derivatives, and ferulic, caffeic, and sinapic acids occur in the coats of many seeds. These may inhibit germination of the seed that contains them or, being leached out into the soil, may inhibit neighboring seeds. Another germination inhibitor sometimes present in seeds is abscisic acid, and germination promoters and growth substances such as gibberellins, cytokinins, and auxins may also occur (for more details see Section 2.4).

USEFUL LITERATURE REFERENCES

SECTIONS 1.1 AND 1.2: SOME ADVANCED LITERATURE ON SEEDS AND GERMINATION

Bewley, J. D., and Black, M., 1978, 1982, *Physiology and Biochemistry of Seeds,* Volumes 1 and 2, Springer-Verlag, Berlin (covers all aspects of viability, germination, dormancy, and environmental control).

Côme, D., and Corbineau, F., 1993, *Basic and Applied Aspects of Seed Biology,* Volumes 1–3, Université Pierre et Marie Curie, ASFIS, Paris (presentations from 4th International Workshop on Seeds).

Khan, A. A. (ed.), 1977, *The Physiology and Biochemistry of Seed Dormancy and Germination,* North-Holland, Amsterdam (multiauthor contributions).

Khan, A. A. (ed.), 1982, *The Physiology and Biochemistry of Seed Development, Dormancy and Germination,* Elsevier, Amsterdam (multiauthor contributions).

Murray, D. R. (ed.), 1984, *Seed Physiology,* Volumes 1 and 2, Academic Press, New York (multiauthor work covering selected topics in development, germination, and reserve mobilization).

Roberts, E. H. (ed.), 1972, *Viability of Seeds,* Chapman and Hall, London (mostly viability but includes some physiology).

Simpson, G. M., 1990, *Seed Dormancy in Grasses,* Cambridge University Press, Cambridge (overview of dormancy, especially in wild oats).

Taylorson, R. B. (ed.), 1989, *Recent Advances in the Development and Germination of Seeds,* Plenum Press, New York (NATO Workshop proceedings of the 3rd International Workshop on Seeds).

SECTION 1.3

Goodchild, N. A., and Walker, M. G., 1971, *Ann. Bot.* **35:**615–621 (measurement of germination).

Hewlett, P. S., and Plackett, R. L., 1979, *An Introduction to the Interpretation of Quantal Responses in Biology,* Arnold, London (methods for mathematical analysis).

Janssen, J. G. M., 1973, *Ann. Bot.* **37:**705–708 (recording germination curves).

Richter, D. D., and Switzer, G. L., 1982, *Ann. Bot.* **50:**459–463 (quantitative expressions of dormancy in seeds).

SECTION 1.4

Corner, E. J. H., 1976, *The Seeds of Dicotyledons,* Cambridge University Press, Cambridge (a comprehensive two-volume work on seed anatomy).

Forest Service, U.S. Dept. Agriculture, 1974, *Seeds of Woody Plants in the United States,* USDA, Washington, D.C. (structure and classification aspects).

Rost, T. L., 1973, *Iowa State J. Res.,* **48:**47–87 (grass caryopsis anatomy).

Vaughan, J. G., 1970, *The Structure and Utilization of Oil Seeds,* Chapman and Hall, London (anatomy of oil seeds).

Webb, M. A., and Arnott, H. J., 1982, *Scanning Electron Microsc.* **3:**1109-1131 (mineral inclusions in seeds and seed coats).

SECTION 1.5

Biochemie und Physiologie der Pflanzen, 1988, **183:**99–250 (multiauthor symposium volume on seed proteins).

Borroto, K., and Dure, L., 1987, *Plant Mol. Biol.* **8:**113–131 (structural relationships and evolution of globulins).

Casey, R., Domoney, C., and Ellis, N., 1986, *Oxford Surv. Plant Mol. Cell Biol.* **3:**1–95 (exhaustive review of legume storage proteins and their genes).

Crocker, W., and Barton, L. V., 1957, *Physiology of Seeds,* Chronica Botanica, Waltham, Mass. (includes seed constituent composition).

Derbyshire, E., Wright, D. J., and Boulter, D., 1976, *Phytochemistry* **15:**3–24 (seed proteins).

Dey, P. M., and Dixon, R. A. (eds.), 1985, *Biochemistry of Storage Carbohydrates in Green Plants,* Academic Press, New York (chapters on seed carbohydrates).

Duffus, C. M., and Slaughter, J. C., 1980, *Seeds and Their Uses,* Wiley, New York (economically important seeds).

Earle, F. R., Curtice, J. J., and Hubbard, J. E., 1946, *Cereal Chem.* **23:**504-511 (composition of corn kernel regions).

Frey, K. J., 1977, *Z. Pflanzenzuchtg.* **78:**185–215 (amino acids in cereal proteins).

Jacobsen, J. V., Knox, R. B., and Pyliotis, N. A., 1971, *Planta* **101:**189–209 (protein bodies in barley aleurone layers).

Kreiss, M., and Shewry, P. R., 1989, *Bio-Essays* **10:**201–207 (seed protein structure and evolution).

Krochko, J. E., and Bewley, J. D., 1988, *Electrophoresis* **9:**751–763 (techniques for separation and identification of legume storage proteins).

Lambert, N., and Yarwood, J. N., 1992, in: *Plant Protein Engineering* (P. R. Shewry and S. Gutteridge, eds.), Cambridge University Press, Cambridge, pp. 167–187 (legume storage proteins, structure, uses and genetic engineering).

Larkins, B. A., 1981, in: *The Biochemistry of Plants. Proteins and Nucleic Acids,* Volume 6 (A. Marcus, ed.), Academic Press, New York, pp. 449–489 (seed storage proteins: review).

Lott, J. N. A., 1981, *Nord. J. Bot.* **1:**421–432 (protein bodies and inclusions: review).

Miflin, B. J., and Shewry, P. R., 1981, in: *Nitrogen and Carbon Metabolism* (J. D. Bewley, ed.), Nijhoff Junk, The Hague, pp. 195-248 (seed storage proteins: review).

Miller, E. C., 1931, *Plant Physiology,* McGraw–Hill, New York (seed constituent composition).

Parker, M. L., 1981, *Ann. Bot.* **47:**181–186 (rye endosperm structure).
Payne, P. I., and Rhodes, A. P., 1982, in: *Encyclopaedia of Plant Physiology,* New Series, Springer, Berlin, Volume 14A, pp. 346–369 (cereal storage proteins: review).
Richardson, M., 1991, *Methods Plant Biochem.* **5:**259–305 (enzyme inhibitors as seed storage proteins).
Shewry, P. R., and Tatham, A. S., 1990, *Biochem. J.* **267:**1–12 (cereal protein relationships and evolution).
Weber, E. J., 1980, in: *The Resource Potential in Phytochemistry. Recent Advances in Phytochemistry,* Volume 14, Plenum Press, New York, pp. 97–137 (composition of corn kernels).
Weber, E., and Neumann, D., 1980, *Biochem. Physiol. Pflanzen* **175:**279–306 (protein bodies and phytin).
Winton, A. L., and Winton, K. B., 1932, *The Structure and Composition of Foods,* Volume 1, Wiley, New York (review).

Chapter 2

Seed Development and Maturation

2.1. EMBRYOGENY AND STORAGE TISSUE FORMATION

2.1.1. Morphological Changes

Before we can consider the specific physiological and biochemical processes intimately involved in seed development, it is necessary to review briefly the morphological and anatomical aspects of embryo and storage tissue formation. The variations in patterns of development are numerous, so we will present here only a generalized picture of what occurs in a conifer (gymnosperm), a dicot, and a monocot angiosperm.

In conifers, the egg nucleus lies within the female gametophyte and is fertilized by a gamete released from the pollen tube. The resultant zygote then divides to produce several free nuclei, around which cell walls are laid down to form the proembryo (Fig. 2.1a, b). Subsequent cell divisions give rise to the embryonal and suspensor cells (Fig. 2.1c, d). These may develop to form a single embryo and an elongated suspensor, but often in the conifers there is cell separation to form four embryos (polyembryony) (Fig. 2.1d). Only one of these develops further and the others degenerate. Within the seed the central portion of the megagametophyte (the haploid female gametophyte) breaks down to form a cavity into which the embryo expands (Fig. 2.1e). Storage reserves—oil, starch,

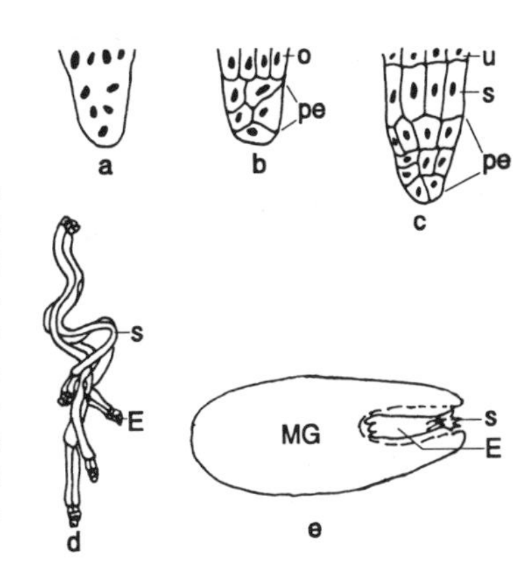

Figure 2.1. Embryo development in conifers. (a) Free nuclear stage. (b) Proembryo (pe) showing internal division of cells, with open tier (o) cells above. (c) Three-tiered mature proembryo (pe) beneath the suspensor tier (s), which elongates to form the primary suspensor, and the upper tier (u), which degenerates. (d) Polyembryonic stage. Elongated suspensor (s) bearing the embryonal masses of cells (E). (e) The single maturing embryo (E) elongates into a cavity in the megagametophyte (MG); the suspensors (s) degenerate.

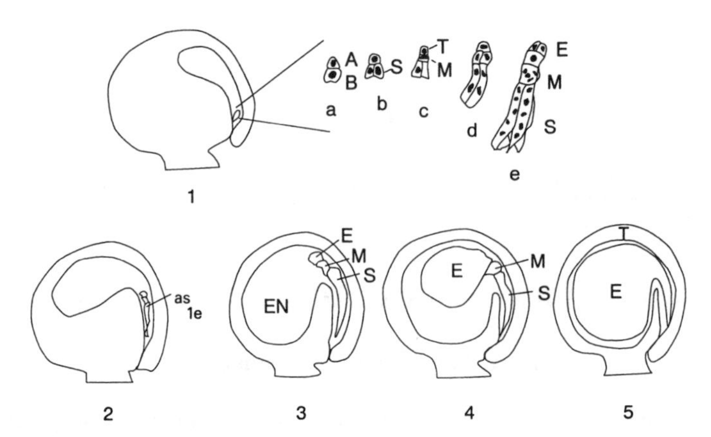

Figure 2.2. (Left) Embryo development in pea, *Pisum sativum.* (1) Early divisions of the zygote. The first division (a) is to form an apical (A) and a basal (B) cell. The basal cell divides transversely (b) to form two suspensor (S) cells, and the apical cell divides (c) to form a middle (M) cell and a terminal (T) cell. The suspensor (S) cells divide to become four, elongate (d, e), and become multinucleate; the middle cells undergo free nuclear division, and the embryonic cells (E) commence cell division. (2–5) The suspensor elongates and then embryo growth occurs into the embryo sac, to eventually utilize the reserves and to occlude the endosperm (EN) and the suspensor/middle cell (S, M). The mature embryo (E) occupies the whole of the inner region of the seed and is surrounded by the testa (T). Based on Marinos (1970) and Maheshwari (1950). (Right) Scanning electron micrographs of the developing embryo of alfalfa (lucerne), *Medicago sativa.* The multicellular globular embryo (g) in (A) is initially smaller than the suspensor (s). It is formed by mitotic divisions of the embryonic cells (E) depicted in the left panel (e). Further cell divisions result in a larger globular embryo (B, C). Cotyledon initiation starts at the early heart stage (D) when a depression (d) is evident in the globular mass of cells. (E) Late heart stage, showing the beginning of cotyledon (c) elongation. The embryo undergoes extensive elongation and expansion during the torpedo stage (F) and the hypocotyl (h) radicle (r) regions are defined. The embryo approaches morphological maturity at the late cotyledon stage (G), and the suspensor is now a diminutive structure. (H) Removal of a cotyledon at stage G reveals the presence of the plumule (p). Bar = 30 μm. From Xu and Bewley (1992).

and protein—are deposited within the persistent parts of the megagametophyte (Fig. 1.3), to be used after germination.

The feature of fertilization in angiosperms is the participation of two male nuclei. One nucleus released from the pollen tube fuses with the egg nucleus to form the diploid zygote, and the other fuses with two polar nuclei to produce a triploid nucleus. Double fusion is almost unique to angiosperms but it has now been found to occur in *Ephedra* (a gnetophyte). In this plant, however, fusion is with one other haploid nucleus so there is no triploid product. Unlike in the gymnosperms, there is no free nuclear stage in angiosperms; the first division results in an axial (distal or apical) cell and a basal cell. In dicot seeds, the

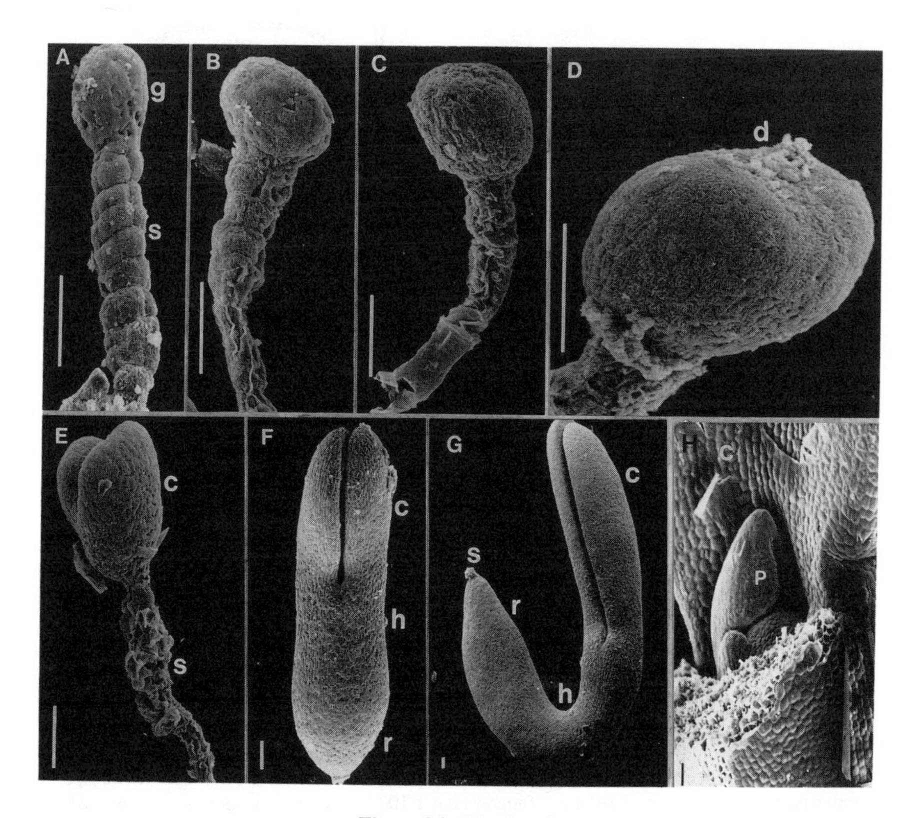

Figure 2.2. *(Continued)*

suspensor is formed from the basal cell, which may also contribute to the embryo; most of the embryo is derived from the axial cell. In monocots, the basal cell does not divide but forms the terminal (haustorial) cell of the suspensor; the embryo and the other few suspensor cells are produced from the axial cell. By definition, the embryos of the mature seeds of the dicots possess two cotyledons, whereas there is only one in monocots. The single cotyledon of the Gramineae is reduced to the absorptive scutellum (Fig. 1.3). Mature gramineous embryos also include a specialized thin tissue which covers the radicle (coleorhiza) and one which is around the plumule and covers the first foliage leaf (coleoptile) (Fig. 1.3). The pattern of development of the embryo and seed of a nonendospermic legume is shown in Fig. 2.2, and of a gramineous monocot in Fig. 2.3i (see legends for details).

The true endosperm, the cells of which are triploid, is found only in the angiosperms; it is derived from the triple fusion nucleus. Two types of en-

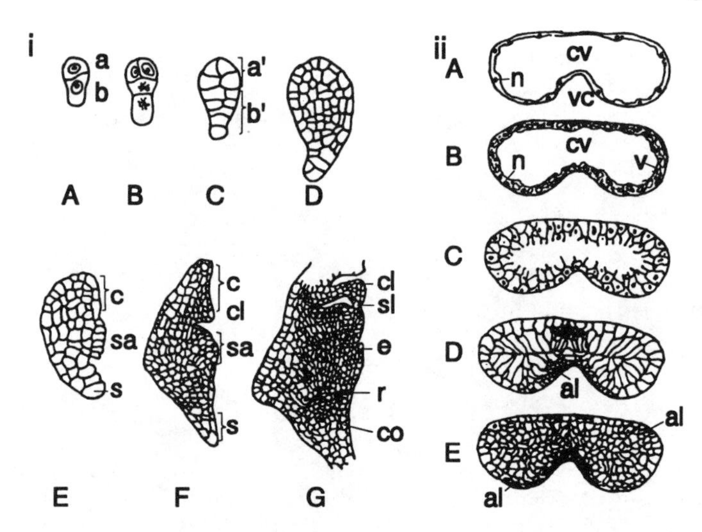

Figure 2.3. (i) Embryo development in a cereal. (A) First division produces an apical (a) and a basal (b) cell, both of which undergo divisions (B and C) to form derivatives, a′ and b′, respectively. After further division (D–F), regions that give rise to the short suspensor (s), cotyledon (c), stem apex (sa), and coleoptile (cl) are delimited. In the nearly mature embryo (G), these regions plus the coleorhiza (co), epiblast (e), stem apex and first leaf (sl), and root initials (r) are in evidence. After Johansen (1950). (ii) Endosperm development in a cereal, barley. Development in the midregion of the embryo sac results in (A) nuclear (n) divisions without cell wall formation (the nuclear endosperm); nuclei in the cytoplasm surround a central vacuole (cv). The ventral crease (vc) is evident. (B) Nuclei undergo further mitotic divisions, and vacuoles (v) appear in the cytoplasm. (C) Endosperm cell walls appear; division and cell wall formation proceed into the central vacuole. (D) Cellularization of the endosperm continues. A large number of cells are formed (about 100,000) which expand as storage reserves (starch and protein) are deposited within them. First cells of the aleurone layer (al) form over the crease area. (E) Aleurone layer differentiation spreads laterally from the central crease to join with other localized regions of formation of this layer. Simplified from Olsen *et al.* (1992).

dosperm development have been noted: (1) nuclear, where the endosperm undergoes several free nuclear divisions prior to cell-wall formation (e.g., apple, wheat, squash), and (2) cellular, where there is no free nuclear phase (e.g., *Magnolia, Lobelia*). Both types occur in monocots and dicots. A less common form of development, found only in some monocots, is the helobial type, where free nuclear division is preceded and followed by cellularization. Nutrients are drawn from adjacent tissues into the endosperm during its development, and new products are also laid down therein. Thus, the growing embryo become enveloped in, or intimately associated with an available food source upon which it can draw during its maturation and subsequent germination/growth stages. In grasses, e.g., cereals, there is limited utilization of the endosperm during maturation, but the part that is depleted forms the intermediate layer which lies between the starchy endosperm and the scutellum of the mature embryo (Fig. 7.1). Barley is used as an example of endosperm development in cereals (Fig. 2.3ii). In nonendospermic dicot seeds, the endosperm reserves are depleted and occluded by the developing embryo (Fig. 2.2) and then reorganized in the cotyledons, which act as the source of stored reserves for the germinated embryos. The endosperm is retained as a permanent storage tissue in the endospermic dicot seeds (e.g., castor bean, Fig. 1.3), and in endospermic legumes (e.g., fenugreek) this storage tissue may be surrounded by an aleurone layer. The nucellus develops into the perisperm in a few nonendospermic species (e.g., *Coffea, Yucca*), but this is rare.

There are many species- or family-specific variations on these generalized patterns of development related, for example, to differences in the origins of the embryogenic tissues from cells of the proembryogenic mass, the extent of suspensor development (from one to hundreds of cells), endosperm ploidy (can be $2n$–$15n$), endosperm morphology (presence or absence of an aleurone layer, number of cell layers thick), and the extent to which the shoot apex (plumule) is developed at maturity—in some species several leaves may be present, whereas in others the apex is quite rudimentary and not even leaf primordia are evident in the mature embryo.

The outer structures of the ovule, the integuments, undergo marked reorganization during seed maturation to form the seed coat (testa). Seed coat structure and thickness are extremely variable between different species, and sometimes even in the seeds of the same species grown under different environmental conditions. Seed coats are reviewed briefly in Chapter 1, but the space available here precludes a more extensive discussion of their development and morphology.

2.1.2. Regulation of Development

Novel morphogenetic events in the embryo and endosperm occur early after fertilization to define the major regions of the embryo, the suspensor, the extraembryonic storage tissues, and the enclosing structures. The embryonic axis becomes established at this time, and the morphogenetic programs to establish the next generation of vegetative plant, albeit in miniature, are initiated. The last three-quarters of embryo and seed development is devoted to the elaboration of structures that appear during the earlier stages, and to the deposition of storage reserves. Interactions between the different regions of the developing embryo and the surrounding tissues must play a defining role in morphogenetic programming. It has been estimated that about 20,000 genes are expressed during embryogenesis, and at least 20% of these are likely to contain information that will specifically affect some aspect of development. Almost nothing is known about the regulation of seed formation and early development, and it will be difficult to elucidate which genes are important, and how they function, given that the complex morphological changes take place within the enclosed environment of the ovule.

We will see later in this chapter that several genes have been identified whose expression is specifically associated with embryogenesis, particularly those involved in the synthesis of stored reserves. Defects, or mutations, of these genes result in a depleted or modified reserve composition, and these can result in changes in the appearance of the mature seed (e.g., modifications to the *R* locus in peas result in a change from a round to a wrinkled seed: see Section 2.3.1). But identifying genes which control morphological development of the seed presents a much greater and more demanding challenge. In recent years, mutants have been sought with obvious impairments in normal seed development so they can be screened for defects in genes with important developmental functions. Mutagenesis studies on maize have yielded several *defective kernel* (*dek*) mutants, in which both embryo and endosperm are defective, and they have provided insights into the genetic programs governing embryogenesis. In *emb (embryo-specific)* mutants of maize, embryo development is severely disrupted, whereas that of the endosperm appears to be undisturbed.

Fifty-one such mutants have been identified affecting different stages in embryo development (Fig. 2.4). It should be possible eventually to isolate the normal genes, clone them, and gain some insight into their products, thus initiating a molecular and biochemical understanding of the regulation of embryogenesis. Embryo-lethal mutants have been identified in several other flowering plants such as *Arabidopsis*, with the blocks to development varying from the very early proembryo stages, to late maturation.

Using developmental mutants is a new approach to unraveling the mysteries related to the regulation of embryo and seed development. The potential of this approach is still to be fully realized, but it offers exciting prospects and is likely to command its place in the seed literature in the upcoming years.

2.2. SOURCE OF ASSIMILATES FOR GRAIN AND SEED FILLING

The extent to which filling occurs in crops harvested for their seed is an important agronomic consideration, and much research has gone into determining the factors which influence yield. This is a large and intensively studied area of crop physiology, which will be given only cursory coverage here. Breeding for higher-yielding seed crops has been undertaken for many years, but factors such as resistance to diseases, tolerance of stresses, water-use efficiency, self-shading, translocation of C- and N-containing compounds, etc., also enter into consideration when trying to produce the ideal line. Here we will briefly outline the source of photosynthate for the storage materials laid down within the seed,

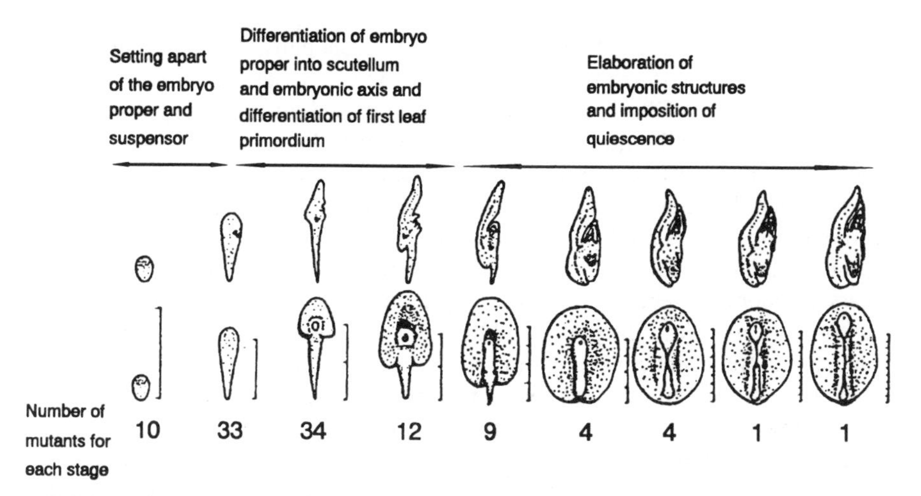

Figure 2.4. Embryo-specific mutants of maize. The stages of development affected by the mutations are shown. Mutants do not develop beyond the particular stage. Although there are 51 separate mutants, many block variably over two or three neighboring stages which is why the numbers add up to more than 51. Each division on the scale bars represents 0.5 mm. After Sheridan and Clark (1993).

their translocation from the vegetative plant to the seed, and the effects of the environment on seed yield.

2.2.1. Cereals

Development of the cereal grain can be divided into two stages: grain enlargement and grain filling. Enlargement is the result of cell division (see Section 2.1) followed by an influx of water which drives cell extension. Filling occurs as the reserves, starch and protein, are deposited within the endosperm. Carbohydrates such as free sugars, starch, and other polysaccharides reach a maximum concentration in the vegetative parts of the parent plant around the time of anthesis (dehiscence of the anthers), after which they start to decrease. Some of this stored carbohydrate is translocated to the growing grain, and although estimates are variable, it may provide up to 15–20% of the final dry weight of the grain. The contribution to grain filling by current photosynthesis in different plant parts is related to their potential photosynthetic activity, their longevity during grain ripening, and the light environment in the crop canopy. Cereals can accumulate up to 90% of their final nitrogen before anthesis also, and this is remobilized to the developing grains as the parts senesce; the final grain N content is to a large extent a consequence of the efficiency of this remobilization. Typically, high-grain-N cultivars have the most effective nitrogen redistribution from vegetative to reproductive tissues.

The photosynthetic rates exhibited by different parts of the plant are quite variable from crop to crop. In wheat (Fig. 2.5A) and barley, net photosynthesis in the flag leaf and ear is relatively high, and these regions provide the major nutrient source to the grain, although each may make quantitatively different contributions in different crops and cultivars. In oats and rice, the flag leaf and penultimate leaf appear to be of equal importance in supplying assimilates for grain filling. Sugars produced by the leaves above the ear in corn (Fig. 2.5B) are translocated efficiently into the kernel, but translocation from leaves below the ears is poor. In general, then, the upper, or uppermost, leaves direct their assimilates mostly to the grains and the stem, and the lower leaves to the roots and tillers. But even within the ear itself the distribution of assimilates may vary. In wheat, for example, assimilates from the flag leaf are distributed preferentially to the lower central spikelets (the first to anthese), and the setting, and hence yield, of the grains in the upper florets may be reduced by the more rapid development of grains from the first flowers to reach anthesis. On the other hand, photosynthates produced by various parts of the ear, e.g., the awns, tend to supply the grains in closest proximity.

The longevity of the green tissue in the various plant parts during the ripening period may be a factor in grain yield. Rice leaves remain green almost until grain maturity, whereas the ear turns yellow fairly early during ripening. In wheat, however, the leaves become yellow (i.e., senesce) before the ear does, and the latter may be the more important in providing photosynthates at the late stages of maturity. At this time, however, respiration by the grain may exceed the net import of sugars and other substrates, resulting in a small loss in grain weight.

The light environment impinging on the crop canopy plays a major role in determining the real photosynthetic activity of a given plant part. The developing ears of barley and wheat are fully exposed to sunlight and can realize their full photosynthetic potential. In contrast, the ears of certain improved rice cultivars have the tendency to bend and are positioned below the flag leaf; hence, they are heavily shaded and make an insignificant contribution to the total assimilates utilized in grain filling. It is notoriously difficult to estimate precisely the photosynthetic contribution by different parts of the plant in a crop canopy under field conditions, but factors such as leaf area index, leaf age and longevity, angle of incidence to the sun, shading, and ear structure (awns or awnless) all play a role, as well as environmental variables like temperature and nutrient and water availability.

2.2.2. Legumes

Substantial differences exist between legume species as to the sources of C and N utilized for filling of the seed. Before flowering, legume plants utilize net photosynthate for respiration by the roots and nodules, and in the investment

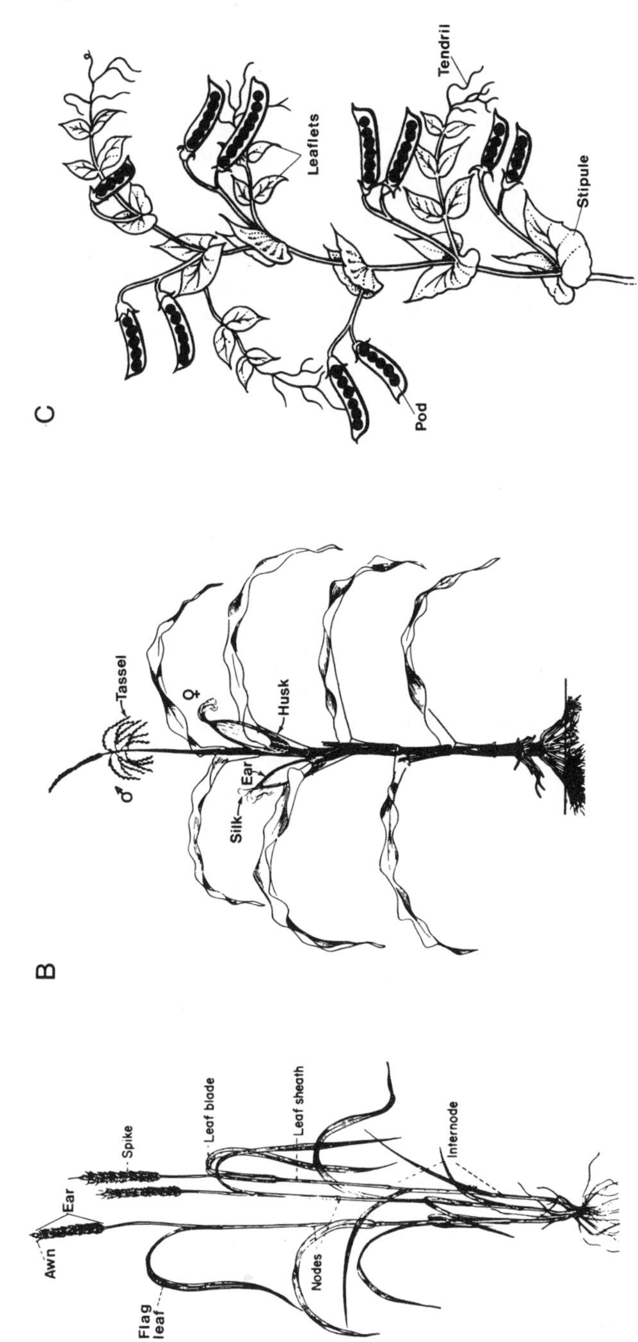

Figure 2.5. Mature (A) wheat and (B) maize (corn) and (C) pea plant at fruiting stage.

of C into dry matter of new leaves and roots. After flowering, some legumes (e.g., lupin, *Lupinus albus*) continue to allocate photosynthate to the roots and nodules, and the requirements for pod and seed (fruit) production are met by remobilization of C- and N-containing compounds in the vegetative parts of the shoot. In cowpea (*Vigna unguiculata*), by contrast, seed set leads to an abrupt decline in allocation of photosynthate to the below-ground parts, progressive abscission of leaves, and translocation of photosynthate to the developing fruit. About 75% of the nitrogen accumulated in the lupin seed is derived from symbiotic N-fixing activity after anthesis, but, in contrast, seeds of cowpea gain 69% of their N requirement from nitrogen fixed before anthesis, which is remobilized to the fruit from protein reserves in the leaflet. Other legumes show variations of these options, e.g., field peas assimilate most of their C and N during flowering and early fruiting, whereas in mung bean this is extended through seed filling.

Generally, the C assimilated before flowering is proportionately much less readily available to the developing fruits than is the N assimilated before flowering. Thus, storage reserve formation in the seed depends heavily on assimilates formed during fruiting itself, and seed yield is likely to be extremely sensitive to adverse environmental factors which reduce photosynthesis.

The pea pod, which is derived from the ovary wall, more or less completes development before swelling of the enclosed seeds commences, i.e., prior to deposition of the major storage reserves within the cotyledons. Although green, the seeds themselves photosynthesize only weakly and lose as much CO_2 by respiration as they gain by photosynthesis. The major source of carbon for the developing pea seed varies with time of development (Fig. 2.6). The pod and the adjacent leaflets (Fig. 2.5C) supply about two-thirds of the C required by the seeds borne at that node. In late development, e.g., 28 days after flowering (Fig. 2.6), seeds acquire most of their C from sources outside the blossom node; at this time the fruit may draw upon C remobilized from senescing tissues of the shoot and roots. Other legumes exhibit different patterns in relation to their sources of C during fruit set, for in some, e.g., cowpea, there is early senescence of the subtending leaflets and the filling seed relies on sources other than its blossom node for most of its assimilates.

2.2.3. Translocation of Assimilates into the Developing Seed

Sucrose is the form of sugar translocated from the sites of photosynthesis to the developing seed. The major pathway for its long-distance transport from the vegetative parts is via the phloem. Developing cereal grains have no direct vascular connections with the parent plant, and a short-distance transport mechanism operates to move the assimilates from the vascular tissues to the region of

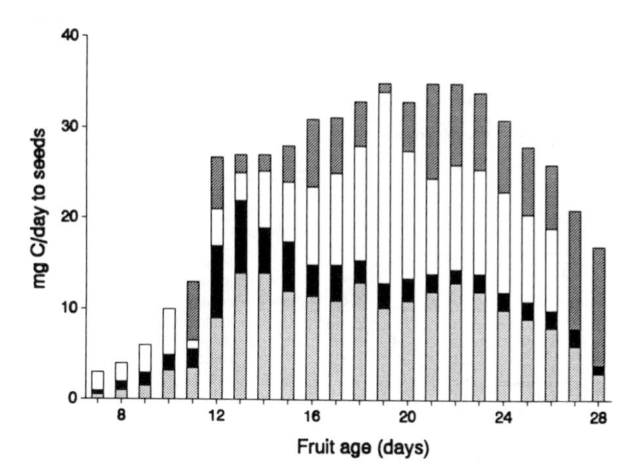

Figure 2.6. Source of carbon to the developing pod located at the lowest reproductive node of the pea plant (*Pisum arvense*). Contributions in mg C per day by the leaflet at the same node ⊡, the stipules ■, the pod □, and from outside of the node at which the fruit is borne ▨, over the time of development of the fruit. Reviewed by Pate (1984).

reserve deposition, the endosperm. In temperate cereals such as wheat and barley, assimilates are supplied via the vascular tissue running in the furrow (crease) and must first pass through the funiculus–chalazal region, then through the nucellar projection, and finally through the aleurone layer before entering the starchy endosperm (Figs. 2.7A and 2.8). Rice grains do not have a crease and nutrients are transported into the developing grain in a single vascular bundle embedded in the pericarp. In maize, and several other tropical cereals (e.g., sorghum, millets), assimilates are unloaded from the phloem terminals located at the base of the grain, the pedicel, where specialized transfer cells facilitate movement from this maternal tissue into the base of the developing endosperm (Fig. 2.7C). Transfer cells have ingrowths of the cell wall and hence increased surface area of the plasmalemma for the absorption or export of solutes.

In barley and wheat, the assimilates unloaded from the phloem along the length of the furrow pass symplastically (i.e., within cells connected by plasmodesmata) through maternal tissues to the cells of the nucellar projection (Fig. 2.8). There, transfer cells redirect assimilates to the apoplast (extracellular regions), possibly utilizing solute pumps in their membranes for active transport, although the operation of a passive, turgor-driven mechanism has also been proposed. The assimilates now diffuse to the endosperm and are taken up into the outer layer; in these regions of the grain the cells of the aleurone layer are modified into transfer cells. Transport now occurs symplastically to the starchy

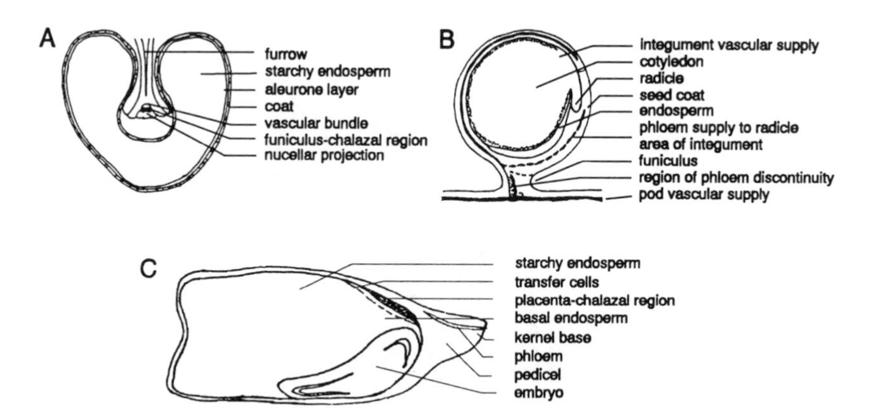

Figure 2.7. (A) Cross section through a developing wheat grain at midpoint between apex and base to show the relationship between the vascular tissue and the starchy endosperm, between which the short-distance transport system operates. (B) Midsagittal section of a developing seed of the garden pea to show the vascular supply to the developing ovule and in the integuments (seed coat). Heavy dashed line indicates the relative position of the phloem supply to the radicle area. Stippled areas are the probable positions of transfer cells in the funiculus and cotyledonary epidermis. (C) Longitudinal section through a developing maize grain. Here the short-distance transport system is through the kernel base (fruit stalk) and pedicel (maternal tissue in which the phloem ends) to the basal part of the starchy endosperm. Adapted from (A) Sakri and Shannon (1975); (B) Hardham (1976). (C) Shannon (1972).

endosperm cells, where the assimilates are used for the synthesis of reserves. Assimilates entering the pedicel of maize are transferred symplastically from the phloem to the placenta–chalazal region, which is a maternal tissue. There they are passively transferred to the apoplast, and diffuse apoplastically to the endosperm transfer cells (Fig. 2.7C), where they reenter the symplast and are translocated throughout the endosperm.

Invertases are present within the apoplast of the pedicel at the base of a maize grain (Fig. 2.7C), which cleave sucrose to glucose and fructose; these products are taken up actively into the endosperm cells and resynthesized to sucrose prior to utilization in starch biosynthesis, and other metabolic pathways. This hydrolysis of sucrose in the maternal tissue is an integral part of the transport mechanism of maize, and mutants (*miniature-1*) lacking invertase exhibit impaired phloem unloading and aberrant development of both the endosperm and the pedicel itself. One possible reason for the damage to the pedicel is that, because of the lack of invertase, sucrose accumulates and upsets the cells osmotically. In wheat and barley, sucrose itself is absorbed from the maternal tissues into the developing endosperm. The uptake of amino acids into developing cereal grains has received less attention, but it is generally accepted that

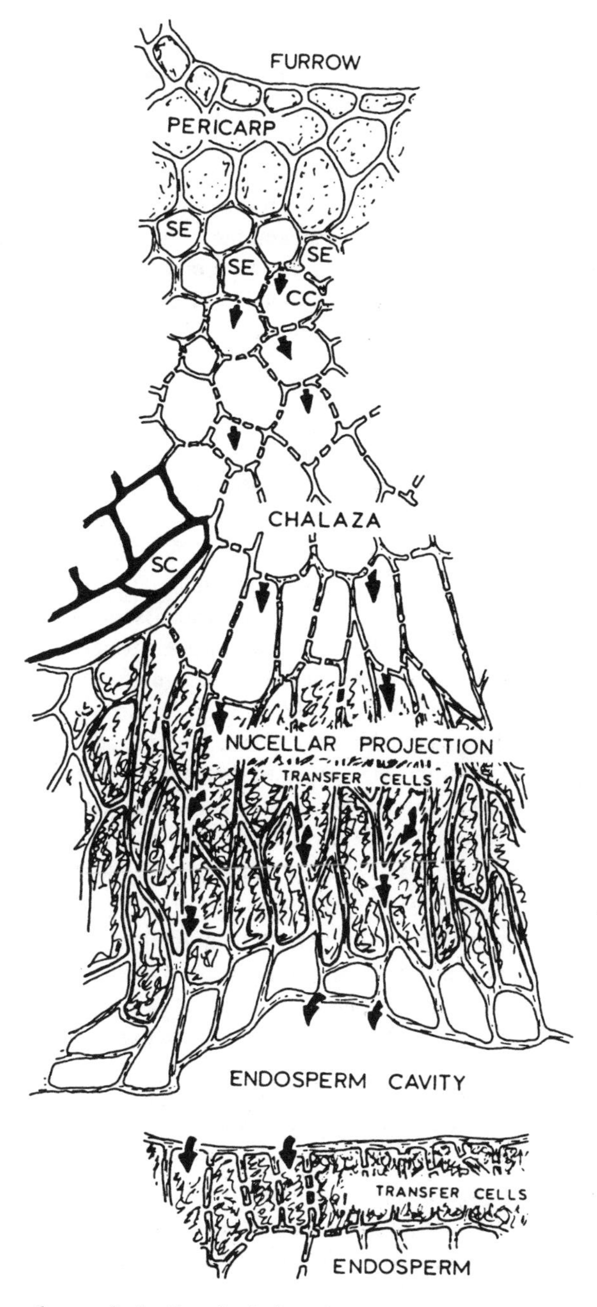

Figure 2.8. The pathways of unloading of assimilates from the vascular phloem (SE, sieve elements) in the furrow of barley or wheat. CC, companion cells; SC, seed coat. Arrows show direction of assimilate transport. After Thorne (1985). Reproduced with permission from Annual Reviews, Inc. © 1985.

asparagine and glutamine are the important translocated forms from the parent plant. Formation of other amino acids must occur within the developing grains themselves. The amino group of the amide-containing amino acids (i.e., glutamine and asparagine) provide the nitrogenous component for newly synthesized amino acids, and carbon skeletons are furnished by the translocated carbohydrates. The rate of storage protein synthesis might be controlled by the rate of delivery of glutamine and asparagine to the ear or the rate of their conversion to other amino acids in the endosperm, or both. Transport of assimilates into the developing cereal grains usually ceases when they become swollen with the deposited reserves, and the regions in which the phloem is located are crushed and nonfunctional.

During the early stages of development the (nonendospermic) legume seed obtains its nutrients from the liquid surrounding the embryo in the embryo sac (Fig. 2.2), referred to as endosperm or nucellar secretion. But assimilates required later for reserve deposition in the cotyledons are translocated from the parent plant. This is facilitated by a vascular strand which branches from the vascular tissue running through the pod and then passes through the funiculus, into the integuments (incipient seed coat) (Fig. 2.7B). Passage of assimilates through the funiculus, and from the seed coat into the cotyledons, by diffusion, is aided by the presence of transfer cells. The phloem in the seed coat, through which the assimilates are distributed, may consist of only one or two strands (e.g., pea), or may exhibit an extensive, reticulate network (e.g., soybean). There are no symplastic connections between the seed coat (maternal tissue) and the embryo, and the assimilates pass from phloem in the coat into the apoplastic space between the two generations. They are then taken up by the embryo, redistributed symplastically, and utilized in reserve synthesis. There are few, if any, transfer cells in the outer layers of the seed coat or embryo (this varies with species of legume), and sucrose is not hydrolyzed during its passage from the coat to the seed.

About 85% of the C translocated in the phloem of the seed coat is in the form of sucrose. The major forms of N in the phloem are asparagine and glutamine, although in soybean and cowpea about 10–15% is ureides. Enzymes to convert ureides to asparagine and glutamine are present in the seed coat of these species. The relative amino acid composition of phloem sap reaching the developing pea seed shows an increasing predominance of asparagine with increasing time after flowering (Fig. 2.9, A). Glutamine import declines with time. The relative abundance of amino acids present in the seed coat phloem is affected by the form in which N is acquired by the plants; those acquiring N by fixation in the nodules import more asparagine than those utilizing nitrate (Fig. 2.9A). The composition of amino acids reaching the developing seed itself is different from that entering the coat. Alanine and threonine are present in greater concentrations than asparagine in the liquid secreted into the apoplast by the seed coat (Fig. 2.9B). Thus, enzymes to convert amino acids are present in the seed

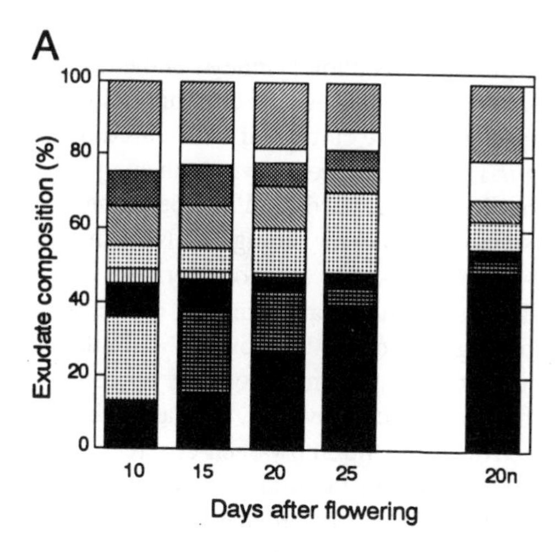

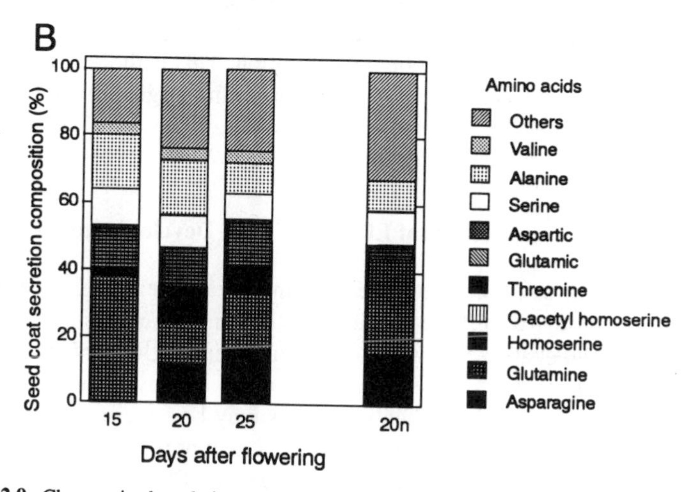

Figure 2.9. Changes in the relative amino acid composition of (A) the phloem sap arriving in the seed coat through the funiculus during development of fruits of pea (*Pisum sativum*), and of (B) the liquid secreted by the seed coat and available to the developing embryo. Plants were fed with nitrate during their growth except those marked (n) which obtained their N by fixation in root nodules. After Rochat and Boutin (1991). Reproduced by permission of Oxford University Press.

coat; asparaginase, for example, which removes the amide group from asparagine, is very active in pea seed coats. These observations point to the complexities of N metabolism in legumes, which varies, for example, with species, N-nutrition, and environmental factors, as well as with time of development. Only the briefest of discussions has been possible here.

In some legumes, e.g., certain cultivars of soybean and field and garden pea, the translocated sucrose produced by photosynthesis in the leaves and pods may be stored temporarily as starch in the pod prior to remobilization and transfer to the developing seed. In species that are cultivated for their edible pods, e.g., runner bean (*Phaseolus vulgaris*), the fleshy pod serves both as a photosynthetic capsule for recycling CO_2 respired by the seeds and as a permanent depository for ostensibly seed-bound sugars. Temporary storage of N in the pod wall is frequently in the form of *O*-acetylhomoserine and homoserine.

In recent years it has been suggested that hormones, particularly abscisic acid, play some role in the regulation of the flow of assimilates into the developing seed, e.g., by controlling phloem unloading or controlling sink strength. The current weight of evidence, however, is against this possibility. Experiments have been carried out using the empty-seed-coat technique in which the contents of a developing seed within the pod, for example, are removed, leaving the attached seed coat which can be filled with solutions of various osmotic strengths. Continued import of sucrose and amino acids from the seed coat into the interior (i.e., when intact, the embryo) requires the presence within the coat of solutes at a high osmotic concentration. This is thought to maintain a low turgor in the seed coat symplast, thus permitting turgor-driven flow from the source, the leaves.

2.2.4. Environmental Effects on Seed Development

Responses to environmental stress during seed development are diverse and complex, although the effects are generally deleterious and result in a decline in seed number and quality. Stresses such as water deficits, low or high temperature, nutrient deprivation, and shading can occur at any time during seed development. Sometimes more than one stress may be experienced by plants, e.g., high temperature and water stress, and the effect of these on seed development may be magnified and be greater than the sum of the individual stresses. Duration of the stress is also an important consideration as well as the time (stage) of occurrence during seed development. Bearing in mind all these variations and permutations in the stresses that plant may experience, it is not surprising that the huge literature on this subject contains quite a number of seemingly contradictory conclusions. A few generalizations are made here, but there will be exceptions.

In cereals, water stress during development of the inflorescence reduces the number of primordia that are produced, resulting in a reduction in the total number of grains that can be formed. If the stress occurs at the time of anthesis and fertilization, pollen production is affected; and the ability to form receptive stigmas (silks), e.g., in maize, can also suffer. Cell division and cell enlargement following fertilization determine the size and storage capacity of the grain, and both can be adversely affected by water and heat stress. Duration of grain filling is shortened at high (over 30°C) temperatures in wheat; this is probably a consequence of accelerated development, and since the rate of grain filling does not increase proportionately to the larger number of cells produced there is diminished grain weight (yield) at maturity. As the temperature experienced by an ear of wheat increases from 15 to 30°C, the proportion of protein storage reserves laid down increases relative to starch deposition, although the synthesis of both is reduced. This is because the formation of one type of starch grain is sensitive to higher temperatures and the decline in its production results in a diminished starch content in the endosperm. A similar sensitivity appears to occur under conditions of water stress. The occurrence of water stress at certain stages of development of wild oat grains can lead to a reduction in their dormancy at maturity.

A similar range of effects of stress during the development of seeds of dicot crops (e.g., legumes) has been reported. Water stress during the early stages of soybean seed development results in a decrease in the number of pods per plant because of the induction of abortion and abscission when the pod fails to expand. Stress at later stages can affect the photosynthetic activity of the vegetative plant, and reduced production of assimilates for the developing seed results in reduced yield. Seed viability and vigor may also be adversely affected. Low-temperature stress during development results in both quantitative and qualitative changes in the lipid reserves of some oil-crop seeds, and a nonlethal frost during late stages of development of rapeseed (canola) prevents the seed coat from losing its green color. Chlorophyll contaminates the oil during commercial extraction, and processing to remove it reduces profitability.

2.3. DEPOSITION OF RESERVES WITHIN STORAGE TISSUES

Most mature seeds contain at least two or three stored reserves in appreciable quantities (Table 1.2), and to a large extent they are generally synthesized concomitantly during seed development. But for the sake of clarity and convenience we have chosen to discuss singly the synthesis and deposition of the major reserves—carbohydrate, triacylglycerols, and protein—and the important minor reserve, phytin.

An interesting question, however, is what determines the types and proportions of the reserves laid down. Since seeds produce at least two kinds of reserves even in the same tissues, or in some cases three in different tissues, they obviously possess multiple biochemical capacities for synthesis. Processes must operate to allocate incoming assimilates to different reserve components in precise quantities and ratios. One feature which must contribute to this is that synthesis takes place in different cellular compartments, e.g., starch and fatty acids in plastids, and proteins in the cytosol and endoplasmic reticulum, so once the carbon from sucrose has been partitioned its destiny is more or less fixed; however, we know very little about the control of such partitioning. Further, it is not known how a plastid regulates the allocation of its imports to the synthesis of fatty acids or of starch. In the maize embryo, for example, the plastids are concerned largely with the production of fatty acids but in the endosperm they make starch. And even in one organ, such as the cotyledons of an oil-storing embryo, starch may be formed in the plastids early in seed development and fatty acids later, though possibly not in the same plastids.

The relative activities of different enzymes obviously must be involved in such phenomena and in all probability the differential expression of genes for them. We have yet to learn about the regulatory mechanisms involved. Endogenous factors must be operative but the environment, also, can be influential, for example in determining the relative amounts of starch and protein in certain oil seeds.

As a prelude to our consideration of the changes which take place in the seed as its reserve content is increasing, let us put these events into perspective with relation to the overall changes which occur during seed development. In Chapter 1 we saw how the seed grows from a single fertilized egg (zygote) to a multicellular embryo by cell division and differentiation. These events, collectively termed histodifferentiation, along with early cell expansion are marked by a rapid increase in whole-seed fresh weight and water content (Fig. 2.10). A period of rapid gain in dry weight then follows as a result of the synthesis and deposition of the stored reserves; cells expand to accommodate these reserves. The whole-seed fresh weight remains relatively stable, although the seed loses water as this is displaced by the accumulating insoluble reserves within the cells of the storage tissues. The decline in water content slows as the seed approaches its maximum dry weight. Finally, as the seed undergoes maturation drying and approaches the quiescent stage, when it may be shed from the plant, there is a period of fresh weight loss accompanied by a rapid decline in whole-seed water content. The duration of each of the major phases of development (Fig. 2.10) varies from several days to many months, depending on species and prevailing environmental conditions.

2.3.1. Starch Synthesis

We should recall that starch exists as amylose and amylopectin. Synthesis of the amylose occurs first, followed by modification to produce the amylopectin component. Sucrose, the sugar translocated to the seed, is the substrate for starch

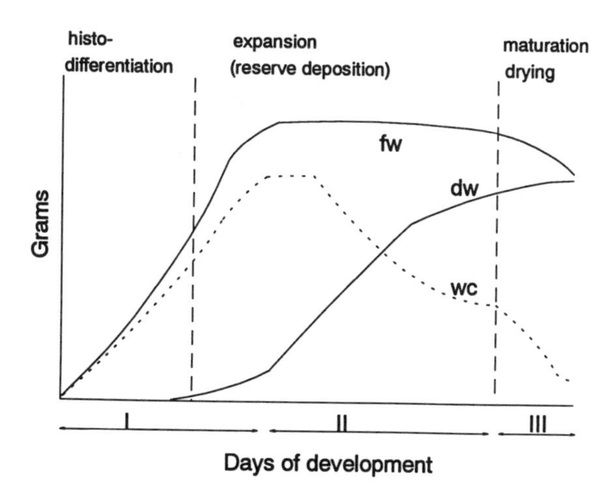

Figure 2.10. Changes in whole-seed fresh weight (fw), dry weight (dw), and water content (wc) during development. Three major phases are noted: I, rapid gain in fresh weight because of cell division and early expansion; II, rapid gain in dry weight because of cell enlargement to accommodate reserve deposition; III, loss of fresh weight as the seed undergoes maturation drying.

formation. First it is converted to fructose and UDPG1c (uridine diphosphoglucose) by sucrose synthetase (sucrose-UDP glucosyltransferase) as follows:

$$\text{Sucrose} + \text{UDP} \rightleftharpoons \text{Fructose} + \text{UDPG1c}$$

The fructose is phosphorylated to Fru-6-*P* by a hexokinase, which is changed to Glc-6-*P* by hexose phosphate isomerase. The UDPG1c is also converted to Glc-1-*P* by activity of the enzyme UDPG1c pyrophosphorylase to produce G1c-1-*P* which is changed by action of a mutase to Glc-6-*P*.

$$\text{UDPGlc} + PP_i \rightleftharpoons \text{Glc-1-}P + \text{UTP}$$
$$\uparrow\downarrow$$
$$\text{Glc-6-}P$$

These steps in the biosynthesis of starch are completed in the cytoplasm (Fig. 2.11). The Glc-6-*P* now enters the starch-synthesizing plastid, the amyloplast, by the activity of a translocator in the membrane of this organelle, and is utilized for conversion to starch. The first change is isomerization to Glc-1-*P*, which is then converted to ADPG1c, another sugar nucleotide, by ADPG1c pyrophosphorylase as follows:

$$\text{Glc-1-}P + \text{ATP} \rightleftharpoons \text{ADPGlc} + PP_i$$

The ADPGlc donates its glucose to the nonreducing end of a small glucose primer [α(1→4)-linked chain of glucoses], thus increasing its chain length by one unit. The process is repeated until the amylose molecule is completed, the enzyme involved being ADPGlc-starch synthetase (ADPGlc-starch glucosyltransferase), as follows:

$$\text{ADPGlc} + \text{Glc}_{(n)} \rightleftharpoons \text{Glc}_{(n+1)} + \text{ADP}$$

There is some debate as to whether there exists an alternative synthetic pathway involving phosphorylase enzymes, as follows:

$$\text{Glc-1-}P + \text{Glc}_{(n)} \rightleftharpoons \text{Glc}_{(n+1)} + P_i$$

There is no evidence that amylose is synthesized in this manner, although phosphorylases might play a role in the formation, from maltose, of the small glucose primer molecules required for ADPGlc starch synthetase action.

As shown in Fig. 1.5, the two forms of starch are the straight-chain amylose [comprised of α(1→4)-linked glucose residues] and the branched-chain amylopectin [linear amylose chains with branches α(1→6)-linked by glucosidic bonds]. Therefore, the presence of a branching enzyme to forge the α(1→6) links is also required. This enzyme introduces short chains of α(1→4)-linked glucose

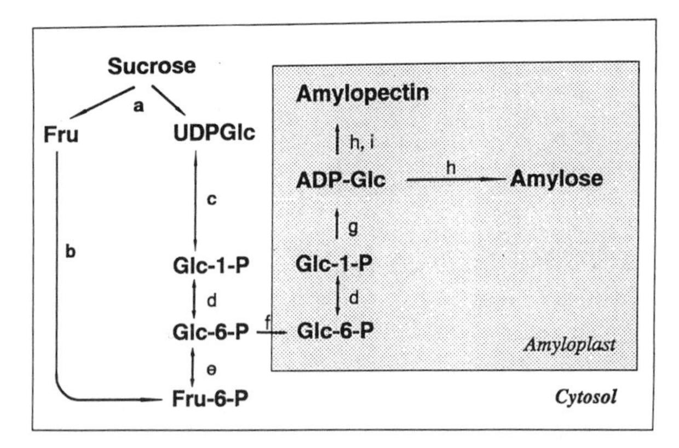

Figure 2.11. Location of enzymes of starch synthesis, based on the studies on the cotyledons of developing pea seeds. a: Sucrose-UDP glucosyltransferase; b: hexokinase; c: UDPGlc pyrophosphorylase; d: phosphoglucomutase; e: hexose phosphate isomerase; f: hexose phosphate translocator; g: ADPGlc pyrophosphorylase; h: starch synthetase; i: starch-branching enzyme. Based on Smith and Denyer (1992).

Table 2.1. Some Mutants of Maize, Characteristics of Their Sugars and Starch, and Possible Relationship to Synthetic Enzyme Activities[a]

Mutant (gene)	Response	Cause	Water-soluble polysaccharides (%)	Starch (%)
Amylose extender (ae)	Fewer branch points in amylopectin	Absence of one of the branching enzymes (IIa)	—	—
Sugary[b] (su)	Smaller, more highly branched, and soluble amylopectin (phytoglycogen)	Altered activity of a branching enzyme (I)	36	30
Shrunken-2 (sh_2)	Reduced starch deposition	Reduced ADPGlc pyrophosphorylase activity: large subunit of enzyme absent	2	25
Waxy (waxy)	Little amylose production, mostly amylopectin	No granule-bound starch synthetase: no protection of amylose from branching enzymes	—	—

[a]Normal maize contains 1.3% water-soluble polysaccharides and 65% starch.
[b]*su-1* is the standard type sweet corn.

residues to amylose; these chains then have more glucose units added onto the nonreducing end by the synthetase enzyme.

Our understanding of the role of enzymes in starch synthesis has been greatly aided by the discovery that certain mutants, particularly of cereals, exhibit variations or defects in their starch-synthesizing pathways. Some of these are listed for maize in Table 2.1. A particularly interesting mutant of starch synthesis causes the wrinkled pea-seed character, used by Gregor Mendel in deriving his laws of inheritance. There are several striking differences between the starch composition of the normal round (*RR*) and wrinkled (*rr*) pea seed (Table 2.2), especially in relation to the much reduced amylopectin content in the latter. The reduction in this branched form of starch in the wrinkled pea is accompanied by a reduction in the activity of the final enzyme in the pathway of starch synthesis, the starch-branching enzyme (Fig. 2.11). Activities of other enzymes in the pathway are not significantly affected. The use of antibodies which specifically detect the major isoform of starch-branching enzyme (SBEI) in pea, has shown that this enzyme is present in developing round peas, but not in wrinkled ones (Fig. 2.12). Failure of the wrinkled peas to produce the enzyme occurs because its mRNA is not transcribed during development; this, in turn, is because the gene that encodes it has been disrupted by a transposon-like element of 800 base pairs (Fig. 2.12B) and is not transcriptionally functional. The round pea has no such insertion in the SBEI gene. Since the wrinkled pea does synthesize some amylopectin (Table 2.2), a second form of starch-branching enzyme (isoform SBEII) must exist even in wrinkled peas, but its activity is considerably below that of SBEI during seed development.

An obvious question is how can the lower activity of one isoform of one enzyme (SBEI) in the starch synthetic pathway lead to such a profound morphological change as the wrinkling of the mature pea seed? The action of the branching enzyme normally results in the formation of many new nonreducing ends in the growing starch polymer. A decrease in the number of nonreducing ends means that these sites of action for starch synthetase (Fig. 2.11) are less abundant, and hence starch synthesis declines. The consequence of this slowing down of the starch synthetic pathway is an accumulation of sucrose in the cells of the cotyledons. In fact, the sucrose content of wrinkled peas is nearly twice that of its round counterpart (Table 2.2). Thus, the wrinkled pea contains less starch to fill up the cells of the cotyledons at late maturity, and more sucrose. The latter, being an osmoticum, will draw water into the cells, giving the wrinkled seed a higher water content (Table 2.2). The final stage of maturation of the pea seed is drying. Prior to this desiccation the wrinkled seeds look very similar to the round seeds, but as water is lost from the former it loses a greater proportion of its volume than the latter, and wrinkling of the overstretched testa occurs because it cannot shrink to the same extent as the embryo, particularly the cotyledons, it encloses.

In nonphotosynthetic, starch-storing tissues, e.g., cereal endosperms, starch synthesis is initiated in the proplastids with the formation of one (rye, wheat, and maize) or more (rice, oats) small starch granules, which at first occupy only a small volume of this organelle. There is then a gradual increase in size of the granule (or granules) until the mature amyloplast is completely filled. Characteristically, the mature starch grains appear to be composed of concentric rings or shells. Each ring probably represents one day's accumulation of starch in the grain. Some cereals, e.g., barley, contain both large and small amyloplasts

Table 2.2. Compositional Differences between Round (*RR*) and Wrinkled (*rr*) Pea Seeds[a]

	Round	Wrinkled
Starch content (% dry wt)	51	29
Amylose (% dry wt)	15	21
Amylopectin (% dry wt)	35	8
Sucrose (% dry wt)	4.1	7.4
Water content of seed prior to maturation drying (%)	60	74

[a]From data summarized by Wang and Hedley (1991).

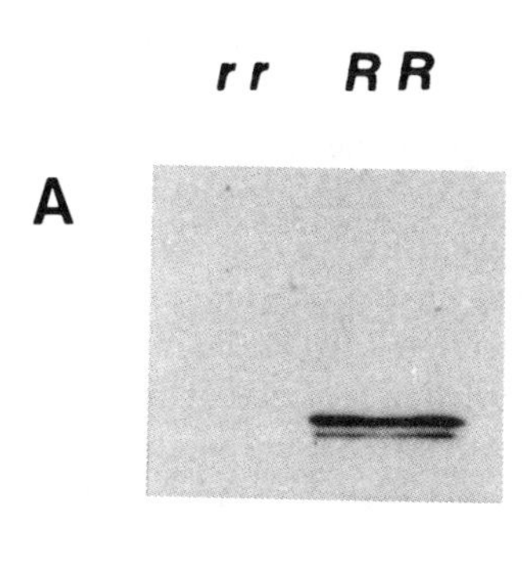

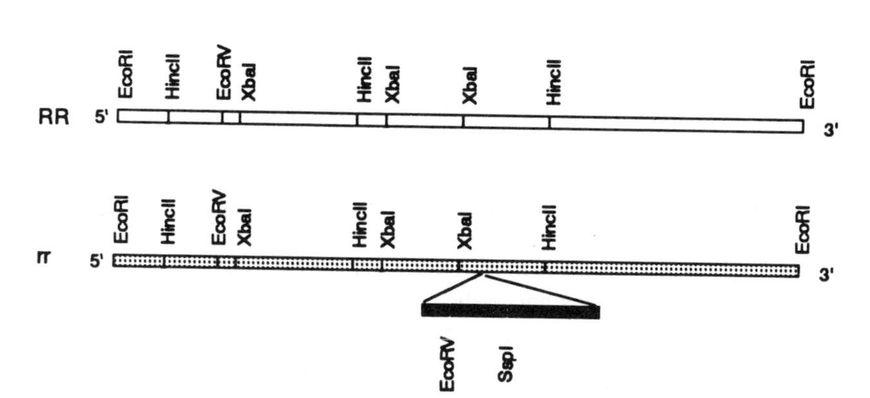

Figure 2.12. (A) Western blot analysis of proteins extracted from round (*RR*) and wrinkled (*rr*) pea seeds during development, identifying SBEI using a specific antibody prepared against the purified 114-kDa enzyme. Note the absence of this enzyme from the wrinkled pea. (B) Restriction maps of the genomic region of pea DNA from which SBEI is transcribed. Note the region of insertion of an 800-bp element (transposon—solid black region) in the gene of the wrinkled pea (*rr*). This prevents proper transcription of the gene. Sites where restriction enzymes cut are shown. After Bhattacharyya *et al.* (1990).

in the mature endosperm. It appears that the large ones (A type) are initiated early after anthesis and the smaller ones (B type) only later, by evagination or constriction of existing amyloplasts. They greatly outnumber the A type (about 10-fold) but account for only 30% of the accumulated starch. The presence of both large and small amyloplasts serves effectively to pack the mature cells of the starchy endosperm and thus, along with the protein bodies, occlude the living contents (Fig. 1.6A).

The amyloplasts of the cotyledons of many legume species, including pea, develop from chloroplast-like plastids. Single starch grains arise between the stacks of thylakoids in the plastids at the time when reserve deposition commences, and they increase in size as starch is deposited. The internal membranes are pushed against the outer membrane, and lose their integrity during maturation drying. Cells located toward the inside of the cotyledons of pea contain larger starch grains; those toward the outside not only contain fewer and smaller starch grains, but also have small chloroplasts in which no starch is deposited. These account for the pale (pea) green color of the mature seed.

A different pattern of starch accumulation occurs during development in many oil seeds such as the legume, soybean, and rapeseed. In these, there is an initial increase in starch content in the cotyledons during development, followed by a decline, so that in the mature seed little starch remains. This is because the starch is remobilized in the later stages of development to provide carbon skeletons for oil and protein synthesis. Presumably the products of starch hydrolysis supplement the restricted flow of assimilates coming into the cotyledons from the vegetative plant.

2.3.2. Deposition of Polymeric Carbohydrates Other Than Starch

Starch is not the only carbohydrate stored in cereals and legumes, and even where it is present, it may not be the major form of stored polymeric sugar. In cereals, the walls of the dead cells of the mature starchy endosperm are comprised of considerable quantities of hemicelluloses and of glucans containing $\beta(1\rightarrow3)$ and $\beta(1\rightarrow4)$ links, although in minor amounts in relation to the amount of starch stored within the cells. In the garden pea, some 35–45% of the seed dry weight is attributable to starch, but in the field pea hemicelluloses in the cell walls eventually make up about 40% of the seed dry matter and starch only some 25%. Hemicellulose synthesis continues after starch synthesis has ceased. Unfortunately, little is known about the synthesis of these cell-wall components in developing seeds.

In certain seeds, including some of the endospermic legumes, the major storage product is deposited in the cell walls as hemicellulose (Section 1.5.1). In species such as fenugreek (*Trigonella foenum-graecum*), ivory nut (*Phytelephas macrocarpa*), coffee (*Coffea arabica*), and date (*Phoenix dactylifera*), the hemicelluloses—largely galactomannans—may be deposited in such large quantities in the cells of the storage organs that the living contents are occluded by the inward thickening of the cell wall. Galactomannan synthesis starts at an early stage of development in fenugreek in the endosperm cells next to the embryo, followed by cells farther away, so that the cells adjacent to the embryo are the first to become filled while those on the outer periphery of the endosperm fill last. No filling occurs in a single outer layer of cells, the aleurone layer (Fig. 7.6) which remains as the only living cells of the mature endosperm. Its role in galactomannan mobilization following germination will be discussed in Chapter 7 (Section 7.3.2). The formation of galactomannan might occur in vesicles of the rough endoplasmic reticulum (RER), the polymer then being secreted through the plasmalemma and into the surrounding cell wall.

The source of carbohydrate for galactomannan synthesis is sucrose entering the endosperm from the maternal tissues. After hydrolysis and phosphorylation, the hexoses are converted to Man-1-*P* and then to the nucleotide sugar, GDP-Man (Fig. 2.13). A membrane (RER)-associated enzyme, GDP-mannose–dependent mannosyltransferase, then transfers Man residues to a linear $(1\rightarrow4)$-β-linked mannose primer, to form the growing backbone chain of the galactomannan polymer. Simultaneously, another membrane-associated enzyme, UDP-galactose–dependent galactosyltransferase, transfers Gal residues as $\alpha(1\rightarrow6)$-linked unit side chains, at intervals which are generally species-specific, to the most recently transferred Man units on the growing chain. Gal cannot be transferred to preformed Man chains, and thus the activities of the mannosyltransferase and galactosyltransferase increase in parallel during galactomannan

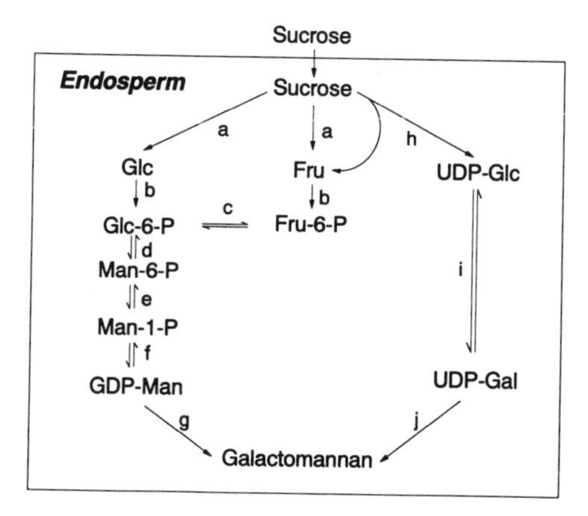

Figure 2.13. Proposed pathway for the synthesis of galactomannan in the endosperm of developing endospermic legumes. Enzymes: a: invertase; b: hexokinase; c: hexose phosphate isomerase; d: phosphomannoisomerase; e: phosphomannomutase; f: GTPMan-1-*P* guanylyltransferase; g: GDPMan-linked mannosyltransferase; h: sucrose synthetase; i: epimerase; j: UDPGal-linked galactosyltransferase. It has not been determined if the cleavage of sucrose by invertase occurs outside or inside the cells of the endosperm.

deposition (Fig. 2.14). Usually, then, the Man/Gal ratio in galactomannan is constant throughout seed development. Any deviations from this occur because of an increase in α-galactosidase activity at late stages of development, which removes some Gal side chains. In senna *(Senna occidentalis)*, for example, there is an increase in Man/Gal ratio from 2.3 to 3.3 over the last 5–10 days of endosperm development.

2.3.3. Triacylglycerol Synthesis

There is considerable variation in the major fatty acid constituents of stored triacylglycerols in seeds. Some examples are presented in Table 1.4, although the diversity is much greater than outlined there. For the sake of brevity and clarity we will not dwell on this diversity but will restrict ourselves to highlighting some of the important facets of triacylglycerol (also called fats, oils, or neutral lipids) biosynthesis in developing seeds.

Sucrose entering the developing oil seed is used mainly for the synthesis of storage triacylglycerols and protein, though why it should be channeled largely into the synthesis of triacylglycerol and not into starch is not clear. The question is relevant because there is ample evidence that oleogenic seeds accumulate starch about midway through their development. Later, however, the starch is mobilized, possibly to be used for triacylglycerol synthesis (though this has not been demonstrated), and the production of triacylglycerols from the imported sucrose becomes predominant.

In broad terms, triacylglycerol synthesis can be considered in three parts: (1) the production of the glycerol backbone, (2) the formation of fatty acids (or fatty acyl moieties), and, (3) the esterification of glycerol with fatty acid components to give triacylglycerols. It is worth noting at this point, however,

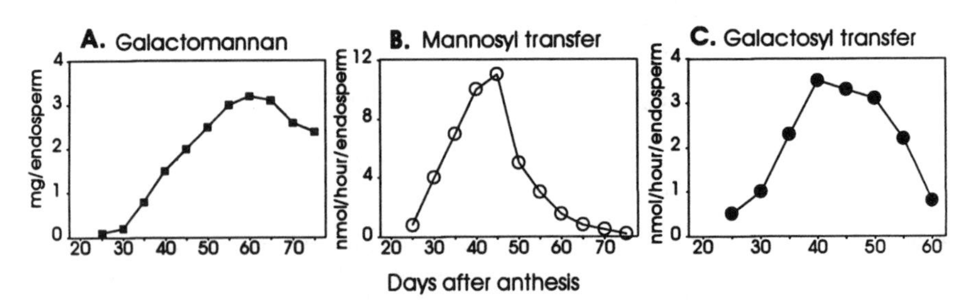

Figure 2.14. (A) Galactomannan deposition in developing endosperms of fenugreek is accompanied by concomitant increases in the activities of (B) GDPMan-linked mannosyltransferase and (C) UDPGal-linked galactosyltransferase. After Edwards *et al.* (1992).

that neither free glycerol nor free fatty acids are involved, but rather glycerol-3-phosphate and fatty acid linked with coenzyme A or an acyl carrier protein.

A generalized scheme for triacylglycerol biosynthesis and the localization of the various steps is shown in Fig. 2.15. Sucrose is translocated into the developing seed and converted to hexose phosphates and triose phosphates by the reactions of glycolysis. One of the latter, dihydroxyacetone phosphate, is reduced in the cytosol to yield glycerol-3-phosphate, which is later acylated. Fatty acid synthesis occurs in the plastids utilizing acetyl-CoA which can be generated by glycolytic reactions in this organelle. Activities of glycolytic enzymes such as enolase and pyruvate kinase in the plastids of oil-storing seeds, for example the endosperm of castor bean, are known to increase during the period of triacylglycerol accumulation. The starting compounds for glycolysis

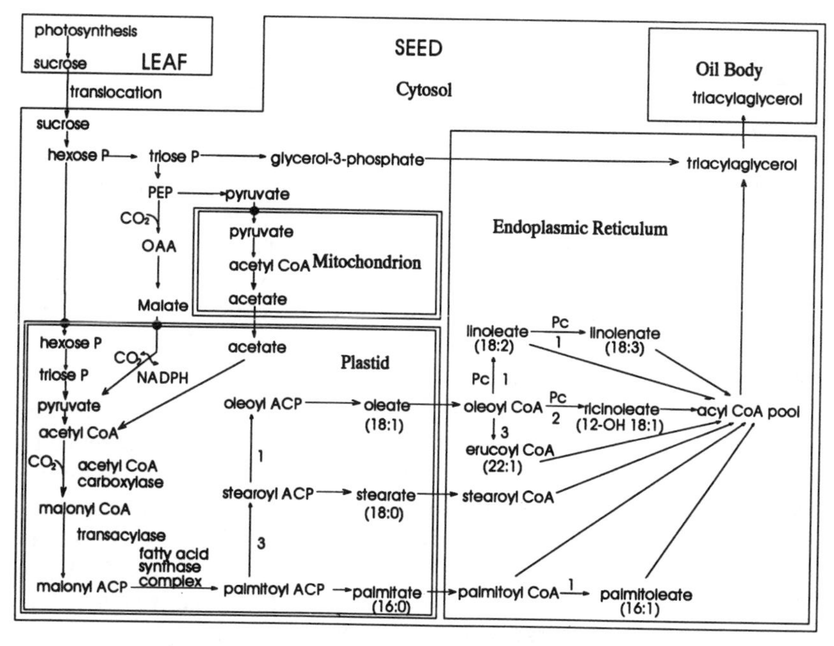

Figure 2.15. Scheme for the synthesis of fatty acids in seeds. Synthesis involves three cell compartments: the cytosol, the plastid, and the endoplasmic reticulum, the latter becoming modified to form the oil bodies. The mitochondrion might also participate. Numbered steps require the following enzymes: 1, desaturase; 2, hydroxylase; 3, elongase. Polyunsaturation and hydroxylation takes place on acyl groups incorporated temporarily into phosphatidylcholine (Pc). Mammalian tissues are not able to desaturate oleate to linoleate: hence, this fatty acid, an essential part of the diet, must be obtained from plant sources. Based on Simcox *et al.* (1977), Stumpf (1977), and Murphy *et al.* (1993).

are held generally to be hexose phosphates which enter the plastids by the activity of specific translocators situated in the outer membranes. Plastids from developing rapeseed embryos contain all of the enzymes for the glycolytic conversion of hexose phosphates, at activities which fit well with the rate of acetyl CoA and fatty acid formation. It has been shown in castor bean endosperm that malate formed in the cytosol can be imported into the plastids where it is decarboxylated to pyruvate, thence to acetyl-CoA. The malate-to-pyruvate conversion, by the malic enzyme, generates NADPH, a reductant used for fatty acid biosynthesis (the other reductant, NADH, and ATP, are produced by plastidial glycolysis). Other possible sources of acetyl-CoA in the plastid are free acetate, produced from acetyl-CoA in the mitochondria, or made available by an acetyl carnitine/carnitine shuttle which is known to operate in chloroplasts of some species. Whether these are important in oil-seed plastids is not clear.

Acetyl-CoA is carboxylated to yield malonyl-CoA, the malonyl residue of which is transferred to acyl carrier protein (ACP) forming malonyl-ACP (Fig. 2.15). There is then a condensation between the acetyl moiety of acetyl-CoA and the malonyl moiety to form acetoacetyl-ACP with the release of CO_2. By two reduction steps, using NADPH and NADH, and a dehydration, the acetoacetyl-ACP is converted to a fully reduced 4C acyl residue. This condenses with another malonyl moiety (attached to ACP), releases a CO_2 molecule, and undergoes reduction and loss of water to produce a fatty acyl moiety, now 6C long. This sequential addition of 2C units continues through 16C (16:0, palmitoyl-ACP) to reach stearoyl-ACP (18:0). Free stearic or palmitic acids may be released, or stearoyl-ACP may be aerobically desaturated to yield oleoyl-ACP, from which free oleate (18:1) may be derived. The conversion to the acyl-ACPs is catalyzed by a complex of enzymes, the fatty acid synthetase complex, consisting of six enzymes arranged around ACP molecules, to which are attached the intermediates en route to acyl formation.

The three fatty acids, palmitate, stearate, and oleate, leave the plastid and are linked with CoA while closely associated with the endoplasmic reticulum. The unsaturated acids, linoleic (18:2) and linolenic (18:3) are formed from the oleoyl residue of oleoyl-CoA after the acyl group has been transferred into a phospholipid, phosphatidylcholine. The unsaturated acyl groups are released from the phospholipid as their CoA derivatives, to enter the acyl-CoA pool. Hydroxylation to form ricinoleate also occurs in the phosphatidylcholine form. The acyl-CoA pool furnishes acyl groups for condensation with glycerol-3-phosphate, eventually to form the triacylglycerols (Fig. 2.16).

Since the fatty acid composition of the triacylglycerol is highly characteristic of a particular species, mechanisms must exist to determine the proportion of the different acyl-CoA types that are formed, to be available for triacylglycerol synthesis. One question that could be posed in this connection relates to why acyl chain elongation does not always continue all the way to oleoyl, leaving no

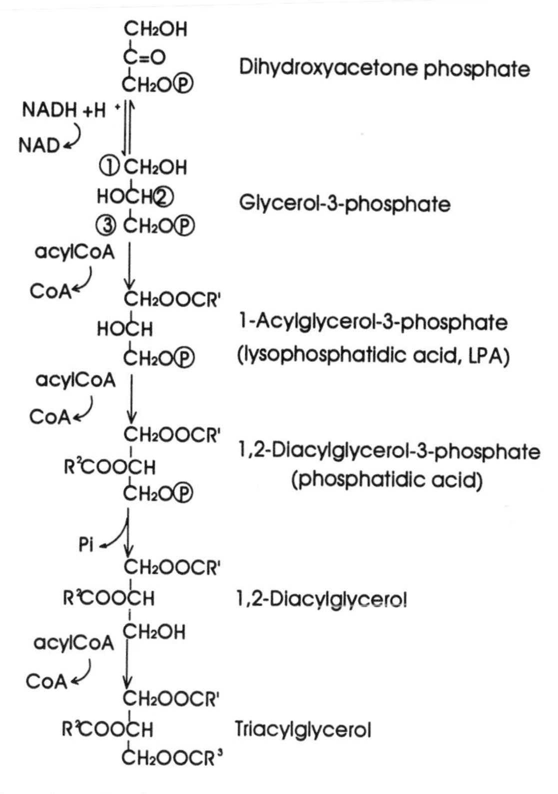

Figure 2.16. The pathway for triacylglycerol synthesis. Acylation is catalyzed by acyltransferases which show preferences for different acyl groups. For example, in developing cotyledons of cocoa seed the enzyme acylating glycerol-3-phosphate prefers palmitic and excludes stearic acid, whereas the second acylation, of LPA, selects unsaturated fatty acids. Hence, the unique composition of triacylglycerol, especially the fatty acids in the 1 and 2 positions, could be determined this way. There is less selection at the 3 position whose fatty acids reflect those available in the acyl CoA pool.

palmitoyl or stearoyl. The answer seems to be that the thioesterases, the enzymes which hydrolyze the acyl-ACPs to give the free fatty acids, and therefore in effect stop further elongation, show strong preference for particular acyl-ACP types. The thioesterases from developing rapeseed, for example, are much more active against oleoyl-ACP than against palmitoyl- and stearoyl-ACP, while the enzymes in palm are more active against palmitoyl-ACP than against the other two. These facts would partly explain the predominance of longer-chain fatty acids in rapeseed and the high proportion of palmitate in palm oil. Using this knowledge about the preferences of different thioesterases it is now possible to alter the fatty acid composition of triacylglycerol. The lauroyl-ACP thioesterase from *Umbellularia californica* (the California bay) has been isolated, purified and the cDNA of the enzyme prepared. Introduction of the cDNA into rapeseed plants induced the formation of high amounts of lauric acid (12:0) in the seeds of the transgenic plants, a fatty acid which is not typical of this species.

Crambe, a species of Cruciferae rich in erucic acid, illustrates many of the principles of seed oil accumulation (Fig. 2.17). During the initial or lag phase of seed development, little or no storage triacylglycerol is present, owing to the absence of the appropriate enzymes. Some galactosylacylglycerols and phosphoacylglycerols, rich in linolenic acid (18:3), can be detected at this time; these are components of the membranes of the cotyledon chloroplasts. In the second phase, some 8–30 days after flowering, there is a gradual accumulation of triacylglycerol, accompanied initially by a transient rise in linoleic (18:2) and oleic (18:1) acids and then a steep increase in the amount and proportion of erucic acid (22:1). In the final stage of seed development, later than 30 days from flowering, triacylglycerol accumulation ceases, and presumably the enzymes involved are destroyed and their synthesis stops.

There are several variations to this generalized pattern of triacylglycerol synthesis. In safflower, for example, there is a temporary accumulation of oleic acid (18:1) at the time when triacylglycerol synthesis commences, presumably

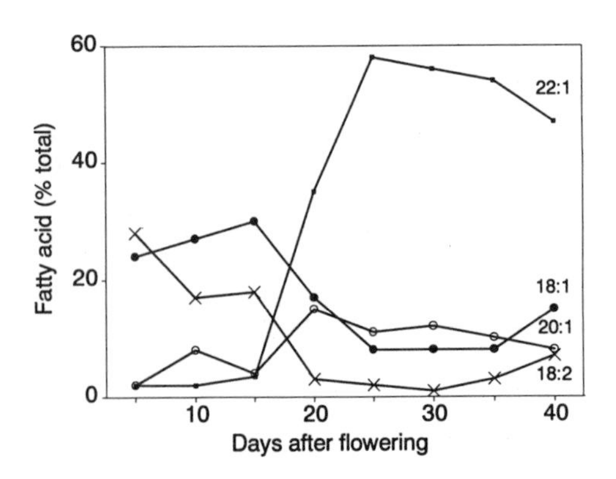

Figure 2.17. Changes in the composition of some fatty acids in *Crambe abyssinica* seed storage oils during development. After Appleby *et al.* (1974).

because the rate of its synthesis initially exceeds the capacity of the desaturase enzymes to produce linoleic acid (18:2). Later the proportion of oleic acid falls as the major storage triacylglycerol is formed. Some attention has been paid to the synthesis of the unique wax esters of jojoba (*Simmondsia chinensis*) seeds. The stored products are liquid waxes comprised of oxygen esters of long-chain fatty acids and alcohols, and they are of great importance because they are difficult to synthesize commercially; the only other natural source of such waxes is the endangered sperm whale. Because jojoba wax is liquid, it is useful for the manufacture of extremely high-pressure lubricants such as those now used in the power transmissions of heavy-duty vehicles. Unlike many of the other oil-bearing plants of agronomic importance, the jojoba plant is a low shrub and only yields seed after taking several years to reach maturity. Studies on the seed show that wax synthesis reaches its maximum potential when the seed is about half its mature weight, and that the desirable composition of waxes is reached about 20 days before the completion of development.

The characteristic fatty acid composition of plant triacylglycerols is largely genetically determined through the different enzymes participating in fatty acid and triacylglycerol synthesis, although environmental factors may result in modifications during seed development. Seed oils of plants grown in cool climates generally tend to be more unsaturated than those grown in warm climates. The major influence is usually on the predominant fatty acid of the seed, but not all oil-rich species, or lines, are affected by growth temperature in this respect. The precise nature of a climatic influence on fatty acid composition in seeds remains to be elucidated, although activity and synthesis of enzymes such as oleate desaturase are increased at low temperature in developing sunflower seeds. This results in an increase in the linoleate/oleate ratio in phosphatidylcholine and triacylglycerols.

Much progress is now being made concerning the biochemistry and molecular biology of fatty acid and triacylglycerol synthesis which will help to clarify some of the previous uncertainties and which also will open the way toward the modification of triacylglycerol production in seeds by genetic engineering. Component enzymes of the fatty acid synthetase complex have been clarified. Mention has already been made of the specificity of acyltransferases and that the genes for some of these enzymes have been cloned and used in plant transformation experiments. Desaturase genes have also been cloned and antisense technology has been used to interfere with desaturation processes, for example in transformed rapeseed. In this case, the proportion of oleic acid in the triacylglycerol has been reduced, with a concomitant increase in stearic acid. In some species, mutants have been produced for various steps in fatty acid or triacylglycerol synthesis which will facilitate the isolation of particular genes. Hence, the technology is becoming available for engineering the fatty acid composition of the seed oils.

Regulation of triacylglycerol synthesis awaits clarification. Important in this respect is the elucidation of its tissue and temporal specificity, i.e., why triacylglycerol is laid down only in certain seed tissues (e.g., cotyledons) and at specific times. The regulation of expression of genes coding for the various enzymes involved is important in both of these. There is some evidence from work on somatic embryos of several species, on wheat zygotic embryos, and from mutants of *Arabidopsis* that the growth regulator abscisic acid and the osmotic properties of the embryo are concerned in the regulation of the synthesis of triacylglycerol and the types of fatty acids therein.

The reserve triacylglycerols within the cells of storage organs are confined to organelles called oil (fat or lipid) bodies. The development of these bodies has been the subject of considerable controversy over the years, but current evidence points clearly to their origin in the ER. Newly formed triacylglycerol accumulates between the two layers of the ER double membrane, leading to its swelling (Fig. 2.18). When the oil-filled vesicle reaches a critical size, it may bud off completely (Fig. 2.18, type A), bud off with a little ER still attached (type B), retain multiple contacts with the ER system (type C), or pinch off as an independent "microsome" (type D)—in the membranes of which are deposited fats. Since the storage triacylglycerols accumulate in the membrane between the phospholipid layers, the outer structure surrounding the mature lipid body is a half-unit membrane.

Associated with the phospholipid moiety of the membrane of the oil body is a unique class of protein, the oleosins. These are proteins of low molecular mass (15–26 kDa) which are structurally incorporated into the membrane of the oil body (Fig. 2.19B) during its ontogeny; they are synthesized on ER-bound polysomes as the triacylglycerol is being deposited within the bodies (Fig. 2.19A). In the scutellum of maize, expression of the oleosin gene may be upregulated by abscisic acid. Oleosins possess three structural domains (Fig. 2.19B): (a) an amphipathic domain of 40–60 amino acids near the NH_2-terminal end, which is associated with the oil body surface; (b) a hydrophobic domain of 68–74 amino acids which penetrates into the oil body and lies within the triacylglycerol matrix; and (c) an amphipathic α-helical domain of 33–40 amino acids situated at or near the COOH-terminus, and which also protrudes from the oil body surface. The difference in molecular masses of the various oleosins (isoforms), which vary between, and even within, species, is the result of extension of the COOH- and NH_2-terminal (a and c) amino acid domains. The oleosins serve a structural role in stabilizing the half-unit phospholipid membrane that surrounds the oil body, and provide the surface with a net negative charge, which by electrical repulsion prevents these organelles from coalescing and aggregating. Oleosins may also protect the phospholipids from hydrolysis by cytoplasmic phospholipases, and provide a binding site for lipases involved in postgerminative triacylglycerol mobilization.

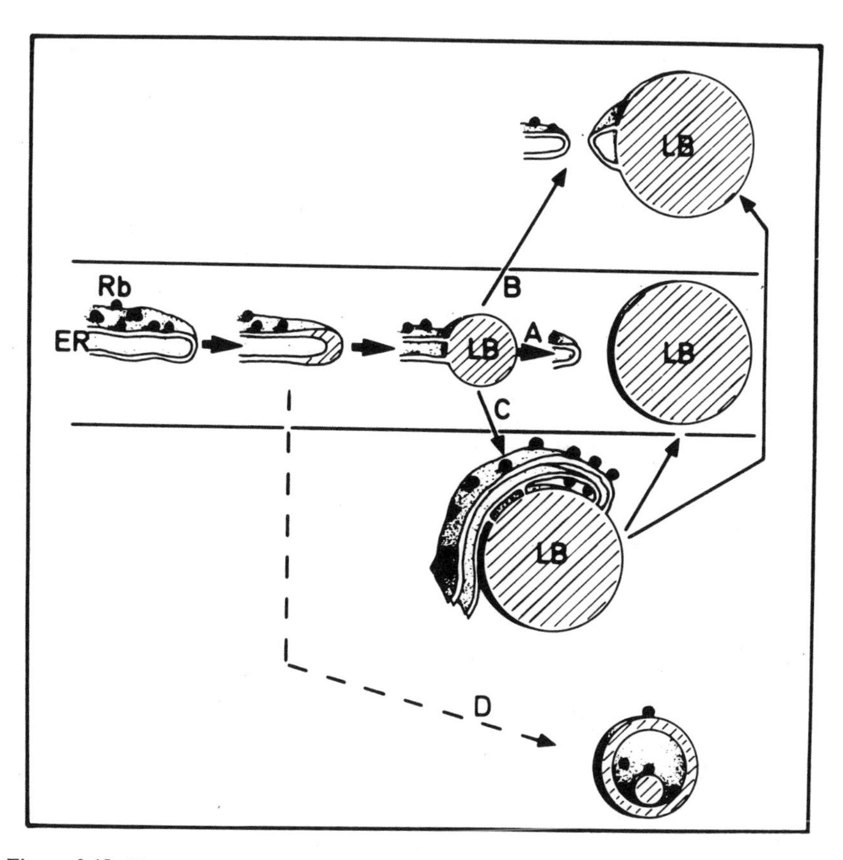

Figure 2.18. The current hypothesis for the development of oil bodies (LB) from endoplasmic reticulum (ER). Several alternatives exist: (A) oil bodies pinch off the ER without a remaining piece of membrane, e.g., broad bean (*Vicia faba*) and pea (*Pisum sativum*); (B) a small piece of membrane remains attached to the oil body, e.g., watermelon (*Citrullus vulgaris*); (C) ER strand maintains contact with the nascent oil body, e.g., pumpkin (*Cucurbita pepo*) and flax (*Linum usitatissimum*); (D) vesicles of oil-rich membranes develop from ER vacuoles by continued oil incorporation into the membrane middle layer, e.g., watermelon. Rb, ribosomes. After Wanner *et al.* (1981).

2.3.4. Storage Protein Synthesis

In general, the syntheses of the various storage proteins of a particular seed are initiated at similar but not identical times during development, and proteins are accumulated at about the same rate. There are some exceptions, however, and in some species the appearance of new storage proteins during late development (even during the early stages of drying) has been noted. Moreover, the

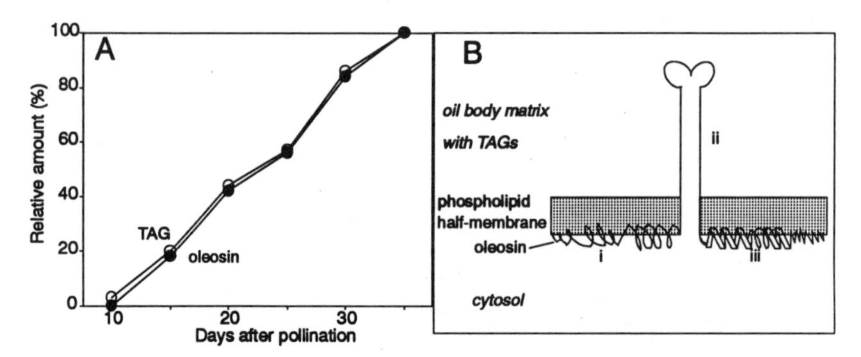

Figure 2.19. (A) The accumulation of triacylglycerols and oleosins in the embryo (scutellum) of maize during kernel development. (B) A model of the 18-kDa maize oleosin associated with oil bodies. The protein is associated with the surface of the bodies in amino-acid domains i and iii, and with the stored triacylglycerols (TAGs) in domain ii. A, after Tzen *et al.* (1993); B, based on Huang (1992). Reproduced with permission from Annual Reviews, Inc. ©1992.

subunit composition of some proteins synthesized throughout development may change during the later stages; e.g., in soybean the α and α′ subunits of conglycinin appear 15–17 days after flowering, but the β subunit appears only 5–7 days later (Fig. 2.20D). The final amount of stored protein varies markedly, and usually there is a characteristic major reserve protein within any one species (Table 1.5). The quantitative and qualitative changes occurring during protein deposition within a cereal grain, a legume, and an oil-storing dicot are illustrated in Fig. 2.20. Although the basic mechanism of protein synthesis is the same in cereals and dicots, the mode of deposition within the protein bodies shows some considerable variations; hence, these two groups of seeds will now be considered separately.

2.3.4.1. Cereals

In barley, wheat, and maize, the major, highly insoluble prolamin storage proteins are synthesized on polysomes closely associated with the ER (Fig. 2.21). The newly synthesized proteins pass through the ER membrane into the lumen where, because of their hydrophobic nature, they aggregate into small particles. These are brought together by cytoplasmic streaming to form larger aggregates, which eventually become protein bodies. In wheat, two different types of protein body accumulate independently within the developing endosperm cells: low- and high-density bodies. Gliadins are present in both types. The light protein bodies are formed early during development in a manner analogous to that in legume seeds (Section 2.3.4.2) and the more dense ones, which are ER-derived, later.

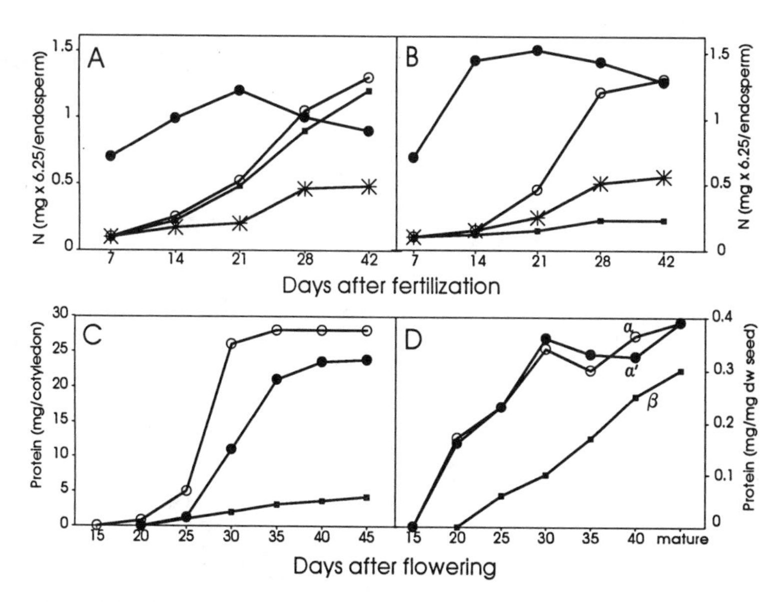

Figure 2.20. (A,B) Changes in the endosperm protein fractions during kernel development of normal Bomi barley (A) and the high-lysine barley mutant Risø 1508 (B). ●, Albumins (plus free amino acids); *, globulins; ■, hordeins (prolamins); O, glutelins. (C) The accumulation of vicilin (O), legumin (●), and albumins (■) in developing cotyledons of broad brean *(Vicia faba).* (D) Accumulation of α, α′, and β subunits of the 7 S storage protein β-conglycinin in developing soybean seeds. A and B, after Brandt (1976); C, after Manteuffel *et al.* (1976); and D, after Gayler and Sykes (1981).

Aggregation of proteins inside the lumen of the ER puts a strain on the membrane and causes it to rupture. The membrane may re-form free of the protein aggregate, and the protein "body" itself is obviously not bounded by an outer membrane. In other cereals, such as millet, rice, maize, and sorghum, where the protein bodies remain as distinct membrane-bound entities even in the mature seed, presumably the strain on the ER is less severe because it distends but does not rupture. Alternatively, the ER itself may be more elastic in nature.

Cereal grains also sequester other storage proteins in addition to the prolamins (e.g., glutelins or globulins, depending on the species) in bodies within the starchy endosperm. The formation of these "crystalline" protein bodies, as well as the protein bodies within the aleurone layer, generally occurs in the same manner as in the cotyledons of legume seeds (Section 2.3.4.2). The high-molecular-weight glutenins of wheat, however, are present in ER-derived protein bodies.

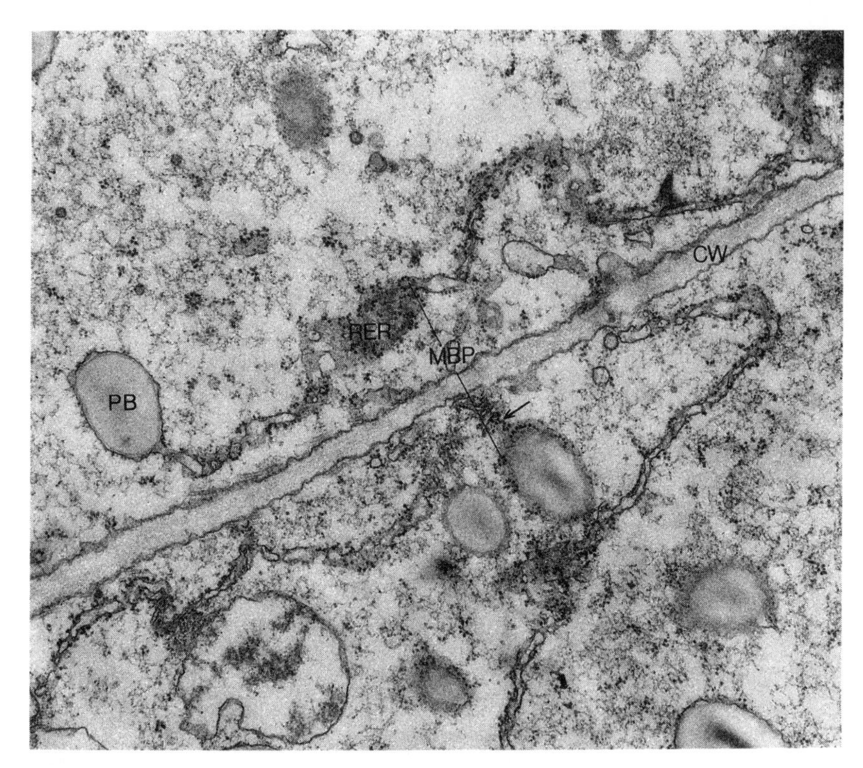

Figure 2.21. Electron micrograph of a developing cell of the starchy endosperm of maize to illustrate the formation of protein bodies (PB) from rough endoplasmic reticulum (RER) and the proximity of membrane-bound polyribosomes (MBP). Note the continuity between the protein body membranes and extended RER cisternae (arrow). CW, cell wall. By Larkins and Hurkman (1978).

Isolation of messenger RNAs for storage proteins from the polysomes associated with the ER [rough endoplasmic reticulum (RER)], and their subsequent translation *in vitro* has been achieved for developing cereal grains. The *in vitro*-synthesized products strongly resemble the prolamin storage proteins characteristic of the species, but their mass is slightly greater (by about 2 kDa) than that of the native protein. This is because within the cells of the developing endosperm itself (i.e., *in vivo*) the newly synthesized protein is modified (shortened) during its passage through the ER membrane into the lumen. This translocation across the membrane is facilitated by the presence of a hydrophobic signal (transit) peptide at the amino end of the nascent protein chain. This signal peptide allows for the passage of the protein across the membrane, but it is removed by proteolytic activity at the same time (Fig. 2.22). This cotranslational modification or shortening of the storage protein occurs *in vivo*, but not *in vitro* in the absence of a membrane preparation. However, as shown by Larkins and coworkers (1979), the cotranslational processing of storage proteins can be achieved even when the mRNA for zein is introduced into an animal cell. They injected this message into oocytes of *Xenopus* and found that not only were all zein components completely synthesized, but also that they were processed correctly and secreted into membrane vesicles!

In the maize endosperm, different classes of zein are deposited within the protein bodies as their development proceeds. In the newly forming bodies only β- and γ-zeins are present, but later the interior of the protein body fills with α-zein, and the β- and γ-zeins are displaced to form a continuous thin layer around the periphery of the protein body at maturity. A few patches of β- and γ-zein occur embedded within the α-zein matrix in interior regions of the mature protein body.

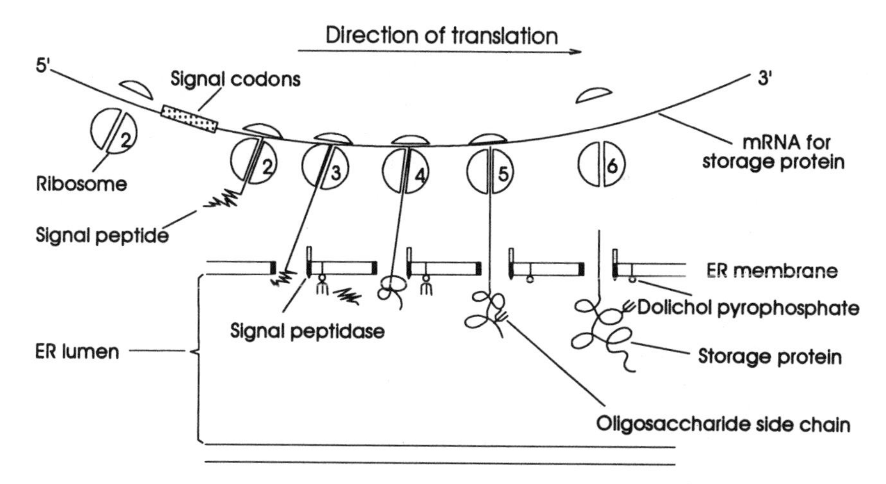

Figure 2.22. A simplified representation of the translocation of storage proteins across the membrane of the rough endoplasmic reticulum. Several stages of the translation of a messenger RNA for a storage protein on a membrane-associated polysome are shown. The polysome contains six ribosomes. The two near the start (5′) end of the mRNA are not yet bound to the membrane. The nascent protein chain "grows" in a passage through the large (lower) ribosomal subunit. The signal peptide is indicated as a jagged line at the amino-terminus of the nascent protein (see second and third ribosomes), or as being cleaved (i.e., cotranslationally) from the protein as translation proceeds by a signal peptidase and located in the membrane on the inner (luminal) face of the ER (see membrane between ribosomes 3 and 4). Glycosylation of certain proteins occurs in the ER lumen by transfer of oligosaccharides from dolichol pyrophosphate carrier molecules located on the inner face of the membrane (transfer occurs between ribosomes 4 and 5). Based on Blobel *et al.* (1979) and Johnson and Chrispeels (1987).

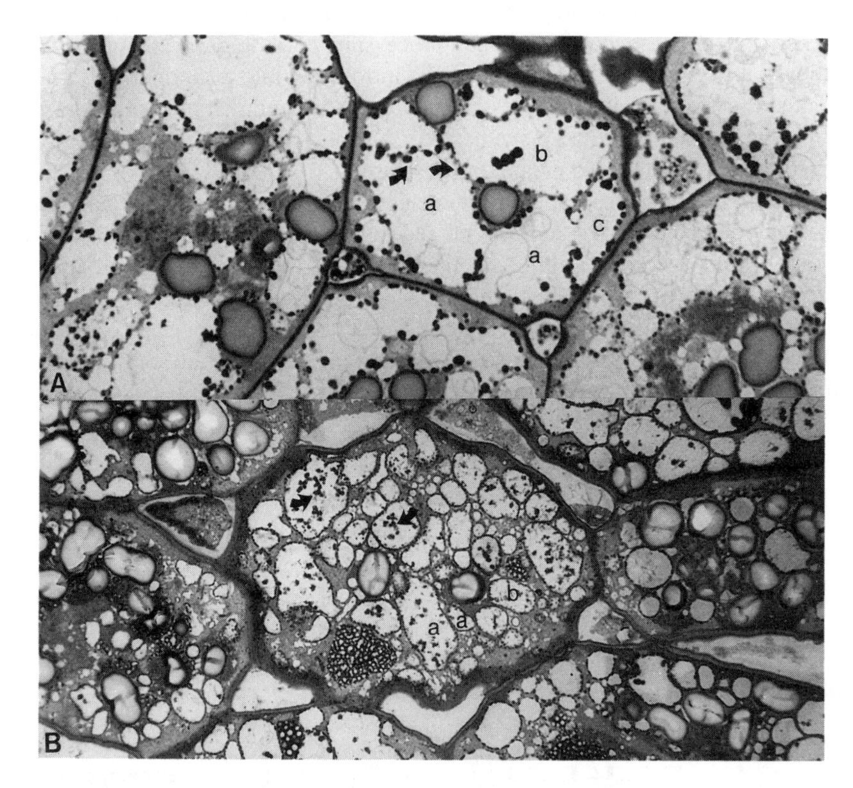

Figure 2.23. Light micrograph of cells of pea cotyledons (A) 12, (B) 15, and (C) 20 days after flowering. (A) A large vacuole is present in the cells. Serial sections through the cell (see original reference) show that (c) is an extension of vacuole (a), which is also continuous with (b). Arrows show proteins deposited in vacuole. (B) Several discrete vacuoles are in evidence, which are irregular in shape; e.g., serial sections through the cell show that (a) is a single vacuole, but it is distinct from (b). Arrows indicate protein deposits. (C) Discrete, spherical protein bodies are in evidence; (a), (b), and (c) are not connected to each other. An indication of the vacuole or protein body diameter, in relation to the diameter of the developing cotyledon cell, can be obtained from the following data:

Days after flowering	Cell diameter (μm)	Vacuole/protein body diameter
10	42	39
12	81	11
15	84	3
20	99	1

Note the large reduction in vacuole size as they fragment to form protein bodies. Original data in Craig *et al.* (1979, 1980).

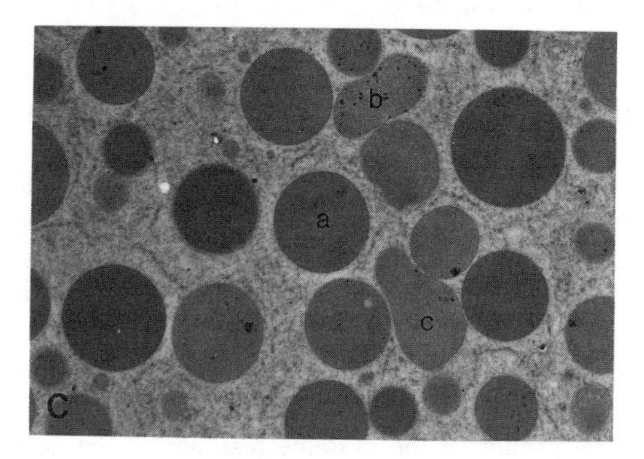

Figure 2.23. (*Continued*)

2.3.4.2. Dicots

The formation of protein bodies in the cotyledons and persistent endosperms of the dicots studied to date follows a distinctly different pattern from that for the prolamin-containing protein bodies in the cereal endosperm. This has been illustrated particularly well for the developing cotyledons of garden pea (*Pisum sativum*) (Fig. 2.23). Early during development, around 12 days after flowering, the cotyledon cells contain one or two large vacuoles, which occupy most of the volume of the cell. Some protein deposition within the vacuole has commenced at this time (Fig. 2.23A). By day 15 many of the vacuoles are considerably smaller and more spherical in shape (Fig. 2.23B), and by day 20 these have become filled with storage protein and form the distinct and discrete protein bodies (Fig. 2.23C). The pattern of development, then, is gradual fragmentation over a few days of a highly convoluted central vacuole, and this leads to the formation of the smaller protein bodies. This sequence of events occurs in the cells of other developing legume and nonlegume cotyledons; it also takes place in the developing endosperm of the oil-storing seeds, e.g., castor bean, except that here vacuolar subdivision occurs before any protein deposition takes place. During the late stages of development of some dicot seeds, e.g., soybean, following the major deposition of reserves, some protein bodies may be formed directly by expansion of the RER cisternae (as in cereal endosperms).

Although deposition of proteins is within the vacuoles/protein bodies, their synthesis occurs on the RER, and there is vectorial transport to the storage organelles. This has been demonstrated in cotyledons of legumes, e.g., soybean. The subunits of glycinin, a 12 S legumin-type protein (Section 1.5.3), are composed of an acidic and a basic polypeptide, which are synthesized sequentially on the same mRNA template (Fig. 2.24). The primary translation product (preprolegumin) contains the signal peptide at the amino-terminal end, the acidic and basic polypeptides joined by a short linker sequence, and a short (penta-) peptide at the carboxy-terminus. In other legumes the linker sequence and the carboxy-terminal peptides may be absent from the preprolegumin. The glycinin is now further processed in the lumen of the RER as the amino- and carboxy-terminal peptides are cleaved off. The protein begins to fold and assemble within the RER into its correct three–dimensional (tertiary and quaternary) structure, an event which is required before the protein can be targeted and transported to its final destination. Proteins within the ER membrane facilitate this process, including a binding protein (BiP) which promotes correct folding of the storage polypeptides, and protein disulfide isomerase (PDI) which forges disulfide bonds using the -SH groups of the cysteine residues of the acidic and basic subunits. The result is the production of prolegumin which is translocated to the protein body, via the Golgi apparatus, for the final stage of maturation, the assembly into hexamers. This occurs only after removal in the protein body of the pentapeptide linking the two subunits by a thiol-containing proteinase (otherwise only trimers can be formed). The six subunits which compose the legumin holoprotein can now assemble, and they become associated into their correct configuration by surface charge interactions.

2.3.4.3. Modifications to Synthesized Storage Proteins, Their Sorting and Targeting to Protein Bodies

Many of the storage proteins in mature seeds are glycoproteins, having one or more oligosaccharide side chains covalently linked to asparagine (Asn) residues in the constituent polypeptides. Examples include the vicilins of legume seeds (e.g., β-conglycinin of soybean) and the lectins of *Phaseolus vulgaris* (phytohemagglutinins); legumins are not glycosylated. The oligosaccharide side chains are of two major types: (1) simple or high-mannose oligosaccharides composed exclusively of mannose (Man) and *N*-acetylglucosamine (GlcNAc), usually in a 5–9:2 ratio, and (2) complex or modified oligosaccharides which are often rich in Man, but contain other residues in addition to GlcNAc, e.g., fucose (Fuc), xylose (Xyl), and galactose (Gal) (Fig. 2.25).

The initial glycosylation of storage proteins is a cotranslational event. High-mannose chains ($Glc_3Man_9[GlcNAc]_2$) are synthesized on carrier molecules of dolichol pyrophosphate (a lipid) via ER-associated glycosyltransferases. As they are inserted into the ER lumen, they are transferred intact to Asn residues occurring in the sequence: –Asn–amino acid–threonine (or serine), on nascent polypeptide chains (Fig. 2.22). Removal of the Glc moieties and one of the four Man residues occurs, leaving mature high-mannose chains attached to the polypeptide. Complex oligosaccharide side chains on glycoproteins are derived

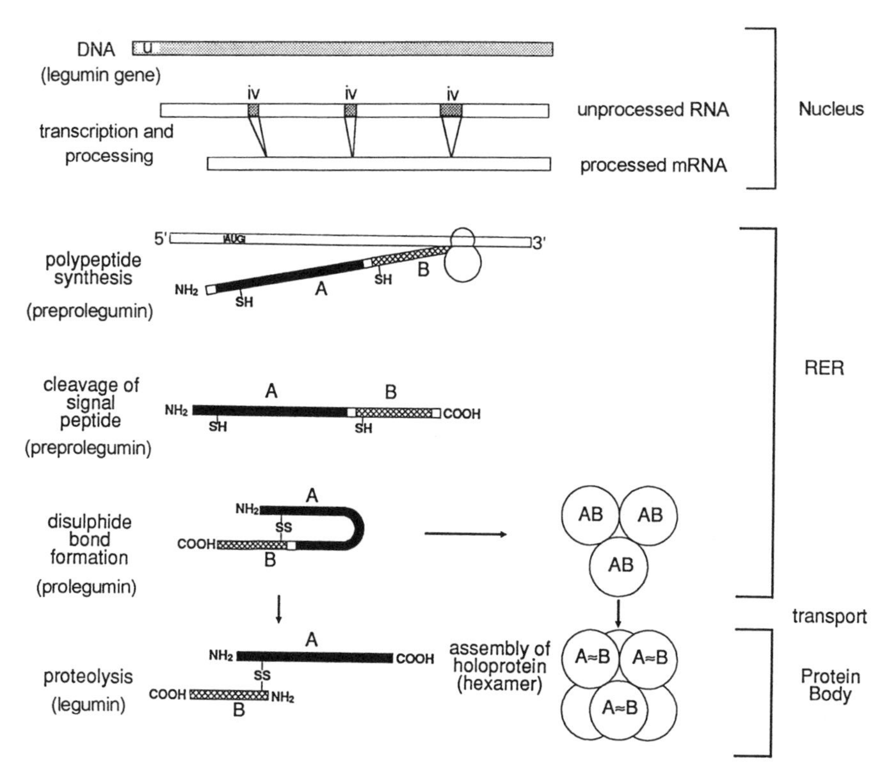

Figure 2.24. A representation of the synthesis of the legumin (glycinin) in soybean cotyledons. The gene for legumin possesses an upstream sequence (u) of several hundred nucleotides which controls its tissue-specific and temporal expression. The first transcriptional product is an unprocessed mRNA which contains nontranslated intron regions (iv), about 1150 bases altogether, and which are spliced out to produce the mature processed legumin mRNA, which is 1455 bases long. (The mRNA is polyadenylated at the 3′ end.) This mRNA is transported from the nucleus to the rough endoplasmic reticulum (RER) where it is translated. The mRNAs for the acidic (A) and basic (B) subunits are joined by codons for a 4-amino-acid linker sequence on the primary translation product (preprolegumin). The mRNA also contains a code for the signal peptide at the amino-terminal end and a pentapeptide at the carboxy-terminal end. The primary product (prolegumin) is processed to yield the A and B subunits joined by the linker sequence and disulfide bonds (S-S); the signal peptide and the pentapeptide are cleaved off. In the final step of processing the mature protein is formed as the linker sequence is removed and the acidic and basic subunits are joined only by the disulfide bonding. The mature acidic subunit contains 278 amino acids (approx. 40 kDa) and the basic subunit 180 amino acids (approx. 20 kDa). Assembly of the subunits within the protein bodies yields the mature hexameric legumin holoprotein. Based on Krochko and Bewley (1989). Reprinted with permission of Longman Group UK. See also Dickinson *et al.* (1989) for details and Bednarek and Raikhel (1992) for a general review.

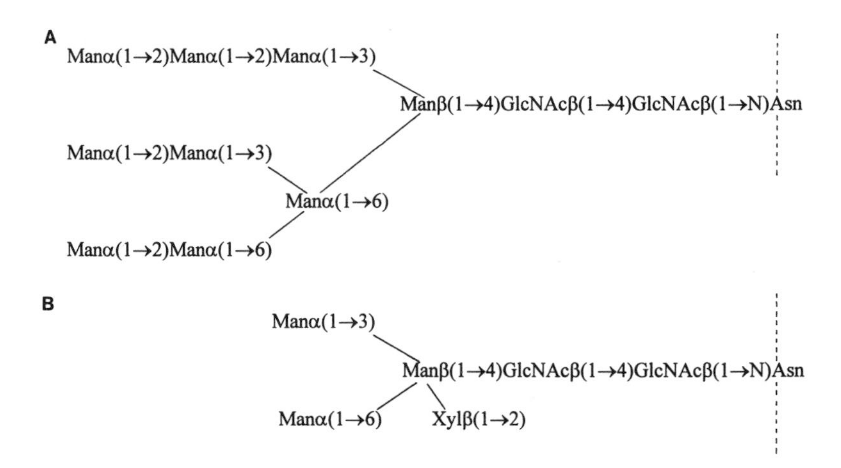

Figure 2.25. The glucan side chains of glycoproteins. (A) A typical Asn-linked simple or high-mannose oligosaccharide side chain (soybean agglutinin). (B) Probable structure of one version of a modified or complex oligosaccharide side chain of phaseolin.

from high-mannose chains following covalent linkage to the polypeptide. The modifications involve several steps, including the removal of several terminal Man residues and the addition of other sugars. Processing of high-mannose side chains to complex side chains does not occur for all glycoproteins, as some have only high-mannose side chains, but many have these and complex side chains on the same polypeptides. The lack of processing of high-mannose side chains may be because they are hidden within the tertiary structure of the polypeptide within the storage protein and are not accessible to the modifying enzymes, α-mannosidase and glycosyltransferase.

While we have a good understanding of the sequence of events involved in glycosylation and modifications of the side chains, the necessity for glycosylation is unclear. Inhibition of protein glycosylation by tunicamycin has no effect on the association of storage protein subunits, the binding ability of lectins, or the transport and targeting of storage protein polypeptides to the protein bodies. The current suggestion is that glycan side chains promote correct protein folding and oligomer formation in the ER.

The synthesis of storage protein precursors and their mode of processing show considerable variation between species, and only a few such examples will be presented. In cotton, the principal storage proteins appear to be synthesized initially as 69- and 60-kDa precursors (Fig. 2.26A). The larger precursor (a) undergoes only a small modification, initially, perhaps by cotranslational re-

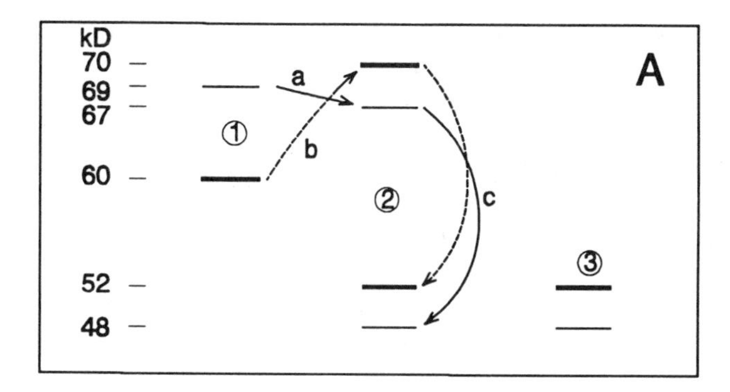

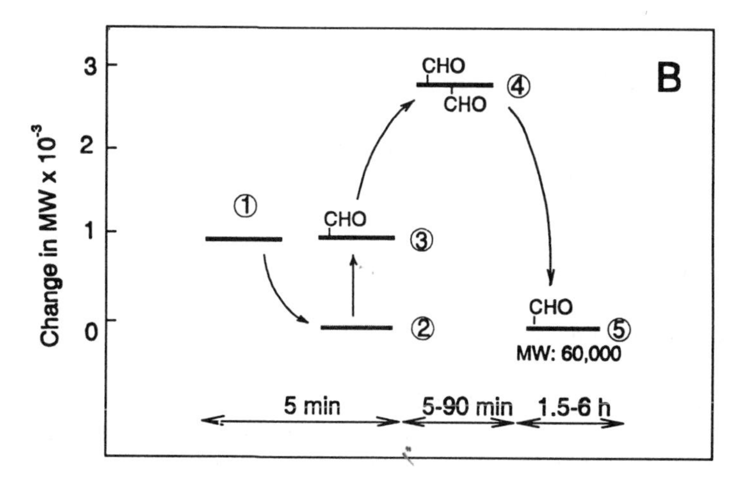

Figure 2.26. (A) The proposed pattern of processing of the principal storage protein in developing cotton *(Gossypium hirsutum)* cotyledons. (1) Precursors. Event (a), cotranslational loss of signal peptide; (b), cotranslational loss of signal peptide and increase in molecular mass owing to glycosylation. (2) Processed products. Event (c), *in situ* cleavage. (3) Molecular mass of abundant storage proteins in the mature seed. (B) Schematic outline of the sequence of events involved in the synthesis of the α subunit of 7 S β-conglycinin storage protein of soybean. (1) Primary translation product of membrane-bound polysomes. This protein is cotranslationally cleaved to form (2) and cotranslationally glycosylated to form (3). After posttranslational glycosylation to produce (4), this is subjected to proteolysis and/or deglycosylation to form the α subunit (5). Time from initial synthesis to completion of processing is approximately 6 h. The final 7 S α subunit is approximately 60 kDa. A, after Dure and Galau (1981); B, after Sengupta *et al.* (1981).

moval of the signal peptide as it passes through the ER membrane. The molecular mass of the smaller precursor increases to about 70-kDa owing to glycosylation—a signal peptide might be lost too. Both precursors are then cleaved further, slowly, and perhaps within the protein bodies, to yield the mature 52- and 48-kDa storage proteins (Fig. 2.26A). The peptides that are cleaved off to produce the proteins from their precursor are very large (15–20 kDa), much larger than any signal peptide, and they may also accumulate within the protein bodies. Comparable posttranslational changes in molecular size during the processing of the 7 S, vicilin-type α subunit of the soybean storage protein are shown in Fig. 2.26B.

In developing endosperms of castor bean, the main storage protein, the 11 S crystalloid complex, is formed by posttranslational cleavage of a 50-kDa precursor into polypeptides of 30 and 20 kDa which become linked by disulfide bridging (as for legumin in soybean, Fig. 2.24). Formation of the matrix protein agglutinins in the castor bean is considerably more complex, however. Agglutinin type I is comprised of two polypeptides of 37 and 31 kDa in the mature state, but they are encoded as 59- and 35-kDa precursors. Cotranslational processing cleaves 1.5- to 2.0-kDa segments (signal peptides), and there is also cotranslational modification of the large-molecular-mass precursor as it increases to 66–69 kDa by glycosylation. Posttranslational cleavage yields the authentic glycosylated 37-kDa agglutinin and 34-kDa protein. The precursor of the smaller agglutinin chain is trimmed to about 32 kDa by cotranslational removal of the signal peptide; subsequently, posttranslational trimming, i.e., removal of small segments of peptides, yields the 31-kDa protein.

As illustrated in Fig. 2.27, storage proteins which are synthesized on the RER are transported to the Golgi apparatus via tubular smooth ER connections. The storage proteins are sorted and packaged into Golgi-derived vesicles and transported to the vacuole/protein body compartment. Fusion of the vesicles with the tonoplast discharges the storage protein into the vacuolar lumen. Evagination of the tonoplast around concentrations of newly deposited storage proteins results in the formation of discrete protein bodies. The processing of the storage proteins during transport from their site of synthesis to their site of deposition has been discussed above; now we must consider how the storage proteins become specifically targeted to the vacuole. For this to occur, the protein must be sorted into Golgi vesicles that are destined for transportation to the vacuole. The determinants for this sorting and targeting are found within the structure of the protein itself, as a contiguous amino acid sequence, or a structure consisting of several amino acids located in separate domains. The regions of the protein which determine that it will be sorted/targeted to the vacuole differ among proteins (Fig. 2.28). For some, e.g., barley lectin, a glycosylated propeptide at the carboxy-terminal end is a necessary and sufficient determinant for it to be sorted and translocated to the vacuole in developing barley embryos. During transport to, or after deposition within the vacuole, the propeptide is removed to

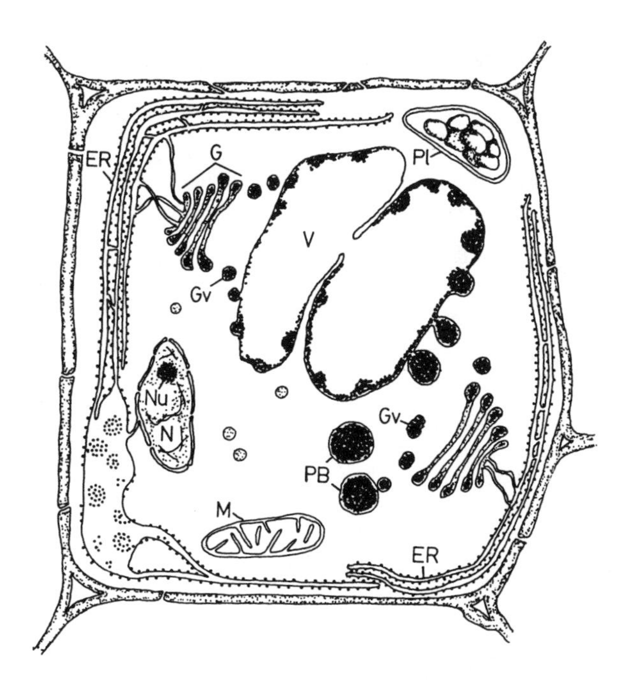

Figure 2.27. The synthesis of storage proteins and their sequestering within the vacuole/protein body, as occurs typically in the storage parenchyma cells of the cotyledons of legume seeds. ER, endoplasmic reticulum; G, Golgi apparatus; Gv, Golgi-derived vesicle; M, mitochondrion; N, nucleus; Nu, nucleolus; Pl, plastid; V, vacuole. After Bewley and Greenwood (1990).

yield the mature lectin. Using genetic engineering techniques, the DNA sequence for the COOH-terminal propeptide of barley lectin was fused to the gene of a protein which is normally secreted from cells, cucumber chitinase. When the resulting chimeric gene was expressed in transgenic tobacco,the fusion protein was redirected to the vacuole, confirming that amino acids within the COOH-terminal region are the vacuolar sorting determinants. There does not appear to be any common sequence of amino acids into the COOH-terminal extensions of the several proteins whose sorting is determined by this region; the only common feature recognized to date is a richness in hydrophobic amino acids.

Other proteins contain an amino-terminus propeptide as a sorting determinant, although no storage protein has been identified in this group. A third class of proteins contains the vacuolar sorting information within the mature protein itself, e.g., phytohemagglutinin and legumin. This information probably consists of a linear sequence of amino acids from two or more regions of the polypeptide chain that lie close together in the three–dimensional structure of the protein.

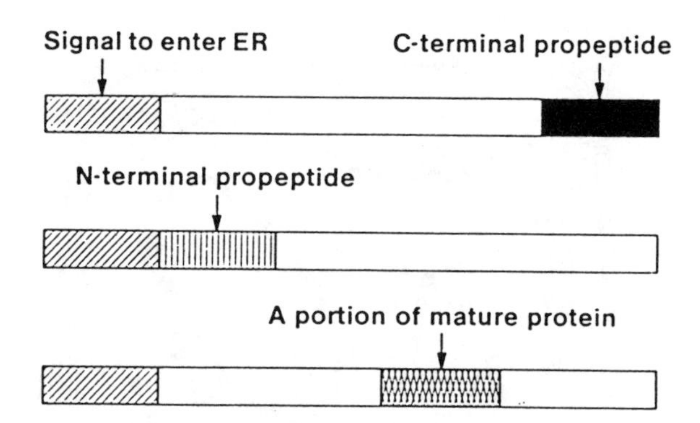

Figure 2.28. Three different regions in which the vacuolar sorting and targeting information occur in plant proteins: in the cleaved carboxy-terminal or amino-terminal propeptide, or internally within the mature protein. The signal peptide for insertion in the ER is shown for each protein. After Chrispeels and Raikhel (1992). See also Holwerda *et al.* (1992), Bednarek and Raikhel (1992), and Nakamura and Matsuoka (1993).

Again, no common sequence of amino acids has been identified in the internal region of these proteins.

Some storage proteins, especially the prolamins in the endosperms of cereal grains, are not targeted to the vacuole but are retained within the ER. Proteins that are ER-resident frequently contain a COOH-terminal tetrapeptide histidine (or lysine)–aspartic acid–glutamic acid–leucine (HDEL or KDEL), which is the ER-retention signal, but this sequence is not present in the cereal storage proteins. Wheat gliadin may contain two sorting signals, one near the COOH-terminal end of the protein which promotes exit of the protein from the ER to the Golgi, and another near the NH_2-terminus which retains the protein and initiates protein body formation within the ER. Sorting of the protein may be determined by a balance between these two opposing signals. In maize kernels, zeins may be retained within the ER via associations with proteins (e.g., BiP, Section 2.3.4.2) that do have an ER-retention signal.

Our knowledge of the mechanisms of protein transport and retention is still at a fairly rudimentary stage. They are the subject of intense interest, however, and although the area of research is a technically difficult one, we can expect progress over the next several years.

2.3.4.4. Seed mRNA Content and Storage Protein Synthesis

In the pea, synthesis of the storage proteins in the cotyledons occurs mostly during the cell expansion phase of growth, following cell division, and ceases as the seed completes its maturation drying (Fig. 2.29). The content of both DNA and RNA in the cotyledons increases during the cell expansion phase. Synthesis of RNA, the bulk of which is ribosomal RNA, during the first half of this phase is to be expected because the expanding cells are commencing the large-scale synthesis of storage and metabolic proteins. The continuing synthesis of DNA in cells that are no longer dividing is less easy to explain. It has been suggested that polyploidization makes available extra copies of the genes for storage proteins. But the weight of evidence is against this possibility; in the dicots and cereals studied so far there is no evidence for any selective amplification of the genes for storage proteins during seed development. Most storage protein genes are either single-copy genes, or there are no more than four or five copies per haploid genome. Thus the extra DNA production may merely have a storage function as a reserve of deoxynucleotides, to be utilized following germination to support the extensive cell divisions in the growing embryo.

The amount of mRNA for storage proteins increases during development, to reach a peak at about the time of maximum protein deposition (e.g., in pea, Fig. 2.30). Then, as the seed begins to mature and dry out, the amount of mRNA declines; storage protein synthesis declines likewise. This pattern of events suggests a simple transcriptional control of protein synthesis: essentially, the amount of storage protein produced is determined by the amount (and stability) of its mRNA present. This probably holds true for most proteins, but there may be exceptions. In soybean and some other legumes, for instance, the subunit

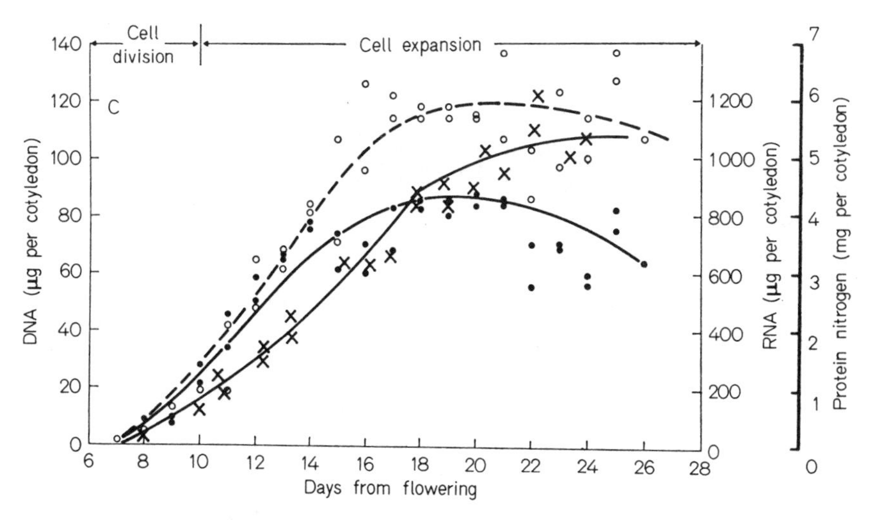

Figure 2.29. Changes in DNA (O), RNA (●), and protein (×) content during pea cotyledon development. After Millerd and Spencer (1974).

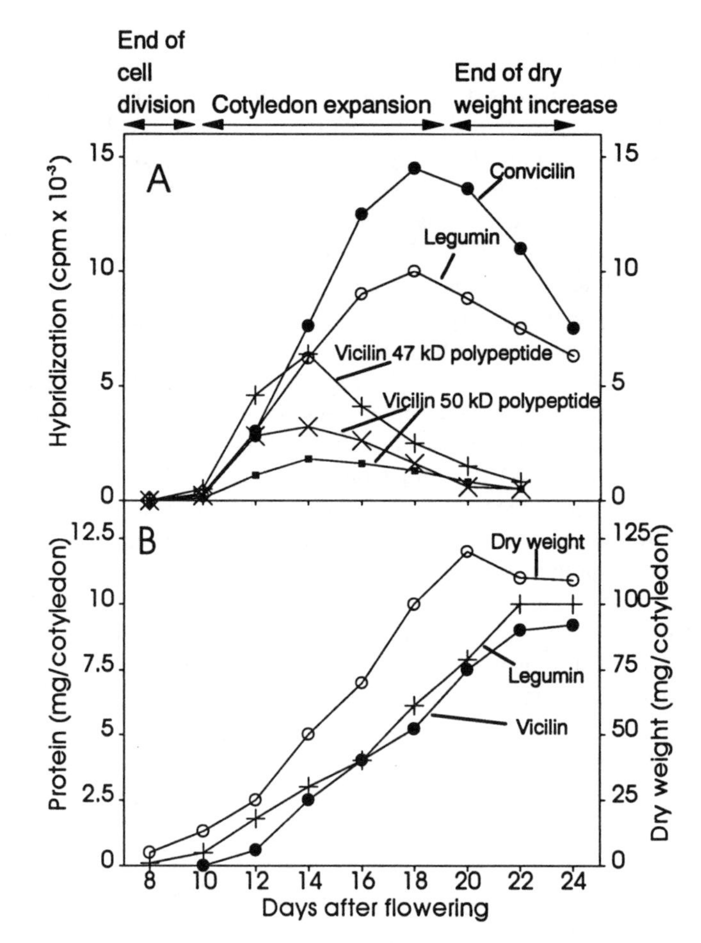

Figure 2.30. (A) Messenger RNAs for different storage proteins in pea *(Pisum sativum)* cotyledons during seed development. RNA from cotyledons of seeds of different ages was hybridized with the appropriate cDNA probes. After Gatehouse *et al.* (1986) and Boulter *et al.* (1987). (B) Accumulation of legumin and vicilin in pea cotyledons during seed development. The changes in dry weight are also shown. Redrawn from Gatehouse *et al.* (1982). (Note that the timing of the events shown in this figure is not strictly comparable with that in Fig. 2.29 as the plants in the two cases were not kept under identical conditions during seed development.)

composition for the 7 S storage protein (conglycinin) may change during development. Also, in soybean, the β subunit of the 7 S protein is only synthesized several days after two of the α subunits (Fig. 2.20D), yet the mRNAs for all of the α and β subunits of the soybean 7 S proteins are present at the earliest times, when only the α subunits are being synthesized (i.e., at 18–20 days). This suggests that there is in operation a translational control mechanism which, early during development (before 28 days), prevents the use of the mRNAs for synthesis of the β subunit and also of the mRNAs for the changed α subunit (α^{o}) produced only at later stages of development.

In some cases, posttranscriptional processes appear to affect mRNA stability. In pea, for example, the transcription rates for the production of different legumin mRNAs are similar, but the actual mRNA content of one of them is much higher; and a similar situation exists with regard to two hordein groups in barley. This suggests that transcripts of certain reserve protein genes are very labile within the developing seeds. Interorgan differences in stability also exist, for example in the case of soybean lectin. Here, although the transcription rate of the gene is 10-fold higher in the embryo than in the root of the parent plant, the protein in the embryo reaches a concentration about 2000-fold higher than in the root. The components of the seed which affect mRNA stability are known in some cases. In the wrinkled pea seed, for example, where amylopectin production is reduced, but sucrose content is relatively high (Section 2.3.1), the latter carbohydrate in some way affects the stability of legumin mRNA and the legumin content is thereby lowered. Stability of mRNA can also be affected by environmental factors. A good example is the response of pea legumin mRNA to the availability of sulfur in the parent plant's environment (see Section 2.3.5.3).

2.3.5. Regulation of Storage Protein Synthesis

Several important points emerge from the pattern of protein deposition exemplified by Fig. 2.30, which is typical of developing seeds. First, synthesis of each reserve protein occurs over a discrete period and not throughout the whole of development. This is the case for the different reserve components in a gross sense (e.g., legumin versus vicilin in peas) and also for different polypeptide subunits, e.g., those of soybean β-conglycinin. The same is true regarding the mRNAs of the protein components and, therefore, for the expression of the appropriate genes. Second, reserve protein is typically produced only in the seed and characteristically in certain parts of the seed—cotyledons or scutellum of the embryo, or endosperm. We can recognize, therefore, strong *temporal* and *spatial* (tissue specific) regulation of synthesis of reserve proteins and hence of the expression of their genes.

To understand the mechanism of regulation, the genes must first be isolated, cloned and the base sequences of the coding and flanking regions determined. We have seen in Chapter 1 that most of the storage proteins are encoded by families composed of many genes which may be located on different chromosomes. Notwithstanding this complexity, cDNA and genomic DNA of numerous storage protein genes of different species have been isolated, cloned, and sequenced; and later, a limited number of examples will be considered to illustrate the main findings. But, before then, an outline summary will be given of the approach which is taken to elucidate, at the molecular level, regulatory processes in storage protein production and in other aspects of seed physiology and biochemistry.

2.3.5.1. Investigating Gene Expression in Seeds

Some of the basic procedures are summarized in Fig. 2.31. Both cDNA and genomic DNA of the gene(s) under investigation can be prepared. The former is particularly useful for obtaining sequence data, i.e., the base composition from which the amino acid sequence of the protein can be inferred. It is also important as a probe since it hybridizes specifically with the appropriate mRNA and hence, after labeling (radioactive or other) and further treatment, it can reveal the presence of the message and, with certain qualifications, the extent of transcriptional expression of the gene. Genomic DNA consists of the coding region of the gene (exons, and introns, where present) and the 3′ and 5′ flanking regions. The latter is upstream of the transcription start, contains promoters and other regulatory sequences such as enhancers, and is predominantly responsible for regulation of transcription of the coding region of the gene; it is, therefore, the main focus of attention in studies of the regulation of gene expression.

Restriction enzymes cleave the genomic DNA at known sites so that the 5′ flanking (upstream) region and the coding sequence can then be handled separately. The coding sequence can be combined with the 5′ upstream regulatory region of another gene (e.g., another storage protein) or with promoters which give constitutive expression (i.e., in all plant tissues not just seeds), for example the 35 S promoter of cauliflower mosaic virus (CaMV) (Fig. 2.31, 2). The 5′ upstream region of the genomic DNA can be ligated with other coding sequences, perhaps of other storage proteins, but more commonly reporter genes are used in these constructs (Fig. 2.31, 1a). These reporter genes are for certain enzymes [e.g., β-glucuronidase (GUS), chloramphenicol acetyl transferase (CAT)] that are normally not found in plants and which can be assayed by a relatively simple color reaction, frequently histochemically, so that activity in particular tissues can be made visible. This procedure tells us if the 5′ upstream regulatory region contains tissue (e.g., endosperm, cotyledons) or temporal (e.g., mid-development) specificity. To locate this regulation more precisely within

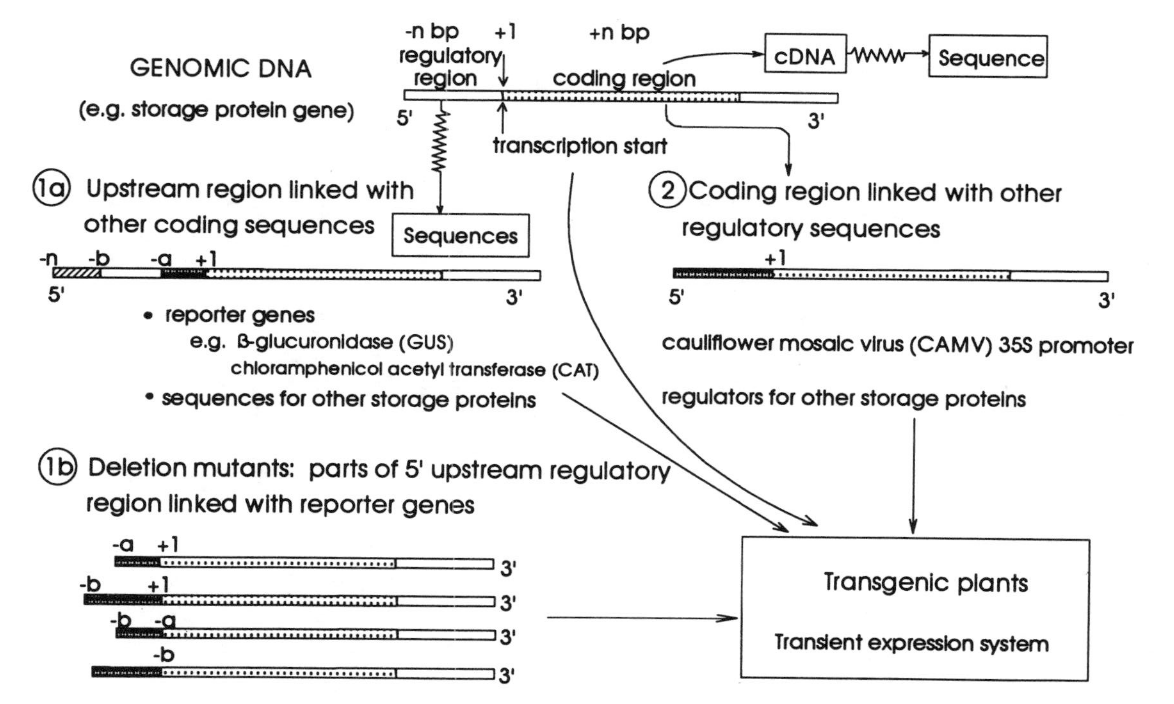

Figure 2.31. Scheme showing the main procedures used to elucidate the regulation of gene expression in seeds. The transcription start site is numbered +1. Base pairs (bp) upstream are given negative numbers; those downstream of the start site, positive numbers.

the 5′ upstream region, it is cleaved into smaller sequence groups, each one of which is fused to a reporter gene. Such sequences are referred to as deletion mutants (Fig. 2.31, 1b) since portions of the regulatory region have been deleted. Of course, the base sequence of the 5′ upstream region is analyzed to discover the composition of the regulatory zone(s).

The ability of these constructs to support gene expression is determined by introducing them into an appropriate expression system, either transgenic plants (e.g., tobacco, petunia) or protoplasts (transient expression systems) each of which offers different advantages. For example, the former is used to follow gene expression within seed or other plant tissues throughout their development. The latter, on the other hand, might be preferred because expression can be obtained within just a few hours; and they are also particularly useful for studying the effects of plant hormones since such substances can be manipulated relatively easily in a protoplast culture but not in the whole plant.

The regulatory sequences of the 5′ upstream region (known as *cis*-acting elements) in many cases function to control transcription by interacting with nuclear proteins that bind specifically to them—DNA-binding proteins or *trans*-acting factors. Once the regulatory sequences have been identified, the proteins binding to them can be isolated by various biochemical procedures.

2.3.5.2. Expression of Storage Protein Genes

Knowledge that has emerged from application of the procedures outlined above can be summarized as follows:

1. Storage protein genes of several species (all crop plants) can be inserted into plants of other species. In these transgenic plants, the genes are expressed in the normal manner, i.e., almost exclusively in developing seeds and at characteristic times.
2. Within the 5′ upstream regions of genes of similar proteins (e.g., legumins, prolamins, or vicilins), similar base sequences (conserved regions) can be recognized which may be involved in particular aspects of regulation, such as to confer tissue and temporal specificity, or hormonal sensitivity.
3. The 5′ upstream region contains several sequences which act as enhancers to upregulate the transcription of the coding region.
4. Nuclear proteins binding to specific sequences of the 5′ upstream region have been isolated. These proteinaceous *trans*-acting factors interact with *cis*-acting elements to regulate transcription of the coding region, i.e., the storage protein. Such interactions are part of the mechanism of tissue and temporal specificity and other aspects of regulation.

5. Expression of some storage protein genes is regulated by plant hormones, particularly abscisic acid, and by other factors, osmotic or nutritional. In several cases the appropriate response elements have been identified in the 5′ upstream promoter region.

We will now consider a few examples to amplify these points.

Some storage protein genes that have been incorporated into transgenic plants are shown in Table 2.3. Favored species for transformation, in most cases by means of vectors derived from the crown gall bacterium, *Agrobacterium tumefaciens*, are tobacco, petunia, and *Arabidopsis*, in which regeneration of transformants is relatively easy. The protein is detected immunologically, which enables its production, location, and subcellular targeting and packaging to be followed, and the mRNA is estimated by means of the appropriate cDNA probe, or more rarely by *in vitro* translation. In almost all cases the storage protein gene (or the reporter gene) is expressed in the transgenic plant in the same storage site as in the parent, i.e., virtually exclusively in the endosperm or cotyledons of the seed. For example, phaseolin, which is produced only in developing seeds of *Phaseolus vulgaris*, and mostly in the cotyledons (though a little occurs in the endosperm), is formed uniquely in the seeds of the transgenic tobacco, mostly

Table 2.3. Some Storage Protein Genes That Have Been Transferred into Transgenic Plants

Protein[a]	Parent species	Transformant
7 S proteins		
α′ Subunit, β conglycinin	*Glycine max*	Petunia
β Subunit, β conglycinin	*Glycine max*	Tobacco, petunia
Vicilin	*Pisum sativum*	Tobacco
Convicilin	*Pisum sativum*	Tobacco
Phaseolin	*Phaseolus vulgaris*	Tobacco, sunflower
11 S proteins		
Legumin	*Pisum sativum*	Tobacco
Legumin	*Vicia faba*	Tobacco
Legumin*	*Vicia faba*	*Arabidopsis*
Helianthinin	*Helianthus annuus*	Tobacco
Glycinin*	*Glycine max*	Tobacco
HMW glutenin*	*Triticum aestivum*	Tobacco
LMW glutenin*	*Triticum aestivum*	Tobacco
BI hordein*	*Hordeum vulgare*	Tobacco
Zein*	*Zea mays*	Tobacco
Zein[b]	*Zea mays*	Petunia

[a]In all cases the entire gene was inserted except in those marked * where constructs of the 5′ upstream region and reporter genes were used.
[b]The coding region of zein was fused with the β-phaseolin promoter.

in the cotyledons with just a trace in the endosperm. It is also packaged in protein bodies as in the parent seed. On the other hand, the reporter genes GUS or CAT linked with the 5′ upstream region of the high molecular-weight (HMW) glutenin of wheat are expressed specifically in the tobacco seed endosperm, the same site where HMW glutenin is produced in the wheat grain. And in all cases, the patterns of expression show evidence of developmental, temporal regulation, i.e., the timing of appearance of the mRNA and of the protein is comparable in the transgenic seeds with that in the parent species. Thus, information resides in the storage protein gene not only for the amino acid composition of the protein but also for the determination of specific tissue and temporal expression. The latter information lies in a 5′ upstream region, since in gene constructs this imposes an expression pattern an the reporter genes similar to that of the native protein in seeds on the parent.

The 5′ upstream regions of similar genes in different species contain similar information. In the various legumin genes, for example, it imposes on the coding region closely similar tissue and temporal expression. Is this because the 5′ upstream regulatory regions of the genes share common structural features? This is indeed the case, for there are base sequences in the 5′ regions which are identical, or nearly so: these are called conserved sequences. Each of the legumin, vicilin, and prolamin genes of different species, respectively, possesses conserved sequences in its 5′ upstream region—the so-called legumin, vicilin, and prolamin boxes (Table 2.4). All three boxes are different, so no common sequence appears to be involved in the three cases. In addition to these boxes, common smaller sequences are evident (for example, the so-called RY repeats) which occur as multiple copies in most of the storage protein genes of legumes, both the legumins and the vicilins. There are also species-specific sequences, for example in several different genes in soybean.

In promoter-deletion studies, portions of the 5′ upstream region are fused to reporter genes and the constructs used to make transgenic plants. Studies of this kind identify more precisely which part of the 5′ region regulates tissue and temporal gene expression. It turns out that discrete regions of the 5′ upstream region are implicated in these effects. For example, in the wheat low-molecular-weight glutenin promoter the DNA between −160 bp and −320 bp (from the transcription start) is necessary to obtain a high amount of expression in the tobacco seed endosperm of the linked reporter gene, CAT. And in the case of one of the *Vicia faba* legumin genes, the linked reporter gene is highly expressed only when, in the constructs, sequences of 5′ upstream promoter (regulatory) region of at least about −1000 bp from the transcription start are used (Fig. 2.32). More detailed analysis shows that within the first 566 bp of this region there are regulatory enhancer elements which function only when key parts of the

Table 2.4. Common Sequences in the 5′ Upstream Region of Some Storage Protein Genes

Name	Base sequence[a]	Example	Position[b]
Legumin box	5′-TCCATAGCCATGCATGCTGAAGAATGTC-3′	*Vicia faba*, legumin	–89 bp
		Pisum sativum, legumin	–89 bp
		Helianthus annuus, helianthinin	–80 bp
		Glycine max, glycinin	–50 bp
Vicilin box	5′-GCCCACCTA/TTTTC/TGTTC/TAC/TTTCAACACNCGTCAANNTNCAT-3′	*Phaseolus vulgaris*, phaseolin	–96 bp
		Glycine max, conglycinin	–76 bp
		Pisum sativum, vicilin	–98 bp
Cereal (prolamin) box	5′-TGTAAAG-3′	*Zea mays*, zein	–332 bp
		Triticum aestivum, gliadin	–318, –510 bp
		Hordeum vulgare, hordein	–300 bp

[a]Bases: A, adenine; C, cytosine; G, guanine; T, thymine; N, variable. A/T indicates adenine or thymine, etc.
[b]Base pair (bp) number from the transcription initiation site (+1) where the 5′ upstream regulatory or promoter sequence commences, i.e., at the 3′ end of this sequence (see Fig. 2.31).

legumin box are also present, namely the sequence CATGCATG around position –95 bp from the transcription initiation site (see Table 2.4); this sequence is also involved in promoting high levels of expression of 11 S glycinin genes. It seems that the CATGCATG motif within the legumin box modulates gene expression, not necessarily in an "on–off" fashion, while other elements farther upstream are also important for maximum expression.

Some of these and other *cis*-acting elements interact with the DNA-binding, *trans*-acting proteins. Such nuclear proteins have been isolated from developing seeds of several species and in some cases the DNA sequences to which they bind have been identified. It is interesting that in the case of the legumins no protein has been found which binds to elements within the legumin box itself, though, as we have seen, these elements certainly are implicated in the regulation of gene expression. Recently, the analysis of DNA-binding proteins interacting with the 5′ promoter of β-phaseolin has been extremely

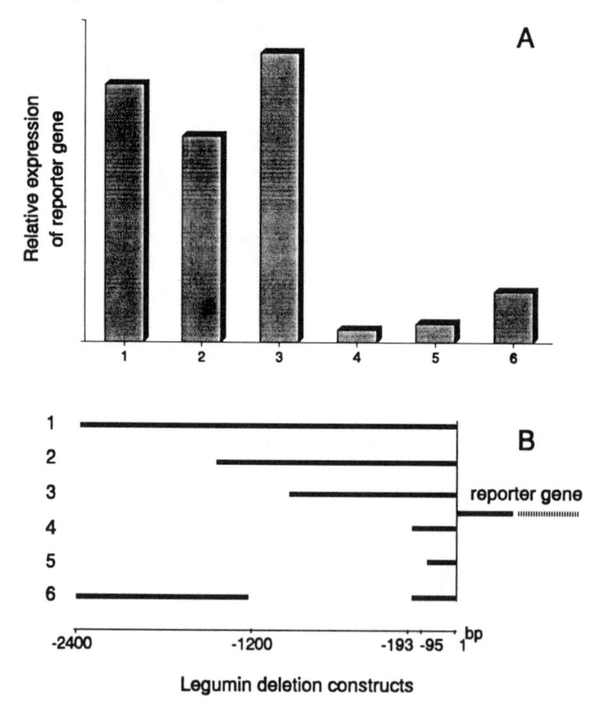

Figure 2.32. Promoter deletion studies on a legumin gene from *Vicia faba*. In (A) the relative extent of expression of the reporter gene neomycin phosphotransferase II is shown for constructs 1–6. (B) Shows the composition of the constructs, and the lengths of the 5′ upstream promoter fragments after deletion of selected regions (2–6). Expression is in transgenic tobacco. Adapted from Baumlein *et al.* (1991).

revealing. Multiple binding sites interact with four discrete proteins. The binding sites have different roles (Fig. 2.33), but one that contains the sequence CACGTG is thought to be the major element conferring seed-specific and temporal expression.

A highly interesting finding is that in several species some of the DNA-binding proteins are present only in seeds and only at times when storage protein gene expression is initiated. Hence, the ultimate control of expression of the storage protein genes might lie in the production of the appropriate binding proteins at the right time and place. Further, these *trans*-acting proteins must be very similar or even identical in different species since the patterns of gene expression governed by them in seeds on the parent plant and in transgenics are so close.

Some of the genes encoding the *trans*-acting proteins have been isolated and cloned. These genes can, of course, mutate and when they do the regulatory system for the control of storage protein synthesis is affected. Certain of the well-known mutants which have an aberrant storage protein content of their seeds are in fact mutants of genes for DNA-binding proteins (e.g., *opaque–2* in maize). This is discussed further in Section 2.3.5.6.

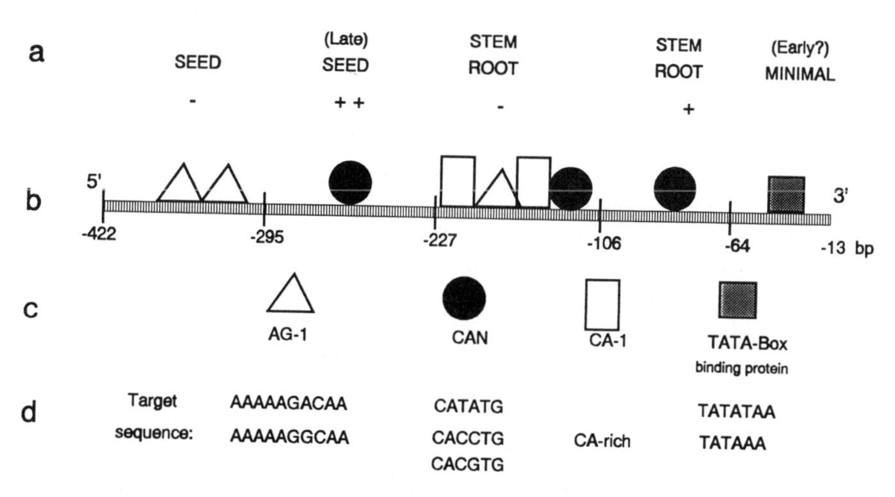

Figure 2.33. Regulation of the β-phaseolin gene. Tissue and temporal regulation is indicated in line a. Sites on the 5′ upstream promoter region (line b) are concerned with downregulation in the seed (–), strong upregulation in the seed later in development (late seed ++), down-regulation in the stem and root (–) of the vegetative plant, moderate up-regulation in the stem and root (+), and minimal early regulation (the low amount of gene expression that occurs early in development). Four DNA-binding proteins are indicated (line c), the regulatory regions on the 5′ promoter region to which they bind (line b) and the target sequences (beneath each protein, line d) in those regions. Adapted from Kawagoe and Murai (1992).

2.3.5.3. Nutritional Effects on Storage Protein Synthesis

It is not surprising that the nutritional status of the plant, for example the availability of nitrogen and sulfur—important elements in protein—should affect the production of storage protein. Shortage of nitrogen might lower the total quantity of protein but sulfur deficiency has discrete effects on specific proteins. Particularly well-documented examples are legumin and albumin in pea. Mild or severe sulfur deprivation (plants supplied with 4 and 2 ppm sulfate, respectively) leads to reductions in accumulation of legumin of 80 and 100% while sulfur-poor vicilin is unaffected, or even increases. The content of legumin mRNA in developing seeds is much lower under low sulfur because of its instability under these conditions. Sulfur stress similarly affects pea legumin in transgenic tobacco. Therefore, the same sulfur-sensitive mechanism affecting the transcripts operates in seeds of both the parent and the transgenic host. This mechanism is not yet understood.

2.3.5.4. Abscisic Acid and the Osmotic Factor

The possibility that deposition of protein and other seed reserves is affected by abscisic acid (ABA) arises out of the observation that in many species the period of high ABA content of developing seeds approximately coincides with the greatest rate of reserve synthesis. But the connection between ABA and storage protein accumulation is more firmly established by the behavior of isolated embryos, and by the properties of certain mutants.

Embryos isolated from developing seeds generally germinate precociously when placed on water or a simple nutrient medium. The cessation of normal embryogenetic development in most cases is accompanied by a halt in storage protein synthesis, but both development and production of storage proteins are maintained when ABA is included in the medium. In one species which illustrates this, rapeseed, the synthesis of mRNAs for the proteins napin and cruciferin (and, of course, synthesis of the proteins themselves) declines when isolated embryos are deprived of ABA. This shows that the growth regulator acts at the gene level, and this also occurs in several other species. Three proteins in developing wheat embryos respond similarly: wheat germ agglutinin, a storage globulin, and the Em (early methionine-labeled) protein (Table 2.5). Interestingly, all three are induced in wheat seedlings by treating them with ABA. Effectiveness of ABA on developing embryos generally depends on their age, with sensitivity to the regulator decreasing as embryos approach maturity.

It could be argued that these effects of ABA on storage (and other) proteins are really indirect, that they simply result from the blockage of germinative growth and thus synthesis of storage protein can continue. Later, we will consider points for and against this view, but in the present context we should note that ABA can affect storage protein accumulation in isolated cotyledons where

Table 2.5. Some Proteins Whose Synthesis Is Promoted by ABA during Seed Development[a]

Species/tissue	Protein
Wheat embryo	Wheat germ agglutinin
	Em protein
	Globulin
Soybean cotyledon	β subunit of β-conglycinin
Rapeseed embryo	Napin
	Cruciferin
French bean embryo	Storage protein
Barley endosperm	Proteinase/α-amylase inhibitor
Sunflower embryo	Helianthinin
Arabidopsis embryo	2 S, 12 S storage protein
Cotton embryo	LEA proteins
Maize embryo	Globulin

[a]In most cases the appropriate mRNA is also ABA-regulated.

germinative growth in the strict sense does not take place. Cultured, isolated soybean cotyledons show a reduced capacity for synthesis of the β subunit of β-conglycinin and its mRNA, which is restored by application of ABA. The production of the β subunit and its mRNA is lowered even further by treating cotyledons with fluridone, an inhibitor of carotenoid and ABA synthesis, but this again is completely overcome by addition of ABA to the medium. Similar effects occur regarding storage protein accumulation in isolated *Vicia faba* cotyledons.

Somatic embryos produced by *in vitro* culture of calluses, cell suspensions, microspores (in anthers), and other plant material show qualitative patterns of reserve accumulation similar to that of normal, zygotic embryos. In somatic embryos of rapeseed, the production of napin and cruciferin is enhanced by the inclusion of ABA in the culture medium. A putative plant hormone, jasmonic acid (which is chemically rather similar to ABA), has similar effects. This substance occurs naturally in both zygotic and somatic rapeseed embryos, so possibly it participates in the regulation of storage protein synthesis in the developing seeds.

Some of the most compelling evidence implicating ABA in the regulation of storage protein accumulation in developing seeds comes from studies on mutants, particularly of maize and *Arabidopsis*. Developing embryos of wild-type maize contain a globulin, but this protein and its mRNA are virtually absent from two mutants, *vp1* (insensitive to ABA) and *vp5* (deficient in ABA). Globulin production is restored by exposing *vp5* embryos to ABA but, not surprisingly, *vp1* mutants do not respond similarly. Thus, the production by developing maize embryos of normal amounts of globulin requires both ABA (the eventual product of *Vp5* action) and sensitivity to it (the effect of *Vp1*). On the other hand, ABA-deficient mutants of *Arabidopsis* (*aba*) and tomato (*sit*) show essentially normal production of storage protein during embryo development. The situation is similar in *Arabidopsis* mutants that have much reduced sensitivity to ABA (*abi1*, *abi2*, *abi3*), but we should note that these contain abnormally high amounts of ABA. Both of these mutant groups in *Arabidopsis* are somewhat "leaky" and therefore recombinants have been constructed in which insensitivity accompanies ABA deficiency. Developing seeds of these recombinants do not produce the 12 S and 2 S storage proteins (and they also show other aberrations that will be discussed in a different context). A recently isolated mutant of *Arabidopsis* has a new allele of *abi3* (designated *abi3–3*) which confers virtually complete insensitivity to ABA. Formation of 12 S and 2 S storage proteins in the developing seeds is dramatically reduced. These findings strongly suggest that accumulation of storage protein in *Arabidopsis* depends on the action of endogenous ABA. They also show that even in developing seeds that have a very low ABA content, but are extremely sensitive to it, ABA-regulated storage protein synthesis can still proceed. Perhaps this is also the situation in the ABA-deficient *sit* mutant of tomato. Vice versa, production of storage protein can continue in *Arabidopsis* seeds having a low sensitivity to ABA if the content of this hormone is so high as to be able to trigger the response.

Before concluding this section it is important to record that there are cases which provide no support for a role for ABA in storage protein production. For example, isolated embryos of pea, barley and cotton show no response to ABA with regard to storage protein synthesis, though other proteins may be induced. Fluridone-treated alfalfa and pea seeds have much reduced ABA contents but the accumulation of storage protein is not deleteriously affected. There may be important species differences, therefore, concerning the participation of ABA. Alternatively, if the sensitivity of some developing seeds to ABA is extremely high, even the small amounts of the hormone present after exposure to fluridone may be enough for ABA-regulated processes to proceed. This scenario would be similar to the situation in the *Arabidopsis* mutants discussed above.

Many of the physiological and biochemical effects brought about in developing embryos by ABA can also be induced by low osmotic potentials. Precocious germination of isolated embryos is prevented, while deposition of reserve proteins in many species is enhanced when high osmolarities are maintained in the culture medium. This presumably reflects the situation in the intact, developing seed where the osmotic potentials are relatively low, for example about −1.5 MPa in alfalfa, barley, and wheat.

But there are cases where ABA and low osmotic potentials act differently. In alfalfa, for example, only in culture media of a low osmotic potential do isolated embryos maintain the same pattern of proteins synthesized in intact seeds; inclusion of ABA does not have the same effect even though precocious germination is inhibited. Low osmotic potentials also prevent precocious germination of isolated maize embryos but globulin synthesis in the embryos occurs only in the presence of ABA. These observations argue strongly against one concept that has been propounded, that osmotic "stress" induces the formation of ABA. If this were true, the two factors would be expected to have the same effect. Indeed, there is biochemical evidence that high osmolarities do not lead to an elevated ABA content in developing embryos as they do, for example, in seedlings. We should note, also, that sensitivities to osmotic potential and ABA vary individually according to age of the developing embryo. Precocious germination of older embryos of alfalfa and other species is readily arrested by osmotica but not so much by ABA. It is likely that ABA and the osmotic factor interact throughout development of the seed. One possibility is that osmotic stress increases sensitivity to ABA, which appears to be the case with respect to ABA-regulated synthesis of Em protein in wheat (see above).

The mechanism of action of the osmotic factor is not the same as that of ABA, at least as far as precocious germination is concerned. High external osmolarities prevent osmotic water uptake so germinative cell elongation cannot take place. The effect of ABA, on the other hand, is not osmotic but appears to be on the cell wall, inhibiting wall yield and loosening which are requisites for cell expansion. As for reserve protein, it is not clear if the osmotic factor operates at the same level of molecular regulation as does ABA (see next section) but this is likely to be so.

2.3.5.5. Molecular Action of ABA

ABA is known to affect the synthesis of reserve and other proteins by regulating transcription: we shall concentrate here on this mechanism. There is evidence, however, that stabilization of mRNA as well as translational and posttranslational effects can also occur, but little detail is known.

Detailed investigations have been carried out on genes for several different seed proteins, e.g., helianthinin (a globulin) of sunflower, β-conglycinin (a vicilin) of soybean, and the Em protein (a LEA, see Section 3.3.2) of wheat. The general approach is the one outlined in Section 2.3.5.1, but in the present context the use of transient expression systems (protoplasts from barley aleurone layers or rice cell suspensions) has been particularly important. This technique allows a more rapid evaluation than that offered by transgenic plants, though the latter have also been employed.

Our present understanding of the regulation of gene expression by ABA can be summarized as follows: (1) The 5′ upstream promoter regions contain *cis*-acting sequences of three kinds: enhancer sequences, seed-specific sequences, and ABA-responsive elements (ABREs). (2) *Trans*-acting factors (DNA-binding nuclear proteins) are produced in the seed which bind to the

ABRE, the consequence of which is an upregulation of transcription of the coding region (gene). We will now consider these points in more detail.

The enzyme GUS is produced in a transgenic tobacco host by constructs consisting of the 5′ upstream promoter regions of wheat Em protein and sunflower helianthinin, for example, and the GUS reporter gene. The gene is expressed specifically in the tobacco seed, probably in response to the native ABA. A further regulation by ABA is shown when developing or mature seeds are exposed to the hormone, when more GUS expression occurs. When constructs containing the Em promoter region are introduced into protoplasts of barley aleurone tissue or rice suspension cells, GUS is formed specifically in response to ABA. In this case, a region extending from −554 bp to +92 bp is sufficient to support the ABA effect. The DNA between −554 bp and −168 bp has three regions rich in the bases adenine and thymine (A, T) which promote a high expression rate of the coding region; further, the region −168 bp to −106 bp contains the ABRE and sequences conferring seed specificity (see Fig. 2.34). The 5′ promoter of helianthinin also has seed-specific elements and ABREs. In both wheat and sunflower the ABREs contain the sequence motif CACGTGGC. Interestingly, this same or very similar motif is found in regulatory elements of many plant, yeast, and mammalian promoters (e.g., see the CAN binding site, Fig. 2.33).

Several nuclear proteins in wheat embryos and rice suspension cells can bind to parts of the 5′ upstream promoter of Em. These *trans*-acting factors interact with sequences in the AT-rich region and with the core motif in the ABRE. Such interactions are involved in the regulation by the 5′ promoter of the transcription of the coding region (e.g., the Em protein, helianthinin, or other ABA-regulated protein). The amino acid sequence of some of these *trans*-acting proteins, deduced from their cDNAs, shows that leucine-zipper regions are

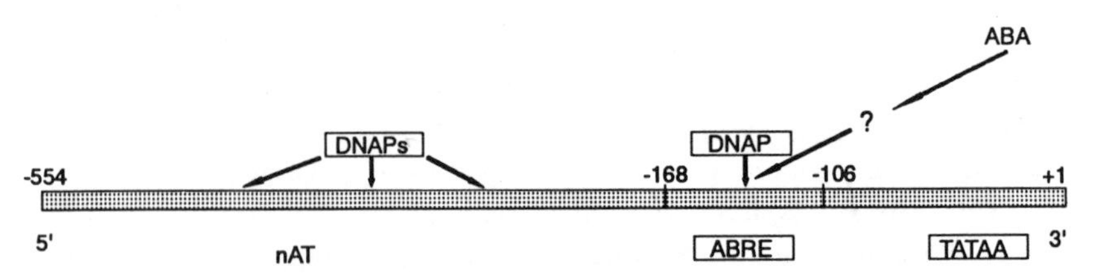

Figure 2.34. The ABA-sensitive 5′ upstream promoter region of the wheat Em gene. A 554-bp fragment of the promoter is shown. This contains an enhancer region (−168 to −554) rich in adenine and thymine (nAT), a region (−106 to −168) in which the ABA-response element (ABRE) is located, and the region containing the TATA box. DNA-binding proteins (DNAP) complex with segments of the AT-rich region and with the ABRE. ABA may be linked to the transcriptional complex via an unknown coupling element. Adapted from Quatrano *et al.* (1993).

present, containing heptad repeats of the amino acid leucine. This is also the characteristic of other transcription factors from plants, yeasts, and humans.

Some very challenging questions now arise. The ABREs contain the same base motif as in other plant (and animal) regulatory elements. The *trans*-acting proteins which bind to them are very similar to other transcription factors. What, then, is the nature of the specific response to ABA and the seed specificity of these genes? The answer is not known and there is still much room for informed speculation. For example, the base neighbors of the ABRE core motif may be involved in determining the binding of some specific subclass of *trans*-acting protein to the ABRE itself. There is no evidence that ABA interacts directly with the transcription factor–ABRE complex and, therefore, other components must be involved in linking ABA action with this complex. The identity of these components remains to be elucidated (Fig. 2.34).

2.3.5.6. Mutants, Protein Content, and Nutritional Quality of Seeds

Over the past decade or so there has been an enormous increase in interest in the nature and synthesis of storage proteins in developing seeds, particularly with a view to improving their nutritional quality. A vast proportion of all human food is derived directly or indirectly (by feeding to livestock) from seeds. Yet, as noted earlier, cereal seeds are deficient in certain essential amino acids (lysine, threonine, and tryptophan), and so are legume seeds (in the S-containing amino acids, methionine and cysteine). Monogastric animals, such as human beings, pigs, and poultry, do not possess the anabolic processes necessary to synthesize these essential amino acids from other C and N sources. Strategies have been developed, or are being researched, to overcome these deficiencies in seed proteins.

In cereals, the prolamins are the major storage proteins; they are high in proline and amide and low in lysine (Section 1.5.3). Spontaneous mutants of some of the cereals have been isolated, e.g., the *opaque–2*, *opaque–7*, *floury-2*, *brittle*, and *shrunken* mutants of maize, which produce less of the prolamin, zein (Table 2.6), and more of the usually minor storage proteins which have a relatively high lysine content. The mutations affect different zeins. *Opaque–2* virtually eliminates the 22-kDa zein, *opaque–7* the 19-kDa class, while *floury-2* affects both kinds. The *opaque–2* mutation is of a gene which encodes a transcription factor (DNA-binding protein) which participates in regulating the expression of the 22-kDa zein gene (see Section 2.3.5.2). In one *opaque–2* mutant, the lesion occurs in the codon for arginine, in the domain which binds to DNA in the zein 5′ upstream promoter region. Instead, lysine is substituted, the effect of which is to demolish the binding capacity of the transcription factor.

The increased nutritional value of these high-lysine mutants to human beings and to experimental animals is well recognized, but, unfortunately, there

Table 2.6. Distribution of Protein Fractions in Endosperms of Normal and *Opaque-2* Maize[a]

	Percentage of total protein	
Protein fraction	Normal	*Opaque-2*
Albumins (plus free amino acids)	3.7	14.7
Globulins	1.7	4.4
Prolamins (including zein)	54.2	25.4
Glutelins	40.4	55.5
% lysine in total protein	1.6	3.7

[a]After Jiminez (1966), quoted in Weber (1980).

are several negative aspects associated with the mutant seeds. As a consequence of the reduced prolamin content, the kernels are more brittle than the normal type and tend to shatter easily during storage; also, the baking properties of the mutant maize meal are poor. The most serious deficiency, however, is that the mutations lead to a severe disruption in carbohydrate metabolism (i.e., starch synthesis), and yield may be decreased by 10–60% (Table 2.1). So far, then, attempts to breed low-prolamin cultivars of maize have been unsuccessful, and little more success has been achieved with other cereals. Natural and induced (by using radiation) mutants of barley have been produced, but none is available as a commercial line. A promising sorghum mutant (P721 *opaque*) has been produced by mutation breeding, which leads to a decline in yield by only 10–15%. Other types of mutants have been sought also, e.g., those with an enhanced ability to synthesize free amino acids, owing to changes in the regulatory mechanisms operating the biosynthetic pathways. There is some evidence that when the lysine (in cereals) or methionine (in legumes) content of seeds is increased, these amino acids appear in proteins with greater frequency. Certainly, supplementing soybean cotyledons developed in culture with methionine can enrich the storage protein fraction in this amino acid by more than 20%. This has led to attempts to increase the content of S-containing amino acids in legume seed proteins by increasing the application of sulfate to the parent plant during seed development.

2.3.6. Phytin Deposition

Although phytin is an important reserve compound in seeds as a source of phosphate and mineral ions (Section 1.5.4), the pathway for phytic acid biosynthesis and its intracellular site are poorly understood. It has been suggested that phytic acid is synthesized within the protein bodies, where it is sequestered, but on the basis of available evidence this is unlikely. Studies on the developing castor bean endosperm using electron microscopy have led to the hypothesis that phytin is synthesized in association with the cisternal endoplasmic reticulum (CER) (Fig. 2.35, Step I), prior to being packaged into transport vesicles (Step II) which migrate to the vacuolar membrane, with which they fuse (Steps III and IV). The phytin particles are released into the lumen of the vacuole (Steps IV and V) and condense therein to form the globoid (Step IV). It is possible that the Golgi apparatus is involved in the packaging and formation of the transport vesicles, as for proteins (Section 2.3.4.3), but this has not been confirmed. During seed development the deposition in vacuoles, which become protein bodies, of phytin and storage proteins coincide. The pathway of phytic acid biosynthesis is unclear but it is likely that *myo*-inositol-1-*P* is synthesized from Glc-6-*P*, and then a further five phosphates added, with ATP as the donor, to form *myo*-inositol hexaphosphoric acid (mIP_6, phytic acid). No intermediates between the mono- and hexaphosphorylated forms of *myo*-inositol have been detected. The association of ions with mIP_6 presumably occurs randomly by the attraction of the metallic cations to the strong negative charges on the exposed phosphate groups.

2.4. HORMONES IN THE DEVELOPING SEED

While the growing seed is accumulating its major storage reserves, changes are also occurring in its content of other important chemical substances, the growth regulators or hormones—auxin, gibberellins, cytokinins, and ABA (Fig. 2.36). These substances are thought to play important roles in the regulation

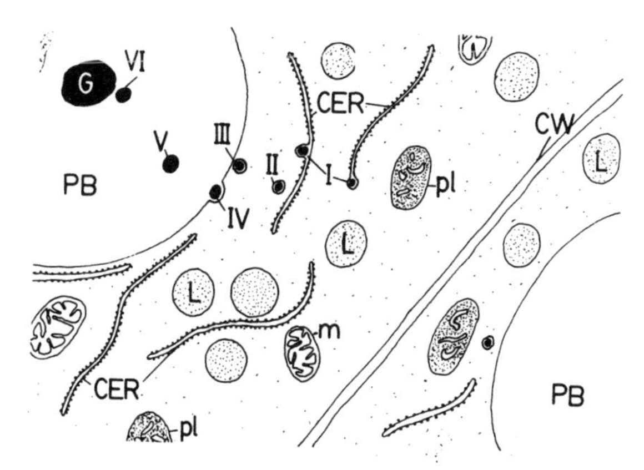

Figure 2.35. A diagrammatic representation showing the site of phytin synthesis and mechanism of deposition of phytin in protein bodies of developing endosperm cells of castor bean. Roman numerals I–VI show phytin at various stages of migration from the CER to the protein body globoid. CER, cisternal endoplasmic reticulum; CW, cell wall; G, globoid; L, oil body; m, mitochondrion; PB, protein body; pl, plastid. After Greenwood and Bewley (1984).

of certain aspects of seed growth and development, as we have seen in the case of ABA. They are also involved in fruit growth and certain other physiological phenomena. Immature seeds or grains were the first higher-plant sources of most of the known plant hormones, and they continue to attract the attention of research workers studying biosynthesis, metabolism, and chemistry of these regulators. We will first briefly discuss the kinds of hormones present in developing seeds, including aspects of their biosynthesis and metabolism, and then go on to consider their possible functions.

2.4.1. Composition and Location

2.4.1.1. Auxins

The major auxin in developing seeds in indoleacetic acid (IAA) (Table 2.7). This is formed in the seed tissues following the normal biosynthetic route

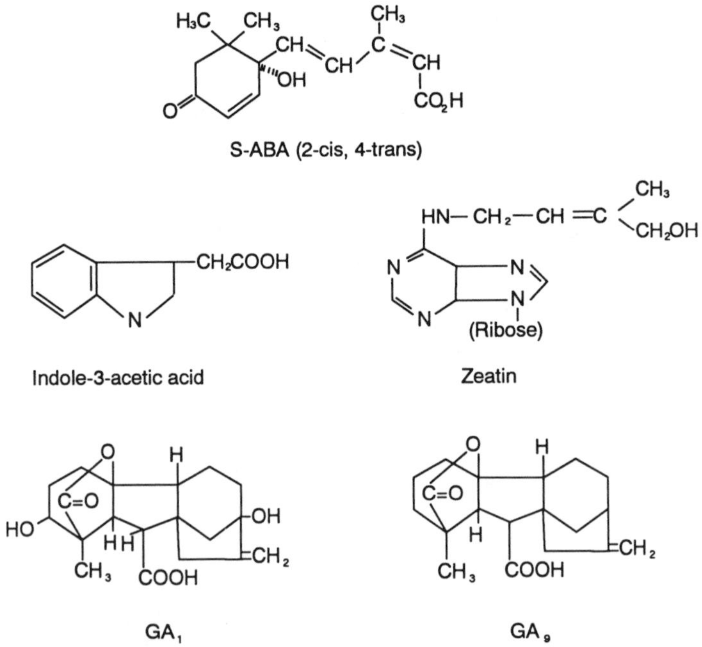

S-ABA (2-cis, 4-trans)

Indole-3-acetic acid

Zeatin

GA_1

GA_9

Figure 2.36. Growth regulators (hormones) in immature seeds. In ribosyl zeatin, ribose is present as shown. In the ribotide, the ribose is phosphorylated. Two gibberellins commonly found in developing seeds are shown.

from tryptophan and is not derived from the parent plant. Free IAA is widely found, and, in addition, various forms of so-called bound auxin are abundant. In immature kernels of maize, for example, IAA arabinoside, IAA *myo*-inositol, and IAA *myo*-inositol arabinoside are present. These compounds are considered to be precursors of IAA which, following germination, is liberated from them enzymatically and is probably transported to the coleoptile tip of the growing seedling. In maize and other cereals, free IAA and its bound forms are extractable from the endosperm. In dicots too, such as the pea, auxin is first found in the endosperm and only later is it detectable in the embryo itself, after the endosperm has been resorbed (Fig. 2.37). The pattern of free IAA content in the pea is typical of that in developing seeds, in that the concentration rises to a peak value and then diminishes, with relatively little remaining at maturity. Variations on this theme are common: in apple, for example, there are two peaks in auxin concentration, the first coinciding with the change in the endosperm from a coenocytic to a cellular structure, and the second with the formation of new endosperm cells. The final decline in free IAA is the result of metabolic conversion to bound forms and other products.

Table 2.7. Growth Regulators (Hormones) of Some Immature Seeds[a]

Species	Auxin	GA	CK	ABA
Prunus cerasus	IAA	—	Zeatin	Present
Malus spp. (apple)	IAA	GA_4, GA_7, GA_9, GA_{12}, GA_{15}, GA_{17}, GA_{20}, GA_{24}	Zeatin Ribosyl zeatin Zeatin ribotide	Present
Pisum sativum	IAA	GA_9, GA_{17}, GA_{19}, GA_{20}, GA_{29}, GA_{38}, GA_{44}, GA_{51}, GA_{53}	—	Present
Zea mays	IAA "Bound" IAA	GA_1, GA_4, GA_8, GA_9, GA_{17}, GA_{19}, GA_{20}, GA_{29}, GA_{34}, GA_{44}, GA_{53}	Zeatin Ribosyl zeatin Zeatin ribotide Ribosyl zeatin glucoside Dihydrozeatin riboside	Present
Triticum aestivum	IAA	GA_{15}, GA_{17}, GA_{19}, GA_{20}, GA_{24}, GA_{25}, GA_{44}, GA_{54}, GA_{55}, GA_{57}	Ribosyl zeatin	Present
Gossypium hirsutum	—	GA_1, GA_3, GA_4, GA_7, GA_9, GA_{13}	Isopentenyl adenine Isopentenyl adenosine	Present

[a]Absence of an entry against a species does not mean that no hormones of that kind exist in the seed but that chemical characteristics have not been recorded.

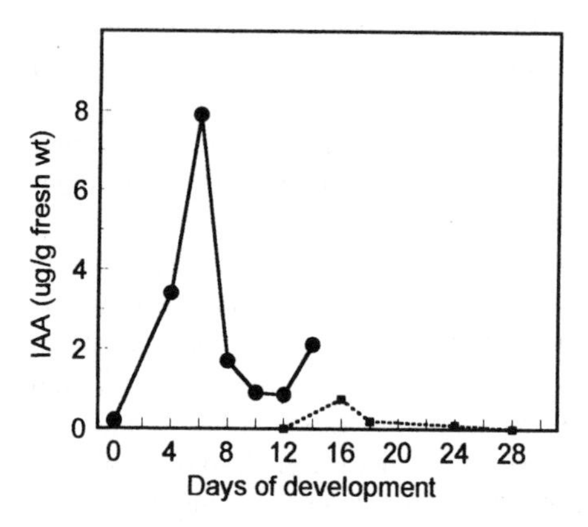

Figure 2.37. Auxin (IAA) in developing pea seeds (cv. Alaska). - - - -, IAA in embryo; ——, IAA in endosperm. After Eeuwens and Schwabe (1975).

2.4.1.2. Gibberellins

By the end of 1992, over 80 gibberellinss (GAs) were known, more than half of which had been found in developing seeds (see Fig. 2.36 and Table 2.7 for some examples). In addition, many GA conjugates have been identified, such as the polar, water-soluble glucopyranosides and glucopyranosyl esters. Because developing seeds are so rich in GAs they are often subjects for studies of GA biosynthesis and metabolism. For example, work on cell-free systems from immature seeds of *Cucurbita maxima, Marah macrocarpus,* and *Pisum sativum* has shown that GA synthesis follows the route: acetyl coenzyme A → mevalonate → isopentenyl pyrophosphate → chain elongation and ring closure → *ent*-kaurene → kaurenoic acid → GA_{12} aldehyde → gibberellins. Developing seeds, or cell-free embryo extracts, very actively interconvert GAs, which is why any one species contains several different kinds. In dwarf peas, for example, two conversion pathways have been identified: (1) GA_{12} aldehyde → GA_{53} →GA_{44} → GA_{19} → GA_{20} (→ GA_{29} → GA_{29} catabolite) → GA_1 → GA_8, and (2) GA_{12} → GA_{15} → GA_{24} → GA_9 → GA_{51} → GA_{51} catabolite. All of the GAs in the pathway are biologically active, except GA_{29}, GA_{51}, and their catabolites. Hence, the interconversions can lead ultimately to the biological inactivation of the endogenous GAs. At the early stages of seed development, however, the major GAs are the active ones, and the inactive ones are formed toward the end of seed maturation (Fig. 2.38). Clearly, if the endogenous GAs were detected solely on the basis of their biological activity, a broad peak would be found at days 18–22, owing to the combined effect of GA_9 and GA_{20}. This probably explains why in the early work on seed GAs, which relied on bioassay for detection of the hormones, single peaks of activity were generally found during seed development in many species. Part of the drop in the free GA content as seeds mature is the result of the formation of various conjugates, the glucosyl esters and glucosides, and GA catabolites, but the fate of the remainder is unknown. The various GAs may be distributed unequally in the seed. In pea, some of the GAs occur at very different concentrations in the testa, cotyledons, and axis; and in maize the GA_1 content of the embryo is about 40 times higher than in the endosperm. One possible cause is that the GAs are transported around the seed and may be subjected to differential metabolism at various sites.

2.4.1.3. Cytokinins

The first cytokinin to be identified in higher plants was zeatin, from developing maize kernels (Fig. 2.36). The chemically characterized cytokinins in immature seeds (see Table 2.7) are substituted adenines: zeatin, isopentenyl adenine, and their derivatives. The latter are various glycosylated forms containing ribose (e.g., ribosyl zeatin, zeatin ribotide, isopentenyl adenosine), glucose (e.g., zeatin glucoside), or both sugars (e.g., ribosyl zeatin glucoside). Cytokinins also occur in certain types of transfer RNA from which they can be liberated by hydrolysis.

It is not clear where the seed cytokinins are synthesized. Evidence points to two possible sources: the parent plant, for example, in the roots, and the seeds or fruits themselves; in some cases (e.g., tomato) the former seems the most

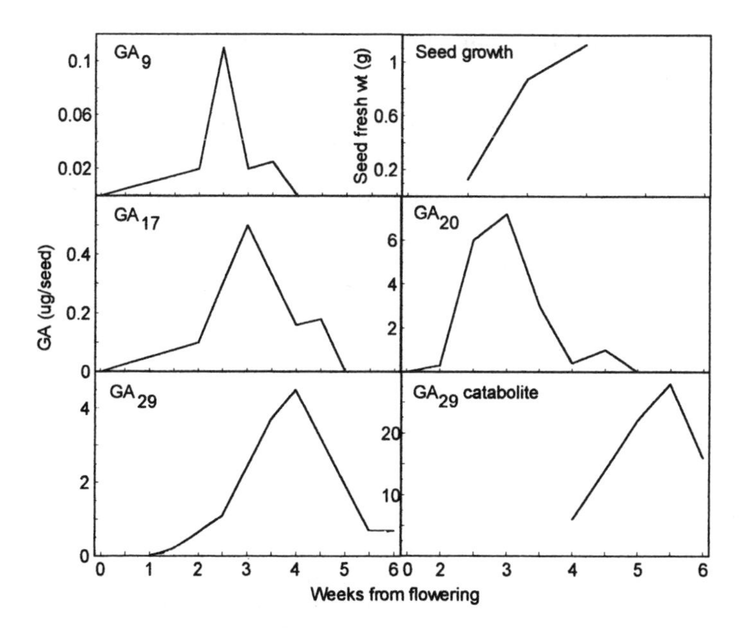

Figure 2.38. Gibberellins in developing pea seeds (cv. Progress no. 9). After Sponsel (1980).

likely source. The amount of cytokinins increases markedly during seed development, particularly while the seed tissues are growing, and then declines with maturation (Fig. 2.39). This temporal distribution is consistent with the suggested role for cytokinins in the control of seed growth, which we will consider below.

2.4.1.4. ABA

ABA has been isolated from immature seeds of many species (see Table 2.7). The free form of the inhibitor can occur at relatively high concentrations, especially in legumes, e.g., about 2 mg/kg fresh weight in soybean (approximately 10 μM), but more commonly the content is between 0.1 and 1 mg/kg fresh weight. Bound forms, the glucosyl ester and glucoside, are also common, and again at relatively high concentrations in legumes. Both free and bound ABA are located in various parts of the seed: the embryo, the endosperm, and the enclosing tissues. Metabolites of ABA, phaseic and dihydrophaseic acid, have also been reported, at high concentrations in legumes.

Like the other growth regulators in immature seeds, ABA rises in concentration during seed development, reaches one or two peaks, and generally then declines rapidly at about the time of seed drying (Figs. 2.40 and 3.9).

2.4.2. Possible Roles of Seed Hormones

The endogenous growth regulators in developing seeds may be involved in several processes, as follows: (1) Growth and development of the seed, including the arrest of growth prior to seed maturation. (2) The accumulation of the storage reserves. (3) Growth and development of the extraseminal tissues.

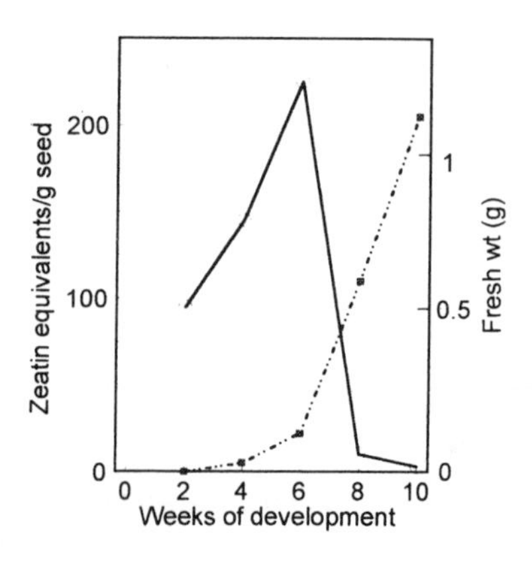

Figure 2.39. Cytokinin in developing seeds of *Lupinus albus.* ——, Cytokinin; ---- growth of seed as increase in fresh weight. Cytokinin was determined by bioassay (soybean callus growth). Amounts are expressed as zeatin equivalents (i.e., amount of zeatin having comparable activity). Adapted from Davey and Van Staden (1979).

(4) Storage for later use during germination and early seedling growth. (5) Various physiological effects on tissues and organs close to the developing fruit.

2.4.2.1. Seed Growth and Development

Most of the evidence linking the growth regulators with seed development comes from correlations between regulator content and embryo growth. The highest concentrations of the biologically active GAs (GA_9, GA_{20}) of dwarf pea seeds, for example, occur during the time of maximum growth of the developing embryo (Fig. 2.38), and in the embryo of *Phaseolus coccineus*, the biologically active GA_{20} accompanies the early stages of growth. A further interesting feature of the latter species is that the suspensor may supply GA_1 to the embryo in the earliest phases of development. Very young, excised embryos do not develop further in culture if deprived of the suspensor, but development continues when GA is added to the medium. Since the suspensor contains relatively high amounts of GA_1 (approximately 1 μg/100 mg tissue), it seems likely that is normally the source of GA for the embryo. Evidence both for and against the possibility of a role for GA in seed development comes from studies of mutants. Normal seed development occurs in GA-deficient mutants of *Arabidopsis* and tomato, which suggests no special role for GA in developing seeds. But in the GA-deficient pea mutant lh^i in which seed GA contents are much reduced, there is a tendency to seed abortion, implying that GA is essential for proper seed development (Fig. 2.41).

The period of cell division and enlargement in the seed (in both the embryo and endosperm) is also the period when the cytokinins are at their highest concentration. Cytokinins are known to promote cytokinesis (cell division) in certain plant tissues, and it is likely that they have this role in the developing seed too. Since it contains cytokinins, the suspensor has been suggested as a

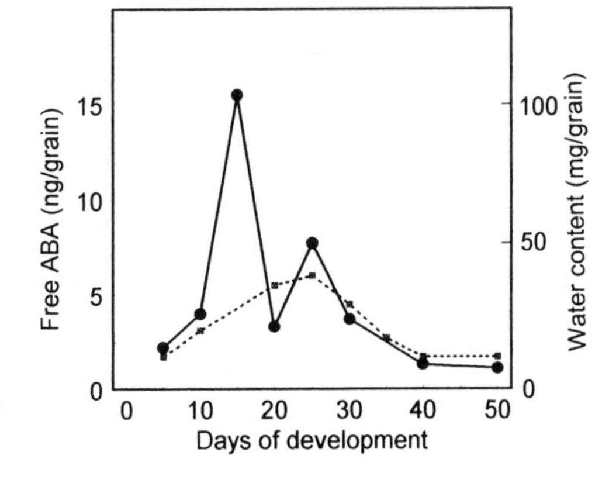

Figure 2.40. ABA in developing wheat grains (*Triticum aestivum* cv. Sappo). Total, extractable free ABA in the whole grain is shown (——), together with water content (- - - -). From data of Mitchell (Ph.D. thesis, 1980).

supplier of these hormones to the young embryo, for example in *Phaseolus coccineus* and *Lupinus albus*.

As might be expected, the inhibitor ABA is associated more with the arrest of embryo growth than with its promotion. There is evidence, in many cases (barley and rapeseed for example), that normal embryogenetic development can occur in the presence of ABA, but that germination and growth (i.e., the longitudinal extension of the axis) cannot. ABA, therefore, is a factor that prevents the embryo from passing directly from embryogenesis to germination while still on the parent plant, without an intervening rest period (see Chapter 3).

Correlations between ABA content and germinability of the developing embryo are indicated in several species. Immature seeds in the early stages of development generally cannot germinate, but the ability to do so is acquired at ages closer to full maturity. ABA content is higher in young, nongerminable seeds than in older ones, e.g., in several cereal grains and legume seeds. In the dwarf bean, when young seeds first become able to germinate they do so with a

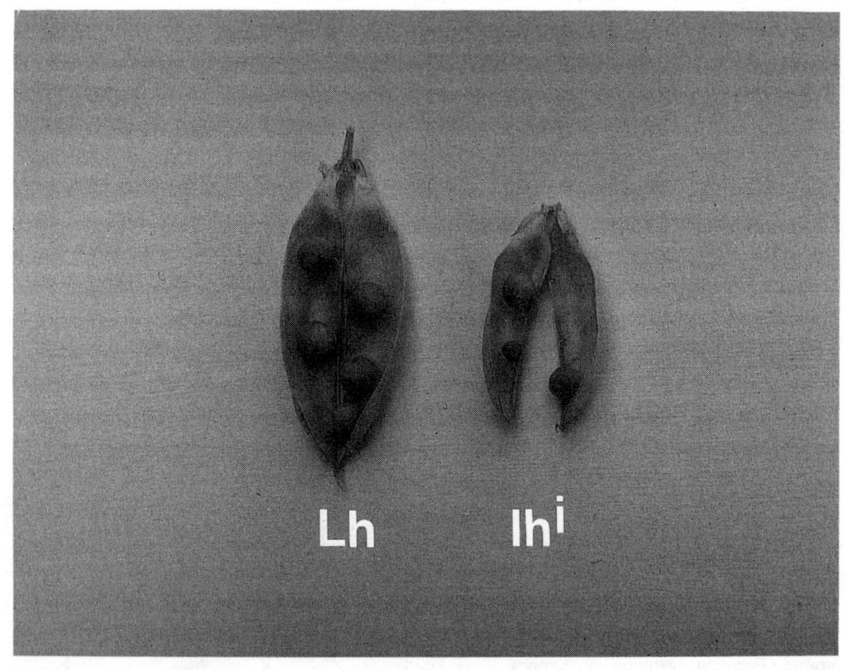

Figure 2.41. Seed abortion in a GA-deficient mutant. Pods with seeds are shown of wild-type (*Lh*) pea (cv. Torsdag) and the GA-deficient lh^{i} mutant (right) containing only two normal seeds, and one aborted seed (left, center). GA content of seeds (ng/g fresh wt): wild type, GA_1 28.7, GA_{20} 19.3; lh^{i} mutant, GA_1 2.4, GA_{20} 9.1. From Swain *et al.* (1993); photograph courtesy of S. M. Swain.

long delay or lag period, but the lag period becomes progressively shorter as the seeds mature; when they are 7–8 weeks old, they germinate within only a day or so after removal of the pod. This decline in the length of the lag period is accompanied by a steep decrease in seed ABA content (Fig. 2.42), and so it seems, therefore, that germinability is related to the ABA concentration within the young seed.

Further evidence suggesting that premature germination is suppressed by ABA comes from cases of vivipary, where there is virtually uninterrupted progression from embryogenesis to germination of the seed on the parent plant. The embryos of most viviparous mutants of maize (e.g., *vp5*) have ABA concentrations 25–50% of those in the normal, nonviviparous wild type. One viviparous mutant (*vp1*), however, shows the same order of ABA concentration as in the wild type, but in this case the embryo of the mutant is much less sensitive to its own ABA. Relative insensitivity to ABA appears to be the case in mangroves (e.g., *Rhizophora mangle*) where vivipary is a normal occurrence. Much higher concentrations of ABA than are normally required are needed to inhibit the growth of excised embryos of this species. Embryos of mutants of other species in which the ABA content or the sensitivity to ABA are lowered or virtually abolished also show precocious germination while still in the fruit. This happens in *Arabidopsis*, for example, in the double mutant *aba1, abi3–1* (ABA deficient and insensitive) and in the ABA-deficient *sitiens* mutant of tomato, in the latter especially in the older fruits. Manipulation of ABA content by biochemical means, by the inhibitor of ABA synthesis fluridone, also induces some precocious germination of developing embryos, e.g., in maize. See Sections 2.3.5.4 and 3.3.2 for more detailed discussions.

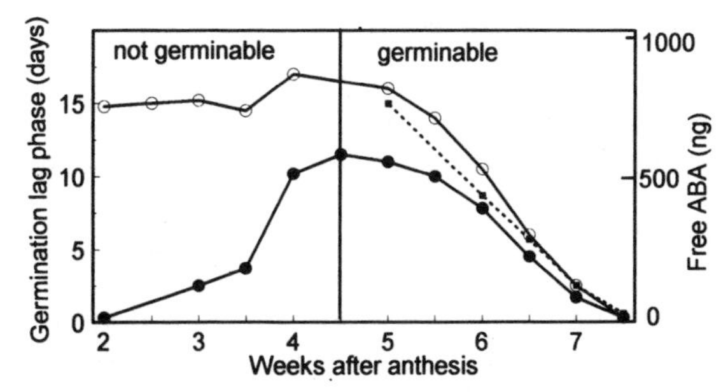

Figure 2.42. Germinability and ABA content of developing seeds of dwarf bean *(Phaseolus vulgaris)*. Seeds were removed from the pod at intervals during development and tested for germinability. Note that for the first $4\frac{1}{2}$ weeks seeds were not germinable. The time taken for radicle emergence to begin in germinable seeds was recorded (the lag phase) (■). ABA is expressed as ng/seed (●) and ng/g seed (O). Adapted from Van Onckelen *et al.* (1980).

Despite the evidence presented, however, it is still too soon to conclude that the control of germinability of the developing seed universally rests with ABA. There are several cases where no clear correlations can be found between ABA content and changes in the ability of young seeds to germinate. The osmotic factor also plays an important role, probably jointly with ABA. And so while we have a tantalizing glimpse of the regulatory system that may operate to control germinability of the immature seed, we still need to know more about several aspects—particularly how sensitivity to ABA might change during seed development—before we can fully assess its plausibility. It is known, for example, that in alfalfa seeds, sensitivity of the embryo to ABA declines progressively during development.

A large component of growth of the developing seed is, of course, associated with the laying down of storage reserves, and hormones have been implicated in this process. It has been suggested, for example, that the cytokinins in the liquid endosperm participate in the mobilization of assimilates to that tissue. As we saw earlier, ABA participates in the regulation of storage protein synthesis. Application of ABA to isolated cotyledons of *Phaseolus vulgaris* and soybean promotes the synthesis of storage protein (see Section 2.3.5.4). In grapes, the accumulation of sugars is enhanced, and grain filling in wheat may be promoted by the endogenous ABA but the evidence is still inconclusive. ABA has been suggested to participate in seed filling in pea. Developing seeds of *Arabidopsis* mutants (especially the ABA-insensitive ones) have reduced contents of certain storage proteins and triacylglycerols. There is also good evidence that triacylglycerol synthesis in wheat embryos is dependent on ABA. On the other hand, there is no reduction in storage protein and carbohydrates in the ABA-deficient *sitiens* mutant of tomato.

2.4.2.2. Fruit Growth and Development

The growth of fruit flesh is in many cases linked to the activity of the developing seeds. For example, fruit size in certain cucurbits (e.g., melon) is positively correlated with seed number. In the botanically false fruit, the strawberry (where the pips or so-called seeds on the fleshy receptacle are the true fruits), the growth of the flesh is greatly reduced when the young, developing seeds are removed. The "seeds" are relatively rich in auxins, and application of auxins to a "deseeded" strawberry to a large extent restores flesh growth. It seems probable, therefore, that flesh growth is promoted by auxin from the developing seeds. The strawberry is a convenient fruit with which to do experiments of this kind because the "seeds" are borne externally. But although they are more difficult, similar experiments have been carried out on true fruits, such as the pea, where the seeds are, of course, internal. By inserting a fine needle though the pod wall (the ovary wall) the seeds inside 2- to 3-day-old pea fruits can be punctured and killed. The pod of fruits treated in this way fails to grow, but virtually normal growth is restored by application of GA or the synthetic auxin, naphthylacetic acid (Fig. 2.43). It has not been established that seeds younger than about 10 days contain GA, but older pea seeds do have GA_9, GA_{20}, and GA_{29}, and these GAs, especially GA_{20}, are active in promoting the growth of seedless pods. An early interpretation of this finding was that GAs (and possibly auxin) from the developing seeds control growth of the ovary wall, but there is now evidence that, instead, the seeds enable the pericarp to produce its own GAs, possibly by supplying precursors.

2.4.2.3. *Hormones for Germination and Growth*

We have seen that seeds of several species, e.g., maize, contain auxin conjugates, many of which furnish IAA on hydrolytic cleavage. There is no evidence that the IAA participates in the germination process or in radicle emergence, but later events—seedling growth—might be controlled by the hormone. Thus, IAA from the conjugates in the endosperm of germinated maize

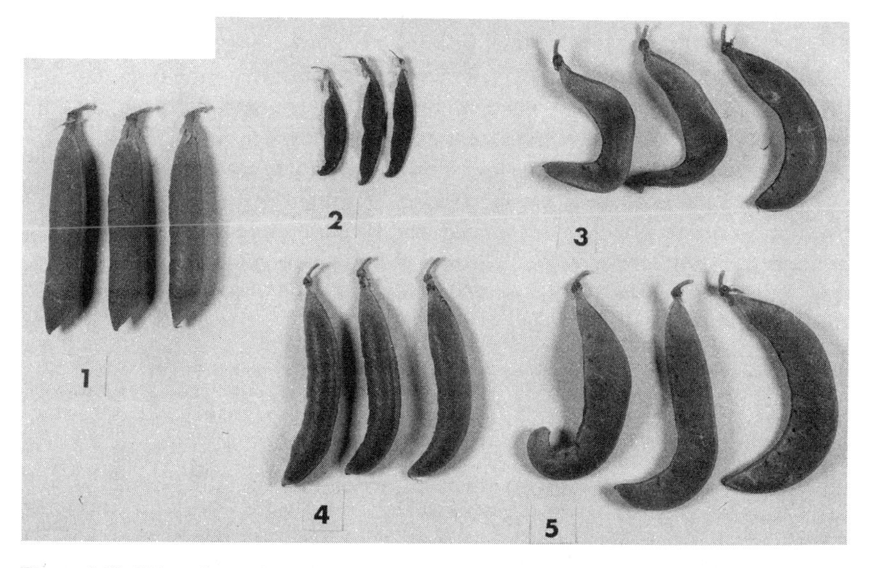

Figure 2.43. Effect of growth regulators on the growth of the ovary wall in peas (*Pisum sativum* cv. Progress no. 9). (1) Control (seed alive). (2–5) Seeds killed. (2) No treatment. (3) Naphthylacetic acid treatment (100 μg/pod). (4) Gibberellic acid (GA_3) treatment (100 μg/pod). (5) Naphthylacetic acid plus GA_3 treatment (each at 100 μg/pod). Growth regulators were applied as solutions in 50% aqueous ethanol. The solvent alone was added to (1) and (2). From Sponsel (1982). Photograph kindly supplied by the author.

(and probably other cereals) is translocated to the tip of the coleoptile from which position it regulates growth of the cells of the elongating zone of that organ.

Radioactively labeled GA_1 fed to developing pea seeds is incorporated into various conjugates, probably glucosides. When the seeds are mature, dried, and then set to germinate, a proportion of the radioactive GA is liberated. GA_{20} is conjugated by maturing maize grains and is liberated, together with GA_1, on germination. Thus, it seems that the conjugates may be laid down in the developing seed as a store of GA for future use.

2.4.2.4. Other Effects of the Seed Hormones

When young seeds inside a pea pod are killed (Section 2.4.2.2), abscission of the fruit stalk is promoted. Treatment of the seedless pod with naphthylacetic acid or GA (including the native GA_9, GA_{20}, and GA_{29}) stops the formation of the abscission layer. One of the effects of these seed hormones, therefore, might be to inhibit the abscission zone. On the other hand, abscission of fruits and neighboring leaves has been suggested to be stimulated by ABA from developing seeds, and although some evidence favors this view, the experimental findings are, on the whole, inconclusive.

An interesting effect which has been attributed to the seed GA occurs in apple. Some cultivars are biennial bearers, i.e., they flower and consequently bear fruit only once every 2 years. Now, flower buds are normally initiated on the spurs while the fruit is developing, and they remain dormant until the following year; so the flowers appearing in one year are those formed in the previous year. In the biennial bearers, flower buds fail to form while fruit is also developing, and hence a flowering season is omitted. Removal of developing fruits allows flower buds to be initiated, but the inhibitory effect of fruits can be mimicked by application of GA. Since developing apple seeds are rich in GAs (see Table 2.7), these regulators are thought to be the cause of flower bud suppression.

USEFUL LITERATURE REFERENCES

SECTION 2.1

Castle, L. A., and Meinke, D. W., 1993, *Semin. Dev. Biol.* **4**:31–39 (embryo-defective mutants, especially in *Arabidopsis*).

Johansen, D. A., 1950, *Plant Embryology. Embryology of the Spermatophyta,* Chronica Botanica, Waltham, Mass. (developmental patterns).

Lindsey, K., and Topping, J. F., 1993, *J. Exp. Bot.* **44**:359–374 (embryogenic development patterns).

Maheshwari, P., 1950, *An Introduction to the Embryology of Angiosperms,* McGraw–Hill, New York (developmental patterns).

Marinos, N. G., 1970, *Protoplasma* **70:**261–279 (embryogenesis of the pea).

Meinke, D. W., 1991, *Plant Cell* **3:**857–866 (developmental mutants and embryogenesis).

Olsen, O.-A., Potter, R. H., and Kalla, R., 1992, *Seed Sci. Res.* **2:**117–131 (cereal endosperm development).

Raghavan, V., 1986, *Embryogenesis in Angiosperms. A Developmental and Experimental Study,* Cambridge University Press, Cambridge (zygotic and somatic embryogenesis).

Sheridan, W. F. and Clark, J. K., 1993, *Plant J.* **3:**347–358 (embryo mutants in maize).

Singh, H., 1978, *Embryology of Gymnosperms,* Borntraeger, Berlin (structure and development).

The Plant Cell, 1993, **5:**1139–1488 (special review issue on plant reproduction, including articles on seed development, embryogenesis).

Xu, N., and Bewley, J. D., 1992, *Plant Cell Rep.* **11:**279–284 (SEM of alfalfa embryo development).

SECTION 2.2

Barneix, A. J., Arnozis, P. A., and Guitman, M. R., 1992, *Physiol. Plant.* **86:**609–615 (N accumulation in wheat grains).

Hardham, A. R., 1976, *Aust. J. Bot.* **24:**711–721 (vascular transport in pea seeds).

Jenner, C. F., Ugalde, T. D., and Aspinall, D., 1991, *Aust. J. Plant Physiol.* **18:**211–226 (starch and protein accumulation in wheat endosperm, and stress effects).

Johnson-Flanaghan, A. M., and McLachlan, G., 1990, *Physiol. Plant.* **80:**460–466 (frost and green seed coats in canola).

Miller, M. E., and Chourey, P. S., 1992, *Plant Cell* **4:**297–305 (sucrose uptake in maize and invertase mutant).

Pate, J. S., 1984, in: *Seed Physiology. Volume I, Development* (D. R. Murray, ed.), Academic Press, New York, pp. 41–82 (C and N translocation in legume plants).

Rochat, C., and Boutin, J.-P., 1991, *J. Exp. Bot.* **42:**207–214 (amino acid composition of assimilates in seed coats and embryos).

Sakri, F. A. K., and Shannon, J. C., 1975, *Plant Physiol.* **55:**881–889 (sugar translocation into wheat grains).

Shannon, J. C., 1972, *Plant Physiol.* **49:**198–202 (sugar translocation into maize kernels).

Tanaka, T., Minamikawa, T., Yamauchi, D., and Ogushi, Y., 1993, *Plant Physiol.* **101:**421–428 (proteins and proteinases in legume pods).

Thorne, J. H., 1985, *Annu. Rev. Plant Physiol.* **36:**317–343 (unloading of C and N assimilates into seeds).

Westgate, M. E., and Peterson, C. M., 1993, *J. Exp. Bot.* **44:**109–117 (stress effects on soybean pod development).

Wolswinkel, P., 1992, *Seed Sci. Res.* **2:**59–73 (transport of nutrients into developing seeds).

SECTION 2.3. CARBOHYDRATES

Bhattacharyya, M. K., Smith, A. M., Ellis, T. H. N., Hedley, C., and Martin, C., 1990, *Cell* **60:**115–122 (wrinkled pea mutant and defective starch synthesis).

Buttrose, M. S., 1960, *J. Ultrastruct. Res.* **4:**231–257 (cereal starch granule formation).

Edwards, M., Scott, C., Gidley, C., and Reid, J. S. G., 1992, *Planta* **187:**67–74 (galactomannan biosynthesis).

Preiss, J., 1991, *Oxford Surv. Plant Mol. Cell Biol.* **7:**59–114 (starch synthesis and its regulation in plants).

Reid, J. S. G., 1985, in: *Biochemistry of Storage Carbohydrates in Green Plants* (P. M. Dey and R. A. Dixon, eds.), Academic Press, New York, pp. 265–288 (galactomannans in seeds).

Smith, A. M., and Denyer, K., 1992, *New Phytol.* **122:**21–33 (starch synthesis in developing pea seeds).

Wang, T. L., and Hedley, C. L., 1991, *Seed Sci. Res.* **1:**3–14 (metabolism and pea seed development).

SECTION 2.3. TRIACYLGLYCEROLS

Appleby, R. S., Gurr, M. I., and Nichols, B. W., 1974, *Eur. J. Biochem.* **48:**209–216 (triacylglycerol synthesis in *Crambe*).

Garces, R., Sarmiento, C., and Mancha, M., 1992, *Planta* **186:**461–465 (temperature affects oleate desaturase in developing sunflower seeds).

Holbrook, L. A., Magus, J. R., and Taylor, D. C., 1992, *Plant Sci.* **84:**99–115 (regulation of triacylglycerol biosynthesis in microspore–derived embryos).

Huang, A. H. C., 1992, *Annu. Rev. Plant Physiol. Plant Mol. Biol.* **43:**177–200 (oil bodies and oleosins).

Murphy, D. J., Rawsthorne, S., and Hill, M. J., 1993, *Seed Sci. Res.* **3:**79–96 (review on triacylglycerol formation in seeds).

Rodriguez-Sotres, R., and Black, M., 1993, *Planta* **192:**9–15 (regulation of triacylglycerol synthesis in embryos—ABA and osmotica).

Simcox, P. D., Reid, E. E., Canvin, D. T., and Dennis, D. T., 1977, *Plant Physiol.* **59:**1128–1132 (proplastids in developing castor bean endosperm).

Slabas, A. R., and Fawcett, T., 1992, *Plant Mol. Biol.* **19:**169–191 (review article on the molecular biology of lipid biosynthesis in plants).

Stumpf, P. K., 1977, in: *Lipids and Lipid Polymers in Higher Plants* (M. Tevini and H. K. Lichtenthaler, eds.) Springer-Verlag, Berlin, pp. 75–84 (lipid biosynthesis in seeds: a review).

Stymne, S., and Stobart, A. K., 1987, in: *The Biochemistry of Plants, Vol. 9* (P. K. Stumpf, ed.) Academic Press Inc., pp. 175–214 (review article on triacylglycerol biosynthesis).

Tzen, J. T. C., Cao, Y.-Z., Laurent, P., Ratnayake, C., and Huang, A. H. C., 1993, *Plant Physiol.* **101:**267–276 (oleosins and oil bodies from diverse species).

Wanner, G., Formanek, H., and Theimer, R. R., 1981, *Planta* **151:**109–123 (oil body formation).

SECTION 2.3. PROTEINS

Altschuler, Y., Rosenberg, N., Harel, R., and Galili, G., 1993, *Plant Cell* **5:**443–450 (N- and C-terminal regions of gliadin, and sorting).

Baumlein, H., Boerjan, W., Nagy, I., Panitz, R., Inze, D., and Wobus, U., 1991, *Mol. Gen. Genet.* **225:**121–128 (upstream sequences regulating legumin gene expression in transgenics).

Bednarek, S. Y., and Raikhel, N. V., 1992, *Plant Mol. Biol.* **20:**133–150 (synthesis, processing, and transport of storage proteins).

Bewley, J. D., and Greenwood, J. S., 1990, in: *Plant Physiology, Biochemistry and Molecular Biology* (D. T. Dennis and D. H. Turpin, eds.), Longman Scientific and Technical, Harlow, pp. 456–469 (storage protein synthesis and deposition in seeds).

Blobel, G., Walter, P., Chang, C. N., Goldman, B. H., Erickson, A. H., and Lingappa, V. R., 1979, in: *Symp. Soc. Exp. Biol.,* **33:**9–36 (signal hypothesis and protein passage through membranes).

Boulter, D., Evans, I. M., Ellis, J. R., Shirsat, A. H., Gatehouse, J. A., and Croy, R. R. D., 1987, *Plant Physiol. Biochem.* **25:**283–289 (storage protein gene expression in *Pisum sativum*).

Brandt, A., 1976, *Cereal Chem.* **53:**890–901 (protein synthesis in high-lysine barley).

Chrispeels, M. J., 1985, *Oxford Surv. Plant Mol. Cell Biol.* **2:**43–68 (posttranslational modifications to vacuolar storage proteins).

Chrispeels, M. J., 1991, *Annu. Rev. Plant Physiol. Plant Mol. Biol.* **42:**21–53 (sorting and secretion of proteins).

Chrispeels, M. J., and Raikhel, N. V., 1992, *Cell* **68:**613–616 (amino acid domains for protein sorting and targeting).

Craig, S., Goodchild, D. J., and Hardman, A. R., 1979, *Aust. J. Plant Physiol.* **6:**81–98 (vacuole changes and protein deposition in pea).

Craig, S., Goodchild, D. J., and Hardham, A. R., 1980, *Aust. J. Plant Physiol.* **7:**327–337 (vacuole changes and protein deposition in pea).

Dickinson, C. D., Hussein, E. H. A., and Nielsen, N. C., 1989, *Plant Cell* **1:**459–469 (assembly of glycinin).

Dure, L. III, and Galau, G. A., 1981, *Plant Physiol.* **68:**187–194 (processing of cottonseed proteins).

Gatehouse, J. A., and Shirsat, H. A., 1993, in: *Control of Plant Gene Expression* (D. P. S. Verma, ed.), CRC Press, Boca Raton, Fla., pp. 357–375 (seed storage protein gene expression).

Gatehouse, J. A., Evans, I. M., Bown, D., Croy, R. R. D., and Boulter, D., 1982, *Biochem. J.* **208:**119–127 (storage protein synthesis in developing pea seeds).

Gatehouse, J. A., Evans, I. M., Croy, R. R. D., and Boulter, D., 1986, *Philos. Trans. R. Soc. London* **314:**367–384 (expression of storage protein genes in developing legume seeds).

Gayler, K. R., and Sykes, G. E., 1981, *Plant Physiol.* **67:**958–961 (conglycinin synthesis in soybean).

Holwerda, B. C., Padgett, H. S., and Rogers, J. C., 1992, *Plant Cell* **4:**307–318 (targeting of a protein through an N-terminal domain).

Itoh, Y., Kitamura, Y., Arahira, M., and Fukazawa, C., 1993, *Plant Mol. Biol.* **21:**973–984 (*cis*-acting regions of soybean 11 S globulin and DNA-binding proteins).

Johnson, K. D., and Chrispeels, M. J., 1987, *Plant Physiol.* **84:**1301–1308 (processing oligosaccharide side chains).

Kawagoe, Y., and Murai, N., 1992, *Plant J.* **2:**927–936 (spatial and temporal specificity of β-phaseolin gene expression).

Kermode, A. R., 1994, in: *Mechanisms of Plant Growth and Improved Productivity* (A. S. Basra, ed.), Dekker, New York, in press (regulation of protein synthesis posttranslationally).

Krishnan, H. B., Franceschi, V. R., and Okita, T. W., 1986, *Planta* **169:**471–480 (protein body formation in rice grains is as in legume seeds).

Krochko, J. E., and Bewley, J. D., 1989, *Electrophoresis* **9:**751–763 (storage protein synthesis and analytical methods).

Larkins, B. A., and Hurkman, W. J., 1978, *Plant Physiol.* **62:**256–263 (zein deposition in protein bodies).

Larkins, B. A., Pedersen, K. Handa, A. K., Hurkman, W. J., and Smith, L. D., 1979, *Proc. Natl. Acad. Sci. USA* **76:**6448–6452 (synthesis of zein in *Xenopus* oocytes).

Lending, C. R., and Larkins, B. A., 1989, *Plant Cell* **1:**1011–1023 (location of zein classes in maize protein bodies).

Manteuffel, R., Muntz, K., Puchel, M., and Scholtz, G., 1976, *Biochem. Physiol. Pflanzen* **169:**595–605 (DNA, RNA, and protein accumulation in *Vicia faba*).

Millerd, A., and Spencer, D., 1974, *Aust. J. Plant Physiol.* **1:**331–341 (RNA and nuclei in pea cotyledons).

Nakamura, K., and Matsuoka, K., 1993, *Plant Physiol.* **101:**1–5 (protein targeting to the vacuole, update).

Quatrano, R. S., Marcotte, W. R., and Guiltinan, M., 1993, in: *Control of Plant Gene Expression* (D. P. S. Verma, ed.), CRC Press, Boca Raton, Fla., pp. 69–90 (ABA regulation of gene expression).

Rerie, W. G., Whitecross, M., and Higgins, T. J. V., 1991, *Mol. Gen. Genet.* **225:**148–157 (developmental and environmental regulation of pea legumin genes).

Rubin, R., Levanony, H., and Galili, G., 1992, *Plant Physiol.* **99:**718–724 (protein body formation in wheat endosperm).

Schmidt, R., 1993, in: *Control of Plant Gene Expression* (D. P. S. Verma, ed.), CRC Press, Boca Raton, Fla., pp. 337–355 (*opaque–2* and zein gene expression).

Sengupta, G., Deluca, V., Bailey, D. S., and Verma, D. P. S., 1981, *Plant Mol. Biol.* **1:**19–34 (posttranscriptional processing in soybean).

Shirsat, A. H., 1991, in: *Developmental Regulation of Plant Gene Expression* (D. Grierson, ed.), Blackie, Glasgow and Chapman and Hall, London, pp. 153–181 (gene expression in developing seeds).

Shotwell, M. A., and Larkins, B. A., 1991, in: *Molecular Approaches to Crop Improvement* (E. S. Dennis and D. J. Llewellyn, eds.), Springer-Verlag, pp. 34–62 (genetic engineering for seed protein quality).

Weber, E. J., 1980, in: *The Resource Potential in Phytochemistry. Recent Advances in Phytochemistry,* Volume 14, Plenum Press, New York., pp. 97–137 (corn mutants).

Zheng, Y., He, M., Hao, S., and Huang, B., 1992, *Ann. Bot.* **69:**377–383 (ER-derived protein bodies in late developing soybean seeds).

SECTION 2.3. NUTRITION, PHYTIN

Greenwood, J. S., 1989, in: *Recent Advances in the Development and Germination of Seeds* (R. B. Taylorson, ed.), Plenum Press, New York, pp. 109–125 (phytin synthesis and deposition).

Greenwood, J. S., and Bewley, J. D., 1984, *Planta* **160:**113–120 (EM of phytin deposition).

SECTION 2.4

Black, M., 1991, in: *Abscisic Acid Physiology and Biochemistry* (W. J. Davies and H. G. Jones, eds.), Bios Scientific Publishers, Oxford, pp. 99–136 (ABA involvement in seed development and maturation).

Davey, J. E., and Van Staden, J., 1979, *Plant Physiol.* **63:**873–877 (cytokinins in developing lupin seeds).

Eeuwens, C. J., and Schwabe, W. W., 1975, *J. Exp. Bot.* **26:**1–14 (growth regulators in developing pea seeds).

Groot, S. P. C., van Yperen, I. I., and Karssen, C. M., 1991, *Physiol. Plant.* **81:**73–78 (precocious germination, reserve deposition in a tomato mutant).

King, R. W., 1982, in: *The Physiology and Biochemistry of Seed Development, Dormancy and Germination* (A. A. Khan, ed.), Elsevier, Amsterdam, pp. 157–181 (abscisic acid in seed development).

Koornneef, M., Hanhart, C. J., Hilhorst, H. W. M., and Karssen, C. M., 1989, *Plant Physiol.* **90:**463–469 (precocious germination, reserve protein accumulation in *Arabidopsis* mutants).

Oishi, M. Y., and Bewley, J. D., 1990, *Plant Physiol.* **94:**592–598 (fluridone-induced precocious germination in maize).

Ooms, J. J. J., Leon-Kloosterziel, K. M., Bartels, D., Koornneef, M., and Karssen, C. M., 1993, *Plant Physiol.* **102:**1185–1191 (precocious germination in ABA mutants of *Arabidopsis*).

Ozga, J. A., Brenner, M. L., and Reinecke, D. M., 1992, *Plant Physiol.* **100:**88–94 (effect of developing pea seeds on pericarp GA).

Sponsel, V. M., 1980, in: *Gibberellins—Chemistry, Physiology and Use,* British Plant Growth Regulator Group Monograph 5 (J. R. Lenton, ed.), Wessex Press, Wantage, U.K., pp. 49–62 (gibberellins in developing pea seeds).

Sponsel, V. M., 1982, *J. Plant Growth Regul.* **1:**147–152 (seed and hormonal control of growth in peas).
Swain, S. M., Reid, J. B., and Ross, J. J., 1993, *Planta* **191:**482–488 (seed abortion in a GA mutant of pea).
Takahashi, N., Phinney, B. O., and MacMillan, J. (eds.), 1991, *Gibberellins,* Springer-Verlag, Berlin (articles on GA in seeds).
Ueda, M., Ehman, A., and Bandurski, R. S., 1970, *Plant Physiol.* **46:**715–718 (bound auxin in maize).
Van Onckelen, H., Caubergs, R., Horemans, S., and DeGreef, J. A., 1980, *J. Exp. Bot.* **31:**913-920 (ABA in developing bean seeds).
Xu, N., Coulter, K. M., and Bewley, J. D., 1990, *Planta* **182:**382–390 (ABA sensitivity and effects on precocious germination).

Chapter 3

Development—Regulation and Maturation

We have seen in the previous chapter that seed development commences with the formation of the single-celled fertilized egg and (generally) terminates when the seed is mature. Between these events there occur many morphological, cellular, and biochemical/synthetic changes which are regulated in a coordinated manner so that the progeny of a particular species are phenotypically more or less identical. Development proceeds in an environment in which seeds are hydrated, yet they do not germinate. What, then, prevents seeds from germinating during development? And how are the controls which maintain seeds in a developmental mode eventually overcome to permit germination? Research into these questions has increased greatly since the first edition of this book appeared, and while much remains to be learned, enough is known to warrant a chapter dealing exclusively with this topic.

3.1. GERMINABILITY DURING DEVELOPMENT

Seeds do not usually germinate during development on the parent plant, but undergo a process of maturation, generally including desiccation, before being shed. Yet maturation is not an obligatory process for the acquisition of germinability, as there is ample evidence that the seed, or at least the embryo therein, is capable of germinating during development. In many species, removal of the seed from its surrounding structures and its placement on water does not lead to germination; however, if the embryo is dissected from the seed, it germinates well. Developing alfalfa seeds, for example, do not germinate when excised from the silique (pod) (Fig. 3.1A) until late during development, a time when the seeds are undergoing maturation drying. Isolated embryos, however, will germinate from an early stage of development, if placed on a nutrient medium (Fig. 3.1A). It takes a longer time for very young embryos to germinate than older ones, and the former are more dependent on the nutrient medium for survival than the latter, which germinate when placed on water (Fig. 3.1A). Presumably this is because early during development the embryos initially lack sufficient nutrients to support their continued development to a germinable stage, and also lack the nutrients and stored reserves to support germination and postgermination growth.

In contrast to this, some seeds, such as those which develop in a fully hydrated environment, e.g., intact seeds of tomato (Fig. 3.1B) and muskmelon, germinate when placed on water after removal from the ripening fruit. Germination of isolated embryos can occur at an earlier stage of development if they are on a nutrient medium. Thus, constraints on the germination of alfalfa during development are imposed by the surrounding seed coat and endosperm, which must be removed to allow the embryo to germinate. At early stages of development, the embryos of tomato seeds appear to be similarly constrained, but as development proceeds germination is prevented by the fruit tissues only, and removal of seeds from this environment is sufficient to allow germination to occur.

During development there is commonly an increase in the ABA content of a seed, both within the embryo itself and in the surrounding structures. It is generally accepted that ABA plays an important role in preventing germination during seed development, that it declines during late maturation (drying), and that at the same time there is frequently also a decrease in sensitivity to its presence. The application of ABA to isolated developing embryos prevents their germination, and they remain in a developmental mode. The importance of ABA, and sensitivity of the seed to it has been elegantly demonstrated using mutants which either fail to synthesize ABA (ABA-deficient mutants, *aba*) or fail to respond to ABA (ABA-insensitive mutants, *abi*) (Section 3.3.2). Developing seeds of wild-type *Arabidopsis* do not germinate on water after being dissected

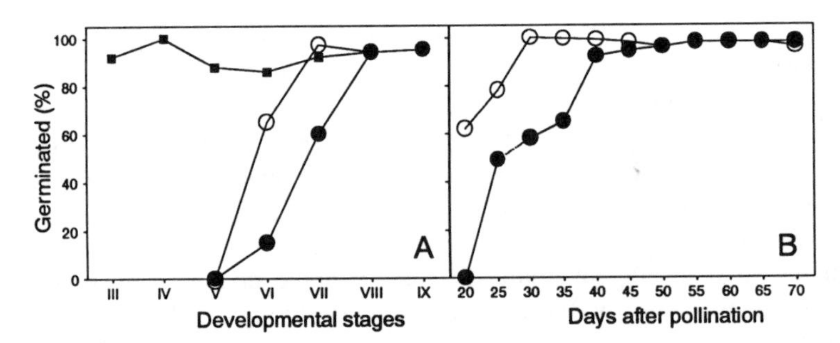

Figure 3.1. (A). Germination of seeds and embryos of alfalfa during their development. At various stages of development, seeds were removed from the pod and placed on water (●), or embryos were dissected from them and placed on a nutrient medium (■) or water (○). (B) Germination of tomato seeds removed from the fruit during development and placed on water for 7 days (●), and germination of embryos excised from the seeds placed on a nutrient medium (○). Tomato seeds take about 60 days after pollination (DAP) to reach maturity. A, after Xu *et al.* (1990); B, after Berry and Bewley (1991).

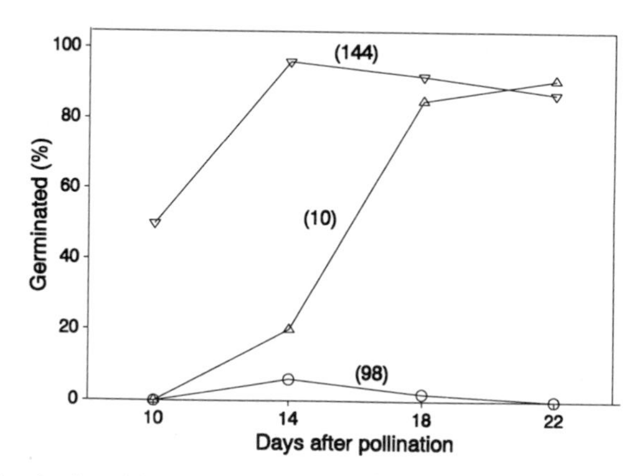

Figure 3.2. Germination of developing *Arabidopsis thaliana* seeds on water after removal from the silique at various times after pollination. O, Wild type; △, ABA-deficient mutant (*aba*); ▽, ABA-insensitive mutant (*abi*). The numbers in parentheses show the concentration of ABA in each type in ng/g fresh wt at 14–16 DAP. After Karssen *et al.* (1987).

out of the silique (Fig. 3.2). However, seeds of the mutant lacking ABA germinate precociously under identical conditions, as do those of the mutant that contains somewhat more ABA than the wild type, but which are insensitive to it. Similar mutants occur in other species, including tomato; these can show a similar behavior to *Arabidopsis*.

The prevention of germination of seeds developing within fleshy fruits seems not to be related exclusively to ABA, even though it is present in the tissues surrounding the seed. Rather, the low water potential of the surrounding fruit tissues, by putting the seeds under an osmotic stress, restricts their ability to germinate. Mutants which germinate precociously appear to be less sensitive to the osmotic constraints imposed by the fruit tissues.

In the subsequent sections of this chapter we will examine further the role of ABA and osmotic restraints on the maintenance of seed development, and how these are overcome to permit germination.

3.2. VIVIPAROUS MUTANTS AND PRECOCIOUS GERMINATION

Mutations that affect seed development provide a valuable means to study the genes that control the complexity of events involved in this process. Particularly interesting in this respect are the mutants which result in developmental arrest and permit germination to occur without the requirement for maturation to be completed. These are the *viviparous (vp)* mutants, whose seeds complete germination precociously, i.e., while still on the parent plant. Such mutants have been reported for several species, including cereals (e.g., maize, wheat, rye, and barley), those producing seeds in fleshy fruits (e.g., lemon, orange, tomato, and melon), and *Arabidopsis* (see also Section 3.1). Vivipary is normal in some seeds, e.g., mangrove (*Rhizophora mangale*), which germinate and produce a long spearlike radicle/hypocotyl before being shed, when they drop from the parent plant and stick into the underlying mud to become established.

In maize, which has been studied the most extensively, at least nine viviparous genes are known, and these can be placed in two mutant classes (Table 3.1). The largest class of mutants includes those whose kernels (embryos and endosperms) are deficient in ABA, and most are deficient in carotenoids also, resulting in a white (rather than yellow) pigmentation. This results from defects in the ABA biosynthesis pathway prior to the step involving the production of carotenoids (Fig. 3.3). Few of these mutants can form viable seedlings since carotenoid deficiency generally leads to photobleaching of chlorophyll in the newly formed leaves, and hence failure of photosynthesis. Because of this lethal condition of the homozygous viviparous mutants, they must be maintained as heterozygotes (i.e., *Vp/vp*). The *vp8* mutation results in a reduction in kernel ABA content, without affecting the amount of carotenoids present in the kernels which remain yellow. The metabolic lesion in this mutant is probably after carotenoid biosynthesis, and possibly in the synthesis of ABA from violaxanthin (Fig. 3.3). The other class of mutants is represented by a single locus (*vp1*). This mutant accumulates appreciable quantities of ABA (Table 3.1) but has a low sensitivity to it; carotenoids also accumulate and the homozygous recessive condition is nonlethal.

Table 3.1. Viviparous Mutants of Maize and Their ABA and Carotenoid Content[a]

Genotype	ABA content (ng/embryo)	Carotenoid
Class I: ABA-insensitive		
vp1	3.9 (4.8)	+
Class II: ABA-deficient		
vp2	0.2 (3.3)	–
vp5	0.25 (1.9)	–
vp7	0.2 (1.8)	–
vp8	0.2 (1.3)	+
vp9	0.25 (2.6)	–

[a]Information on *w3*, *y9* and *al* mutants not included. In ABA content column the amount present in the wild type is shown in parentheses following that in the mutant. For carotenoid content, + means amount similar to the wild type; – indicates a deficiency. Based on information in Neill *et al.* (1986) and McCarty and Carson (1991).

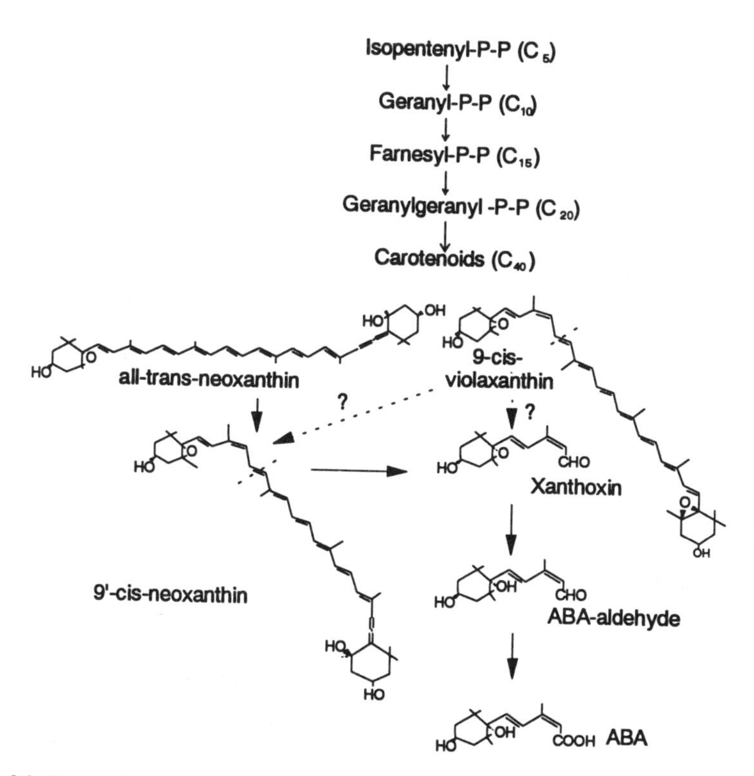

Figure 3.3. Proposed pathway for the biosynthesis of abscisic acid in higher plants. The involvement of 9′-*cis*-neoxanthin is unproven. After Parry and Horgan (1992). Reprinted by permission of Kluwer Academic Publishers.

These *viviparous* mutants thus provide compelling evidence for the role of ABA in maintaining seeds in the developmental mode, and in preventing their germination prior to maturation. Indeed, even if wild-type kernels of maize are sprayed with an inhibitor of ABA (and carotenoid) synthesis, fluridone, at critical stages during their development, precocious (viviparous) germination occurs (Fig. 3.4).

The question of interest, of course, is: what developmental events do the normal *Vp* genes control? So far, most work on this question has been conducted on the *Vp1* gene. In the *vp1*, ABA-insensitive mutant there is an almost complete independence of the embryo from the endosperm, and it is possible to obtain *vp* embryos in a wild-type endosperm, and vice versa. When the endosperm is of the mutant type *vp1*, there is reduced expression of several genes normally active during late aleurone-layer development. Consequently, this layer fails to accumulate anthocyanin, and enzyme

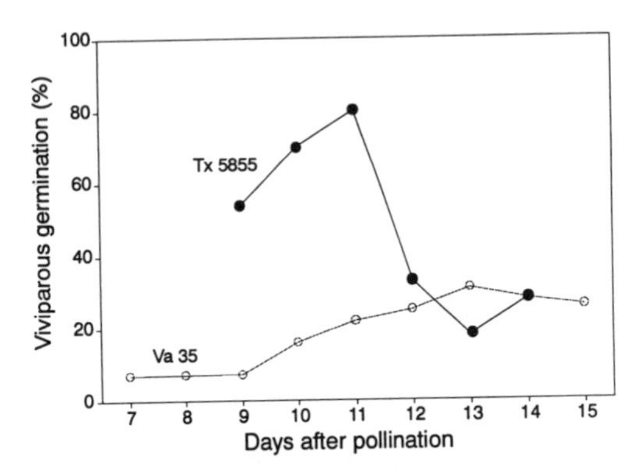

Figure 3.4. The effect of fluridone, an ABA-biosynthesis inhibitor, on viviparous germination of two lines of maize during their development. Developing kernels were sprayed with fluridone at 9–14 DAP (*Tx5855*) or 5–15 DAP (*Va35*) and then allowed to mature to 45–50 DAP before germination was assessed. The effectiveness of fluridone thus varies with time of application during development. After Fong *et al.* (1983).

deficiencies in its biosynthetic pathway, and in other pathways, have been found. These changes are independent of the ABA response. The block to anthocyanin synthesis is seed-specific, for seedlings and vegetative plants of *vp1* mutant accumulate amounts of pigment similar to the wild type.

The effects on seed development of the mutation to the *Vp1* gene are thus pleiotropic, i.e., there are several distinct responses to a change in this single gene. *Vp1* encodes a unique protein of approximately 73 kDa, and the basis for the pleiotropic phenotype may be in its multifunctional nature. The protein may affect the expression of several genes related to development, as a transcriptional regulator. One such gene may be the *C1* gene whose product, the C1 protein, is also a nuclear regulatory (transcription) factor, which activates the transcription of specific enzymes in the anthocyanin synthesis pathway in the maize kernel. The *C1* gene is expressed in wild-type kernels, but not in *vp1* mutant kernels (Fig. 3.5). Since the *Vp1* gene is expressed in the wild-type kernels, but not in the mutant, it is possible that its product, the Vp1 protein, is also a transcriptional regulator—one that turns on the expression of the *C1* gene and the C1 protein. This, in turn, promotes the transcription of genes for enzymes in the anthocyanin synthesis pathway. In the *vp1* mutant kernels, the gene expressing the Vp1 protein is defective, and consequently there is no transcriptional activation of the *C1* gene nor, subsequently, genes related to anthocyanin synthesis. A similar cascade of events might operate for the expression of a gene for an ABA-receptor protein; the Vp1 protein might upregulate such a gene, or that of another protein which in turn is the regulator.

3.3. MATURATION DRYING AND THE "SWITCH" TO GERMINATION

Maturation drying is the normal terminal event in the development of many seeds, after which they pass into a metabolically quiescent state. Seeds may

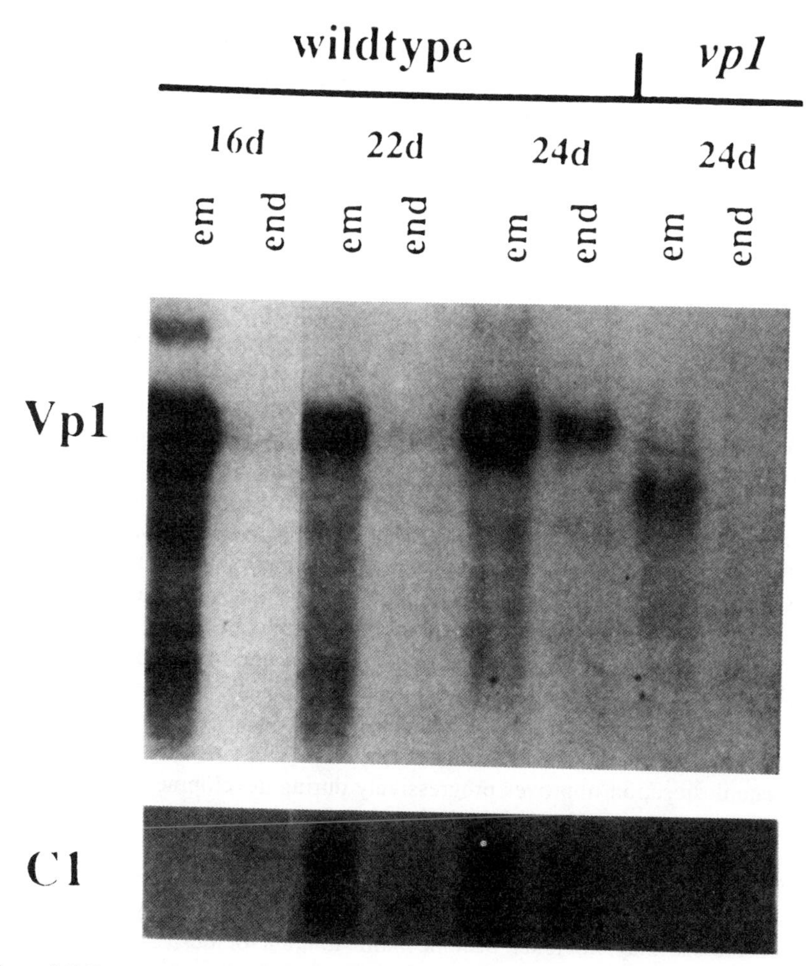

Figure 3.5. Expression of the *Vp1* gene, and the *Vp1*-regulated gene *C1*, during development of wild-type maize kernels, and at 24 DAP in developing kernels of the *vp1* mutant. mRNAs for the *Vp1* and *C1* protein in the embryo (em) and endosperm (end) were detected by Northern hybridization. Note the increase in expression of the *Vp1* and *C1* genes in the endosperm (presumably aleurone layer) of the wild type during development to 24 DAP. In the *vp1* mutant endosperm, the *Vp1* gene is not expressed at 24 DAP, and neither is *C1*, the gene which the Vp1 protein transcriptionally regulates. After McCarty *et al.* (1989).

remain in this dry state from several days to many years and still retain their viability (Chapter 9). Upon hydration under suitable conditions, the seed commences germination—if it is not dormant. There is now substantial evidence that the "switch" from a developmental to a germinative mode is elicited by the intervening maturation drying event.

3.3.1. The Acquisition of Desiccation Tolerance

Seeds are not capable of withstanding desiccation at all stages during their development, but their acquisition of desiccation tolerance is usually considerably earlier than maturation drying itself. In the developing castor bean (*Ricinus communis*), seeds excised from the capsule and placed on water do not germinate until some 50–55 days after pollination (DAP), when maturation drying has commenced (Fig. 3.6). However, if the excised seeds are first desiccated, and then placed on water, germination is achieved as early as 25–30 DAP. Seeds desiccated at 20 DAP will not germinate, nor will they survive, for they have yet to reach the desiccation-tolerant stage of their development. It appears, then, that desiccation-tolerance is achieved within 5 days during development (from 20 to 25 DAP), and a similar rapid acquisition of tolerance occurs in developing seeds of other species (e.g., *Phaseolus vulgaris*, mustard, and rapeseed). However, at early stages of development, survival upon desiccation in castor bean occurs only if the seed is dried slowly, over several days, whereas at later times of development tolerance of rapid desiccation is acquired. Thus, the ability of seeds to tolerate desiccation improves progressively during development. This is probably a consequence of physiological and morphological changes which take place gradually as development proceeds, perhaps including the synthesis of specific protective substances in the later stages (see Section 3.3.2). Another observation that has been made in castor bean (Fig. 3.6) and soybean is that as seeds mature they not only become more tolerant of desiccation, but upon rehydration there is an increased capacity to form normal seedlings. Even though germination of near-mature castor bean seeds (50–55 DAP) occurs, the percentage that establish normal seedlings is low. Drying leads to improved seedling quality following subsequent germination. Seeds of some species, including those of certain Gramineae, may withstand rapid desiccation relatively early during their development. *Avena fatua* (wild oat) grains, for example, can survive air-drying when only 5–10 days old, and germinate upon subsequent rehydration; maturation drying normally occurs some 15–20 days later. The onset of desiccation tolerance in developing seeds may or may not coincide with their, or their embryos', ability to germinate, depending on the species. In barley, for example, isolated embryos can germinate at about 10 DAP, but they cannot withstand desiccation until at least 5 days later (Fig. 3.7).

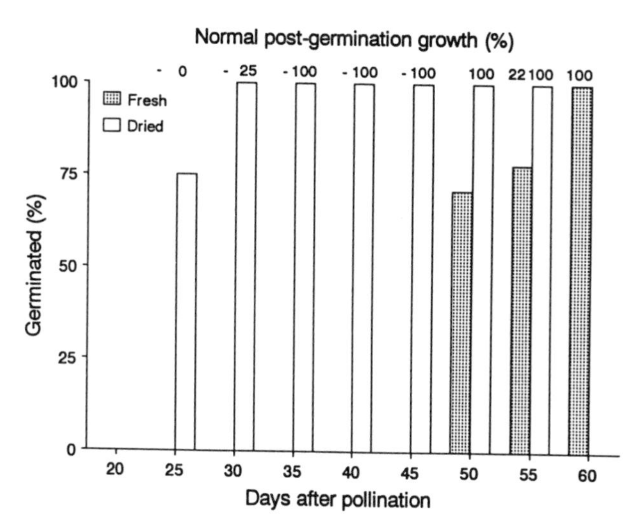

Figure 3.6. Germination of dried or fresh intact seeds of developing castor bean harvested at various days after pollination (DAP) and placed on water. Note that fresh seeds not subjected to premature drying do not germinate until 50 DAP; even these seeds do not achieve normal postgerminative growth (i.e., do not attain the normal radicle lengths of mature germinated seeds). Those seeds dried prematurely from 35 DAP onward exhibit normal postgerminative growth. Drawn from data in Kermode and Bewley (1985).

As we observed in a previous section (3.1), the developing seed is usually incapable of germinating unless first dried, although developing embryos germinate on a liquid medium without prior drying (Fig. 3.1). It would appear, therefore, that drying, like isolation, overcomes restraints imposed on the embryo by the surrounding structures (storage tissue, coat, and parental environment), or by internal factors. That ABA or osmoticum, or both, are important in maintaining seeds in a developmental mode was discussed in previous sections, and it is sufficient to surmise here that desiccation causes their decline and/or negates their effects, perhaps by decreasing the sensitivity of the seeds to these or other agents, thus permitting germination. It is commonly observed, for example, that to inhibit germination of embryos following desiccation a much higher concentration of ABA (even 10^2–10^4 times more) is required than during development, and the germination of mature seeds of some species is virtually unpreventable by ABA.

There may be a role for ABA in the acquisition of desiccation tolerance itself during development. Double mutants of *Arabidopsis thaliana*, which are ABA-deficient and ABA-insensitive, germinate precociously during develop-

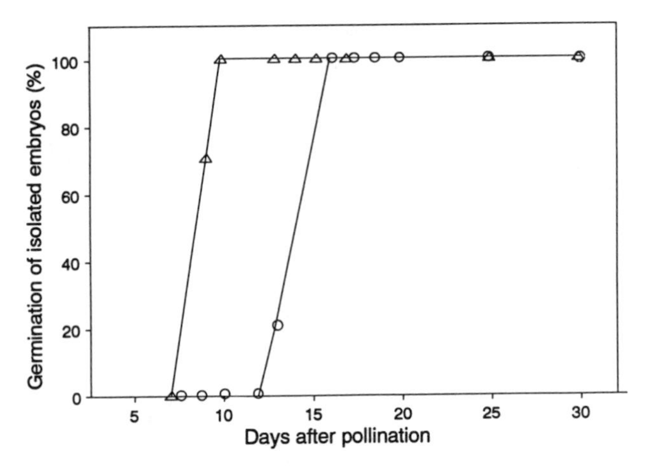

Figure 3.7. Germination of isolated embryos of *Hordeum vulgare* (barley, cv. Aura) after excision and incubation on water (△), and following air-drying before incubation on water (○). After Bartels *et al.* (1988).

ment, but at no time can they withstand being dried. Application of ABA to the developing seed renders them desiccation-tolerant.

3.3.2. Changes in Metabolism Associated with Drying

Several metabolic changes occur in seeds either just prior to or during drying. These changes involve the appearance of two types of products that are thought to have functional significance with respect to the protection of seed tissues against the rigors of desiccation, viz., (1) oligosaccharides and sugars and (2) specific types of proteins.

Sugars. In maturing seeds of several species [e.g., white mustard (*Sinapis alba*), soybean, *Brassica campestris*, maize], the concentrations of certain sugars and oligosaccharides increase in association with the onset of desiccation tolerance and in the early phase of water loss. The disaccharide sucrose and the oligosaccharides raffinose and stachyose, in some cases, are relatively abundant. These compounds generally occur at much lower concentrations in the desiccation-intolerant stage when the monosaccharides glucose, mannose, fructose and galactose predominate. In developing soybean embryos, for example, there is an increase in some sugars and oligosaccharides within embryos induced to become desiccation-tolerant by slow drying, but not in those maintained in the intolerant state (Fig. 3.8).

Strong evidence for the involvement of sucrose and oligosaccharides in desiccation tolerance comes from ABA-insensitive mutants of *Arabidopsis thaliana* (e.g., *abi3-4*, *abi3-5*) and the recombinant ABA-insensitive and deficient ones (*abi3-1*, *aba1*) which, respectively, have much lower or no desiccation tolerance as compared with the wild type. Seeds of these mutants and the recombinants, while having relatively high amounts of total sugars (monosaccharides and disaccharides), have much lower quantities of the oligosaccharides raffinose and stachyose than the wild type. Indeed, the oligosaccharide/monosaccharide ratio in the wild type is about 400 times higher than in the desiccation-intolerant recombinant mutant. It must be noted, however, that the relationship between oligosaccharide (e.g., raffinose) content and the onset of tolerance in the wild type is not firmly established because the temporal correlation is not rigorous.

It seems clear, therefore, that changes in carbohydrate metabolism occur in association with the inception of desiccation tolerance, but relatively little is known about what these changes are. One possibility is that there is a conversion of previously existing monosaccharides to sucrose and/or oligosaccharides. It seems likely that in some cases a substantial part of the carbohydrate for sucrose and oligosaccharide synthesis comes from the breakdown of starch. In maturing seeds of *Brassica campestris*, soybean, and white mustard, there is a steep drop in the amount of starch, coincident with the acquisition of desiccation resistance, just before and during seed drying. Whatever the source of carbohydrate, enzymes for the synthesis of sucrose and oligosaccharides must also be present,

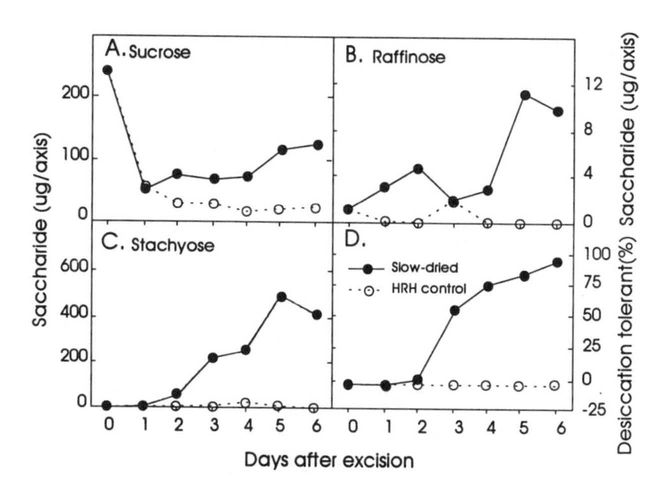

Figure 3.8 Changes in (A) sucrose, (B) raffinose, and (C) stachyose in axes of excised developing soybean seeds during slow drying (●) or maintenance in an atmosphere of high relative humidity (O). Desiccation tolerance is induced by slow drying (D) but not in axes maintained in a hydrated state. After Blackman *et al.* (1992).

and presumably these appear only when maturation drying is imminent. Questions arise as to how this happens. Is it part of an inherent developmental program or does it occur in response to the initial change in water content? Answers cannot yet be given because experimental investigation is still needed. Studies on the ABA-insensitive mutants of *Arabidopsis* strongly imply that the changes in carbohydrate metabolism are ultimately driven by ABA. There is good evidence that at least one carbohydrase, aldose reductase, is an ABA-dependent enzyme produced only in desiccation-tolerant *Arabidopsis* seeds, and it would not be surprising, therefore, to find others having the same features.

The importance of sucrose and/or oligosaccharides partly lies in the role that such substances might play in the promotion of vitrification of water and the consequent protection of cytosolic structures (Section 3.4.1). These compounds might also facilitate the stabilization of lipid and protein in cell membranes. There is good evidence from *in vitro* studies that various disaccharides and oligosaccharides bind to phospholipid by hydrogen bonding between their OH groups and the phosphate of the polar head of phospholipid. Such carbohydrates also prove to be very effective stabilizing agents when tested *in vitro* against several different proteins exposed to dehydration.

Proteins. It has been known for some time that different groups of transcripts and their protein translation products arise at discrete times during seed development and maturation. One group, first identified in cotton embryos, appears late in embryogenesis and accumulates to a relatively high concentration. When discovered, this set was therefore named the late embryogenesis abundant (LEA) proteins, encoded by the *lea* genes. Their amino acid sequence has been deduced from the base composition of their respective cDNAs and other studies have revealed information about their physiological properties. It soon became clear that seeds of many species possess similar proteins, which were later also found in other plant parts. All of these are now referred to as LEAs and are considered to be a highly homologous group. LEAs are strongly hydrophilic proteins, a property that arises from their amino acid composition in which glycine is strongly represented. They are highly stable proteins, not denatured by boiling. Their ability to attract water molecules maintains a water-enriched local environment or in some ways even substitutes for water, and, therefore, at the subcellular level they are thought to play an important role in protection against desiccation.

Two sets of LEA proteins occur in cotton embryos. The first, consisting of 6 members, appears relatively early in development soon after the completion of histodifferentiation; it exhibits two transient peaks and reaches a maximum content approximately 3 days before the seed begins to dry (Fig. 3.9). The second set is larger, with 12 members; transcripts and the proteins are produced in late maturation reaching a peak concentration just before or during seed dehydration. Concentrations are relatively high, as much as 30% of the nonstorage reserve protein, and are substantially higher still at certain cytosolic sites. The mRNA transcripts persist throughout desiccation and in the mature seed but are rapidly degraded on imbibition.

First discovered in developing cotton embryos, LEAs or *lea* transcripts are now known in embryos of many species, such as pea, soybean, rape, carrot, castor bean, *Arabidopsis* and several cereals; the Em proteins of wheat and maize are LEAs, sharing sequence homology with many LEAs from the above-named species. These proteins are in fact not confined to seeds but under certain circumstances also occur in seedlings and adult plants. In these cases transcription of the genes is induced by water stress. Examples are the dehydrins of barley, pea, and maize, and the RAB proteins of rice seedlings. The resurrection plant, *Craterostigma plantagineum*, which can withstand very severe desiccation, also produces LEAs in the early stages of drying. Transcription of some *lea* genes is induced by premature desiccation of cotton embryos in their early maturation phase, which strongly suggests that the normal drying of the seed is a factor which stimulates LEA formation. Hence, in seeds, seedlings, and adult plants, changes in gene expression brought about by drying lead to the production of a set of proteins which are thought to participate in desiccation tolerance by virtue of their special properties of hydrophilicity and stability. The mechanism by which LEAs function is not clear but several possibilities are being investigated, including membrane protein solvation, ion trapping (e.g., phosphate), and stabilization of membranes by acting as a surrogate water film.

There is good evidence that ABA is involved in effects of water stress on vegetative tissues; including the expression of *lea* genes. In barley and maize seedlings, changes in ABA content elicited by water stress correlate well with the appearance of *lea* transcripts and proteins. Maize and tomato mutants which are unable to make ABA (*vp2, vp5*, and *flacca*) do not produce *lea* transcripts upon the imposition of water stress but do so after the application of ABA. On the other hand, seedlings of some ABA-insensitive mutants of *Arabidopsis thaliana* (*abi1, abi2*) succeed in expressing a *lea* gene in response to mild

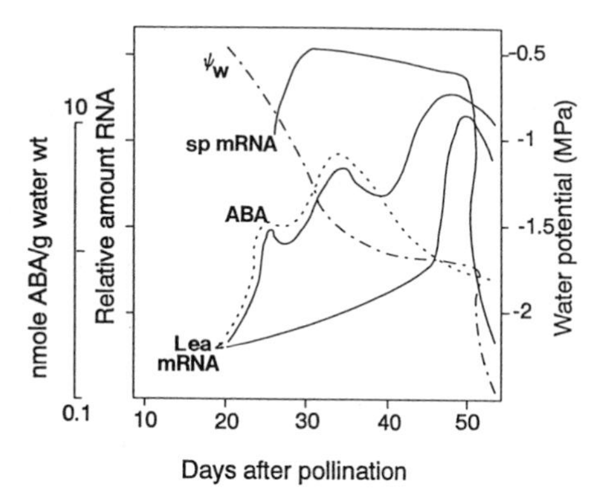

Figure 3.9. Changes in the relative abundances of the two classes of *lea* mRNAs during cotton seed embryogenesis in relation to changes in seed water potential (ψ_w), storage protein (sp) mRNA, and abscisic acid (ABA). Two distinct classes of *lea* mRNAs increase during development. One class increases at about the midpoint of development (25–35 DAP) coincidental with an increase in the growth regulator ABA, and the other class increases at the time of maturation drying (45–50 DAP). After Galau *et al.* (1987) and Hughes and Galau (1989).

dehydration, a finding which is still in need of explanation. But, on the whole, the sum total of evidence does suggest that ABA participates in LEA production in vegetative tissues experiencing water stress. The obvious question now arises as to whether a similar situation exists in seeds.

Certain LEAs can be induced by applied ABA in isolated, immature embryos (e.g., barley), a treatment that also confers upon them resistance to dehydration. During normal seed development, *lea* transcripts and proteins first appear at about the time when the ABA content is at or close to its highest value. But high expression of the second set of *lea* genes occurs at the start of and during maturation drying itself, generally when the amount of ABA has precipitously declined (Fig. 3.9). Arguments can be constructed to explain this lack of correlation, e.g., there is an early regulated, ABA-controlled "switch" that begins to operate only when drying commences; but on the other hand it may really be the case that ABA is not involved in the expression of this second group of *lea* genes. Studies of mutants, however, implicate ABA in some cases, but here again there are discrepancies which require explanations. Seeds of the double mutants of *Arabidopsis* (deficient and insensitive, *aba/abi3*) lack several LEAs and are also desiccation-intolerant. Tolerance to drying is conferred on the developing seeds by treating the parent plants with an analogue of ABA or by exposing excised, developing seeds directly to ABA and sucrose. One particular group 1 LEA, which has homology with Em protein, is much reduced in concentration in seeds of *abi3* and slightly so in *aba*. Seeds of another ABA-insensitive mutant, *abi3-3*, also cannot tolerate desiccation but it is not known if it lacks LEAs, though certainly its production of storage protein is much diminished. In mutants of maize the many *lea* genes appear to be differentially affected. Embryos of the ABA-deficient *vp2* and *vp5* produce the RAB 17 and RAB 28 LEAs, but the latter is absent from those of the ABA-insensitive *vp1*. The maize Em LEA, in contrast, is not found in either *vp5* or *vp1*, but, surprisingly, some LEAs are induced in *vp1* by applied ABA. Some doubt as to the universal involvement of ABA in LEA production is also cast by the fact that in wheat embryos osmotic stress induces Em formation even when ABA production is prevented by an inhibitor; also, some barley *lea* genes respond differently to ABA and water stress. It is possible, therefore, that LEA production is regulated by different mechanisms, involving ABA in some but not in all cases.

Several of the *lea* genes which are responsive to ABA have been investigated at the molecular level to study their mechanism of regulation. Their 5′ upstream regions contain ABA-regulated elements (ABREs) whose sequences have been analyzed; this is discussed in detail in Section 2.3.5.5. The *lea* gene *rab29* in *Arabidopsis* has an additional *cis*-acting element, sensitive to osmotic regulation.

Information is accumulating to suggest that the many different LEAs do not have equal functional significance. In the present context, their possible role as desiccation protectants has been emphasized in which they interact with and stabilize membranes and the proteins therein. Certain LEAs, however, are now suspected to have enzymatic activity as protein kinases. For example, an ABA-sensitive and dehydration-sensitive LEA in wheat seedlings, on the basis of the cDNA structure, has amino acid homology with a serine/threonine type of kinase. At present, the role of a protein kinase in desiccation tolerance is a matter for speculation. Possibly, phosphorylation of proteins is an important occurrence in maturation drying; it is plausible, also, that phosphorylation of *trans*-acting factors (i.e., DNA-binding proteins) takes place in the spectrum of events associated with activation of the *lea* ABREs.

Before closing this section we must raise one important question which is implicit in the above discussion. The acquisition of desiccation tolerance by seeds may involve the production and action of certain sugars, oligosaccharides, and/or LEAs. What, then, is the situation regarding these components in seeds which do not acquire resistance to desiccation, i.e., the so-called recalcitrant seeds? We will turn our attention to this topic in Section 3.5.

3.3.3. Metabolic Changes on Rehydration following Desiccation

Desiccation of developing seeds, whether prematurely or during the final stages of maturation, not only promotes germination on subsequent imbibition, but also results in the cessation of developmentally related synthetic events (e.g., reserve synthesis) and the onset of synthesis associated with germination and postgerminative growth. This has been well illustrated in the endosperm of castor bean, where premature drying leads to a change in the pattern of polypeptide synthesis from one which is distinctly developmental to one which is identical to that found in germinated seeds following normal maturation drying (Fig. 3.10). Premature drying of developing seeds also elicits the production of enzymes required for the mobilization of stored reserves. In soybean and castor bean seeds, for example, active enzymes for the mobilization and conversion of the stored lipid reserves are almost absent during seed development, but increase greatly after premature drying, and to a similar extent as in the germinated mature seed (Fig. 3.11A). Mobilization of the starch reserves in germinated cereal grains requires the production and secretion of α-amylase by the aleurone layer, which occurs in response to the hormone gibberellin (GA) (Chapter 8). The ability of the aleurone layer to produce α-amylase is not normally acquired until the onset of maturation drying, but premature desiccation of wheat, triticale, and maize all induce the ability subsequently to synthesize this enzyme. In developing wheat the effect of drying is to increase the sensitivity of the aleurone layer to GA, which then provokes the synthesis of α-amylase (Fig. 3.11B). This may occur as a result of desiccation-induced membrane changes and alterations to integral

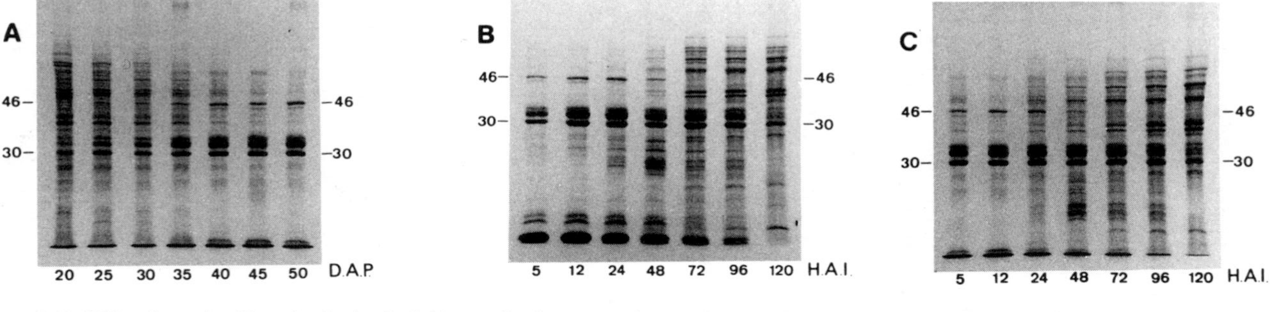

Figure 3.10. SDS–polyacrylamide stained gels of soluble proteins from castor bean endosperms isolated during (A) development, (B) germination and growth, and (C) following premature desiccation and rehydration at 40 days of development. Note that the pattern of proteins produced on rehydration following premature drying at 40 DAP is identical to that produced by mature dry seeds during germination and growth. Germination of the embryos is completed at 24 h after imbibition (HAI). This demonstrates that the "switch" from a developmental to a germinative mode of metabolism is elicited by desiccation. After Kermode and Bewley (1985). Reproduced by permission of Oxford University Press.

hormone receptors. It has been suggested, also, that drying-induced depletion of ABA in the aleurone layer increases the response of this tissue to GA. In maize, however, inhibition of ABA synthesis in the aleurone layer does not permit it to respond to GA without the imposition of a drying treatment.

The changes in protein synthesis from a developmental to a germinative/growth mode are indicative of a switch in genome activity. This results in the permanent suppression of developmental protein synthesis and an induction of germination/growth-related proteins. Drying ultimately affects gene transcription, with the production of mRNAs for developmental proteins being permanently suppressed by premature and maturation drying, and those for germination and growth being switched on following subsequent imbibition. Developmentally related messages frequently decline during drying itself, and any residual messages present in the dry state are presumably degraded on rehydration by the normal turnover processes, and are not replaced because their genes have been off-regulated. Sometimes these developmental genes are expressed transiently during or following germination; either the residual messages that are present following drying are translated, or some developmental genes temporarily retain transcriptional competence.

Clear examples of changes in gene expression resulting from drying are shown in Fig. 3.12A and B. In the castor bean endosperm, storage protein

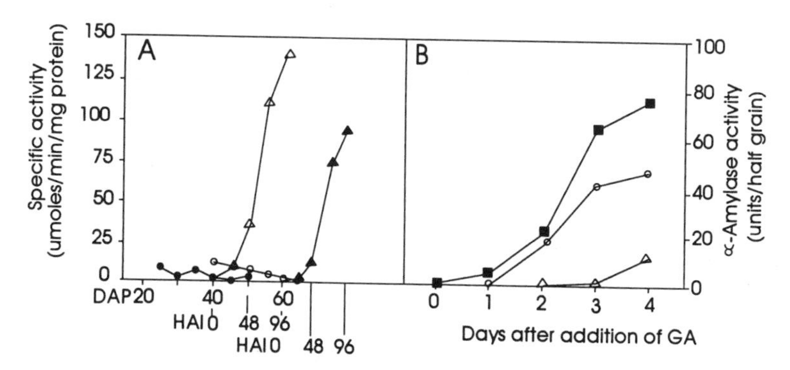

Figure 3.11. (A) Increases in isocitrate lyase activity (an enzyme involved in the conversion of fatty acids to sugars, Chapter 7) in castor bean endosperms elicited by maturation drying (▲) and premature drying at 40 DAP (△). Developing seeds (●) do not produce the enzyme, nor do seeds removed from the capsule at 40 DAP and placed on water (○) for 96 h, i.e., not desiccated. DAP, days after pollination. HAI, hours after imbibition. (B) Time course of α-amylase production by half-grains of mature (■) and immature wheat grains dried prematurely at 25 DAP (○) or maintained in a moist condition in an atmosphere of 100% RH (Δ). The half-grains were incubated in GA; only those subjected to drying produced substantial amounts of the enzyme. From data in (A) Kermode and Bewley (1985), (B) Cornford *et al.* (1986). A Reproduced by permission of Oxford University Press.

messages present during development decline during maturation drying, from 50 DAP to maturity. They are not resynthesized on rehydration. Seeds that are prematurely dried at 40 DAP exhibit a similar decline in storage protein message within the endosperm, so that in the dry state (0 HAI, prem dried/rehy, Fig. 3.12A) little is detectable by Northern blot analysis. There is no increase in this mRNA on rehydration of the seeds. In contrast, the message for an unidentified growth-related protein (D38) is expressed in low amounts during development, but is abundantly present in the endosperms at 48 HAI. Likewise, following premature drying at 40 DAP, transcription of this mRNA is strongly evident by 48 h after rehydration (Fig. 3.12A), showing that it is induced following desiccation.

Expression of α-amylase genes (*Amy 1* and *2*) in the barley aleurone layer in response to GA is upregulated by drying (Fig. 3.12B; see Fig. 3.11B for corresponding changes in enzyme activity). Transcription to produce mRNAs for this enzyme is evident in the isolated aleurone layers from mature, dry grains, incubated for 48 h in the presence of GA; no amylase messages occur in the absence of this growth regulator. Aleurone layers removed from immature, developing grains that are not dried do not produce transcripts for α-amylase, even in response to GA, whereas those that are first dried show a strong positive response (Fig. 3.12B).

How desiccation is able to up- and downregulate genes associated with development and germination is still unknown. In some instances, e.g., α-amylase synthesis in response to GA, it is possible that the effect of drying is indirect and that the induction step is the result of increased sensitivity of a receptor to the growth regulator. This may be the case with other drying-related synthetic events, where increased or decreased sensitivity to ABA caused by desiccation, for example, could result in changes in the expression of specific genes. In other cases, the promoter region of certain genes could be sensitive to the drying event, although it would appear to be easier to invoke the synthesis, release, or activation of desiccation-related repressors and inducers as a primary response, which then interact with appropriate regions of the genome to modulate transcription.

3.3.4. Seed Germination without Prior Desiccation

The development of some seeds occurs within a fleshy fruit, e.g., squashes and tomato, which frequently do not dry out prior to dispersal. When mature, the seeds within a ripe tomato fruit have a water content of approximately 40%. Mature dry seeds of other species at the time of shedding, in contrast, have a 10–15% moisture content. Tomato seeds taken from the fruit during development and placed on water, or a nutrient medium, germinate without any require-

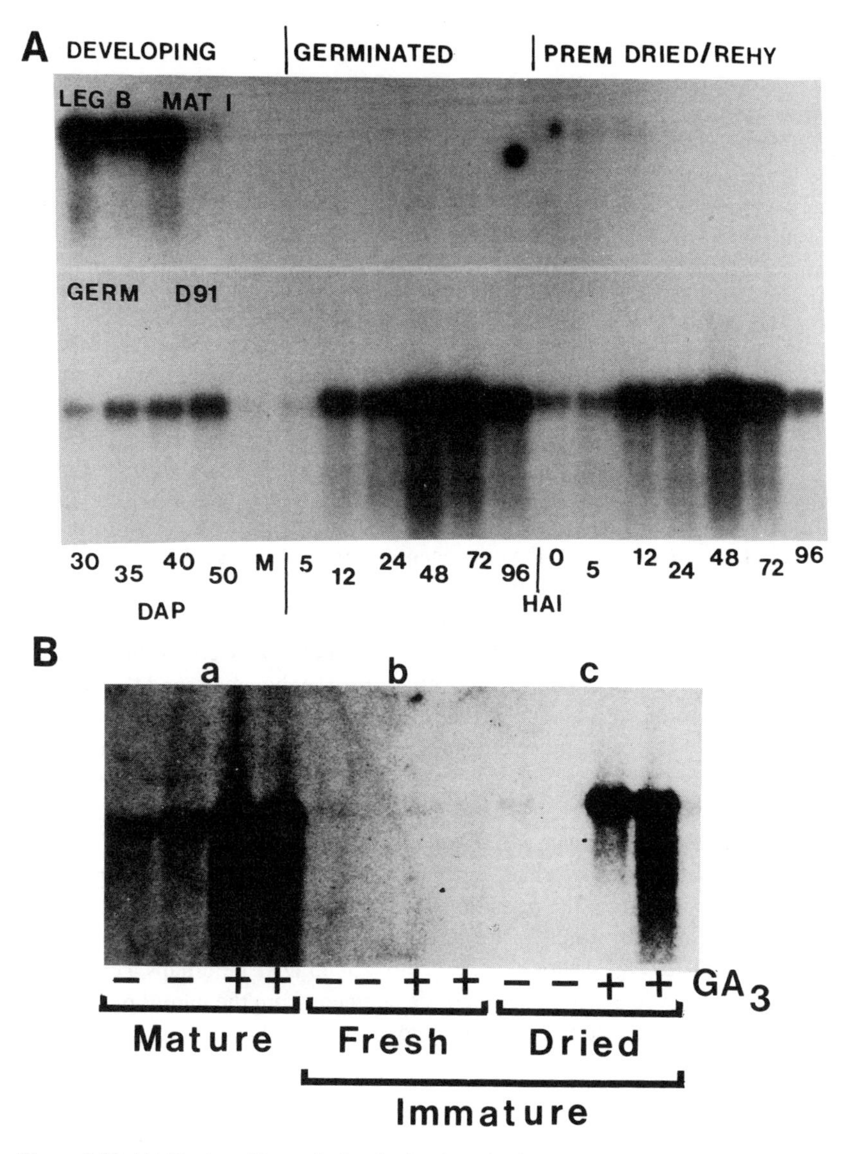

Figure 3.12. (A) Northern blot analysis of a developmental messenger RNA (*Leg B Mat 1)* and a postgermination message (*D91*) in the endosperm during development and germination/growth of castor bean seeds. Mature (M) seeds were germinated for up to 96 h after imbibition (HAI), or at 40 days after pollination (DAP) seeds were prematurely dried, and mRNAs extracted from the dry seed (0 HAI), and from those incubated for up to 96 HAI (prem dried/rehy). (B) Northern hybridization of α-amylase (high pI) cDNA probes with mRNAs extracted from (a) aleurone layers of mature wheat grains, (b) immature grains at 25 days postanthesis (fresh), or (c) immature grains of the same age subjected to drying. The aleurone layers were incubated in the presence (+) or absence (−) of 2×10^{-6} M GA_3, before mRNA was extracted. A, data of D. W. Hughes, G. A. Galau, A. R. Kermode, and J. D. Bewley, published in Oliver and Bewley (1992);. B, from Cornford *et al.* (1986).

ment for drying, so that removal of the seeds from the constraints of the fruit is sufficient (Fig. 3.1B and Section 3.1).

Even seeds which normally dry out at maturity, in which the transition from development to germination is effected by desiccation, will respond similarly if only partially dried. Castor bean again provides an example, and here, when developing seeds at 25–40 DAP are detached and maintained in an atmosphere of high relative humidity, they lose less than 15% of their water content (compared with up to 65% when desiccated), yet they germinate on subsequent return to full hydration. This is accompanied by identical changes in the patterns of protein and mRNA synthesis, and the induction of postgerminative enzymes associated with catabolism, as in seeds undergoing premature or maturation drying. Partial drying of desiccation-sensitive species, *Acer pseudoplatanus* and the aquatic grass *Zizania palustris*, also promotes their germination. This is particularly the case for dormant seeds of *Zizania*, which afterwards require a 100-fold lower concentration of GA to break dormancy and stimulate maximum germination.

In wheat grains, GA-induced synthesis of α-amylase mRNA and protein can occur in detached developing grains maintained under conditions of high humidity. This treatment is not consistently effective, however, and the amount of α-amylase produced is always much lower than that elicited by premature, or natural drying. Moreover, other enzymes which normally increase in response to GA following drying, e.g., phosphatase, proteinase, and ribonuclease, fail to do so after partial drying.

Why partial drying can wholly or incompletely reproduce the effects of desiccation in those seeds which normally undergo maturation drying is not known. Detachment from the parent plant alone is not sufficient, for if developing seeds are placed in water following isolation from the capsule or pod, they will not germinate (e.g., castor bean, Fig. 3.6, 25–45 DAP). Perhaps a combination of some water loss with a period of detachment from the parent plant leads to changes within the seed that are similar to those caused by drying; but by what means is unknown.

3.4. THE DRY SEED

3.4.1. Bound Water and Vitrification

We have considered the metabolic changes associated with, and resulting from maturation drying, but now let us consider the state of the mature dry seed itself. The water content of the seed declines steadily during phase II of development, as the insoluble reserves are laid down, with the final loss of water occurring during phase III, maturation drying (Fig. 2.10). This latter loss probably commences when the funicular region joining the seed to the parent plant starts to shrivel, thus cutting off the supply of water. How water is then lost from the seed is largely unknown, although evaporation through the desiccating outer structures, e.g., pod, capsule, glumes, etc., is a likely route.

With the progressive removal of water from seeds, the forces resisting water loss from the cells increase. At the moisture content which usually exists in the mature dry seed, water exists as "bound" water (a poor term which is losing popularity). This refers to water associated with a macromolecular surface; it has virtually no mobility and is sufficiently structured so that its thermodynamic properties differ from those of free or bulk water. Most significantly, it is not readily freezable, and hence many dry seeds can withstand subzero temperatures, even being plunged into liquid nitrogen. The relationship between water content and survival of freezing is shown in Fig. 3.13; it is evident that when frozen at below approximately 25% water content (percent water on a fresh mass basis) soybean seeds can subsequently germinate, and membrane damage, measured by leakage following thawing and imbibition, is minimal. At higher water contents, freezing is fatal and considerable cellular damage is inflicted.

Essentially, there are three types of bound water. Type 1 is absorbed very tightly onto macromolecules through ionic bonding and the water behaves as a liquid rather than as a solvent. Type 2 has glassy characteristics and is a thin film of water and solutes that coats the surface of macromolecules. Type 3 is less well defined, but it may form bridges over hydrophobic sites and its presence results in changes in the phase behavior of membrane lipids. The energy with which the different types of water are bound varies, with Type 1 being bound most tightly, Type 2 weakly, and Type 3 with negligible energy. This affinity for water of the seed components can be measured. Thus, water in dry seeds is both structured and nonfreezable. In fact it is now recognized that in the dry state the cytoplasm of desiccation-tolerant organisms, such as seeds, exists in a glassy (vitrified) state, even at physiological temperatures. A glass is a liquid of high viscosity, which stops or slows down all chemical reactions requiring molecular diffusion. As such, it traps residual water molecules, and by assuring quiescence and stability over time, it prevents damaging interactions between cell components. Glass formation (vitrification), for example, has been shown in model (*in vitro*) systems to retard or prevent denaturation of proteins, including enzymes, by holding them in their stable, folded state. Another advantage of vitrification is that it suppresses or prevents crystallization of solutes in the cytoplasm. Some solutes aid in the formation of glasses at low moisture contents. For example, the survival of desiccation by numerous animals, and some lower plants and fungi has been attributed to the presence of trehalose. This sugar is absent from seeds but it has been suggested that other sugars or oligosaccharides, e.g., sucrose and raffinose, which can be present in substantial quantities in dry seeds, contribute to glass formation.

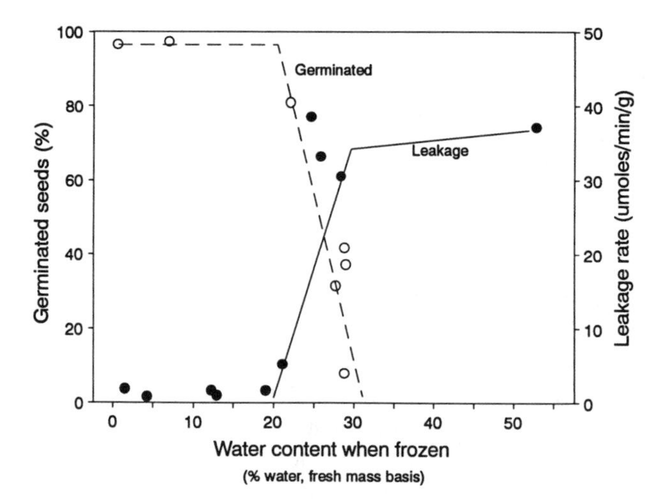

Figure 3.13. The effect of freezing soybean seeds with different water contents at –65°C on seed germination (O) and electrolyte leakage (●) upon subsequent return to 25°C and imbibition on water. After Leopold and Vertucci (1989).

The glassy state is a metastable state, and the rate at which glasses decompose is variable, and depends on the nature of the cellular components (i.e., it will vary between seeds and their tissues), water content, and temperature. The water content in dry, stored corn kernels is usually less than 12% and at these values vitrification will occur between 0 and 25°C At higher water contents, the temperature at which glasses form is much lower, and the presence of free water will result in the highly viscous phase being melted, with resultant freezing damage. Since the glassy state in a seed may gradually decompose, and will be affected by temperature extremes, loss of viability during long-term storage could result.

3.4.2. Metabolism in the Dry State

The term "dry" is a relative one and does not necessarily mean that water is absent. Some water is present in the "dry" seed and the amount and extent of metabolic reactions proceeding in the mature seed will depend on what its water content actually is. As stated in the previous section, those seeds with only bound water, in the glassy state (Type 2), are unlikely to support metabolic reactions. If enzymes are sufficiently hydrated, and in contact with their appropriate substrate (such as in the presence of Type 3 water, or greater hydration), then reactions can proceed, albeit at a low rate. Very little oxygen consumption occurs at below 20% water content in maize, soybean or pea (Fig. 3.14); at 25% water content some contribution to this process might be made by the mitochondria, structurally incomplete though they are (see later). Some oxygen consumption at low water contents might be made by lipoxidases, enzymes which modify or cleave fatty acids oxidatively. Experiments with $^{14}CO_2$ suggest that it is incorporated into organic compounds in dry seeds of charlock (*Sinapis arvensis*), and radioactive ethanol, supplied to dry, dormant wild oat grains in quantities which did not change their water status was observed to be incorporated into sugars, amino acids, and proteins. Probably all seeds can metabolize slowly when maintained under conditions where glass formation is not totally suppressive. Nonenzymatic reactions may also proceed in dry seeds, and free-radical formation increases with decline in moisture content. The deteriorative processes resulting from free-radical activities are particularly important in seeds stored for extended periods of time, and thus will be discussed in Chapter 9.

Several ultrastructural studies have been conducted on dry seeds, but interpretation of the results of some must be treated with caution. In these, aqueous fixatives were used prior to embedding and sectioning: cells imbibe water rapidly from such fixatives, resulting in partial to complete hydration before adequate fixation occurs. This can lead to the formation of artifacts owing to distortion of cellular contents, and the resultant picture obviously is not that of a dry cell. Far better fixation, giving a truer picture of the inclusions within dry cells, can be obtained by using nonaqueous fixatives or osmium vapor, or by freeze-etching. Cells of the coleoptile of dry rice grains (fixed in osmium vapor

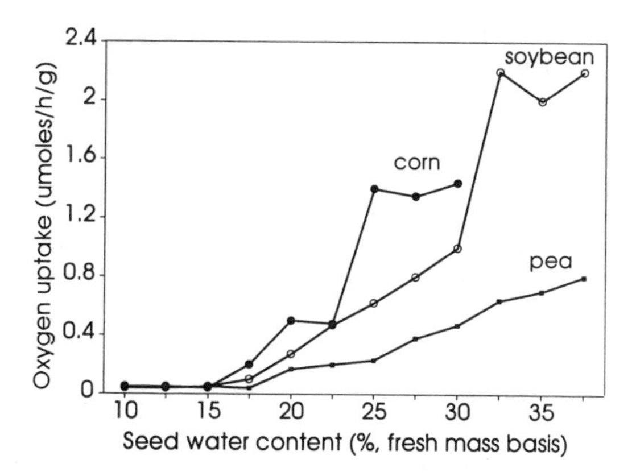

Figure 3.14. Oxygen consumption as a function of corn (●), soybean (O), and pea (■) seed water content. After Leopold and Vertucci (1989).

for 11 months at room temperature) are shrunken and have highly folded walls. Nuclei, plastids, mitochondria, and lipid bodies have irregular outlines because of shrinkage and compression.

All cellular membranes, including the bounding plasmalemma, appear to be intact, however. Early during imbibition the mitochondria have a poorly defined inner structure (Chapter 4). But it is not possible to tell if this is also the case in the dry seed, or if the perturbation of this organelle is caused by the rapid uptake of water (and, presumably, concomitant osmotic changes) immediately on imbibition. From the little work that has been done on the structure of other dry seeds, it appears that they also contain intact organelles with no discernible damage to membranous structures, or their configuration, within the cells.

3.5. RECALCITRANT SEEDS

There is a class of seeds which, at maturity, are not subjected to, nor are capable of withstanding, water losses of the magnitude we have discussed so far. In fact, these so-called recalcitrant or "unorthodox" seeds must maintain a relatively high moisture content in order to remain viable. But even when these seeds are stored under moist conditions, their life span is frequently brief and only occasionally exceeds a few months. Some species that produce recalcitrant seeds are listed in Table 3.2. Included are several large-seeded hardwoods (e.g., species of *Corylus, Castanea, Quercus, Aesculus*; in addition, *Salix, Araucaria* spp., and *Juglans* are recalcitrant) and important plantation crops like coffee, kola nut, cacao, tea, and rubber. Seeds of most aquatic species (e.g., wild rice and some mangroves) also rapidly lose viability in dry conditions (but see below).

The inability to store seeds of recalcitrant species is a serious problem, for while vegetative propagation is possible for some species and is the usual practice, the retention of a viable seed stock is desirable to preserve maximum genetic diversity. Unfortunately, methods of storage other than drying may also be detrimental to recalcitrant seeds; e.g., low-temperature storage is inappropriate for some species, particularly for seeds of tropical plants. Grains of wild rice present a different problem, for although they may be successfully stored under moist conditions at low temperatures, they eventually lose dormancy and sprout.

At present, the optimal storage conditions for recalcitrant seeds can be determined only by trial and error, which is a tedious, time-consuming, and expensive approach. As our knowledge advances, it is apparent that some seeds (e.g., those of *Citrus limon*) that were once thought to be recalcitrant can now be classified as orthodox—either the original method of seed drying or germination testing was at fault. Another example, wild rice (*Zizania palustris*), is recalcitrant only if dried at temperatures below 25°C followed by quick rehydra-

Table 3.2. Some Species That Produce Recalcitrant Seeds and Examples of Appropriate Storage Conditions[a]

Species	Longevity and (% germinated)	Storage conditions	Damaging conditions
Avicennia marina	Few days	Moist	Drying
Corylus avellana (hazel)	6 mo +	1°C in polyethylene bag	Drying
Castanea crenata (Japanese chesnut)	6 mo	0–3°C in ventilated can or polyethylene bag	< 0°C; excessive water or drying
Quercus borealis (red oak)	20 mo + (50%)	5°C in sealed tin	< 20–40 moisture content
Juglans nigra (black walnut)	4 yr	3°C in outdoor pit	Drying
Hevea brasiliensis (rubber)	4 mo (3%)	7–10°C in damp sawdust in perforated polyethylene bag	< 20% moisture content
Cocos nucifera (coconut)	16 mo	Ambient temperatures and high RH%	Drying
Coffea arabica (coffee)	10 mo (59%)	25°C in moist charcoal at 92–98% RH	< 8–35% moisture content and < 10°C
Cola nitida (kola nut)	5 mo (80%)	Ambient, heaped in open and kept moist	Drying
Theobroma cacao (cacao)	8–10 wk	21–27°C in pod, coated with fungicide	< 13°C, drying
Zizania aquatica (wild rice)	14 mo (86%)	1°C in water	Drying at < 25°C

[a]Taken from data in Berjak *et al.* (1989), King and Roberts (1979), and Kovach and Bradford (1992).

tion, but seeds survive and can be stored when first dried at temperatures above 25°C and are subsequently rehydrated slowly. A category intermediate between orthodox and recalcitrant is now recognized (e.g., coffee) in which seeds survive desiccation but become damaged during dry storage at low temperatures. However, much work still lies ahead to establish clearly which species are truly recalcitrant and then to define quantitative relationships between storage conditions and viability. We still do not understand the physiological and biochemical basis for desiccation intolerance in recalcitrant seeds. But information is rapidly becoming available which we will now consider briefly.

In many cases the sensitivity of recalcitrant seeds to drying is so great that just a relatively small change in water content is enough to cause damage. As an example, we can compare the response to loss of water of the recalcitrant *Acer pseudoplatanus* (sycamore) seed with that of its orthodox, close relative *Acer platanoides* (Norway maple) (Fig. 3.15). In the former, a reduction in water content from 60% to 50% lowers germinability by approximately 10%; this continues to fall as seeds dry, and no seed can survive drying to 10% water content.

It should be noted that the ability of *A. platanoides* seeds to withstand drying (i.e., whether or not they behave as the orthodox type) depends on the stage of maturation at which they are dried, a point which applies to other species too. That is, seeds may appear to be recalcitrant if they are dehydrated before they have fully acquired desiccation tolerance. A true recalcitrant seed, of course, never becomes completely tolerant, at least under natural conditions of drying. We make this latter qualification because under certain conditions some recalcitrant seeds do survive dehydration. For example, isolated axes (e.g., sycamore) show more tolerance than do intact embryos or seeds. The rate of drying is also important and in many cases slow dehydration is more damaging than rapid, "flash" drying. This is illustrated by *Landolphia kirkii* (a vinelike shrub of South East Africa) whose seeds suffer lethal damage when the axis water content is reduced slowly to about 0.8 g/g, though the isolated axes themselves are unharmed by flash drying to 0.3 g/g.

How can we explain recalcitrance? To do so, we must first understand the basis of desiccation tolerance and intolerance. We saw in Section 3.4.1 that water in seeds exists in several different states. Very few studies have been carried out to investigate which of the various water fractions are specifically important in the recalcitrance syndrome, but in one case, *Landolphia kirkii*, biophysical evidence suggests that the seeds can survive the loss of "freezable" water only if it is removed very quickly, by flash drying. The loss of "non-freezable" water cannot be tolerated, however, possibly because of the consequent loss of integrity of cellular components.

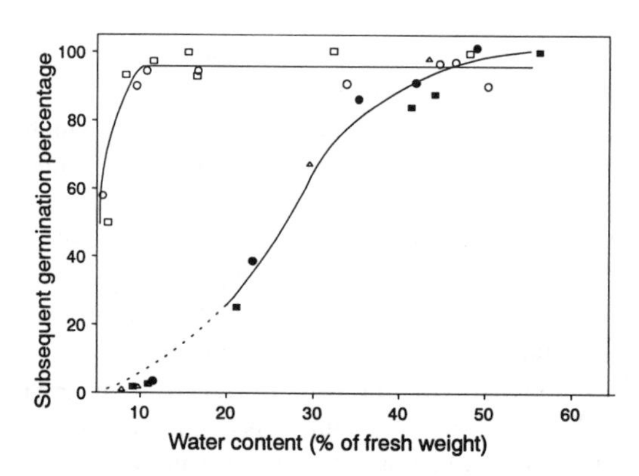

Figure 3.15. Effect of loss of water on subsequent germinability of an orthodox and a recalcitrant *Acer* sp. Mature embryos (△) seeds (□,■) or fruits (○,●) of recalcitrant *Acer pseudoplatanus* (closed symbols) and the orthodox *A. platanoides* (open symbols) were dried to different water contents and their germination was then tested. Embryos of *A. platanoides* were not used because they are difficult to separate from the testa. The fall in germination of *A. platanoides* after drying to below 4% water content is unexpected and unexplained. After Dickie *et al.* (1991).

LEA proteins, sucrose and certain oligosaccharides appear to be associated with the acquisition of desiccation tolerance (Section 3.3.2). One obvious possibility regarding recalcitrance is that the production or effectiveness of these substances is faulty. Information about this is just beginning to emerge but already there are discrepancies which do not allow ready interpretations. Considering the LEAs first, it has been claimed that these are absent from seeds of the recalcitrant mangrove *Avicennia marina* (although the method used for their identification was not stringent), and the ABA content of the developing embryos is extremely low. On the other hand, dehydrins (e.g., RAB 16 mRNA and protein) are formed in wild rice embryos under different conditions which provoke either the orthodox or the recalcitrant type of behavior, and reports are appearing that LEAs occur in recalcitrant seeds of several other species, e.g., English oak (*Quercus robur*) and chestnut (*Aesculus hippocastanum*). Further, during their development, recalcitrant seeds of several species (*Acer pseudoplatanus, Theobroma cacao*, and *Zizania palustris*) contain ABA at concentrations comparable with those in orthodox species. As far as carbohydrates are concerned, sugars, including stachyose, are abundant in *Avicennia marina* seed, and sucrose accounts for 11% of the dry weight of the recalcitrant *Camellia sinensis* (tea) embryonic axis! Clearly, there is no easy explanation of recalcitrance in terms of only one factor—LEA, carbohydrate, or ABA—and it seems likely that a complex interaction may be involved. Much more information is required; we need to know, for example, precisely what the different LEAs are, their functions and their locations in the cells of the embryo. Notwithstanding the present uncertainties, the fact that a kind of recalcitrant behavior can be provoked by interfering with the production of ABA, LEAs, and certain carbohydrates in mutants of *Arabidopsis* (Section 3.3.2) is extremely persuasive of the importance and participation of these factors.

Finally the case of *Avicennia marina* is worthy of mention. Here, the embryo passes from development to germination within the maturing seed. The cells are highly vacuolated, and reserve mobilization and utilization commence before seed shedding. These features of the embryo may render it especially susceptible to dehydration damage. But there is insufficient information to indicate how widespread such a phenomenon might be.

USEFUL LITERATURE REFERENCES

SECTION 3.1

Berry T., and Bewley, J. D., 1991, *Planta* **186:**27–34 (regulation of tomato seed development in the fruit).

Karssen, C.M., Groot, S. P. C., and Koorneef, M., 1987, in: *Developmental Mutants in Higher Plants* (H. Thomas and D. Grierson, eds.), Cambridge University Press, Cambridge, pp. 119–133 (hormone mutants in *Arabidopsis* and tomato).

Welbaum, G. E., Tissaoui, T., and Bradford, K. J., 1990, *Plant Physiol.* **92:**1029–1037 (one of series of papers on muskmelon seed development in the fruit).

Xu, N., Coulter, K. M., and Bewley, J. D., 1990, *Planta* **182:**382–390 (prevention of alfalfa seed germination during development).

SECTION 3.2

Fong, F., Smith, J. D., and Koehler, D. E., 1983, *Plant Physiol.* **73:**899–901 (fluridone and precocious germination of maize).

McCarty, D. R., and Carson, C. B., 1991, *Physiol. Plant.* **81:**267–272 (review of maize *vp* mutants and molecular aspects).

McCarty, D. R., Carson, C. B., Stinard, P. S., and Robertson, D. S., 1989, *Plant Cell* **1:**523–532 (molecular effects of *vp1* mutant on maize development).

Neill, S. J., Horgan, R., and Parry, A. D., 1986, *Planta* **169:**87–96 (ABA and carotenoid content of maize mutants).

Parry, A. D., and Horgan, R., 1992, in: *Progress in Plant Growth Regulation* (C. M. Karssen, L. C. van Loon, and D. Vreugdenhil, eds.), Kluwer, Dordrecht, pp. 160–172 (ABA biosynthesis in plants).

SECTION 3.3

Aldridge, C. D., and Probert, R.J., 1992, *Seed Sci. Res.* 2:199–205 (partial drying of desiccation-sensitive seeds).

Bartels, D., Singh, M., and Salamini, F., 1988, *Planta* **175:**485–492 (desiccation tolerance and germinability of barley).

Bartels, D., Schneider, G., Terstappen, D., Piatkowski, D., and Salamini, F., 1990, *Planta* **181:**27–34 (*lea* genes in a resurrection plant, especially ABA-regulated ones).

Blackman, S. A., Obendorf, R. L., and Leopold, A. C., 1992, *Plant Physiol.* **100:**225-230 (sugars and proteins in maturation of soybean seeds).

Bray, E. A., 1991, *Abscisic Acid Physiology and Biochemistry* (W.J. Davies and H. G. Jones, eds.), Bios Scientific Publishers, Oxford (ABA-induced production of LEAs during drought stress).

Chen, Y., and Burris, J. S., 1990, *Crop Sci.* **30:**971–975 (carbohydrates in maturing maize embryos in relation to desiccation tolerance).

Cornford, C. A., Black, M., Chapman, J. M., and Baulcombe, D. C., 1986, *Planta* **169:**420–428 (α-amylase synthesis, GA response and desiccation in wheat).

Dure, L.S., 1993, in: *Control of Plant Gene Expression* (D.P.S. Verma, ed.), CRC Press, Boca Raton,Fla., pp.325–335 (LEA proteins in higher plants).

Dure, L. S., Crouch, M., Harada, J.J., Ho, T.-H. D., Mundy, J., Quatrano, R. S., Thomas, T., and Sung, Z. R., 1989, *Plant Mol. Biol.* **12:**475–486 (amino acid sequence homologies of different LEAs).

Galau, G. A., Bijaisoradat, N., and Hughes, D.W., 1987, *Dev. Biol.* **123:**198–212 and 213–221 (LEA protein accumulation and ABA).

Hughes, D. W., and Galau, D. W., 1989, *Gene. Dev.* **3:**358–369 (LEA protein accumulation in cotton embryos).

Kermode, A. R., 1990, *Crit. Rev. Plant Sci.* **9:**155–195 (extensive review on desiccation of seeds).

Kermode, A. R., and Bewley, J. D., 1985, *J. Exp. Bot.* **36:**1906–1915 and 1916–1927 (desiccation tolerance and the switch from development to germination).

Kermode, A. R., and Bewley, J. D., 1989, *Plant Physiol.* **90:**702–707 (partial drying and the switch from development to germination).

Koster, K. L., and Leopold, A. C., 1988, *Plant Physiol.* **88:**829–832 (sugars and desiccation tolerance).

Leprince, O., Hendry, G. A. F., and McKersie, B. D., 1993, *Seed Sci. Res.* **3:**275–290 (review on desiccation tolerance).

Meurs, C., Basra, A. S., Karssen, C. M., and van Loon, L. C., 1992, *Plant Physiol.* **98:**1484–1493 (ABA mutants and desiccation tolerance).

Miles, D. F., TeKrony, D. M., and Egli, D. B., 1988, *Crop Sci.* **28:**700–704 (soybean germination and seedling establishment without desiccation).

Mundy, J., and Chua, N. H., 1988, *EMBO J.* **7**:2279–2286 (drought and ABA-induced LEAs in rice seedlings).

Oliver, M. J., and Bewley, J. D., 1992, in: *Water and Life* (G. N. Somero, C. B. Osmond, and C. L. Bolis, eds.) Springer-Verlag, Berlin, pp. 141–160 (review on desiccation tolerance).

Ooms, J. J. J., Leon-Kloosterziel, K. M., Bartels, D., Koornneef, M., and Karssen, C.M., 1993, *Plant Physiol.* **102:**1185–1191 (desiccation tolerance, sugars in *Arabidopsis* mutants).

Roberts, J. K., Desimone, N. A., Lingle, W. L., and Dure, L. S., 1993, *Plant Cell* **5:**769–780 (subcellular location of LEAs in cotton embryos).

SECTION 3.4

Edwards, M., 1976, *Plant Physiol.* **58:**237–239 (dry charlock seed metabolism).

Leopold, A. C., and Vertucci, C. W., 1989, in: *Seed Moisture* (P.C. Stanwood and M.B. McDonald, eds.), Crop Sci. Soc. America, Madison, pp. 51–67 (cellular responses to low water content).

Leopold, A. C., Bruni, F., and Williams, R. J., 1992, in: *Water and Plant Life* (G. N. Somero, C. B. Osmond, and C. L. Bolis, eds.), Springer-Verlag, Berlin, pp. 161–170 (water in dry organisms).

Opik, H., 1980, *New Phytol.* **85:**521–529 (dry seed structure).

Williams, R. J., and Leopold, A.C., 1989, *Plant Physiol.* **89:**977–981 (glassy state in corn embryos).

SECTION 3.5

Berjak, P., Farrant, J. M., and Pammenter, N. W., 1989, in:*Recent Advances in the Development and Germination of Seeds* (R.B. Taylorson, ed.), Plenum Press, New York, pp. 89–108 (basis of recalcitrant seed behavior).

Chin, H. F., and Roberts, E. H. (eds.), 1980, *Recalcitrant Crop Seeds,* Tropical Press, Malaysia (several review articles).

Dickie, J. B., May, K., Morris, S. V. A., and Titley, S. E., 1991, *Seed Sci. Res.* **1:**149-162 (recalcitrance and orthodoxy in two *Acer* species).

Ellis, R. H., Hong, T. D., and Roberts, E. H., 1990, *J. Exp. Bot.* **41:**1167–1174 (intermediate category of storage behavior).

Farrant, J. M., Pammenter, N. W., and Berjak, P., 1993, *Seed Sci. Res.* **3:**1–14 (a detailed consideration of recalcitrance in *Avicennia marina*).

King, M. W., and Roberts, E. H., 1979, *Report for the International Board for Genetic Resources Secretariat,* Rome, (recalcitrant seed storage).

Kovach, D. A., and Bradford, K. J., 1992, *J. Exp. Bot.* **43:**747–757 (drying conditions and recalcitrance in wild rice).

Chapter 4

Cellular Events during Germination and Seedling Growth

When dry, viable seeds imbibe water, a chain of events is initiated which ultimately results in the emergence of the radicle, signifying that germination has been successfully completed. On imbibition, metabolism quickly recommences. Respiration, enzyme and organelle activity, and RNA and protein synthesis are fundamental cellular activities intimately involved in germination and the preparation for subsequent growth. It is not surprising, therefore, that most research into the biochemistry of germination and growth has concentrated on these events in an attempt to elucidate the key processes which lead to the successful completion of germination. Even so, the amount of research on germination is very scanty, especially compared with that on seed development, and only sporadic progress has been made.

4.1. IMBIBITION

Here we will concentrate on the physical and structural changes occurring in normal, viable seeds during the initial phases of imbibition and through germination to seedling establishment. Later in the chapter we will deal with the metabolism that is provoked by hydration.

4.1.1. Uptake of Water from the Soil

The uptake of water by seeds is an essential, initial step toward germination. The total amount of water taken up during imbibition is generally quite small and may not exceed two or three times the dry weight of the seed. For subsequent seedling growth, which involves the establishment of the root and shoot systems, a larger and more sustained supply of water is required.

Several factors govern the movement of water from the soil into the seed, but particularly important are the water relations of the seed and of the soil. Water potential (ψ) is an expression of the energy status of water, and net diffusion of water occurs down an energy gradient from high to low potential (i.e., from pure water to water containing solutes). Pure water has the highest potential, and by convention, it is assigned a zero value. Other potentials, therefore, have positive (i.e., > 0) or negative (i.e., < 0) values. The water potential of the cells in a seed can be expressed as follows:

$$\psi_{cell} = \psi_{\pi} + \psi_{c} + \psi_{p}$$

This means that cell water potential is affected by three components: (1) ψ_{π} the osmotic potential. The concentration of dissolved solutes in the cell determines the osmotic potential—the greater their concentration, the lower is the osmotic (water) potential and hence the greater the energy gradient along which water will flow. Thus, the concentration of solutes in the cell influences water uptake. (2) $\psi_{c,}$ the matric component. This is contributed by the hydration of matrices (e.g., cell walls, starch, protein bodies) and their ability to bind water. (3) $\psi_{p,}$ the pressure potential, which occurs because as water enters a cell the internal pressure builds up which exerts a force on the cell wall. Values for ψ_{π} and ψ_{c} are negative since they have a lower potential than pure water, and ψ_{p} is a positive and hence opposite force. The sum of the three terms, the water potential, is a negative number, except in fully turgid cells where it approaches zero. Water potential can be expressed in terms of pressure or energy, and the *bar* was the unit most frequently used (1 bar = 10^3 dynes · cm^{-2}, 10^2 J · kg^{-1}, or 0.987 atm). The unit megapascal (MPa) is now preferred, and –1 MPa = –10 bar.

The soil also has its own water potential (ψ_{soil}), which is the sum of its ψ_{π}, ψ_{c}, and ψ_{p} although of these only ψ_{c} plays a significant role (except in saline soils where ψ_{π} may be appreciable). The difference in water potential between seed and soil is one of the factors that determines availability and rate of flow of water to the seed. Initially, the difference in ψ between the dry seed and moist soil is very large because of the high ψ_{c} of the dry coats, cell walls, and storage reserves. But as the seed moisture content increases during imbibition and the matrices become hydrated, the water potential of the seed increases (i.e., becomes less negative) and that of the surrounding soil decreases as water is withdrawn. Hence, the rate of water transfer from soil to seed declines with time, more quickly in soils of low water-holding capacity (e.g., sandy soils) (Fig. 4.1). Continued availability of water to the seed depends on the water potential of the zones of soil immediately surrounding the seed and on the rate at which water moves through the soil, i.e., the hydraulic conductivity of the soil. Capillary and vapor movement of water near the seed is influenced considerably by soil compaction (bulk density), which may result in mechanical restraint of the swelling seed and decreased imbibition. Other factors may play a role in determining the rate and extent of water uptake regardless of the difference in ψ between the seed and soil, e.g., the impedance of the soil matrix (caused chiefly by surface and colloidal factors) and the degree of contact of the

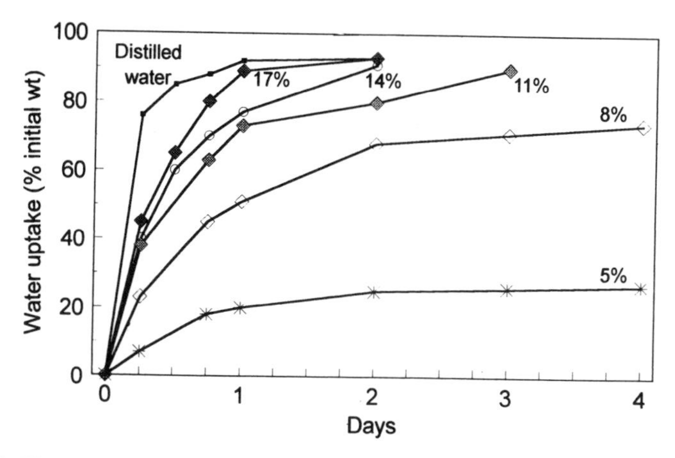

Figure 4.1. Water uptake by chick-pea (*Cicer arietinum*) seeds in distilled water or in soils at various moisture contents (on a percent dry weight basis). After Hadas (1970).

seed with soil moisture (i.e., seed–soil contact). The latter varies with seed size and shape and with the texture and compactness of the soil itself. Small seeds, seeds that produce mucilage, and seeds with relatively smooth coats tend to be the most efficient in absorbing water owing to their greater contact with soil and their larger surface area/volume ratio. The influence of hydraulic conductivity and seed-water contact area on germination at a particular soil water potential varies between soil types, and so the germination response to soil water potentials in sandy soil is markedly different from that in clay soils (Fig. 4.2). Unsaturated sandy soils have a lower hydraulic conductivity than clay soils, and because of the larger particle size of the former, the seed–water contact area is also reduced. Hence, at any given stress, germination is better in clay soils (Fig. 4.2A) than in sandy soils (Fig. 4.2B). This figure also illustrates that different seeds can utilize different soil moisture conditions for successful establishment; e.g., ryegrass can germinate better in lighter soil types than can other grass species. This may be related to variations in the efficiency with which seeds take up water or may be a reflection of the fact that some seeds can complete germination at lower water contents than others.

4.1.2. Water Uptake by Seeds

Under optimal conditions of supply the uptake of water by seeds is triphasic (Fig. 4.3).

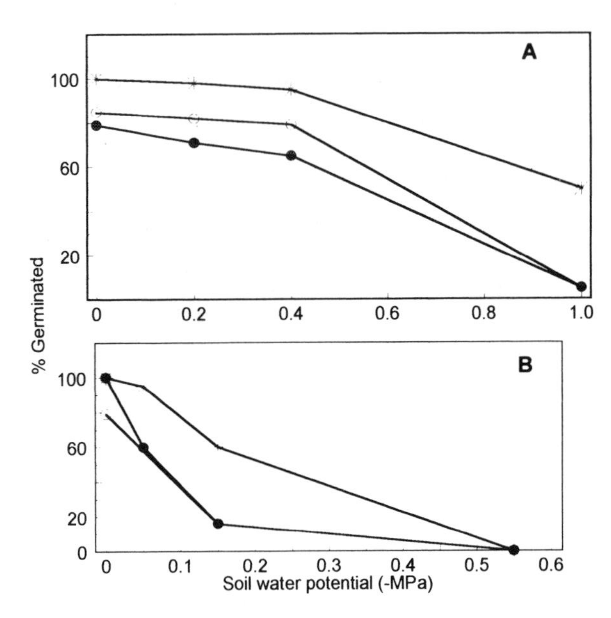

Figure 4.2. The germination of *Themeda australis* (●), *Danthonia* spp. (O), and ryegrass·(*Lolium perenne*) (*) in (A) clay soil and (B) sandy soil at different water potentials. After Hagon and Chan (1977).

Phase I. As noted in Section 4.1.1, the water potential of a mature dry seed is much lower than that of the surrounding moist substrate and can exceed −100 MPa because of its high ψ_c Phase I, or imbibition, is largely a consequence of these matric forces, and water uptake occurs regardless of whether the seed is dormant or nondormant, viable or nonviable. A wetting front is formed as water permeates the seed, and there is an abrupt boundary of water content between wetted cells and those about to be wetted. Moreover, the average water content of the wetted area increases as a function of time. This initial pattern of water uptake is thus marked by three characteristics: (1) a sharp front separating wet and dry portions of the seed, (2) continued swelling as water reaches new regions, and (3) an increase in water content of the wetted area. In some seeds, however, water uptake may be uneven because of certain inherent structural features. In certain legumes (*Vicia* and *Phaseolus* spp., for example), water uptake through the micropyle may be greater than through the rest of the testa, and in some hard-coated seeds of the Papilionaceae (e.g., *Melilotus alba* and *Trigonella arabica*) a plug covering a special opening—the strophiolar cleft—must be loosened or removed before water can enter, and then only through that region. Imbibition is probably very rapid into the peripheral cells of the seed and into a

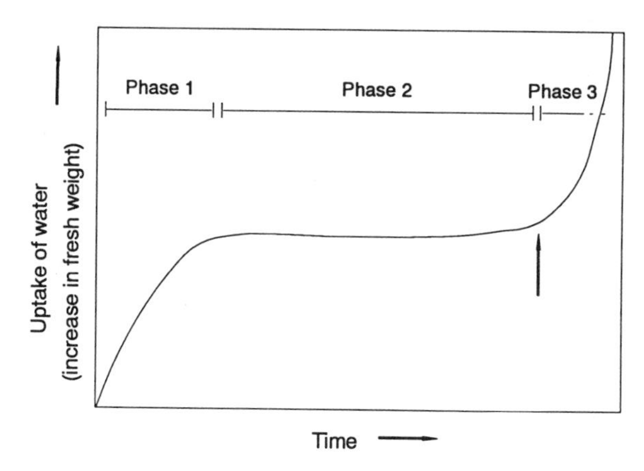

Figure 4.3. Triphasic pattern of water uptake by germinating seeds. Arrow marks the time of occurrence of the first signs of radicle protrusion (i.e., completion of germination).

tissue as small as the radicle. Hence, metabolism can commence during this first phase, within minutes of introduction of the seed to water.

Phase II. This is the lag phase of water uptake when the ψ_c no longer plays a significant role, and the ψ of the seed is largely a balance between ψ_π and ψ_p In this phase, the value of ψ for many seeds probably does not exceed −1 to −1.5 MPa. During this phase major metabolic events take place in preparation for radicle emergence from nondormant seeds; dormant seeds are also metabolically active at this time.

Phase III. Although dormant seeds may achieve phase II, only germinating seeds enter this third phase which is concurrent with radicle elongation. The increase in water uptake is initially related to the changes that cells of the radicle undergo as they extend (Section 4.2), marking the completion of germination. Then water uptake is influenced by decreases in ψ_π resulting from production of low-molecular-weight, osmotically active substances resulting from the postgerminative hydrolysis of stored reserves. Endosperms and nonpersistent (hypogeal) cotyledons do not expand and hence do not achieve phase III of water uptake: eventually their water content declines as degeneration occurs.

The duration of each of these phases depends on certain inherent properties of the seed (e.g., hydratable substrate content, seed coat permeability, seed size, oxygen uptake) and on the prevailing conditions during hydration (e.g., temperature, moisture content, composition of substrate). Different parts of a seed may pass through these phases at different rates; e.g., an embryo or axis located near the surface of a large seed may commence elongation (i.e., enter phase III of water uptake) even before its associated bulky storage tissue has become fully imbibed (i.e., completed phase I). As an example, when the water content of whole dent corn (*Zea mays*) grains reaches 75%, the water content of the embryo on a dry weight basis is 261%, but that of the remainder of the grain is only 50%.

Water may be directed preferentially toward the radicle of the maize embryo during early imbibition. There are structural modifications to the coat adjacent to the radicle which facilitate this. Water then diffuses to the shoot of the embryo. Hydration of the endosperm is slower because water has to penetrate the surrounding pericarp which is not structurally modified to permit its rapid uptake.

In concluding these two sections we can appreciate that the kinetics of water uptake into seeds is influenced by the properties of the seed, as well as by the environment in which it is situated. A water potential gradient between the seed and its surroundings (the soil matrix) is the driving force for water uptake, but the permeability of the seed to water is more important in determining its rate of uptake. Seed permeability is influenced by morphology, structure, composition, initial moisture content, and temperature of imbibition. Water uptake rates are not necessarily affected by these in a predictable manner because of complex interactions. Surprisingly, for example, soybean embryos with a higher initial moisture content imbibe faster than those with a lower content (Fig. 4.4), even though the water potential gradient for the drier seeds is much greater (see legend to Fig. 4.4). This reduced water uptake could occur because the drying of seeds beyond a certain critical water content results in changes in their permeability, by physical or metabolic perturbations, or both. Thus, the kinetics

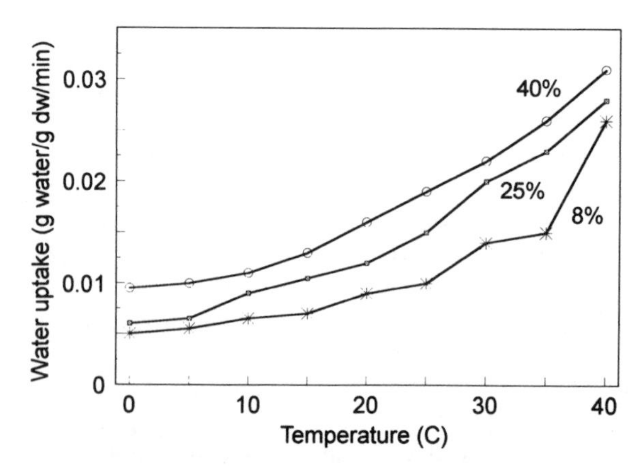

Figure 4.4. Rate of water uptake by soybean embryos of different moisture contents (percent water on a fresh mass basis) over a range of temperatures. The initial ψ of the seed at 8% was approx. –50 MPa, at 25% approx. –10 MPa, and at 40% < –5 MPa. Based on Vertucci and Leopold (1983).

of imbibition are complex and are influenced by interactions within the seed in a manner which is not easy to model. The two major processes involved, wetting and hydraulic flow of water within the seed, exhibit separate dynamics and the variables affecting them change as the seeds pass through phases I–III.

The rate of water penetration into the seed is critical to the success of germination. If water uptake is too slow, then germination is reduced because seeds may deteriorate; if water uptake is too rapid, seeds may suffer excess imbibitional damage, a subject for the next section.

4.1.3. Soaking Injury and Solute Leakage

As seeds start to take up water, there is a release from colloids of adsorbed gases (Section 4.3.2) and a rapid leakage into the surrounding medium of solutes such as sugars, organic acids, ions, amino acids, and proteins. In the field, these solutes might stimulate the growth of fungi and bacteria in the soil, which might invade the seed and lead to its deterioration. Some seeds (e.g., soybean) also leak certain proteinase inhibitors and lectins from their cell walls early upon imbibition, and these may function as protective agents against microbial or insect invasion. Damaged legume seeds, e.g., with cracked seed coats, which often subsequently exhibit poor seedling vigor may extrude starch grains and protein bodies through the coats under pressure when first imbibed. The source of these intracellular substances is the outer layers of the cotyledons, which may blister within minutes of introduction to water.

When imbibed, pea embryos, i.e., seeds from which the testa has been removed, show an immediate and rapid leakage of potassium and of electrolytes, the latter being detected as an increase in conductivity of the surrounding water (Fig. 4.5A). A similar rapid leakage of sugar and protein occurs. This initial leakage lasts for up to about 30 min (Fig. 4.5B) and occurs only from the outermost cell layers of the embryo at this time. Embryos that have imbibed for 60 min, been dried, and then rehydrated show the same rapid initial leakage (Fig. 4.5B). Intact seeds do not leak so copiously into the surrounding medium. In some species this is because the uptake of water is restricted by the enclosing structures (e.g., testa, pericarp), or else they act as barriers to the efflux of solutes, or both. Seeds of certain species e.g., *Phaseolus vulgaris*, pea, and soybean, germinate very poorly and show reduced growth or vigor when hydrated without their testas. In these seeds, then, the coat plays an important protective role during imbibition. Damage to the coat during harvesting or planting could seriously reduce yield and, as noted above, permit extensive structural damage to the cotyledons during early imbibition. Seeds that are hydrated wholly or partially in water-saturated atmospheres before introduction to liquid water, and seeds that do not undergo drying during their final stages of maturation, do not

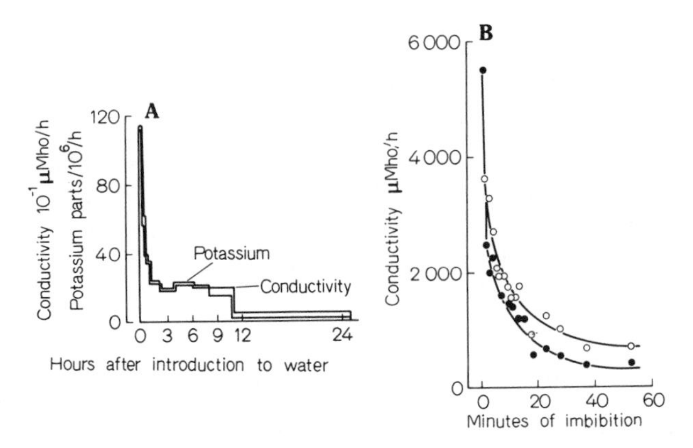

Figure 4.5. (A) Time course of leakage of electrolytes and potassium from pea embryos (i.e., seed minus testa) immersed in water. (B) Electrolyte leakage from pea embryos immersed in water (●) and from embryos removed after 60 min in water, dried, and then returned to water (O). After Simon and Raja Harun (1972).

leak solutes when placed in water. Such seeds do not imbibe, however, since they are already hydrated.

Although some components are leaked from the cell wall, many are lost from within the cells themselves. Hence, the selectively permeable membranes of the tonoplast and plasmalemma that normally retain solutes within cells lose their integrity during drying and do not act as retentive barriers during the initial stages of imbibition. Membranes are composed chiefly of proteins and phospholipids arranged in a fluid bilayer. Individual phospholipid molecules are lined up side by side with their polar groups facing the aqueous phase on either side of the membrane and with their hydrocarbon chains forming the hydrophobic central region (Fig. 4.6). This molecular organization is stabilized by the relationship between membrane components and water. About 25% hydration above fully dried weight is essential for the maintenance of this liquid-crystalline membrane configuration. During maturation drying of the seed it is likely that the phospholipid components of the membranes undergo a phase transition to a gel-like state (Fig. 4.6). When the seeds imbibe, the membranes pass once again from the gel to the liquid-crystalline phase, and in doing so they become transiently leaky. It follows that the extent of leakiness of the membranes will be reduced if the transitions to and from the gel phase can be eliminated. The incorporation of specific molecules into the membrane in the region of the phospholipid head groups prevents the formation of the gel state (Fig. 4.6). It is suggested that in seeds, sugars such as sucrose and the raffinose-series oligosaccharides, stachyose and raffinose, or proteins [e.g,. late embryogenesis abundant

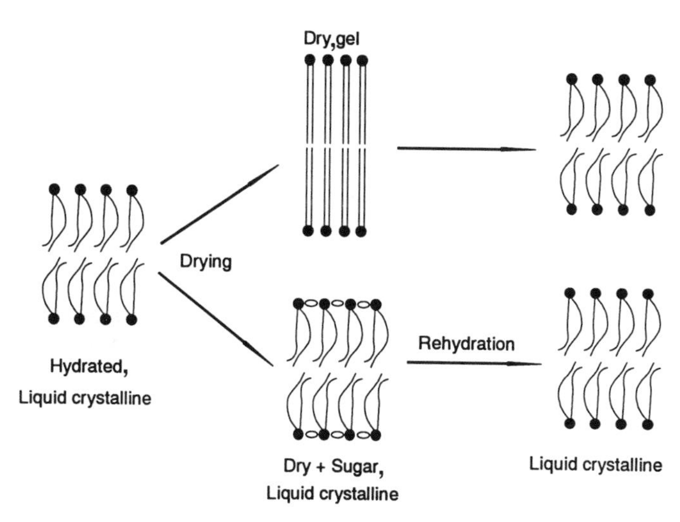

Figure 4.6. Transitions of phospholipid components of the membrane from a liquid-crystalline to a gel phase during drying, and the reformation of the liquid-crystalline state on rehydration. When membranes are protected by sugars, for example, formation of the gel phase on drying is prevented. Based on Crowe and Crowe (1992).

(LEA) proteins] (Section 3.3.2) are synthesized prior to maturation drying to improve the stability of the membranes during subsequent water loss. In developing soybean embryos, for example, there is an increase in some sugars in desiccation-tolerant embryos but not in intolerant ones (see Fig. 3.8). While this observation, and others made on mutant seeds of *Arabidopsis* (Section 3.3.2) do not provide a cause-and-effect link between cellular sugar content, membrane stability, and desiccation tolerance, the possibility is intriguing.

Obviously, there is not a complete maintenance of the liquid-crystalline state during drying because there is transient leakage on rehydration. But the ability of seeds to survive drying may depend on the prevention of extensive gel formation by the phospholipid portion of the component membranes.

4.2. THE COMPLETION OF GERMINATION: RADICLE ELONGATION AND ITS CONTROL

The cell walls of the embryos in mature dry seeds are shrunken, giving them a wavy or folded appearance under the electron microscope. Initially on imbibition, the walls expand a little and straighten as they take up water, but this occurs in nonliving and in dormant seeds, as well as in viable nondormant

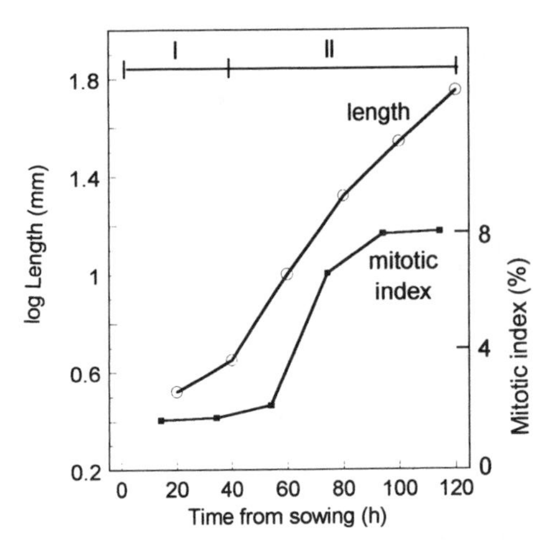

Figure 4.7. Length and mitotic index of roots of germinating and germinated broad bean (*Vicia faba*) seeds. I and II refer to phases of radicle elongation (see text). Based on Rogan and Simon (1975).

ones. For germination to be completed, the radicle must expand and penetrate the surrounding structures; over the years, there has been considerable debate whether this occurs solely by cell expansion, or if cell division is required too. It is now generally conceded that cell division is neither correlated with nor is necessary for radicle expansion. In *Vicia faba* (Fig. 4.7), for example, the radicle increases in length about 18 h before any cell division (increase in mitotic index) takes place. In other seeds, the difference in timing between cell division and cell expansion is much closer, but there is no strong evidence that the former ever precedes the latter during germination, or that they occur simultaneously. From the pattern of cell elongation shown by *Vicia faba* (Fig. 4.7) it is evident that extension of the radicle occurs in two phases. There is initially a slow phase up until about 48 h (I), by which time the root has pierced the testa in about 30% of the seeds. Then a rapid phase ensues (II), which is accompanied later by an increase in the number of cells, and by mitoses in the apical region of the radicle. A similar biphasic mode of radicle elongation has been reported for other seeds.

That initial cell elongation is different from subsequent radicle growth can be demonstrated also by examining the sensitivity of these two events to water stress (Fig. 4.8). Seeds that have not commenced radicle elongation, i.e., not completed germination, can be prevented from doing so by incubation in solutions with a maximum water potential of −1.2 MPa. On the other hand, seeds that have already entered the second phase of radicle elongation, i.e., commenced growth, need lower water potentials (at least −2 MPa) for complete arrest. This implies that the initiation of cell elongation is controlled in a different manner from later events. But what is the nature of this control? What specific events

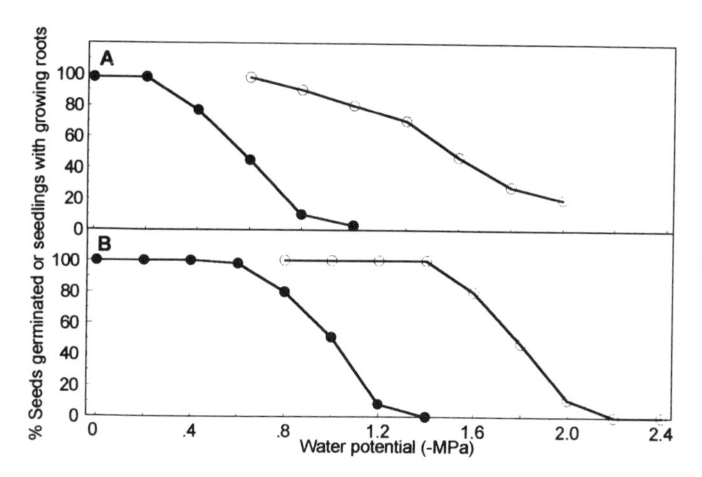

Figure 4.8. Germination (●) and seedling radicle growth after germination (○) of (A) calabrese (*Brassica oleracea* var. *italica*) and (B) cress (*Lepidium sativum*) incubated in solutions of polyethylene glycol at different water potentials. After Hegarty and Ross (1978).

are essential for radicle elongation to commence, and what internal signals are responsible for setting this process in motion?

Possible causes for the commencement of radicle growth are (1) lowering of the osmotic (solute) potential, ψ_π in the radicle cells as a result of solute accumulation, thus increasing water uptake and raising the turgor pressure, causing cell elongation; (2) relaxation (i.e., increased extensibility) of the radicle cell walls, allowing for elongation; or (3) weakening of the tissues surrounding the radicle tip, thus allowing it to elongate—or a combination of these.

Experiments have been conducted on muskmelon (*Cucumis melo*) and tomato seeds, the radicles of which must penetrate the surrounding structures (perisperm/endosperm, or endosperm, respectively) to complete germination. In muskmelon seeds and isolated axes (Fig. 4.9) there is no decrease in ψ_π (rather, there is a slight increase) and no increase in turgor pressure (but a slight decrease) in the time period immediately prior to radicle extension. Nor does cell-wall relaxation result in the formation of a water-potential gradient driving water uptake. Rather, the tissues surrounding the radicle impose a mechanical stress, preventing it from taking up water. Weakening of these tissues opposite the radicle tip is sufficient to release this restraint, permitting the radicle to emerge.

The cause of the weakening of the perisperm/endosperm in muskmelon is not known, but it is possible that hydrolytic enzymes are produced which digest or separate the cell walls in the radicle region. Such is the situation in tomato seeds, where prior to radicle emergence there is an increase in activity of a

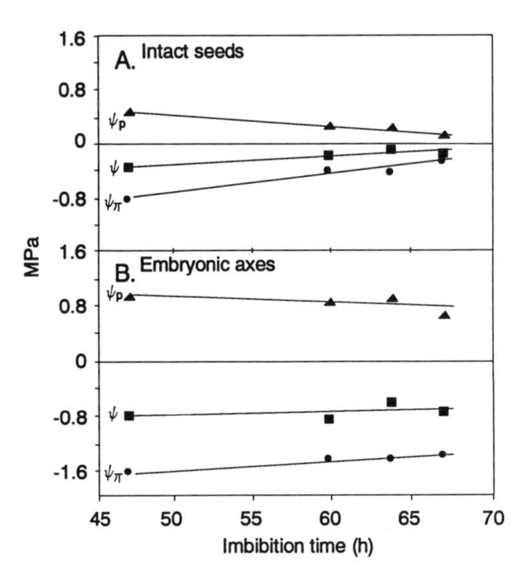

Figure 4.9. Water potential (ψ), osmotic potential (ψ_π), and turgor potential (pressure) (ψ_p) of intact seeds and embryonic axes of muskmelon during phase 2 (Fig. 4.3) of imbibition, just prior to radicle emergence. After Welbaum and Bradford (1990).

mannanase which can hydrolyze the galactomannan-containing cell walls of the endosperm, and thus permit protrusion of the radicle. This increase in mannanase activity is stimulated by gibberellin. In tomato mutants which are deficient in this hormone, there is an increase in enzyme activity only after the seeds (endosperms) are exposed to exogenous GA (Fig. 4.10).

The germination model that is proposed for tomato (Fig. 4.11) is that radicle emergence is determined by a balance between the growth potential or turgor of the radicle, and the yield threshold or mechanical resistance of the endosperm cap. As indicated in Fig. 4.11, a high water potential (ψ) increases

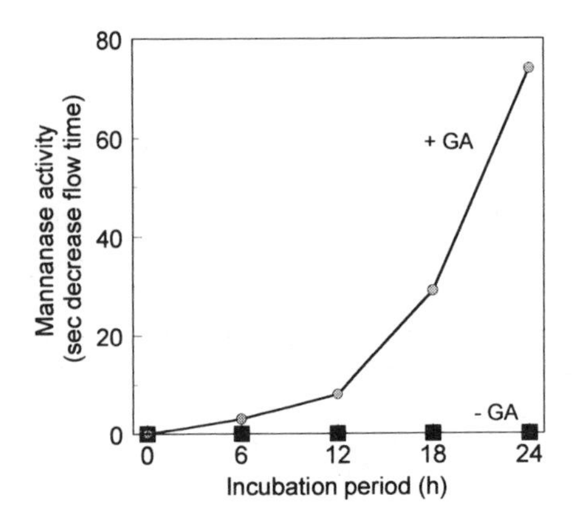

Figure 4.10. Endo-β-mannanase activity in de-embryonated endosperms of a gibberellin-deficient mutant (*gib1*) of tomato, in the presence or absence of GA_{4+7}. After Groot *et al.* (1988).

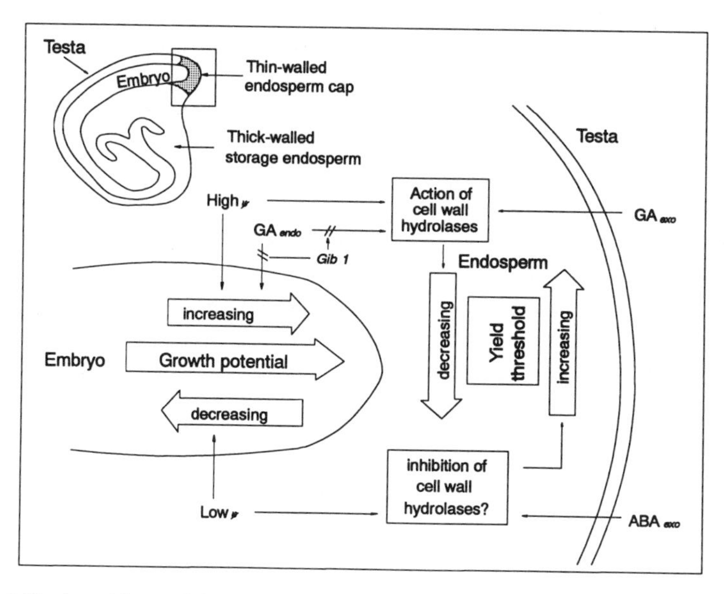

Figure 4.11. A model to explain the germination of tomato seeds in relation to the balance of forces between the growth potential of the radicle and the yield threshold of the endosperm cap which imposes a mechanical restraint on the radicle. The roles played by water potential (ψ) and endogenous (*endo*) or exogenously supplied (*exo*) hormones are noted, and the block to germination of the gibberellin-deficient mutant (*gib1*) is noted. Simplified from Ni and Bradford (1993).

the growth potential of the embryo, presumably permitting the production or activation by GA of the cell wall hydrolases (e.g., mannanase) and thus decreasing the yield threshold of the endosperm close to the radicle tip. A low water potential, e.g., imbibition in an osmoticum, imposes a stress on cell metabolism, perhaps decreasing hydrolase production, and by reducing the turgor of the radicle cells there is a decrease in growth potential. In the *gib1* mutant, endogenous GA is not available to promote the action of cell wall hydrolases, and an exogenous supply of this hormone is needed. Addition of ABA also causes a decline in growth potential of the radicle, perhaps by inhibiting the GA-induced increase in mannanase activity. The actions of low ψ in preventing germination are not mediated through an increase in ABA in the seeds.

The failure of seeds of many species to germinate cannot be the result of mechanical constraints by surrounding structures, for they do not all possess a perisperm or endosperm at maturity, and the testa is frequently too yielding a structure to present a physical barrier to radicle emergence (for some exceptions, see the account on coat-imposed dormancy in Chapter 5). The rapeseed (*Brassica napus*) is such an example, where the fragile testa cracks during imbibition.

Growth of the radicle begins when there is loosening (i.e., increased extensibility) of its walls and the resistance to turgor-driven elongation (i.e., the minimum turgor required for cell expansion) declines. There are significant changes in solute content of the radicle cells prior to the completion of germination. Exogenously supplied ABA prevents radicle emergence of rapeseed by preventing cell wall loosening, thus limiting the growth potential of the radicle.

The basis for the changes in cell wall rigidity as the radicle enters the elongation phase is not known nor is how ABA interferes with the loosening process. One hypothesis is that expanding cells secrete protons (hydrogen ions) into their walls, resulting in their acidification; this leads to a breaking of the hydrogen bonds between adjacent chains of wall carbohydrates, or the low pH optimizes the conditions for limited enzymatic hydrolysis of the wall. The resultant change in wall rigidity, i.e., increased stretchability, thus allows for cell extension. Studies on rapeseed have failed to find any positive correlations between hydrogen ion secretion and ability to complete germination, however. Thus, the nature of cell-wall extension, the event which is essential for the completion of germination, remains unknown.

4.3. RESPIRATION—OXYGEN CONSUMPTION

4.3.1. Pathways, Intermediates, and Products

Three respiratory pathways are assumed to be active in the imbibed seed: glycolysis, the pentose phosphate pathway, and the citric acid (Krebs or tricarboxylic acid) cycle. Glycolysis, catalyzed by cytoplasmic enzymes, operates under aerobic and anaerobic conditions to produce pyruvate, but in the absence of O_2 this is reduced further to ethanol, plus CO_2, or to lactic acid if no decarboxylation occurs. Anaerobic respiration, also called fermentation, produces only two ATP molecules per molecule of glucose respired, in contrast to six ATPs produced during pyruvate formation under aerobic conditions. In the presence of O_2, further utilization of pyruvate occurs within the mitochondrion: oxidative decarboxylation of pyruvate produces acetyl-CoA, which is completely oxidized to CO_2 and water via the citric acid cycle to yield up to a further 30 ATP molecules per glucose molecule respired. The generation of ATP molecules occurs during oxidative phosphorylation when electrons are transferred to molecular O_2 along an electron transport (redox) chain via a series of electron carriers (cytochromes) located on the inner membrane of the mitochondrion. An alternative pathway for electron transport, which does not involve cytochromes, may also operate in mitochondria (see Section 4.4.2). The pentose phosphate pathway is an important source of NADPH, which serves as a hydrogen and electron donor in reductive biosynthesis, especially of fatty acids.

Intermediates in this pathway are starting compounds for various biosynthetic processes, e.g., synthesis of various aromatics and perhaps nucleotides and nucleic acids. Moreover, complete oxidation of hexose via the pentose phosphate pathway and the citric acid cycle can yield up to 29 ATPs.

4.3.2. Respiration during Imbibition and Germination

Respiration by mature "dry" seeds (usual moisture content: 10–15%) of course is extremely low when compared with developing or germinating seeds, and often measurements are confounded by the presence of a contaminating microflora. When dry seeds are introduced to water, there is an immediate release of gas. This so-called "wetting burst," which may last for several minutes, is not related to respiration, but is the gas that is released from colloidal adsorption as water is imbibed. This gas is released also when dead seeds or their contents, e.g., starch, are imbibed.

Keto acids (e.g., α-ketoglutarate, pyruvate), which are important intermediates in respiratory pathways, are chemically unstable and may be absent from the dry seed. A very early metabolic event during imbibition, occurring within the first few minutes after water enters the cells, is their reformation from amino acids by deamination and transamination reactions (e.g., of glutamic acid and alanine).

The consumption of O_2 by many seeds follows the basic pattern outlined in Fig. 4.12 although, as indicated, the pattern of consumption by the embryo differs ultimately from that by storage tissues. Respiration is considered to involve three or four phases:

Phase I. Initially there is a sharp increase in O_2 consumption, which can be attributed in part to the activation and hydration of mitochondrial enzymes involved in the citric acid cycle and electron transport chain. Respiration during this phase increases linearly with the extent of hydration of the tissue.

Phase II. This is characterized by a lag in respiration as O_2 uptake is stabilized or increases only slowly. Hydration of the seed parts is now completed and all preexisting enzymes are activated. Presumably there is little further increase in respiratory enzymes or in the number of mitochondria during this phase. The lag phase in some seeds may occur in part because the coats or other surrounding structures limit O_2 uptake to the imbibed embryo or storage tissues, leading temporarily to partially anaerobic conditions. Removal of the testa from imbibed pea seeds, for example, diminishes the lag phase appreciably. Another possible reason for this lag is that the activation of the glycolytic pathway during germination is more rapid than the development of mitochondria. This could lead to an accumulation of pyruvate because of deficiencies in the citric acid cycle or

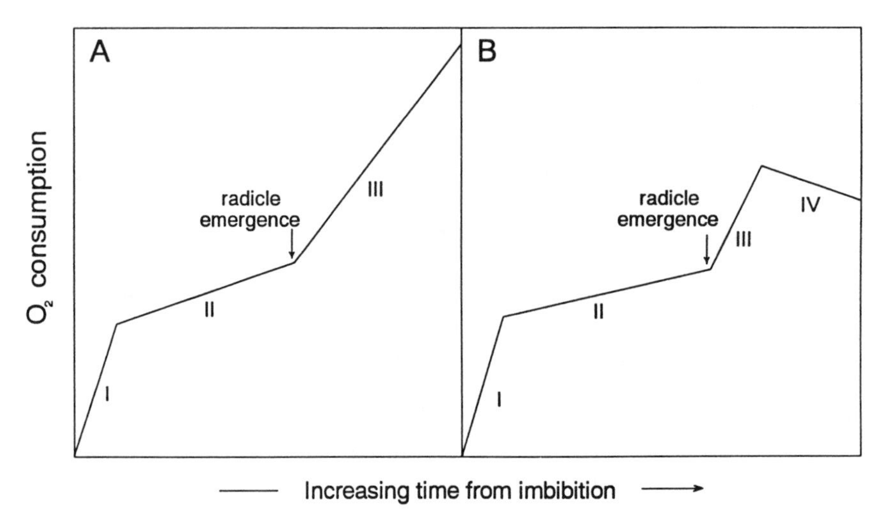

Figure 4.12. Pattern of oxygen consumption by the embryo (A) and storage tissues (B) of seeds during and after germination.

oxidative phosphorylation (electron transport chain); hence, some pyruvate would be diverted temporarily to the fermentation pathway, which is not O_2 requiring.

Between phases II and III in the embryo the radicle penetrates the surrounding structures: germination is completed.

Phase III. There is now a second respiratory burst. In the embryo, this can be attributed to an increase in activity of newly synthesized mitochondria and respiratory enzymes in the proliferating cells of the growing axis. The number of mitochondria in storage tissues also increases, often in association with the mobilization of reserves. Another contributory factor to the rise in respiration in both seed parts could be an increased O_2 supply through the now punctured testa (or other surrounding structures).

Phase IV. This occurs only in storage tissues and coincides with their senescence following depletion of the stored reserves.

The lengths of phases I–IV vary from species to species owing to such factors as differences in rates of imbibition, seed-coat permeability to oxygen, and metabolic rates. Moreover, the lengths of the phases will vary considerably with the ambient conditions, especially the temperature. In a few seeds, e.g., *Avena fatua*, there is no obvious lag phase (II), in oxygen uptake. The reasons for its absence are not known, but it could be because efficient respiratory systems become established early following imbibition, including the development of newly active mitochondria, thus ensuring a continued increase in O_2

utilization. Also, coat impermeability might not restrict O_2 uptake prior to the completion of germination.

During germination a readily available supply of substrate for respiration must be present. This may be provided to a limited extent by hydrolysis of the major reserves, e.g., triacylgylcerols, which are present in almost all parts of the embryo, including the radicle and hypocotyl, although their greatest concentration is in storage tissues. It is important to note, however, that extensive mobilization of reserves is a postgerminative event (Chapter 7). Most dry seeds contain sucrose, and many contain one or more of the raffinose-series oligosaccharides: raffinose (galactosyl sucrose), stachyose (digalactosyl sucrose), and verbascose (trigalactosyl sucrose), although the latter is usually present only as a minor component. The distribution and amounts of these sugars within seeds are very variable, even between different varieties of the same species. To introduce a brief digression, the raffinose-series oligosaccharides have attracted some interest in relation to human nutrition because they are the cause of flatulence in humans and domestic animals. These sugars escape digestion and absorption in the upper gastrointestinal tract but are degraded and fermented by microbes in the colon to yield H_2 and CO_2. The odious and malodorous consequences of the relatively high amounts of stachyose and raffinose in digested (baked) beans are legendary.

During germination, sucrose and the raffinose-series oligosaccharides are hydrolyzed, and in several species the activity of α-galactosidase, which cleaves the galactose units from the sucrose, increases as raffinose and stachyose decline. Although there is little direct evidence that the released monosaccharides are utilized as respiratory substrates, there is strong circumstantial evidence. Free fructose and glucose may accumulate in seeds during the hydrolysis of sucrose and the oligosaccharides, but there is no buildup of galactose (e.g., in mustard, *Sinapis alba*, Fig. 4.13). Hence, it is probably rapidly utilized, perhaps through incorporation into cell walls or into galactolipids of the newly forming membranes in the cells of the developing seedling.

4.4. MITOCHONDRIAL DEVELOPMENT AND OXIDATIVE PHOSPHORYLATION

4.4.1. Site of ATP Production during Early Imbibition

Mitochondria in dry and freshly imbibed seeds are functionally and structurally deficient, and in the past there has been considerable debate as to whether they can effectively conduct oxidative phosphorylation during the first few hours after hydration. Nowadays, however, it is generally accepted that they can, and

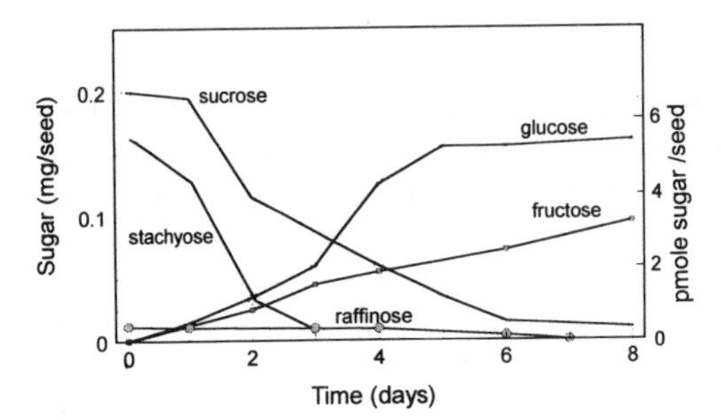

Figure 4.13. Decline in stachyose, sucrose, and raffinose in seeds of *Sinapis alba* accompanied by an increase in glucose and fructose. A separate analysis showed that most of the stored stachyose, sucrose, and raffinose is located in the cotyledons, but that most of the glucose and fructose is found in the root and stem. After Gould and Rees (1964).

mitochondrial oxidative phosphorylation is the major source of ATP from the start of imbibition.

4.4.2. The Route of Electrons between Substrate and Molecular Oxygen

Many studies on the development of mitochondrial activities in seeds have been conducted on the cotyledons of legumes, and for the moment we will have to assume that similar changes occur within the axis. Mitochondria in dry seeds are characteristically poorly differentiated and lack cristae. To understand the changes that occur in the mitochondria during germination, we must first review the major routes by which oxidizable substrates enter this organelle and then how they become involved in oxidative phosphorylation. Glycolysis and various other metabolic pathways produce NADH, from which NAD has to be regenerated. The outer membrane of the mitochondrion is permeable to NADH (called exogenous NADH because it is produced outside of the mitochondrion), and it passes to the intermembrane space (Fig. 4.14). Pyruvate also enters the mitochondrion, but it is transported through the membranes to the inner matrix, which contains the enzymes of the citric acid cycle. This cycle produces NADH also (called endogenous NADH because it is formed within the mitochondrion). Both endogenous and exogenous NADH are cleaved by membrane-bound NADH dehydrogenase complexes, although the released reducing equivalents (as electrons) enter the electron transport chain at different sites. Electrons originating from endogenous NADH are transferred to ubiquinone via a coupled proton efflux site (site I) which permits energy conservation by driving phosphorylation, i.e., ATP is produced. Electrons released from exogenous NADH are not transferred via site I but are passed directly to the ubiquinone pool, thus eliminating the first phosphorylation step. The oxidation of succinate to malate in the citric acid cycle is not NAD linked; succinate dehydrogenase (SDH) is inserted into the inner membrane on the matrix side, and the released electrons are transferred to ubiquinone. There is no site of energy conservation along this pathway (Fig. 4.14).

The electrons are transferred from the ubiquinone pool via cytochromes *b*, c_1, and *c* and finally to O_2 via cytochrome oxidase (aa_3). This part of the chain is very important because it incorporates two sites of proton extrusion and, consequently, of energy conservation (Fig. 4.14). Mitochondria of many higher plant tissues possess an alternative electron transfer pathway which, unlike the cytochrome chain, is insensitive to cyanide (and hence it is often called the cyanide-insensitive pathway). The alternative pathway branches between ubiquinone and cytochrome *b*, and electrons are passed via an "alternative oxidase" system, as yet unidentified, to O_2. There are no sites of energy conservation linked to this pathway, and hence when exogenous NADH and succinate are oxidized by this route, there is no ATP production. When an NAD-linked substrate from the mitochondrial matrix, such as malate and α-ketoglutarate, is oxidized through the alternative pathway, only one ATP is produced (at site I) compared with three (at sites I, II, and III) when electrons pass through the cytochrome pathway (Fig. 4.14).

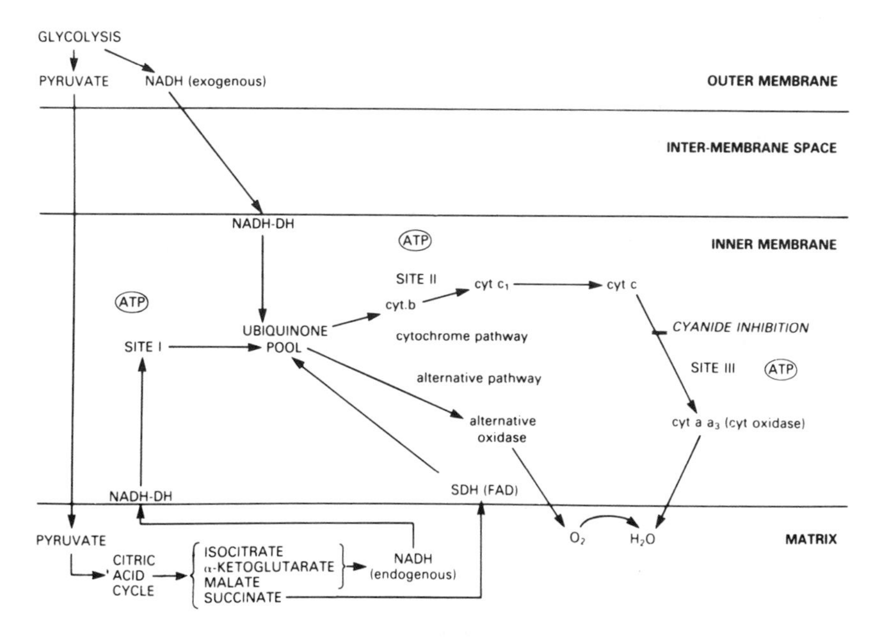

Figure 4.14. The pathways for utilization of reducing power and electrons in plant mitochondria. NADH-DH, NADH dehydrogenase; SDH (FAD), flavoprotein-linked succinate dehydrogenase; sites I–III, sites of proton extrusion where ATP (circled) is produced; cyt, cytochrome.

Isolated mitochondria must be used to study their phosphorylating activity. Their ability to oxidize various substrates and to phosphorylate ADP is monitored using an oxygen electrode, and a typical trace of oxygen consumption by a mitochondrial preparation consuming O_2 and conducting oxidative phosphorylation is shown in Fig. 4.15. Initially the isolated mitochondria are incubated in the absence of both ADP and added substrate (α-ketoglutarate, malate, succinate, or NADH) to test activity driven by the purely endogenous components. A low amount of oxygen consumption may take place: this is state 1 respiration. Then substrate (in this case malate) is added and further O_2 consumption occurs, which is the state 2, or substrate rate, of O_2 uptake. Next, ADP is added to the mitochondrial preparation, and the rate of O_2 uptake increases appreciably as oxidative phosphorylation proceeds (state 3 respiration). The rate of uptake then begins to level off as ADP becomes limiting (state 4). State 3 respiration can be reinduced by adding more ADP, until the available O_2 in the incubation mixture becomes limiting and anaerobiosis (state 5 respiration) is achieved. Two important calculations can be derived from this type of experiment: (1) the ADP/O ratio, which tells us how many micromoles of ADP are used (to make ATP) per micromole of O_2 consumed and is a measure of efficiency of phosphorylation, and (2) the respiratory control ratio (RCR), which is the ratio of state 3 to state 4 respiration and is an indication of how tightly respiration is controlled by ADP availability. If phosphorylation is efficient (i.e., tightly coupled to O_2 consumption) and utilizes the cytochrome pathway, then, in reference to Fig. 4.14, α-ketoglutarate and malate should elicit an ADP/O ratio close to 3, and succinate and added NADH a ratio close to 2. Substantial lowering of these ratios indicates that mitochondria are deficient in their cytochrome pathway and/or phosphorylating ability. A clear transition from state 3 to state 4 is expected in efficiently phosphorylating mitochondria; failure to achieve this transition can result from damage to the inner membrane (allowing leakage of protons from the sites of proton extrusion), or if electron transport is diverted to the nonphosphorylating alternative pathway.

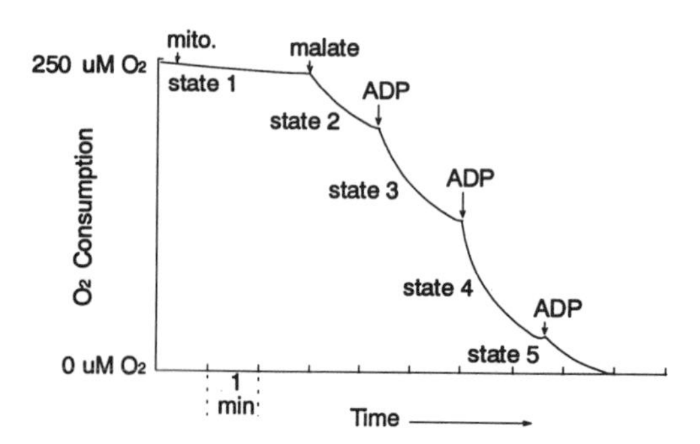

Figure 4.15. Pattern of oxygen consumption by a preparation of mitochondria measured with an oxygen electrode. Arrows indicate the time of addition of the mitochondrial extract (mito.), substrate (malate), and ADP. A well-aerated mitochondrial reaction mixture is 250 μM with respect to O_2; at the end of the experiment all O_2 is depleted.

Now let us consider mitochondrial changes in the imbibed seed in relation to this information.

4.4.3. Mitochondrial Development in Imbibed Seeds

As noted previously, mitochondria in unimbibed embryos and storage tissues are poorly differentiated internally. As might be expected, they exhibit deficiencies in their capacity to utilize reducing power from NADH and succinate dehydrogenase (SDH) and to conduct oxidative phosphorylation. Generally, mitochondria extracted from dry seeds can oxidize supplied succinate and NADH (i.e., can achieve state 3 respiration). Hence, the electron pathways between succinate and ubiquinone, and between exogenous NADH and ubiquinone, are present and are immediately active on hydration. Although these pathways may initially be low in activity, they become increasingly prominent with time after imbibition (Fig. 4.16). In contrast, mitochondria from early-im-

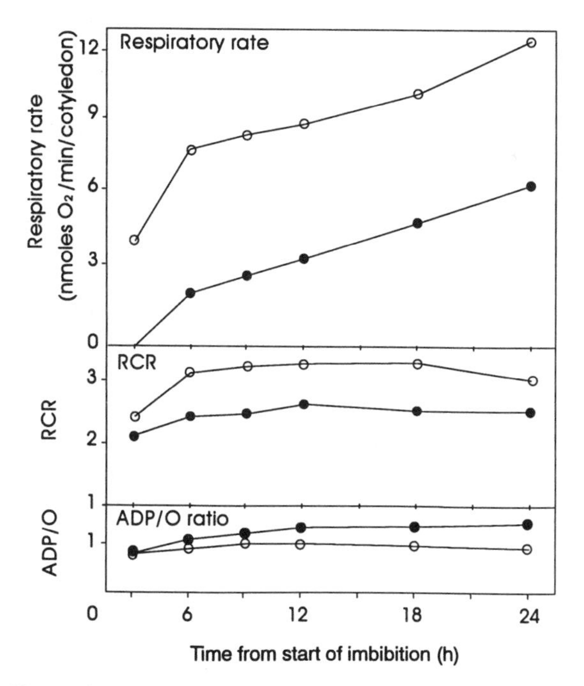

Figure 4.16. Changes in activities of mitochondrial fractions isolated from pea cotyledons (cv. Alaska) over 24 h from the start of imbibition. Substrate: succinate (●); NADH (○). After Morohashi and Bewley (1980).

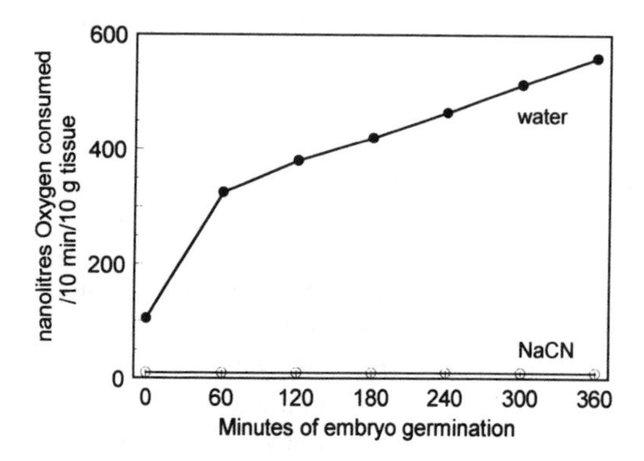

Figure 4.17. Oxygen consumption by maize embryo axes incubated in water or 10 mM NaCN for 6 h from the start of imbibition. Note that those embryos on water only passed through phases I and II of O_2 consumption (see Fig. 4.12). After Ehrenshaft and Brambl (1990).

bibed seeds oxidize malate or α-ketoglutarate only poorly, and the development of oxidizing capacity using endogenous NADH occurs more slowly during germination. During the early stages of germination, mitochondria may not respond clearly to ADP addition when malate and α-ketoglutarate are used as substrate, and a transition from state 3 to state 4 respiration does not occur. Hence, both the RCR and ADP/O ratio are extremely low, indicating a block to the transfer of electrons through the pathway to ubiquinone via site I. Note that the results presented in Fig. 4.17 were obtained using pea cotyledons, and hence the changes that occur are not directly related to germination, which is an axial phenomenon. However, similar changes appear to occur in the axis, although in some species the time required for mitochondria to develop a response to added malate and α-ketoglutarate may be considerably shorter.

Mitochondria from dry and early-imbibed peanut embryonic axes are deficient in cytochrome *c*, and this deficiency remains until about the end of the lag period. But mitochondrial respiration during this time probably still takes place via a cytochrome pathway, since oxidation of succinate is inhibited by cyanide, a specific inhibitor of the terminal oxidation (cytochrome oxidase) reaction (Fig. 4.14). Hence, although there is only a low amount of cytochrome *c* present in mitochondria until germination is completed, it might still be important in preventing completely uncontrolled loss of respiratory substrate, and some oxidative phosphorylation can occur.

Maize embryos exhibit a sharp increase in O_2 consumption very soon after the start of imbibition. This is dependent on the mitochondrial electron transport chain, which is clearly inhibited by cyanide (Fig. 4.17), even within minutes of the start of imbibition. Cytochrome *c* oxidase is conserved in an active form in the mitochondria of mature dry maize embryos, and is responsible for the terminal oxidation reaction on imbibition. New synthesis of this enzyme occurs early during germination, with nucleus-encoded subunits of cytochrome *c* oxidase being synthesized and incorporated into mitochondria within 2 h of imbibition. The counterparts encoded on the mitochondrial genome are synthesized after about 4 h. These events appear to be involved in the biogenesis of new mitochondria, which accompanies germination and subsequent seedling growth.

Cyanide sensitivity of the electron transport chain has been observed in the mitochondria from axes and storage tissues of other seeds also. This argues strongly against the suggestion that the alternative (cyanide-insensitive) pathway is the predominant electron transport pathway during the early stages of germination. It is evident, however, that the alternative pathway can become more active with time after the start of imbibition, and in pea cotyledons (Fig. 4.18) it reaches a maximum several days after SDH and cytochrome oxidase do. In storage tissues, the alternative pathway may play an important role in the reoxidation of NADH produced by such catabolic pathways as β-oxidation and the glyoxylate cycle (see Section 7.5.4). Reducing equivalents are passed to the mitochondria from the glyoxysome and need to be oxidized to maintain an adequate supply of NAD for operation of important enzymes in this latter organelle. But transfer of electrons through the normal cytochrome pathway could lead to excess ATP production and result in a negative feedback and a shutdown of metabolism. Hence, to maintain a high rate of metabolism and an optimum supply of ATP, a balance of electron flow between the cytochrome and alternative pathway is maintained in times of high NADH production.

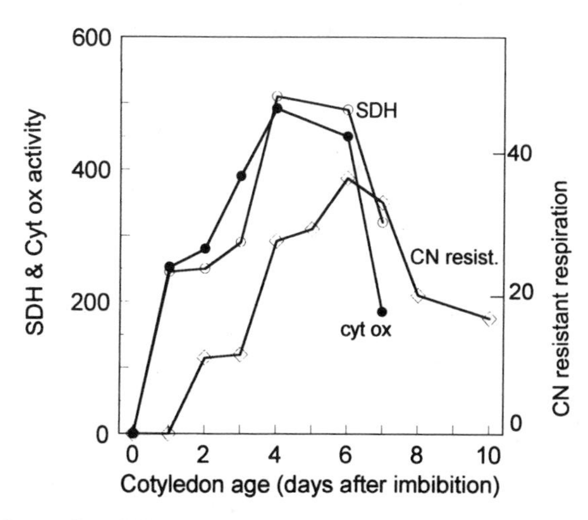

Figure 4.18. Changes in activities of succinate dehydrogenase, cytochrome oxidase (both expressed in nmol substrate reduced/min per mg protein), and cyanide-resistant respiration, with succinate as substrate in mitochondria from pea (cv. Homesteader) cotyledons. After James and Spencer (1979).

A general feature of mitochondrial oxidation in seeds is that it increases with time from imbibition. This may be related to increased efficiency of existing mitochondria and/or an increase in the number of mitochondria. It is now evident that there are two distinct patterns of mitochondrial development in imbibed seeds: (1) repair and activation of organelles already existing within the mature dry seed, as typified by the pea cotyledon and the storage tissues of some other starch-storing seeds, and (2) production of new mitochondria, as typified by the peanut cotyledon and the storage tissues of other oil-storing seeds (Table 4.1). The pattern of development has not been followed in the axes of these or other dicot species during germination. After germination, there must be considerable biogenesis of mitochondria as new cells are formed in the growing axes by mitotic divisions.

In pea cotyledons, some development of mitochondria occurs within the first day of imbibition, and by the end of the second day cristae appear to be more or less fully developed. The structurally deficient mitochondria from pea cotyledons are also enzymatically deficient, having only low amounts of malate dehydrogenase and cytochrome oxidase. SDH is present but, unlike in normal mitochondria, it is not tightly bound to the inner membrane. Incorporation of applied radioactive leucine into mitochondrial proteins can occur within 6 h of the start of imbibition and may continue for at least another 18 h. Inhibition of protein synthesis on cytoplasmic ribosomes prevents this synthesis of mitochondrial proteins, whereas inhibition of protein synthesis on mitochondrial ribosomes does not. Therefore, the new mitochondrial proteins are cytoplasmic in origin. But even in the absence of synthesis of new mitochondrial proteins, cytochrome oxidase and malate dehydrogenase activities increase, and mitochondria have a greater respiratory efficiency. Moreover, SDH activity becomes tightly associated with the inner membrane when both cytoplasmic and mitochondrial protein synthesis are inhibited. Thus, protein synthesis is not a prerequisite for the development of mitochondrial activity. Instead, it appears that

Table 4.1. Seeds Exhibiting Repair or Biogenesis of Mitochondria in Their Storage Tissues following Imbibition, as Related to Their Major Nonprotein Reserve[a]

Repair and activation (starch-storing)	Biogenesis (oil-storing)
Cowpea (*Vigna sinensis*)	Okra (*Hibiscus esculentus*)
Mung bean (*Vigna radiata*)	Pumpkin (*Cucurbita* sp.)
Black gram (*Vigna mungo*)	Cucumber (*Cucumis sativus*)
Egyptian kidney bean (*Dolichos lablab*)	Castor bean (*Ricinus communis*)
Soybean (*Glycine max*)	Peanut (*Arachis hypogaea*)
Garden pea (*Pisum sativum*)	

[a]From data in Morohashi and Bewley (1980), Morohashi *et al.* (1981a), Morohashi (1986).

preformed proteins, both structural and enzymatic, are transferred into preformed, immature mitochondria after imbibition to produce active and efficient mitochondria with complete membrane and oxidative systems.

Mitochondria in peanut cotyledons are difficult to detect early after imbibition, but they increase in quantity and quality (i.e., in internal structure) over the first few days after imbibition. Mitochondrial development is accompanied by an increase in state 3 respiration, ADP/O ratio, RCR, and cytochrome oxidase and other enzyme activities, and by an increased protein-nitrogen and phospholipid content of the mitochondrial membrane fraction. Moreover, there is an increase in incorporation of radioactive leucine into mitochondrial proteins, which ceases when cytoplasmic protein synthesis is inhibited. These observations all support the contention that the increase in mitochondrial activity in peanut cotyledons is the result of *de novo* synthesis of these organelles. This does not preclude the possibility that there is also improvement in the low number of mitochondria preexisting in the dry seed.

As the stored reserves in the cotyledons become depleted, the mitochondria become disorganized and gradually lose their respiratory efficiency, enzyme complement, and activity. For example, in cotyledons of dark-grown Alaska peas there is a marked decline in RCR and ADP/O ratio, although cytochrome oxidase and malate dehydrogenase activities remain fairly constant. Degenerative changes in the inner membrane may occur, for SDH is no longer tightly bound to it in aging cotyledons, and some activity is lost (Table 4.2). The fall in respiratory activity that accompanies the overall decline in stored reserves might not occur uniformly in all parts of dicot cotyledons. In squash (*Cucurbita maxima*) cotyledons, for example, which persist and become photosynthetic, the mitochondria are considerably reduced in numbers in storage cells as the reserves are expended and chloroplasts are formed, but not in those cells close to the veins. Thus, there is a redistribution of respiratory activity in the cotyledons with age, it being fairly uniform during reserve mobilization but becoming more restricted to vascular tissue as the organs assume a photosynthetic role. The vein-cell mitochondria might be important for the vein-loading of catabolites and the subsequent products of photosynthesis. In dark-grown seedlings, which do not become photosynthetic, vein-localization of mitochondria is less distinct.

4.4.4. ATP Synthesis during Germination

Synthesis of ATP during germination has been studied in only a few seed species, but the same general pattern emerges. The amount of ATP in dry seeds is extremely low compared with other adenine nucleotides (ADP and AMP), but on hydration it increases very rapidly (e.g., in lettuce, Fig. 4.19A,B). During the lag phase of O_2 consumption ATP remains fairly constant; a further rise in ATP

Table 4.2. Decline in Respiratory Activities of Mitochondria in Cotyledons of Alaska Pea as a Result of Senescence

Cotyledon age	State 3 respiration + succinate (nmoles O_2 min^{-1} $cotyledon^{-1}$)	Relative activity			
		RCR	ADP/O	Cytochrome oxidase	SDH
1 day	118	2.7	1.75	100	100
5 days	32	1	nr	102	38

[a]The cytochrome oxidase and SDH activities are those associated with the inner membrane. Note that pea cotyledons show signs of senescence 5 days after the start of imbibition. nr, No response to added ADP. From data in Nakayama *et al.* (1978).

occurs after germination. The increase in ATP in the seed is accompanied by an increase in the total pool of adenine nucleotides, which suggests an early requirement for their *de novo* synthesis. The ATP pool during the lag phase is not static, for if ATP synthesis is inhibited, e.g., by placing seeds in an atmosphere of N_2 to prevent terminal oxidation in the mitochondria, then the ATP pool is quickly used up (Fig. 4.19B). The ATP pool returns rapidly to control amounts on reintroduction of the seeds to air. Thus, during the lag phase, although there is no net increase in ATP, the rate of its synthesis is balanced exactly by its rate of utilization.

4.4.5. The Synthesis and Utilization of Reducing Power: Pyridine Nucleotides

Reduced pyridine nucleotides (NADH and NADPH) are essential coenzymes for several key metabolic pathways and hence are required by the germinating seed, the developing seedling, and the storage organs. They participate, for example, in glucose respiration, amination reactions, deoxynucleotide synthesis, and fatty acid catabolism. Metabolic pathways might be controlled by the availability of these coenzymes or by their rates of reoxidation within the cell.

The oxidized pyridine nucleotide NADP is a coenzyme of glucose–6-phosphate dehydrogenase (Glc-6-Pdh), which is a key link between the glycolysis pathway and the pentose phosphate pathway (PPP). Glycolysis requires a supply of NAD for the intermediate step involving glyceraldehyde–3-phosphate dehydrogenase (Fig. 4.20). When NADP is in limited supply, glycolysis proceeds at the expense of the PPP, whereas under conditions of limiting NAD, the reverse is the case. The relative activities of the two pathways have been studied in only a few seeds, and more usually in the cotyledons than in the axis; nevertheless, the following pattern of events emerges.

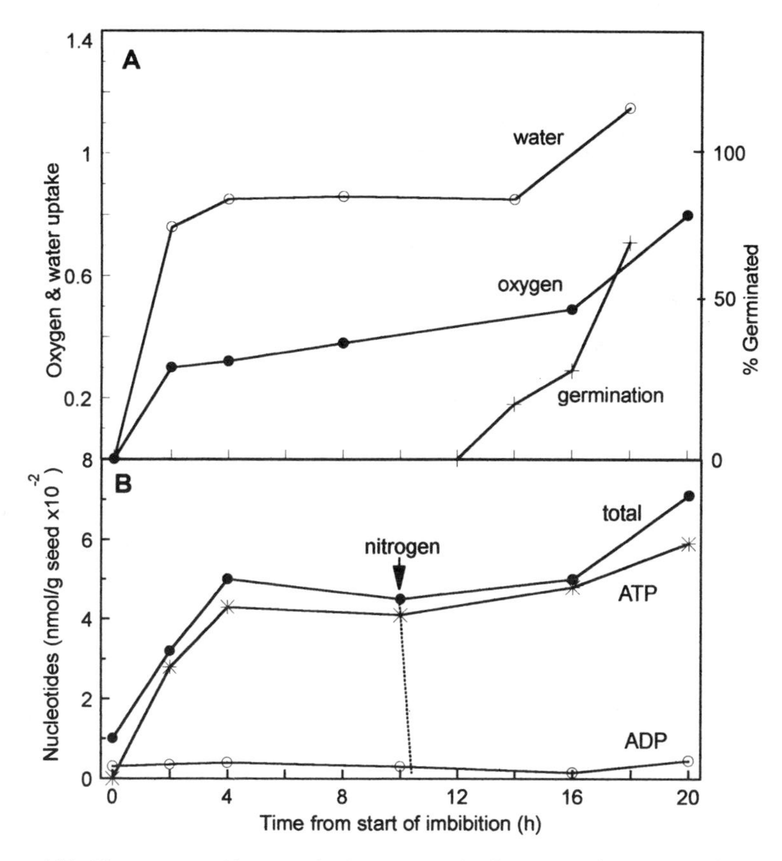

Figure 4.19. Time courses of increases in (A) water uptake, O_2 consumption, and germination and in (B) total adenine nucleotides, ADP, and ATP in lettuce seeds. (A) O, water uptake in g/g seed; ●, O_2 uptake in μmol/min; +, percent germinated. (B) O, ADP; *, ATP; ●, total adenine nucleotides. The arrow indicates time of transfer of the seeds to an atmosphere of nitrogen, and the dotted line the immediate and rapid decline in ATP (to zero within 6 min). Based on data in Pradet *et al.* (1968) and Hourmant and Pradet (1981).

During the first few hours to several days (depending on the species) after the start of imbibition, it appears that the glycolytic pathway predominates, favoring pyruvate formation and some synthesis of ATP (Fig. 4.20). Pyruvate is converted to ethanol under the partially anaerobic conditions that tend to prevail in intact seeds prior to emergence of the radicle. By channeling respirable substrate through the glycolytic pathway under these conditions the cells can to some extent offset the reduced ATP production by mitochondria resulting from natural anaerobiosis due to limited O_2 availability, or from their inherent struc-

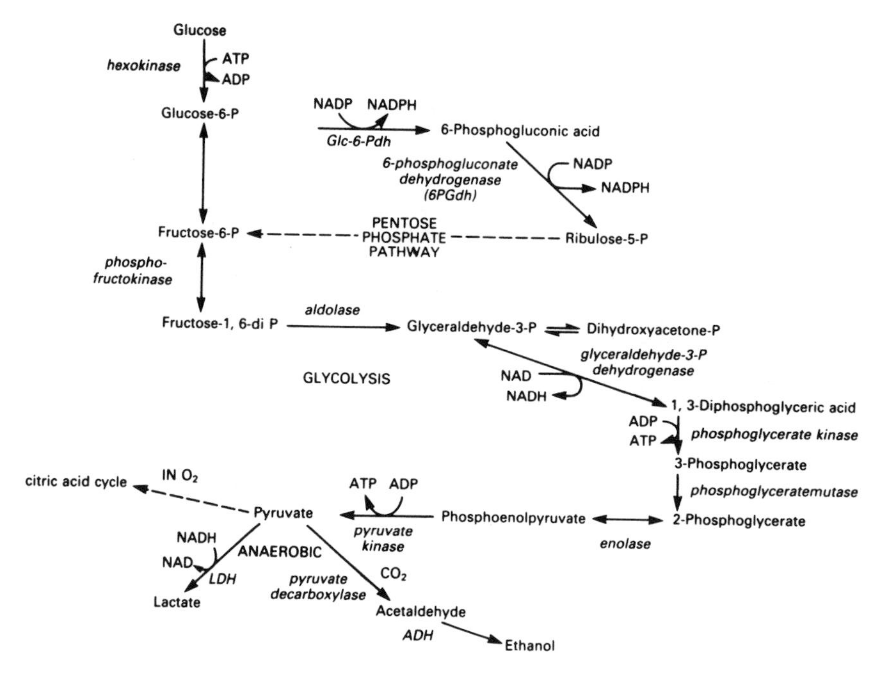

Figure 4.20. Some important steps in glycolysis and the pentose phosphate pathway (PPP) in relation to NADH, NADPH, and ATP production. Enzymes are in italics.

tural or enzymatic deficiencies. Later, the PPP predominates, and as mitochondria become more active and O_2 more available, the seeds enter a period of energy production for increased metabolic activity associated with growth and reserve mobilization, with NADPH production increasing to provide the necessary reducing power. Some indication of the type of changes in status of the glycolytic and PPP pathways is illustrated, using cotton (*Gossypium hirsutum*) as the example, in Fig. 4.21. Aldolase and phosphofructokinase, both glycolytic pathway enzymes, decline in activity in the radicles and to a lesser extent in the cotyledons, during the first 1–3 days after sowing. Glc-6-Pdh, a PPP enzyme, increases in activity after imbibition, particularly in the growing regions. Although activities of extracted enzymes assayed *in vitro* are not always an accurate measure of their *in vivo* activity, the illustrated changes are at least consistent with the generally accepted view that carbohydrate catabolism via the glycolytic pathway precedes that via the PPP. Incidentally, the rise in aldolase activity in the cotyledons after 3 days (Fig. 4.21) could be related to the onset of gluconeo-

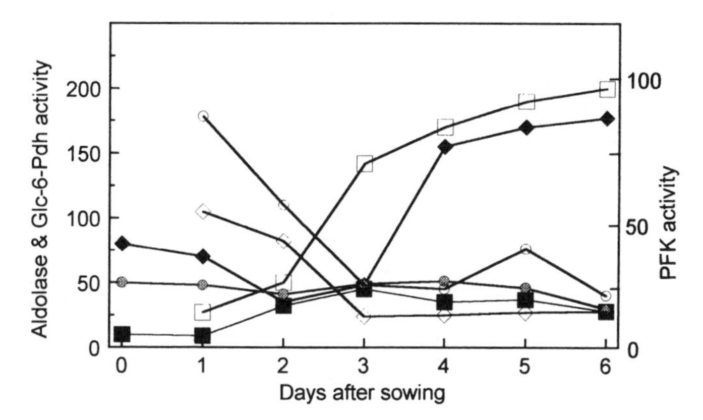

Figure 4.21. Changes in activities (expressed in μmol product/mg protein per min) of the PPP enzyme glucose–6-phosphate dehydrogenase (■,□) and the glycolysis enzymes phosphofructokinase (●,O) and aldolase (♦,◊) in radicles (open symbols) and cotyledons (closed symbols) of cotton seeds. Taken from data in Purvis and Fites (1979).

genesis associated with oil catabolism, which requires the operation of most of the glycolytic pathway in reverse (Fig. 7.9), but not that part involving phosphofructokinase.

4.5. RESPIRATION UNDER ANAEROBIC CONDITIONS

As outlined in Section 4.3.2, many seeds experience conditions of temporary anaerobiosis during lag phase II of respiration (Fig. 4.12). Consequently, both ethanol and lactic acid, products of anaerobic respiration, accumulate within the seed; they are found in different proportions in different species. On penetration of the enclosing structures by the radicle, these anaerobic products decline as they are metabolized under conditions of increasing aerobiosis. Lactate dehydrogenase (LDH) and alcohol dehydrogenase (ADH) are responsible for both the synthesis and the removal of lactate and ethanol, respectively. Under aerobic conditions, the former enzyme converts lactate back to pyruvate, which can then be utilized by the citric acid cycle, and the latter converts ethanol to acetaldehyde, which is oxidized to acetate by acetaldehyde dehydrogenase. The acetate is activated to acetyl-CoA, which can be used in many metabolic processes. Not surprisingly, then, ADH and LDH tend to be present in seeds during germination, and often they are even present in the dry seed, e.g., ADH in pea cotyledons (Fig. 4.22). In some species of seeds the activity of LDH and ADH increases appreciably (tenfold or more) during germination, perhaps because of *de novo* synthesis. After germination has been completed, and when conditions are more aerobic, ADH (Fig. 4.22) and LDH become negligible: loss of both ethanol and lactate from the seed parallels this decline.

This period of natural anaerobiosis in seeds during germination can last from a few hours to several days. Anaerobic conditions can be prolonged by sowing seeds under water. Many species will actually complete germination when submerged, although subsequent growth of the radicle is stunted, and if the germinated seeds are maintained in water they will die. Seeds of certain aquatic species germinate better under conditions of reduced O_2 tension, e.g., *Typha latifolia,* and some, e.g., *Juncus effusus,* will germinate readily even after submergence for up to 7 years. Grains of aquatic species such as rice (*Oryza sativa*) and barnyard grass (*Echinochloa phyllopogon*), a noxious weed in rice fields, germinate and grow under water and show some interesting adaptations to these conditions of reduced oxygen availability. Both rice and barnyard grass will germinate in a totally O_2-free environment, although only the coleoptile elongates; root growth is inhibited, as is emergence of the leaves from the coleoptile (Fig. 4.23). The growth rate of *Echinochloa* under anoxic conditions is about 25% of that in air. The seedlings of both species produce considerable quantities of ethanol (shown for barnyard grass in Fig. 4.24A), much of which (up to 95% in rice) diffuses into the surrounding water. Even so, under anoxic conditions the seedlings may contain close to 100 times more ethanol than the aerated

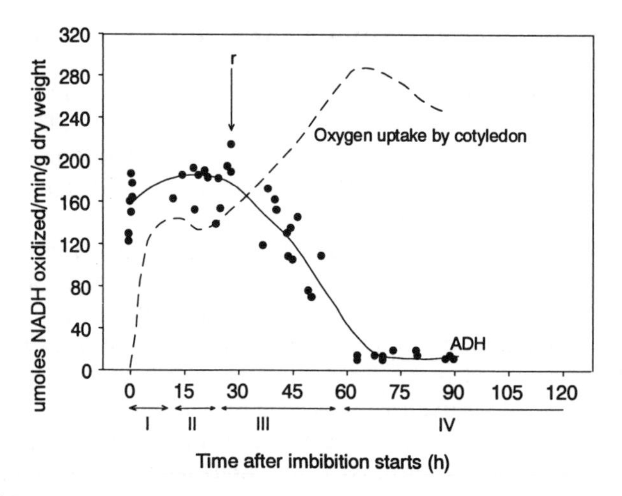

Figure 4.22. Alcohol dehydrogenase activity in, and oxygen uptake by, cotyledons of dark-germinated intact pea seeds cv. Rondo. Arrow (r) marks the time of penetration of the testa by the radicle. After Kollöffel (1968).

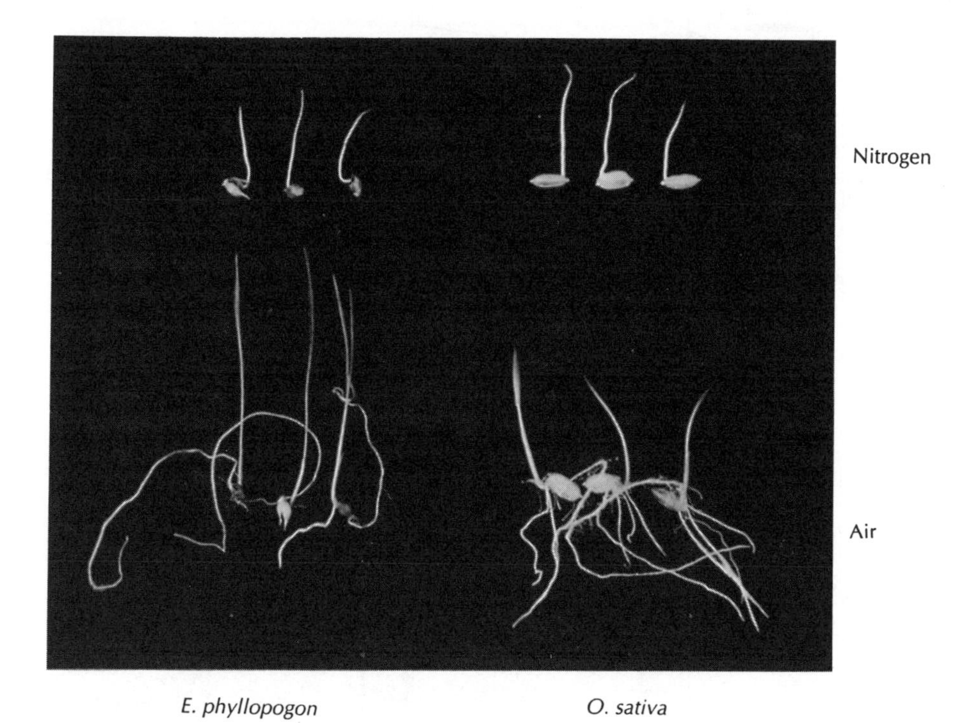

Figure 4.23. Seven-day-old seedlings of (left) barnyard grass (*Echinochloa phyllopogon)* and (right) rice (*Oryza sativa*) imbibed in N_2 (top row) or air (bottom row) since sowing. Note differences in root morphology and extent of leaf emergence from the coleoptile. Courtesy of R. A. Kennedy.

controls. ADH activity rises along with ethanol production (Fig. 4.24B). Increases in lactate occur also, but they are very small in comparison with ethanol. Mitochondrial development in both rice and barnyard grass seedlings under anoxic conditions is not very different from that in fully aerated seedlings. Both exhibit substantial increases in many of the citric acid cycle enzymes under anoxia, for example. Although oxidative phosphorylation obviously does not occur in grains imbibed in nitrogen, the mitochondria develop and maintain their capacity for ATP production in the absence of O_2. On return of seedlings to atmospheres containing O_2, oxidative phosphorylation commences almost immediately. Synthesis of ATP takes place in the absence of O_2 (Fig. 4.25) from the earliest stages of imbibition,

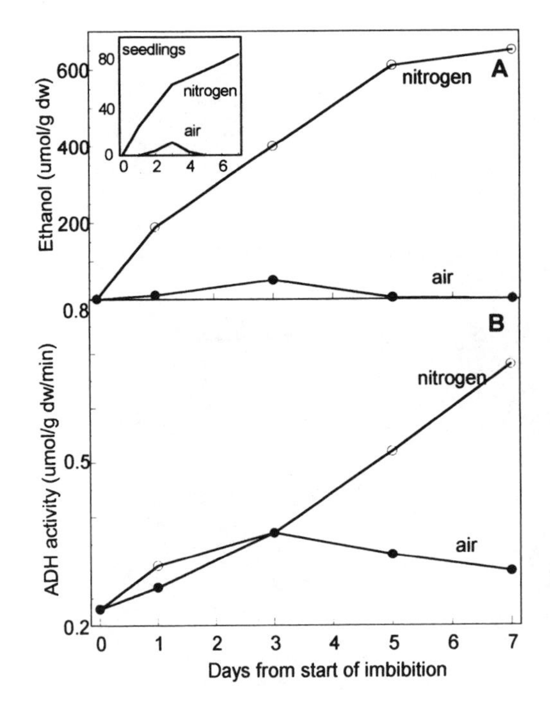

Figure 4.24. (A) Ethanol content of grains of *Echinochloa phyllopogon* imbibed in atmospheres of N_2 or air. The main graph is the sum of the ethanol content of the grains and the surrounding liquid, and the inset is the content of the seedlings alone. (B) Alcohol dehydrogenase activity. Germination was completed (i.e., coleoptile elongation commenced) by the third day from the start of imbibition. After Rumpho and Kennedy (1981).

probably by operation of the glycolytic pathway, terminating in the conversion of pyruvate to ethanol (Fig. 4.20).

4.6. PROTEIN SYNTHESIS DURING GERMINATION AND ITS DEPENDENCE ON mRNA SYNTHESIS

The transition from development to germination necessitates fundamental changes in the control of gene expression within a seed. At some stage the expression of genes coding for reserve proteins, and for enzymes involved in the synthesis of stored material and other activities, has to be switched off, whereas genes coding for enzymes involved in germination, the initiation of axial growth, and subsequent reserve mobilization must be switched on. Concomitant with germination, there must be met the increasing demand for proteins essential to this event and later the synthesis of proteins for seedling growth. In this section we will concentrate on studies using germinating tissues, viz., isolated embryos of cereals or axes of legumes and other dicots. In Section 4.6.4 we will consider changes occurring within the seed storage tissues.

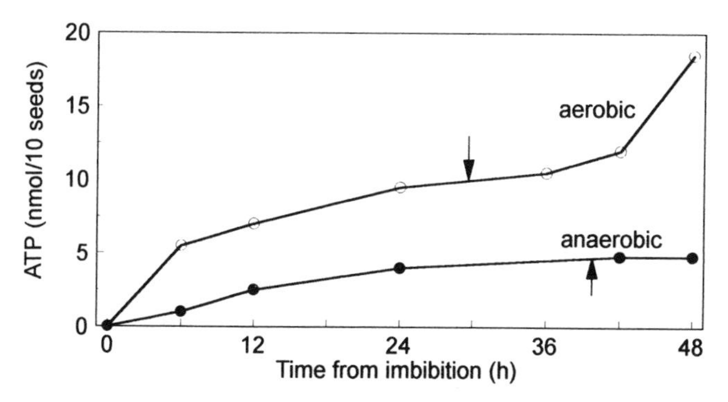

Figure 4.25. ATP content of rice during germination and growth under aerobic (○) and anaerobic (●) conditions. Arrows mark the time of coleoptile expansion (i.e., completion of germination) for each treatment. After Pradet and Prat (1976).

4.6.1. Protein Synthesis in Germinating Embryos and Axes

Polysomes are absent from dry seeds, but they rapidly increase as soon as the cells become hydrated. Accompanying this increase in polysomes is a decline in the number of single (unattached or free) ribosomes as these become recruited into the protein-synthesizing complex. An example of early polysome formation is found in the wheat embryo (Fig. 4.26). Here the extracted ribosomal pellet from the embryo of the mature dry grain is incapable of catalyzing *in vitro* protein synthesis in the absence of added mRNA, because there are no polysomes

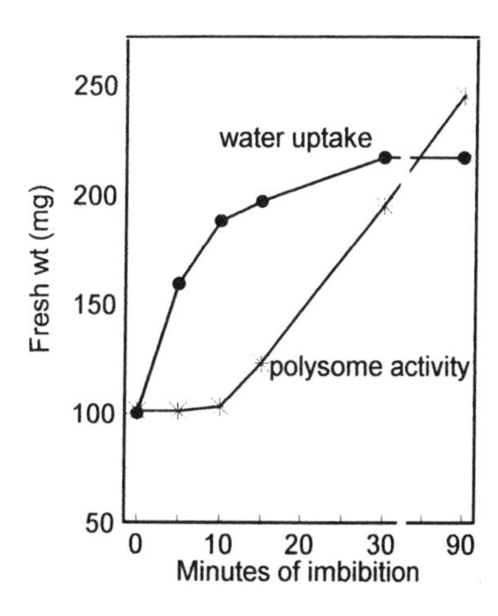

Figure 4.26. Comparison of water uptake and development of polysomal activity assayed *in vitro* during the early stages of imbibition by isolated wheat embryos. After Marcus *et al.* (1966).

present. But after a lag period of only 10–15 min, during which the embryo rapidly takes up water, the protein-synthesizing capacity of the extracted pellet is evident. Thereafter its activity increases as more and more polysomes are formed.

As discussed in Chapter 2, considerable protein synthesis occurs during seed development (see Figs. 2.22 and 2.24 for details on the mechanism of protein synthesis), but this ceases as desiccation proceeds during the final stages of maturation. It is pertinent to ask, therefore, whether some or all of the components involved in protein synthesis during development are destroyed as a consequence of desiccation, or whether they are conserved in the dry seed and then reutilized on subsequent rehydration. A necessary ingredient of many *in vitro* protein-synthesizing systems is a supernatant prepared from mature dry wheat embryos or wheat germ. From this supernatant have been isolated a number of the components including ribosomes, transfer RNA (tRNA), initiation and elongation factors, amino acids, and aminoacyl tRNA synthetases. Many of these components have been extracted from dry embryos, axes, and storage tissues of other species too, and hence there can be no doubt that they are present in a potentially active form in the cells of dry seeds. Moreover, they appear to be present in sufficient quantities for protein synthesis to commence immediately when the seeds are hydrated.

More heatedly debated has been the question as to whether protein synthesis can commence in the imbibing seed without the necessity for any prior RNA synthesis. Although it has been accepted that active ribosomes (containing rRNA) and tRNA are present in the dry seed, the status of translatable mRNA has taken longer to resolve. Isolated wheat embryos synthesize mRNA very soon after they start to imbibe water, and within 30–90 min newly synthesized mRNA becomes incorporated into polysomes. However, this new mRNA synthesis might not be required for the resumption of protein synthesis on imbibition, for during the first hour polysomes are still formed when over 90% of the initial RNA synthesis, including mRNA, is prevented by chemical inhibitors such as cordycepin (Fig. 4.27). Since imbibition of wheat embryos in effective concentrations of this inhibitor does not prevent rapid polysome formation, it is likely that mRNA is conserved in the dry embryo in a suitable condition for the support of early protein synthesis. Similar conclusions have been drawn from studies on embryos and axes from other species of seeds. In wheat embryos, mRNA-processing events (e.g., capping and internal methylation) do not appear to be required for conserved (or newly synthesized) messages to be recruited into polysomes.

Although protein synthesis may proceed in the absence of RNA synthesis during early imbibition, it eventually comes to depend on it—before germination is completed. For example, protein synthesis in radish embryonic axes is insensitive to cordycepin for the first 2–3 h from the start of imbibition (Fig. 4.28A),

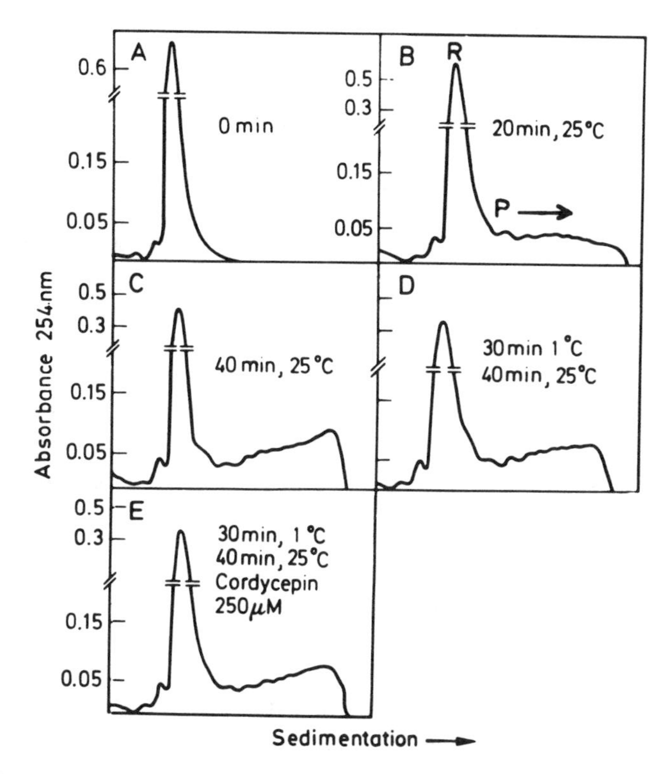

Figure 4.27. Polysome formation during early germination of wheat embryos and the effect of RNA synthesis inhibitors. Isolated embryos were imbibed in water at 25°C for (A) 0 min; (B) 20 min; (C) 40 min; (D) 1°C 30 min, then 40 min at 25°C, both in water; or (E) 1°C 30 min, then 40 min at 25°C, both in cordycepin. Cordycepin inhibited uridine incorporation into RNA by 82%. R, area of single ribosomes; P, area of polysomes. After Spiegel and Marcus (1975).

indicating that new RNA synthesis (presumably including mRNA synthesis) is not required during this period. Poly(A)-containing RNA declines in the embryos germinating on cordycepin during the first few hours, although the total poly(A) content of control embryos increases after an initial decline (Fig. 4.28B). Thus, it appears that there is a loss of conserved mRNA during the first few hours of germination and an increase in new messages, with protein synthesis becoming increasingly dependent on the latter with time of germination.

In pea axes, the change in mRNA status with time is reflected by a qualitative change in the types of proteins being synthesized (Fig. 4.29). Here the *in vitro* translation products of mRNAs extracted from dry pea axes (Fig. 4.29A) and from early imbibed axes (Fig. 4.29B) are quite similar, both quanti-

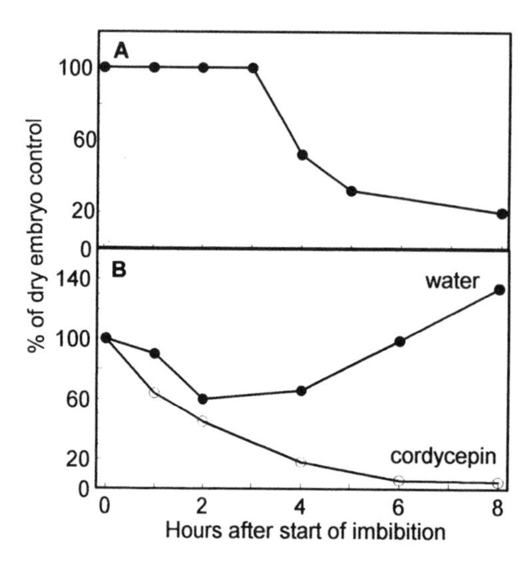

Figure 4.28. (A) Decline in the amount of *in vivo* protein synthesis coded for by mRNA present in the dry embryonic axes of radish (*Raphanus sativus*). New RNA synthesis during the 8-h experimental time period was prevented by cordycepin. (B) Poly(A)-containing RNA (mRNA) content of radish embryos during germination in the presence (○) or absence (●) of cordycepin. After Delseny *et al.* (1977).

tatively and qualitatively. At later times (Fig. 4.29C–F), up to the completion of germination at 16 h, the types of proteins determined by the extracted messages are considerably different from those at earlier times. In particular, many of the messages present in the dry axes are not replaced as germination proceeds. Several studies have now shown that embryos of wheat and other species also exhibit qualitative changes in their mRNA complement within a few hours of the start of imbibition.

There are probably two classes of mRNA present within the dry embryo or axis: (1) Residual mRNAs—these are mRNAs produced during seed development and which are not destroyed during late maturation and desiccation. They are not essential for germination and may be degraded early after imbibition. The early ribosomal protein messages in wheat might fall into this category. Another example is the Em protein (Section 3.3.2). (2) Stored or conserved mRNAs—these are laid down during development so that they are available immediately on hydration and hence can quickly be translated into proteins that are an integral part of germination. These may be subcategorized into mRNAs for (a) enzymes essential for intermediary metabolism, i.e., for essential cellular processes not necessarily unique to germination, and (b) proteins essential for the successful completion of the germination process *per se*, culminating in the elongation of the radicle.

But how, then, do mRNAs come to reside within the dry seed, and how do the cells of the imbibed seed discriminate between mRNAs to be utilized in germination and those to be destroyed? The answers to these questions are by no means clear. One possible explanation is that mRNAs

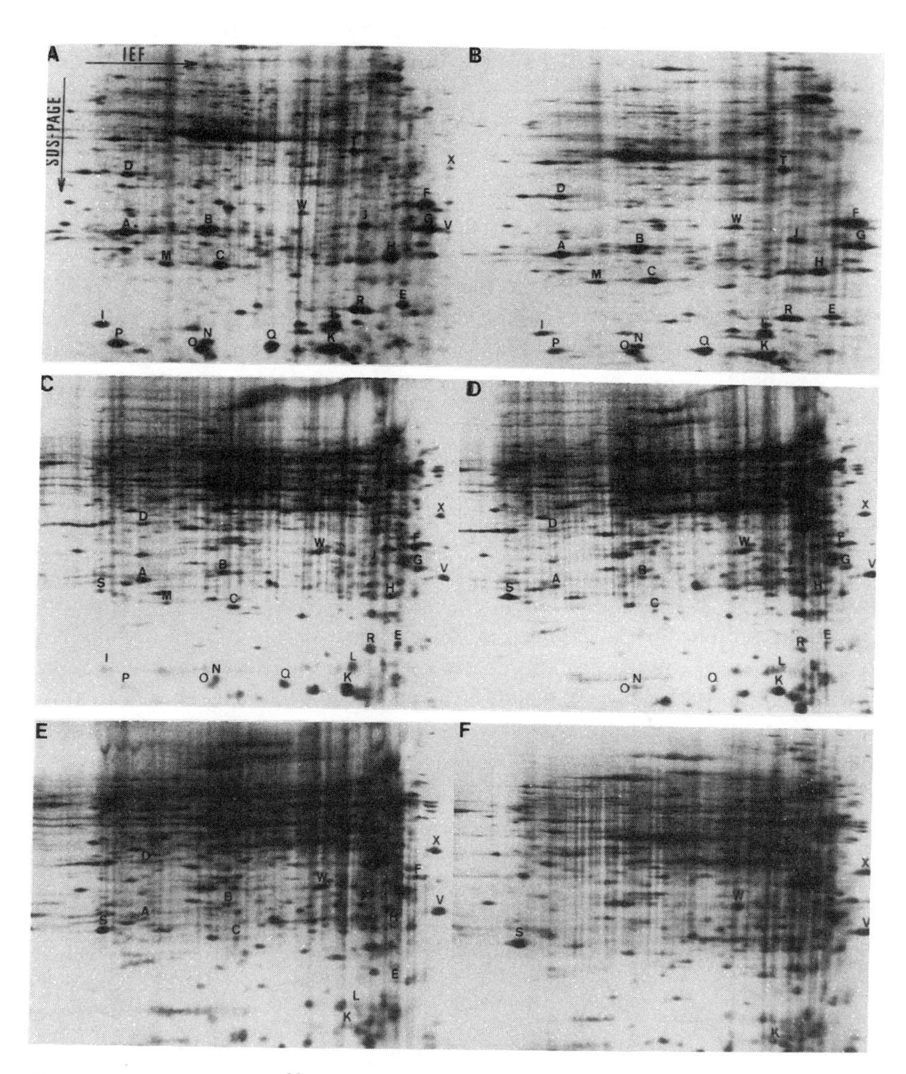

Figure 4.29. Fluorographs of ^{35}S-methionine-labeled *in vitro* translation products of total RNA from pea axes separated by two-dimensional gel electrophoresis. The RNA preparations, which include mRNAs, were obtained from (A) dry axes, and axes imbibed for (B) 30 min, (C) 4 h, (D) 8 h, (E) 12 h, and (F) 16 h. Note the changes in the translation products with time of imbibition, and particularly the decline in those directed by mRNAs present in the dry axes. The mRNAs for proteins S, X, and V increase with time. From Lalonde and Bewley (1986). Reproduced by permission of Oxford University Press.

to be conserved in the dry seed are protected by being stored in the nucleus during maturation, either in their activated (polyadenylated) state or ready to be activated, although the former appears to be the more likely. These stored mRNAs are then released from the nucleus to the protein-synthesizing sites within the cytoplasm when required. There is some evidence that in cotton embryos and rapeseed the subcellular location of polyadenylated messages is within the nucleus. On the other hand, the accumulation of mRNAs in cytoplasmic informosomes [called messenger ribonucleoproteins (mRNPs), which describes more accurately the status of these mRNAs as being associated with proteins and hence being surrounded and protected by them] during development of wheat and rye embryos has been advocated. For example, in dry rye embryos most of the mRNA is in the form of cytoplasmic mRNPs ranging in size from 25 to 104 S. When the protein moiety is removed, the mRNAs are revealed to have sizes from about 7 to 35 S, with the majority being between 15 S and 25 S. In dry wheat embryos, the mRNPs and mRNAs may be considerably smaller.

What are the genes whose expression is initiated after seed rehydration and what are the functions of their protein products? In broad terms they are likely to fall into two groups. The first includes those genes which code for the enzymes and other proteins that are needed for all of the basic activities of living plant cells, for example in respiration, protein and nucleic acid synthesis, membrane synthesis, amino acid metabolism, etc.—so called "housekeeping" proteins. Second, there are those which may be involved specifically in germination processes. Of course, this begs the question: what are these specific processes? None as yet has been identified with any degree of confidence though a few have been proposed. Some metabolic processes have been suggested to be typical of germinating as opposed to nongerminating (dormant) seeds and, therefore, the former might possess distinctive sets of the appropriate enzymes. Also, since germination culminates in the initiation of cell elongation (e.g., in radicle emergence) followed by cell division we might suspect that enzymes which participate therein are included in the products of the specifically expressed genes.

The characterization of the novel transcripts and proteins that appear during germination nevertheless has been very limited. But a protein which has been investigated in depth is one whose formation begins just before radicle elongation in wheat embryos and continues throughout seedling growth—the protein germin. This is a water-soluble homopentameric glycoprotein (approximately 25 kDa) with some unusual properties, being resistant to digestion by pepsin and some other proteolytic enzymes, and remaining intact under certain conditions which normally cause proteins to dissociate. Germin is encoded by a family of genes (about seven copies) of known chromosomal location. Full-length cDNA clones and genomic clones of germin yield sequence data showing that the protein has close homology to an oxalate oxidase isolated from barley. Enzymological and immunochemical methods confirm that germin is, indeed, an oxalate oxidase.

The role that such an enzyme might play in germination and the initiation and maintenance of radicle elongation is not clear, but some highly interesting possibilities have been proposed. In plants, oxalate normally occurs as calcium oxalate (it is present in wheat embryos, for example) which on degradation by the oxidase yields calcium ions and hydrogen peroxide—two substances of potential physiological importance. Calcium can function as a second messenger in several regulatory systems, while peroxide, together with peroxidase, is thought to participate in cell wall metabolism. That a high proportion of germin in wheat embryos is associated with cell walls is consistent with its possible action at, or close to, this site. Since cell extension terminates the germination process, it is a highly attractive hypothesis that an enzyme whose action can lead to biochemical modification of cell walls might be a candidate for a "germination enzyme". Germin is known in several cereals (barley, rice, maize, oats) but its widespread occurrence and significance in germination remain to be established.

To summarize this section, then, most of the essential components of the protein-synthesizing complex are conserved in the dry seed including the vital templates for synthesis, the mRNAs. These mRNAs are in a protected form in the dry seed, perhaps in the nucleus or in the cytoplasm associated with proteins as mRNP particles. Unprotected messages are presumably hydrolyzed during late maturation or early germination. Protein synthesis begins within minutes of hydration of the seed, and some of the conserved mRNA is utilized to direct this. There is an excess of conserved mRNA in relation to the early requirements of the cell, but the mRNAs that are not translated might not be qualitatively different from those that are. Within a few hours there is synthesis of new mRNAs, and as the conserved ones are degraded, protein synthesis, leading to the completion of germination, probably becomes increasingly dependent on them. Some of the newly synthesized mRNAs may code for the same proteins as the conserved messages, whereas others are for distinctly different products, perhaps proteins that are essential for the commencement of radicle elongation. Such proteins remain to be identified, however.

4.6.2. Ribosomal and Transfer RNA Synthesis

Synthesis of rRNA in imbibing wheat embryos has been detected as early as that of mRNA and within the first 1–2 h of imbibition. The new rRNA becomes incorporated into ribosomes soon after its synthesis. Maturation of rRNA in rye embryos also occurs quickly after imbibition, with the following sequence of events taking place:

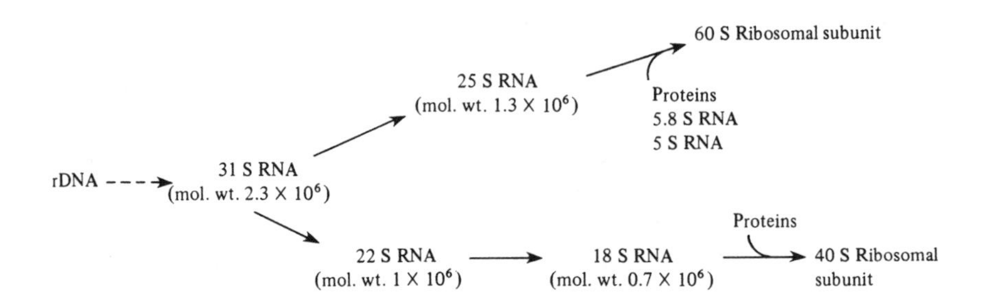

After 40 min from the start of imbibition the 31 S rRNA precursor is evident, and by 60 min it has started to mature into the 25 S and 18 S rRNA of the larger and smaller subunits, respectively. Between the third and sixth hours from the start of imbibition, processing of rRNA is virtually complete and large amounts of the mature rRNAs are present in the embryo. The large 31 S rRNA precursor may accumulate in the nucleolus (the site of the rRNA genes) prior to its migration into the cytoplasm: such appears to be the case in pea embryos. In rye, it is not known when the newly synthesized rRNAs become associated with ribosomes actively participating in protein synthesis. On the basis of studies on other embryos and axes, however, there is little reason to expect that its incorporation into ribosomes is delayed. As yet, it is not known in any seed when and how the low-molecular-weight RNA components (5.8 and 5 S) of the large ribosomal subunit are synthesized.

Amplification of rRNA genes occurs during certain developmental processes in animals, but current evidence is against such an event occurring during seed germination and seedling establishment.

Transfer RNAs and their aminoacylating enzymes are present in dry seeds, and it has been surmised that there is sufficient of each present to support protein synthesis during germination. Even so, tRNA synthesis commences soon after imbibition, e.g., within 20 min in rye embryos and within 1 h in dissected axes of *Phaseolus vulgaris*. It is also possible that a proportion of the tRNAs present in dry seeds are defective at the -CCA end (to which the amino acid becomes attached), and these molecules must be repaired before they can be used in protein synthesis. The enzyme responsible for the repair of tRNA, nucleotidyltransferase, is present in the dry axes of lupin (*Lupinus luteus*) and increases in activity after imbibition. In wheat, however, the low activity present in the dry embryo does not increase until germination is completed.

4.6.3. Enzymes and Precursors of RNA Synthesis

Synthesis of RNA requires the involvement of three RNA polymerases: (1) RNA polymerase I, a nuclear polymerase that transcribes genes coding for 18 and 25 S rRNA; (2) RNA polymerase II, localized within the nucleoplasm and responsible for transcription of mRNA; and (3) RNA polymerase III, found within the nucleoplasm (and possibly the cytoplasm), which transcribes genes coding for tRNA and 5 S RNA. Very little is known about these polymerases in seeds. Dry wheat germ, rye embryos, and soybean axes contain DNA-dependent RNA polymerases, and presumably these enzymes are present in sufficient quantities to catalyze RNA synthesis as soon as is necessary on imbibition. Polymerase activity may increase during germination, but the largest increase probably occurs after germination is completed. This, at least, appears to be the case with RNA polymerases in soybean axes.

The initiation and continuation of RNA synthesis are dependent on a ready supply of ribonucleoside triphosphate precursors. ATP, CTP, GTP, and UTP are present in very low quantities in the dry wheat embryo and all four increase rapidly during the early stages of germination. The rise in the pyrimidine nucleoside triphosphates (UTP and CTP) is slower, however, although they increase substantially by 3 and 5.5 h after the start of imbibition, respectively. Hence, the rate of RNA synthesis at these later times might be controlled by the supply of pyrimidine nucleoside triphosphates alone. But, in truth, the nature of the quantitative and qualitative control mechanisms governing RNA synthesis in germinating seeds is unknown. Many levels of control are possible, e.g., chromatin template availability, polymerase activity, availability of substrates, processing of RNA (particularly mRNA and rRNA), but at the moment there is no substantive evidence for the operation of any of them during germination.

4.6.4. Protein and RNA Synthesis in Storage Tissues

A major event in any storage tissue is the mobilization of stored reserves, which requires the participation of many enzymes, a substantial number of which are synthesized *de novo*. The synthesis of specific hydrolytic enzymes will not be dealt with here, but in Chapters 7 and 8. Instead, we will concern ourselves briefly with the establishment of the protein-synthesizing complex in storage tissues during and after imbibition. Since the cells of the starchy endosperm of cereals are dead at maturity, they are incapable of synthetic processes and hence will receive no attention here; changes in the aleurone layer are considered in Chapters 7 and 8 also.

Total RNA in imbibed storage tissues may increase over several days before eventually declining, e.g., in peanut and cucumber cotyledons and in castor bean endosperms. In other tissues, e.g., pea and broad bean (*Vicia faba*) cotyledons, however, RNA remains constant until there is a decline associated with organ senescence. The RNA content of the megagametophyte of red pine (*Pinus resinosa*) changes little over 2 weeks after initial imbibition, even though radicle emergence occurs on day 5, the cotyledons expand on day 10, and by day 14 the megagametophyte and seed coat are attached to the end of the expanding cotyledons and are about to be shed.

Not all regions of a storage organ necessarily exhibit the same pattern of RNA metabolism. Nuclear RNA appears to increase in the outer storage tissues of field pea (*Pisum arvense*) cotyledons during the first 5 days after imbibition and then declines markedly before any changes occur in the nuclei of cells in the inner storage region. Although RNA synthesis has been detected in cotyledons of rapeseed as early as 2 h from the start of imbibition, it is confined to localized regions. More regions commence RNA synthesis with time. Hence, measurements of RNA synthesis (and also protein synthesis) made on whole storage organs must be regarded as an aggregate of changes occurring at different locations. From the start of imbibition, the increase in RNA synthesis in whole organs might, therefore, reflect an increase in the number of cells commencing synthesis (at least initially) rather than an increased synthetic capacity of individual cells.

Mature dry storage tissues contain ribosomes that can catalyze protein synthesis *in vitro*, and it is safe to assume that other components of the protein-synthesizing complex are also present. Certainly, dry cotyledons of several species contain all of the tRNA species and aminoacyl tRNA synthetases necessary for protein synthesis to commence. As in the axis, protein synthesis commences quickly on imbibition, as indicated by rapid polysome formation and *in vivo* incorporation of radioactive amino acids. Messenger RNAs are present in dry storage tissues of seeds and may become involved in, or be essential for, early protein synthesis, in much the same way as in axes. Many of the enzymes involved in reserve mobilization and utilization are synthesized from mRNAs newly transcribed following germination (Chapters 7 and 8).

4.7. DNA SYNTHESIS AND CELL DIVISION

Expansion of the radicle within the seed occurs initially by cell elongation, and its subsequent emergence through the seed coat may or may not be accompanied by cell division. Hence, cell division is largely a postgermination phenomenon, concerned with axis growth and establishment of the seedling. It may be initiated close to the time of axial elongation, however, and in conifer seeds, it seems to precede radicle protrusion through the seed coat.

The relationship of the status of DNA to the cell cycle is illustrated in Fig. 4.30. After mitosis (M), when chromosomes are in their two-stranded (2C) configuration, there is a period of normal cell growth (Gap 1 or G_1) during which synthetic events, including those for subsequent DNA synthesis, take place. DNA synthesis (S) results in a doubling of the chromosomes to a four-stranded (4C) configuration without cell division, and a second preparative growth period (G_2) occurs before mitosis. The reader will appreciate that if the G_2 phase is long, then so will be the lag between DNA synthesis and cell division; some cells, in fact, synthesize DNA but never divide, whereas in others there may be a pause of many hours between synthesis and division.

If all of the nuclei of a dry embryo contained chromosomes in the 4C state, then DNA synthesis would not be a prerequisite for the first cell division. To date, however, there has been no recorded instance of all nuclei being exclusively 4C, although some species contain both 2C and 4C and others only 2C (Table 4.3). Within a particular species the proportions of 2C and 4C nuclei may vary between parts of the seed. The significance of nuclei in the dry seed being at different stages of the cell cycle is unclear. But in those species whose nuclei are exclusively in the G_1 phase there must presumably be some stringent control of nuclear events to ensure that all nuclei enter the final developmental and drying phases of maturation at the same stages of their cell cycle.

During mitosis in the radicles of germinated onion (*Allium cepa*) seeds the nuclei first enlarge from a diameter of about 7 μm to more than double this size. DNA synthesis commences prior to mitotic cell division, for the former occurs when the root length is 1.4 mm (and the nuclei are 15 μm in diameter), but the latter is not evident until the roots are at least 2.8 mm long. Between DNA

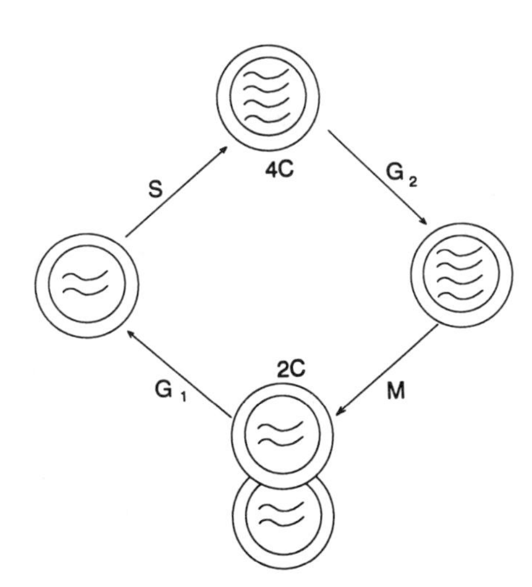

Figure 4.30. Simplified diagram to illustrate the states of DNA during the cell cycle. For an explanation of terms, see text.

Table 4.3. Some Species of Seeds Which in the Mature Dry State Have Embryo or Radicle Nuclei in the 2C, or 2C and 4C, Stage of the Cell Cycle[a]

2C only	2C and 4C
Allium cepa	*Crepis capillaris*
Dactylis glomerata	*Hordeum vulgare*
Festuca arundinacea	*Pisum sativum*
Lactuca sativa	*Triticum durum*
Pinus pinea	*Vicia faba*
Tradescantia paludosa	*Zea mays*
Cichorium endiva	*Phaseolus vulgaris*
Fagus sylvatica	*Spinacia oleracea*

[a]Data from Conger and Carabia (1978) and Bino *et al.* (1993) and references therein.

synthesis and mitosis (S to G_2 phase) there is an interlude of approximately 9 h. Similar observations have been made on other seeds.

Two distinct phases of DNA synthesis may occur during germination. During the first few hours after imbibition of maize embryos there is incorporation of thymidine into DNA, but no increase in the amount of DNA. Later, following germination, there is a second phase of thymidine incorporation into DNA, this time accompanied by an increase in total DNA (Fig. 4.31A). The first increase in DNA synthesis has been attributed to repair of DNA, to restitute damage resulting from maturation drying and subsequent storage in the dry state. But synthesis of mitochondrial DNA, a minor component of the total DNA within the cell though a large contributor to the cytoplasmic pool of DNA, probably accounts for most of the observed thymidine incorporation up to 12 h. DNA synthesis associated with subsequent nuclear division accounts for the second phase, which includes a net increase in this nucleic acid. Interestingly, synthesis of histones, the proteins associated with DNA in the nucleus, increases many hours before nuclear DNA synthesis (Fig. 4.31B). The reason for this is unknown; nor is it known in seeds if the presence of specific histone variants is related to expression of sets of development- or germination-related genes.

In dicot storage tissues, which do not undergo cell division, the amount of DNA may or may not change after imbibition. In some, e.g., field pea (*Pisum arvense*), DNA in the cotyledons remains constant during and after germination of the axis and then declines as senescence sets in. In others, e.g., peanut cotyledons, DNA may double in amount up to the seventh to tenth day after imbibition and then decline again over several days. Part of the increase might be related to synthesis of mitochondrial or plastid DNA, although this probably does not account for more than 1–2% of the total.

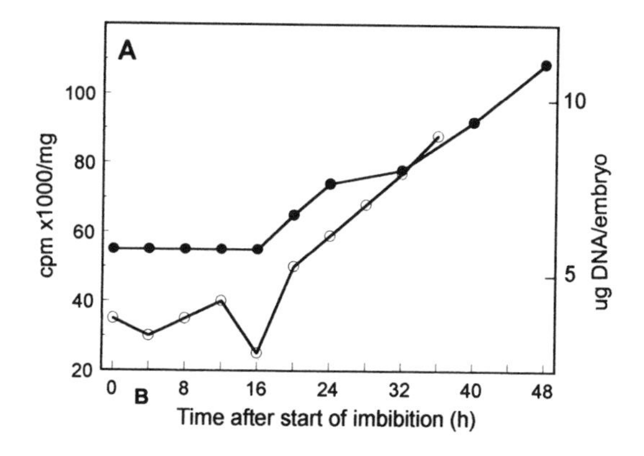

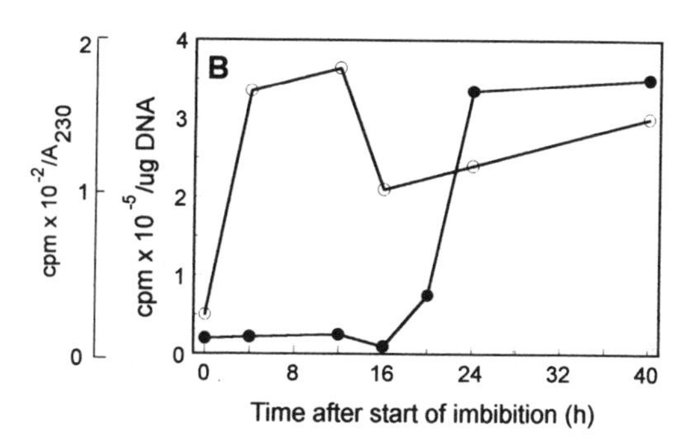

Figure 4.31. (A) Incorporation of thymidine into total cell DNA (O) and DNA content (●) in maize embryos during (0–12 h approx.) and after germination. (B) The synthesis of histone proteins (O) and nuclear DNA (●) during and following germination of maize embryos. A, after Zlatanova *et al.* (1987); and B, after Zlatanova and Ivanov (1988).

4.8. SEEDLING DEVELOPMENT

The first sign that germination has been completed is usually an increase in length and fresh weight of the radicle. In many seeds the radicle penetrates the surrounding structures as soon as growth commences, but in others (e.g., *Vicia faba* and other beans) there is considerable radicle growth before the testa is ruptured. There are some seeds, however, from which the hypocotyl is the first structure to emerge; this occurs in some members of the Bromeliaceae, Palmae, Chenopodiaceae, Onagraceae, Saxifragaceae, and Typhaceae. In some grains (e.g., rice and barnyard grass) germinated under anoxic or hypoxic conditions coleoptile growth precedes that of the radicle (Fig. 4.23); such an unusual

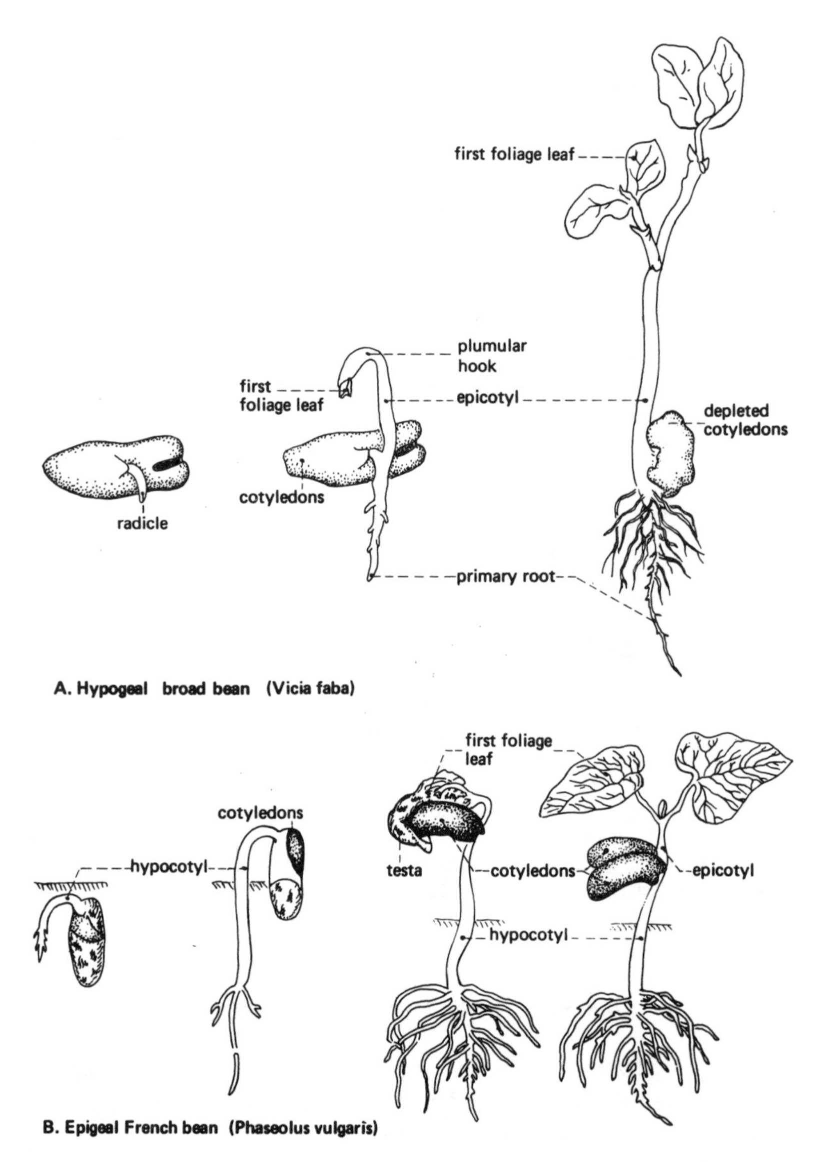

Figure 4.32. The hypogeal and epigeal type of seedling development shown by two species in which the cotyledons are the storage organs.

Table 4.4. Some Species Exhibiting Hypogeal and Epigeal Seedling Growth

	Hypogeal	Epigeal
Endospermic	*Triticum aestivum*	*Ricinus communis*
	Zea mays	*Fagopyrum esculentum*
	Hordeum vulgare	*Rumex* spp.
	Phoenix dactylifera	*Allium cepa*
	Tradescantia spp.	*Trigonella foenum-*
	Hevea spp.	*graecum*
Nonendospermic	*Pisum sativum*	*Phaseolus vulgaris*
	Vicia faba	*Cucumis sativus*
	Phaseolus multiflorus	*Cucurbita pepo*
	Aponogeton spp.	*Sinapis alba*
	Tropaeolum spp.	*Crambe abyssinica*
		Arachis hypogaea

germination pattern is also sometimes found in wheat when grains begin to sprout on the parent plant (Section 9.3).

Seedlings can be conveniently divided into two types on the basis of the fate of their cotyledons following germination: (1) epigeal, in which the cotyledons are raised out of the soil by extension of the hypocotyl and often become foliate and photosynthetic, and (2) hypogeal, in which the hypocotyl remains short and compact, and the cotyledons stay beneath the soil. The epicotyl expands to raise the first true leaves out of the soil (Fig. 4.32, Table 4.4).

In endospermic seedlings showing the epigeal mode of seedling growth, e.g., castor bean (*Ricinus communis*), the endosperm may be carried above ground by the cotyledons as they utilize its food stores. In onion (*Allium cepa*), another epigeal type, the absorptive tip of the single cotyledon may remain embedded in the degrading endosperm, while the rest of the cotyledon turns green (Fig. 4.33). The cotyledon in monocots may become highly specialized for absorption; in the Gramineae, for example, it is modified to form the scutellum, which may become extended as an absorptive haustorium (Fig. 7.2). Highly developed haustorial cotyledons are found in the Palmae. When the small embryo of the date palm (*Phoenix dactylifera*) commences growth, the cotyledon tip enlarges to form an umbrella-shaped body buried within the endosperm, from which it absorbs the hydrolyzed reserves. Likewise, the absorptive cotyledon of the coconut (*Cocos nucifera*) enlarges to invade the endosperm.

An interesting pattern of growth is shown by certain species of *Peperomia* (a dicot) in which one cotyledon emerges from the seed and the other remains as an absorptive organ (Fig. 4.33). It has been suggested that the monocotyle-

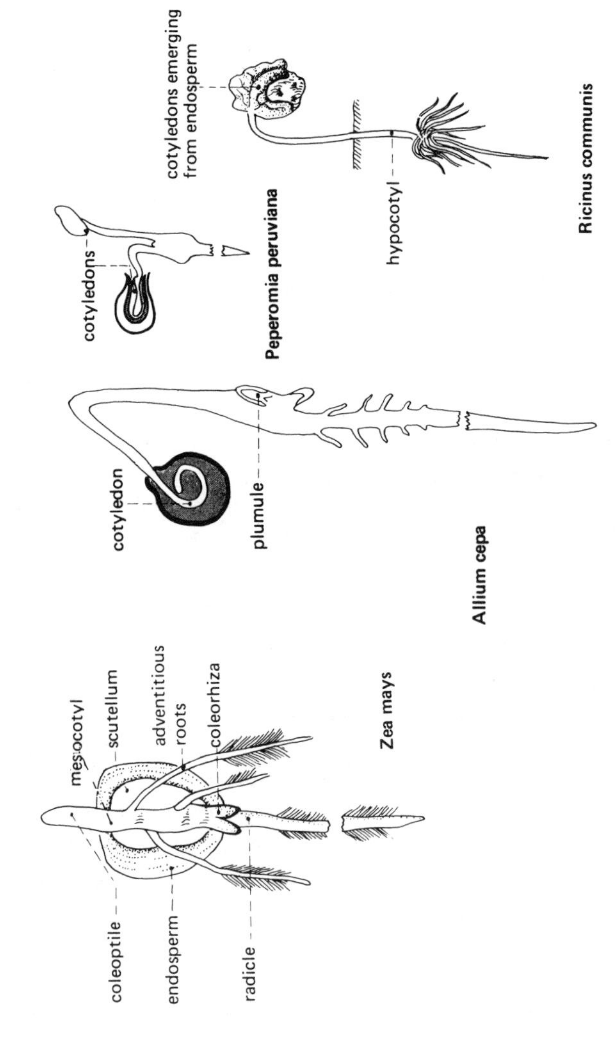

Figure 4.33. Patterns of seedling development and the different roles played by the cotyledons (see text for details). Not drawn to scale.

donous condition evolved from this pattern of emergence, although this remains a matter for conjecture.

USEFUL LITERATURE REFERENCES

SECTION 4.1

Collis-George, N., and Melville, M. D., 1978, *Aust. J. Soil Res.* **16:**291–310 (seed surface–soil interactions).

Crowe, J. H., and Crowe, L. M., 1992, in: *Water and Life* (G. N. Somero, C. B. Osmond, and C. L. Bolis, eds.), Springer-Verlag, Berlin, pp. 87-103 (membrane integrity in dry organisms).

Currie, J. A., 1973: *Seed Ecology* (W. Heydecker, ed.), Butterworths, London, pp. 463–480 (seed–soil interactions).

Finch-Savage, W. E., and Phelps, K., 1993, *J. Exp. Bot.* **44:**407–414 (predicting germination in horticultural seed beds).

Hadas, A., 1970, *Isr. J. Agric. Res.* **20:**3-14 (soil moisture stress and germination).

Hagon, M. W., and Chan, C. W., 1977, *Aust. J. Exp. Agric. Anim. Husb.* **17:**86–89 (soil moisture stress and germination).

Simon, E. W., and Raja Harun, R. M., 1972. *J. Exp. Bot.* **23:**1076-1085 (leakage during imbibition).

Spaeth, S. C., 1987, *Plant Physiol.* **85:**217–223 (extrusion of intracellular substances during imbibition).

Vertucci, C. W., 1989, in: *Seed Moisture* (P. C. Stanwood and M. B. McDonald, eds.), Crop Sci. Soc. America, Madison, Wisc., pp. 93–115 (extensive review on kinetics of seed imbibition).

Vertucci, C. W., and Leopold, A. C., 1983, *Plant Physiol.* **72:**190-193 (dynamics of soybean embryo imbibition).

Waggoner, P. E., and Parlange, J.-Y., 1976, *Plant Physiol.* **57:**153-156 (water diffusivity and imbibition).

SECTION 4.2

Groot, S. P. C., Kieliszewska-Rokicka, B., Vermeer, E., and Karssen, C. M., 1988, *Planta* **174:**500–504 (endosperm cell wall hydrolysis in tomato).

Hegarty, T. W., and Ross, H. A., 1978, *Ann. Bot.* **42:**1003-1005 (water stress, germination and growth).

Ni, B.-R., and Bradford, K. J., 1993. *Plant Physiol.* **101:**607–617 (germination of tomato seed).

Rogan, P. G., and Simon, E. W., 1975, *New Phytol.* **74:**273–275 (root elongation and mitosis).

Schopfer, P., and Plachy, C., 1985, *Plant Physiol.* **77:**676–686 (germination of rapeseed).

Welbaum, G. E., and Bradford, K. J., 1990, *Plant Physiol.* **92:**1046-1052 (imbibition and germination of muskmelon).

SECTIONS 4.3–4.5

Botha, F. C., Potgieter, G. P., and Botha, A. M., 1992, *Plant Growth Regul.* **11:**211–224 (respiration during germination: review).

Ehrenshaft, M., and Brambl, R., 1990, *Plant Physiol.* **93:**295–304 (mitochondrial biogenesis in maize embryos).

Gould, S. E. B., and Rees, D. A., 1964, *J. Sci. Food Agric.* **16:**702–709 (oligosaccharides and respiration).

Hourmant, A., and Pradet, A., 1981, *Plant Physiol.* **68:**631–635 (early oxidative phosphorylation).

James, T. W., and Spencer, M. S., 1979, *Plant Physiol.* **64:**431–434 (cyanide-insensitive respiration in peas).

Kennedy, R. A., Rumpho, M. E., and Fox, T. C., 1992, *Plant Physiol.* **100:**1–6 (anaerobic metabolism in plants).

Kollöffel, C., 1968, *Acta Bot. Neerl.* **17:**70–77 (ADH in peas).

Morohashi, Y., 1986, *Physiol. Plant.* **66:**653–658 (mitochondrial development in starchy and fatty seeds).

Morohashi, Y., and Bewley, J. D., 1980, *Plant Physiol.* **66:**70–73 (pea mitochondrion development).

Morohashi, Y., Bewley, J. D., and Yeung, E. C., 1981, *Plant Physiol.* **68:**318–323 (peanut mitochondrion development).

Nakayama, N., Iwatsuki, N., and Asahi, T., 1978, *Plant Cell Physiol.* **19:**51–60 (mitochondrial degeneration and senescence).

Pradet, A., and Prat, C., 1976, *Etudes de Biologie Végétale,* R. Jacques, Paris, pp. 561–574 (anoxia and ATP in rice).

Pradet, A., Narayanan, A., and Vermeersch, J., 1968, *Bull. Soc. Fr. Physiol. Veg.* **14:**107-114 (adenosine phosphate content of seeds).

Purvis, A. C., and Fites, R. C., 1979, *Bot. Gaz.* **140:**121–126 (glycolysis and PPP).

Rumpho, M. E., and Kennedy, R. A., 1981, *Plant Physiol.* **68:**165–168 (anaerobic metabolism of *Echinochloa*).

Salon, C., Raymond, P., and Pradet, A., 1988, *J. Biol. Chem.* **263:**12278–12287 (fatty acids, carbon fluxes, and respiration).

SECTION 4.6

Bewley, J. D., and Marcus, A., 1990, *Prog. Nucleic Acid Res. Mol. Biol.* **38:**165–193 (gene expression in germination and development).

Bino, R. J., Lanteri, S., Verhoeven, H. A., and Kraak, H. L., 1993, *Ann. Bot.* **72:**181–187 (2C and 4C nuclear complements).

Bryant, J. A., and Dunham, V. L., 1988, *Oxford Surv. Plant Mol. Cell Biol.* **5:**23–55 (review of nuclear DNA replication).

Cheung, C. P., and Suhadolnik, R. J., 1978, *Nature* **271:**357–358 (ribonucleotide triphosphates in wheat).

Conger, B. V., and Carabia, J. V., 1978, *Environ. Exp. Bot.* **18:**55–59 (2C and 4C nuclear complements).

Cuming, A. C., and Lane, B. G., 1979, *Eur. J. Biochem.* **99:**217–224 (mRNA changes during germination).

Datta, K., Parker, H., Averyhart-Fullard, V., Schmidt, A., and Marcus, A., 1987, *Planta* **170:**209–216 (gene expression in germinating soybean).

Delseny, M., Aspart, L., and Guitton, Y., 1977, *Planta* **135:**125–128 (loss of conserved mRNA during germination).

Dziegielewski, T., Kedzierski, W., and Pawalkiewicz, J., 1979, *Biochim. Biophys. Acta* **564:**37-42 (tRNA repair).

Guilfoyle, T. J., and Jendrisak, J. J., 1978, *Biochemistry* **17:**1860–1866 (RNA polymerases).

Guilfoyle, T. J., and Malcolm, S., 1980, *Dev. Biol.* **78:**113–125 (RNA polymerases in soybean).

Ingle, J., and Sinclair, J., 1972, *Nature* **235:**30–32 (rRNA gene amplification).

Lalonde, L., and Bewley, J. D., 1986, *J. Exp. Bot.* **37:**754–764 (changes in mRNA populations during pea germination).

Lane, B. G., 1991, *FASEB J.* **5:**2893–2901 (properties and occurrence of germin).

Lane, B. G., Dunwell, J. M., Ray, J. A., Schmitt, M. R., and Cuming, A. C., 1993, *J. Biol. Chem.* **268:**12239–12242 (germin is an oxalate oxidase).

Marcus, A., Feeley, J., and Volcani, T., 1966, *Plant Physiol.* **41:**1167–1172 (early polysome formation in wheat).

Misra, S., and Bewley, J. D., 1986, *J. Exp. Bot.* **37:**364–374 (new mRNAs associated with germination).

Sen, S., Payne, P. I., and Osborne, D. J., 1975, *Biochem. J.* **148:**381–387 (early RNA synthesis in rye).

Spiegel, S., and Marcus, A., 1975, *Nature* **256:**228–230 (protein synthesis without mRNA synthesis).

Zlatanova, J., and Ivanov, P., 1988, *Plant Sci.* **58:**71–76 (DNA and histone synthesis in maize embryos).

Zlatanova, J., Ivanov, P., Stoilov, L. M., Chimshirova, K. V., and Stanchev, B. S., 1987, *Plant Mol. Biol.* **10:**139–144 (DNA repair and synthesis in maize embryos).

Chapter 5

Dormancy and the Control of Germination

5.1. INTRODUCTION

Whether or not a viable seed germinates and the time at which it does so depend on a number of factors, including those present in the seed's environment. First, the chemical environment must be right. Water must be available, oxygen may have to be present since the seed must respire, and noxious or inhibitory chemicals should be absent. The physical environment, too, must be favorable. The temperature must be suitable and so also, in many cases, must the light quality and quantity. But in many instances all of these conditions may be satisfied and nevertheless the seed fails to germinate. The reason for this, as we have indicated in Chapter 1, is that there exists within the seed (or dispersal unit) itself some block(s) that must be removed or overcome before the germination process can proceed: such a seed is said to be *dormant*. To be released from dormancy, a seed must experience certain environmental factors or must undergo certain metabolic changes. Hence, the control of germination exists at two levels. One—dormancy—is related entirely to the state of the seed itself, and the second involves the operation of environmental factors on both dormancy and germination. We call these the *internal* and *external* controls, respectively. The relationship between dormancy and germination and the points at which control exists are shown in Fig. 5.1.

5.2. INTERNAL CONTROLS

5.2.1. Dormancy—Its Biological Role

Since the function of a seed is to establish a new plant, it seems curious that dormancy—an intrinsic block to germination—should exist! But it may not necessarily be advantageous to a seed to germinate freely, and we shall see that dormancy, in fact, offers considerable benefits. One is that it is a means whereby distribution of germination in time can be achieved. This occurs in three ways.

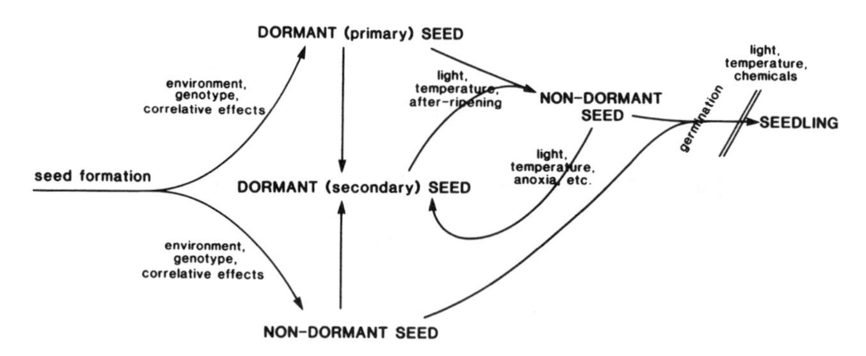

Figure 5.1. Dormancy and germination.

First, seeds are dispersed from the parent plant with different degrees of dormancy, a phenomenon known as polymorphism, heteromorphy, or heteroblasty. Frequently, the variation is reflected in the appearance—color, size, and thickness of coat—of the seeds or dispersal units. In *Bidens bipinnata*, for example, the more dormant, outer cypselas of the inflorescence (a capitulum) are short, brown, and wrinkled, whereas the less dormant, inner ones are longer, black, and smooth. The green seeds produced by *Salsola volkensii* have virtually no dormancy, but the nongreen seeds have dormancy. And in *Chenopodium album* four types of seed can be found—brown or black ones each with reticulate or smooth coats; of these, the smooth, black seeds have the deepest dormancy. Differences such as these, in both appearance and dormancy, are displayed by seeds from the same plant (e.g., *Bidens bipinnata*) or from different plants (e.g., *Chenopodium album*). In the former case, correlative effects (Section 5.4.3) operate to produce the variation, and in the second, both environmental and genetic causes can be traced. When there are polymorphic seeds, germination is spread temporally, with new seedlings emerging at irregular intervals and thus reducing competition and increasing the likelihood that some individuals will survive. Such a temporal distribution clearly can have advantages with regard to the continuation and spread of the species.

Dormancy also provides for the distribution of germination in time through the dependence of dormancy breakage on some environmental factor which itself has a time distribution. For example, seeds are commonly released from dormancy by being chilled, sometimes for several weeks or months at temperatures of 1–5°C. Since such temperatures are found only during the winter, seeds that rely on such a means of dormancy breakage must await the passage of this cold season before they can germinate. The advantage of this strategy is that the young seedling emerges in the spring and establishes itself over the favorable succeeding months; emergence before winter would entail the risk of succumbing to the inclement conditions of that season. Seeds of many species enter a state of dormancy, called secondary dormancy, when they experience conditions unfavorable for germination such as relatively high or low temperatures, but in the succeeding weeks or months they slowly emerge from dormancy. Seeds forced into secondary dormancy by low temperatures (winter) end their dormancy at a different time of year from those made dormant by higher temperatures (summer), and thus the seedlings of the two types emerge at different seasons (see also Section 6.8).

Seed dormancy can also lead to a distribution of germination in space—another aspect of its biological importance. In many cases, dormancy is terminated by light, the most effective wavelengths being in the red region of the spectrum. Those seeds which utilize this mechanism therefore do not germinate in light transmitted through green leaves, since it is poor in the red component, but do so only when unrestricted by leaf canopies. Hence, seedling emergence occurs in open, competition-free situations (Section 6.4). Moreover, germination of such kinds of seeds does not occur at depths in the soil to which suitable light cannot penetrate. This is important for small seeds whose food reserves can support only a very limited amount of seedling growth below ground (Section 6.2.2).

5.2.2. Categories of Dormancy

Dormancy is fundamentally the inability of the embryo to germinate because of some inherent inadequacy, but in many cases it is manifest only in the intact seed, and the isolated embryo can germinate normally. Here, the seed is dormant only because the tissues enclosing the embryo, generally referred to loosely as the seed coat (but often including endosperm, pericarp, or extrafloral organs), exert a constraint that the embryo cannot overcome. This type of dormancy is generally called *coat-imposed dormancy* though a better term may be *coat-enhanced dormancy*. Experimentally, the coats can often be replaced by other kinds of constraints, such as an osmoticum (to create water stress) or artificial wrappings (e.g., moist filter paper). For example, isolated embryos of lettuce germinate when placed on a substratum moistened with water, but they exhibit all of the characteristics of dormant seeds when in contact with 0.3–0.4 M mannitol solution. In contrast to coat-imposed dormancy are those cases where the embryo itself is demonstrably dormant. Removal of the coat does not permit such embryos to germinate normally, and so the block to germination is, in a sense, more profound than in seeds with coat-imposed dormancy. This category of dormancy—*embryo dormancy*—is common in woody species, especially in the Rosaceae, but is sometimes found in herbaceous plants such as some

Table 5.1. Some Species Having Coat-Imposed and/or Embryo Dormancy

Coat-imposed dormancy[a]	Embryo dormancy
Avena fatua— some strains (palea, lemma, pericarp)	*Acer saccharum*
Hordeum spp. (palea, mainly pericarp)	*Avena fatua*—some strains
Betula pubescens (pericarp)	*Corylus avellana*
Peltandra virginica (pericarp)	*Fraxinus americana*
Acer pseudoplatanus (pericarp, testa)	*Hordeum* spp.
	Prunus persica
Phaseolus lunatus (testa)	*Pyrus communis*
Sinapis arvensis (testa)	*Pyrus malus*
Xanthium pennsylvanicum (testa)	*Sorbus aucuparia*
	Syringa reflexa
Lactuca sativa (endosperm)	*Taxus baccata*
Pyrus malus—some cvs. (endospermal membrane)	
Syringa spp. (endosperm)	

[a]Tissues responsible are shown in parentheses.

grasses (e.g., *Avena fatua,* wild oats). Examples of the two categories of dormancy are given in Table 5.1. Both types of dormancy exist simultaneously or successively in some species. In apple seeds, for example, embryo dormancy predominates, but a contribution is made by the covering tissues, the endosperm and testa, and their removal reduces the amount of dormancy-breaking treatment (chilling) that is required (Section 5.5.1.2). Mature sycamore dispersal units (actually fruits, not seeds) possess only coat-imposed dormancy, yet just before the end of their maturation on the plant they have embryo dormancy. And in the grasses *Aristada contorta* and *Bouteloua curtipendula*, dormancy of the seed in the newly dispersed units is so deep that removal of the covering hull has no effect, whereas some months later this treatment promotes their germination. The later dormancy is therefore coat imposed.

In many cases the expression of dormancy shows a strong temperature dependence. For example, grains of several grasses and cereals (wheat, barley, oats) are dormant only at temperatures above a particular value, a condition known as *relative dormancy* (see Fig. 5.11). Frequently, the "critical" temperature shifts with seed age or after treating the seeds in various ways, such as with certain chemicals, e.g., cytokinins.

Seeds are said to have *primary dormancy* when they are dispersed from the parent plant in a dormant state. Here, the dormancy is initiated during seed development. Dormancy can also be induced in mature, nondormant seeds—*induced* or *secondary dormancy*. This sets in when seeds are held under conditions unfavorable for germination, e.g., anoxia, unsuitable temperatures or illumination.

Seeds of several species display more complex patterns in which the parts of the embryonic axis differ in the depth of dormancy. In the so-called *epicotyl dormancy* (e.g., in *Paeonia* spp. and *Lilium* spp.), radicle emergence occurs readily but the epicotyl fails to grow. In *Trillium* spp. and *Caulophyllum thalictroides*, the radicle does have some dormancy but it is less deep than that of the epicotyl, and so the two organs differ in the degree of treatment (chilling) needed to break dormancy: such cases are said to exhibit *double dormancy*.

5.2.3. Mechanism of Dormancy

Embryo and coat-imposed dormancies share one common feature. In both, the embryo is unable to overcome the constraints imposed on it, in the former case by factors within the embryo itself and in the second by the enclosing tissues. So when we attempt to understand the mechanism of dormancy, we have to answer two questions: What is the mechanism of action of the constraints? Why cannot the embryos overcome them? Let us begin with the first question in relation to embryo dormancy.

5.2.4. Embryo Dormancy—The Inherent Constraints

Unfortunately, few cases have been examined in detail, but in those that have, two factors appear to be involved: (1) the cotyledons and (2) germination inhibitors.

Amputation of the cotyledons often allows the embryonic axis of the dormant embryo to germinate and grow. In this way, dormancy is partially or completely broken in *Corylus avellana* (hazel) and *Euonymus europeus* (spindle tree) by excising one cotyledon, and in *Fraxinus excelsior* (ash) by cutting off two. Embryo dormancy in barley can be relieved by removal of the scutellum (which is a modified cotyledon), and dormancy of apple embryos is progressively reduced as increasing amounts of cotyledonary tissue are cut off (Fig. 5.2). Interesting evidence for the inhibitory role of the cotyledons is additionally provided by those species in which the isolated, dormant embryos succeed in making only very sluggish growth, to form a dwarf plant—*physiological dwarfism*. If the cotyledons are excised at an early stage of dwarfism, a normal growth habit is resumed (e.g., in peach). Cotyledon excision also relieves epicotyl dormancy (e.g., in *Viburnum trilobum*).

It appears, therefore, that some inhibitory effect emanates from the cotyledons to block the germination of the embryonic axis. This inhibitory influence also reveals itself in other ways, within the cotyledons themselves. Freshly harvested seeds of *Xanthium pennsylvanicum* (cocklebur) show embryo dormancy: if the cotyledons of these embryos are excised, they do not expand or form chlorophyll when placed on a moist substratum in the light, but after the

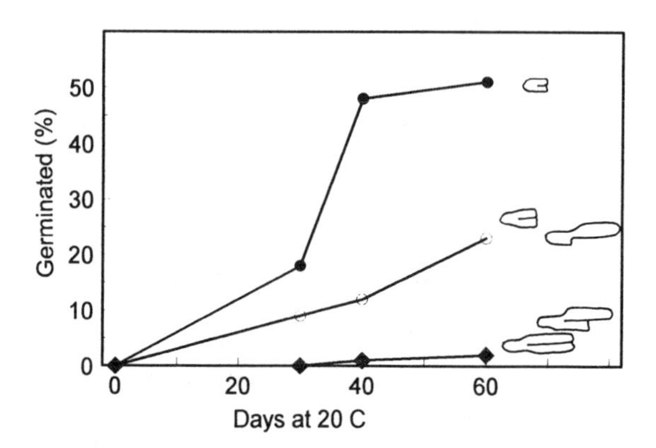

Figure 5.2. Effect of cotyledon removal on embryo dormancy in apple. Portions of cotyledons(s) were removed from isolated, dormant embryos. The treated embryos were placed on moist cotton wool. After Thevenot and Côme (1973).

loss of embryo dormancy the isolated cotyledons enlarge and become green. On the other hand, in some dormant embryos (e.g., apple, pear), the cotyledon in contact with the moist substratum does grow and develop chlorophyll in the light, but the uppermost cotyledon, not in direct contact with the wet substratum, remains small and white. The beneficial effects of contact with a moist substratum suggest that the inhibitory influence can be leached from the cotyledon into the surrounding water, and implicit in this inference is that a chemical inhibitor is involved. There is good evidence that in apple cotyledons the inhibitor is ABA.

Support for the possibility that ABA is also involved in the maintenance of embryo dormancy comes from experiments on embryos of *Taxus baccata* and apple. In both cases, conditions which encourage leaching relieve the dormancy and allow radicle extension to proceed. For example, 2 weeks on a liquid medium permits *Taxus* embryos to germinate, though the same period of time spent on a medium solidified with agar (which would be less likely to encourage the leaching of inhibitor) is less effective (Fig. 5.3A). During the 2 weeks in the liquid medium almost all of the free and bound forms of ABA leave the embryo, a coincidence that prompts the suggestions that ABA, when remaining in the embryo, stops it from germinating. Supporting this contention (but not proving it!) is the finding that addition of ABA to the liquid medium very much reduces the efficacy of the washing treatment (Fig. 5.3B). The cotyledons of apple contain bound forms of ABA from which free ABA is gradually released and transported to the radicle, whose extension is thus inhibited. There is evidence that when the radicle is buried in agar the free ABA declines, and radicle growth ensues.

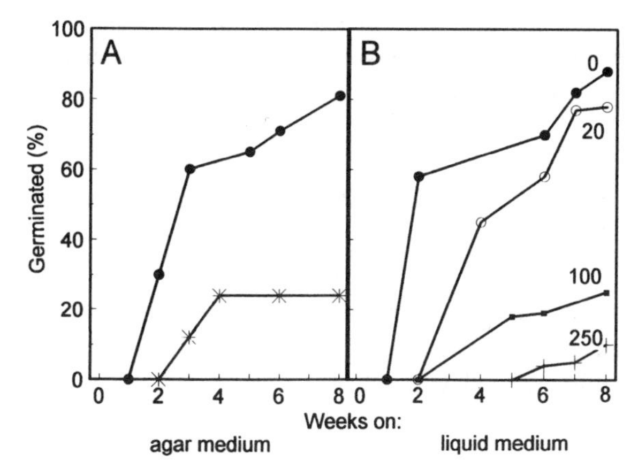

Figure 5.3. The effect of leaching on relief of embryo dormancy in *Taxus baccata*. In (A), embryos were placed on a liquid (●) or solid (*) agar medium, and germination over several weeks at 22°C was followed. In (B), embryos were on a liquid medium supplemented with ABA at the concentrations shown (in μg/liter). After Le Page–Degivry (1973).

An interesting observation on *Helianthus annuus* embryos suggests that here continuous synthesis of ABA (and not release from bound forms) is required for the expression of embryo dormancy. Embryos with established dormancy (more than 22 days old) lose this dormancy when treated with fluridone, the inhibitor of ABA synthesis; the embryos are stimulated to commence radicle growth (Fig. 5.4). The axes themselves are dormant and when excised they too are sensitive to the promotive fluridone treatment.

It may be, therefore, that embryos are held in the state of dormancy by ABA, either generated by the cotyledons, or made within the axis, or possibly

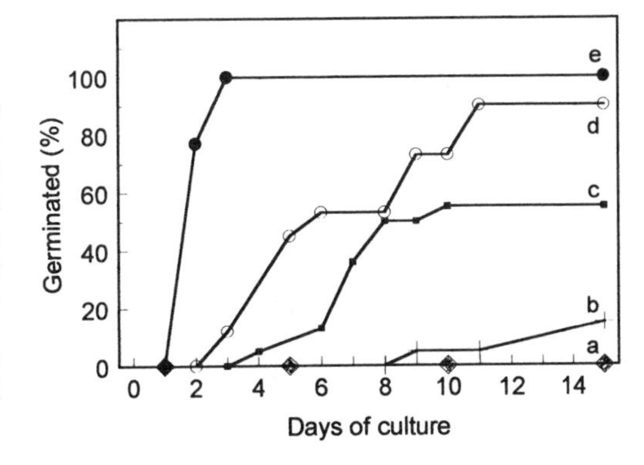

Figure 5.4. Effect of fluridone on the relief of embryo dormancy in *Helianthus annuus*. Embryos (31 days old) were held on agar without (a) or with (b–e) treatment with fluridone. This was applied directly to the embryos as 100 μg/ml in 10% acetone immediately after isolation from the seed (b), after 1 day (c), 3 days (d), or 7 days (e). After Le Page–Degivry and Garello (1992).

both. Dormant embryos of many species contain ABA and/or other inhibitors, but more extensive and detailed investigations are required before we can accept this as a ubiquitous scenario.

5.2.5. Coat-Imposed Dormancy—The Constraints

The following are possible effects of the tissues enclosing the embryo:

1. Interference with water uptake
2. Mechanical restraint
3. Interference with gas exchange
4. Prevention of the exit of inhibitors from the embryo
5. Supply of inhibitors to the embryos

Prevention of germination may be related to the action of one or more of these effects.

5.2.5.1. Interference with Water Uptake

This is a common effect especially in seeds of the Leguminosae, but it is also found in other families such as the Cannaceae, Convolvulaceae, Chenopodiaceae, and Malvaceae. Many species have seeds with extremely hard coats which, by preventing the entry of water, delay germination for many years. For example, about 20% of soaked *Robinia pseudoacacia* seeds remain ungerminated for 2 years because insufficient water reaches the embryo, and 1.5% are in this state for 20 years! It could be argued that these are not really cases of dormancy in the strict sense, because the embryos simply do not have enough of one of the basic requirements for germination, i.e., water; nevertheless, hard-coated seeds, in which water entry is the factor limiting germination, are generally considered under the heading of dormancy.

The testa is generally responsible for impeding water uptake. *Melilotus alba* is an example which illustrates the salient anatomical features of this tissue (Fig. 5.5). Waterproofing can be conferred by several parts of the testa: the waxy cuticle, the suberin, and the thick-walled palisade and osteosclereid layers. All of these contribute to some extent, and, indeed, in some species of Leguminosae the waxy cuticle plays a major role. But careful experiments carried out on several species strongly suggest that the main barrier to water uptake is offered by the osteosclereids, because only when these cells are punctured do most seeds begin to imbibe water (Fig. 5.6). We must remember that the testa in the Leguminosae has the hilum, micropyle, and strophiole (Section 1.4.3). In species with impermeable coats, such as *Phaseolus lunatus*, the micropyle is occluded and the fissure in the hilum remains closed except under very low relative

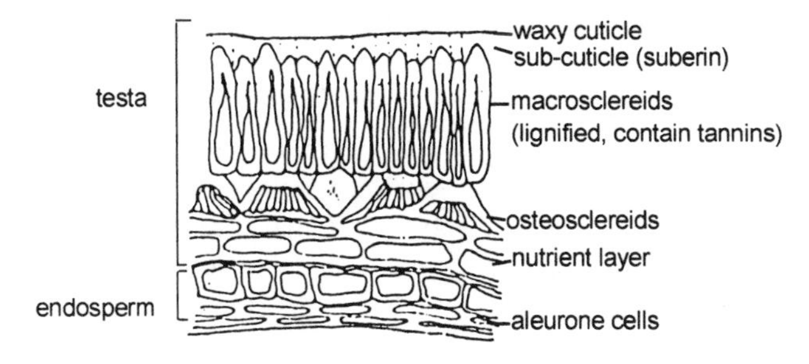

Figure 5.5. A section of the seed coat of *Melilotus alba*. After Hamly (1932).

humidity. The strophiole stays intact unless the seed experiences certain conditions that cause the plug to be ejected, a point we will consider briefly in a later discussion. Thus, the testa is a very effective waterproofing tissue, which, because of this property, can delay germination of hard-coated seeds for considerable periods of time.

5.2.5.2. Mechanical Restraint

The coats of many dispersal units are hard, tough tissues which may be expected to offer considerable resistance to the emergence of the embryo. The hard shells of different kinds of nuts are outstanding examples, and obviously if the embryos cannot generate enough force to penetrate these tissues, they cannot properly germinate. Various tissues surrounding the embryo in other dispersal units are also extremely resistant. A rather good example is provided by dormant

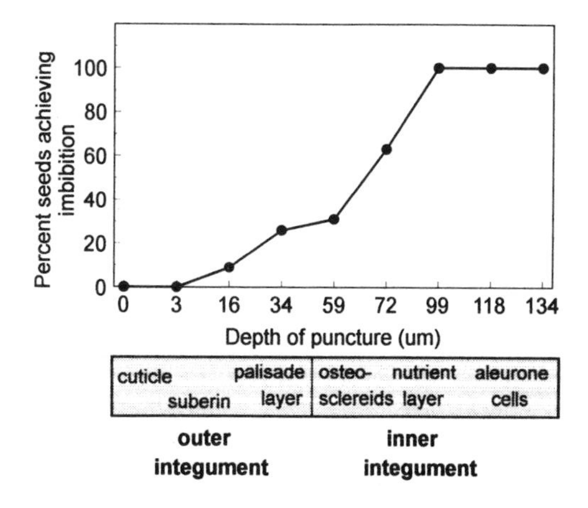

Figure 5.6. The effect of puncturing on seed coat permeability to water. The testae of *Coronilla varia* seeds were punctured to different depths with a fine needle and afterward placed on moist filter paper for imbibition. Adapted from McKee *et al.* (1977).

cultivars of the lettuce "seed" (actually an achenic fruit—a cypsela). Here, the enclosing tissue that is responsible for coat-imposed dormancy is the endosperm; so if the pericarp and testa are both removed, the embryo is still unable to germinate. The embryo is completely enclosed by the endosperm, which is two cells thick, each cell having relatively thick walls composed predominantly of mannans, not cellulose (i.e., a polymer of mannose, not glucose). That the endosperm is a tough, constraining tissue can be revealed by certain treatments with chlorine-generating compounds that do not stop the growth of the embryo but do prevent it from breaking through the endosperm. Because this tissue is so resistant, the embryo grows inside it, developing a contorted radicle/hypocotyl that cannot penetrate the relatively thick, tough tissue surrounding it (Fig. 5.7). But nevertheless, in those species where the force needed to puncture the coat has been determined, there seems to be no clear correlation between the degree of dormancy and the resistance of the coat to puncturing, i.e., coats of dormant seeds are no tougher than those of nondormant ones. Of equal importance to the coat resistance, however, is the thrust generated by the embryo; clearly, the relationship between coat resistance and embryo "push" will determine whether or not the embryo can penetrate the enclosing tissues. Experiments on a few species, notably *Syringa* (lilac) and lettuce, in which isolated embryos are placed in an osmoticum which acts as an external constraint, strongly suggest that before receiving a dormancy-breaking treatment the embryo does not have enough growth potential to pierce the coat, whereas after the termination of dormancy it can generate the necessary force. In this connection we should note that embryos taken from nondormant *Xanthium* seeds generate more than double the growth thrust of embryos from dormant seeds.

It is known in some cases that the tissues restraining the embryo must be weakened chemically before the radicle can emerge. This weakening is caused by enzymes produced under the influence of the embryo: nongerminable, dormant seeds fail to produce these enzymes. In tomato, for example, embryos in dormant seeds do not provoke the appearance of the enzyme mannanase required for the degradation of the cell walls of the endosperm—the tissue which imposes coat dormancy. A similar situation appears to exist in lettuce and celery seeds. The regulation of production of these enzymes that occurs during dormancy breakage and germination is discussed in Sections 4.2 and 5.5.1.8.

5.2.5.3. Interference with Gas Exchange

The several layers of tissue surrounding the embryo might limit the capacity for gaseous exchange by the embryo in two ways. First, entry of oxygen may be impeded; second, escape of carbon dioxide may be hindered. One important consequence could be the inhibition of respiration.

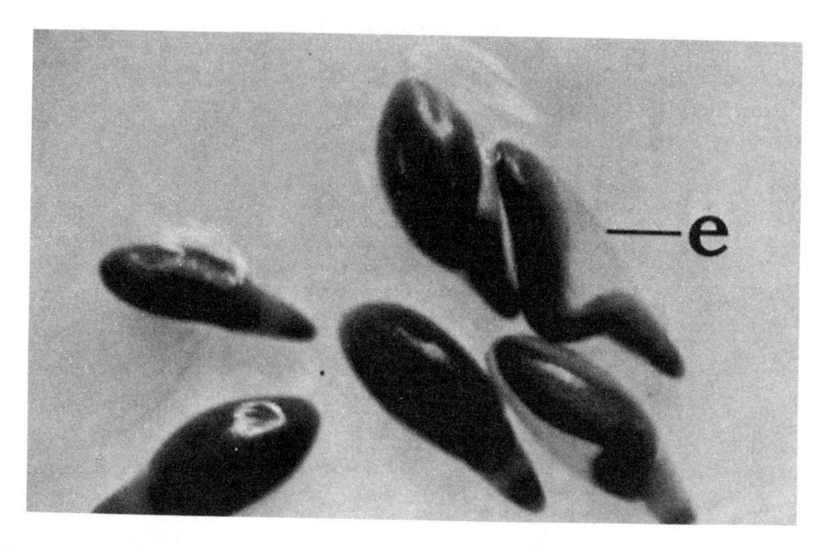

Figure 5.7. Resistance of lettuce endosperms. Lettuce seeds were treated with dichloroisocyanurate which did not stop the growth of the embryo but, for an unknown reason, did prevent it from breaking the endosperm (e). This illustrates that the endosperm is a tough tissue that can even constrain a germinated embryo. By courtesy of A. Pavlista.

The possibility that the seed coat imposes dormancy by affecting gaseous exchange gains support from the fact that in many cases the inhibitory action of the tissues surrounding the embryo is much reduced simply by scratching or puncturing them. Thus, a pinprick through the endosperm near the radicle of the lettuce seed or through the pericarp over the embryo of the intact, dormant wheat grain can cause some germination. Numerous cases of such effects are known (Table 5.2), and it is difficult to understand how such moderate "surgical" operations can interfere appreciably with the mechanical resistance of the coat. Another finding that leads us to suspect that germination in the intact dispersal unit is held in check by insufficient oxygen is that dormancy is frequently overcome by oxygen-enriched atmospheres (Table 5.2). To summarize at this point, it can be suggested that the embryo in the intact, dormant dispersal unit fails to germinate because of the restrictions by the enclosing tissues, particularly on oxygen uptake. Removal, abrasion, or puncturing of these tissues gives the embryo access to oxygen and germination can proceed. To assess the plausibility of this view we must have information relating to the following three questions: (1) What is the permeability of seed coat tissues to oxygen? (2) How much oxygen does the embryo need to support its respiration? (3) Does the impermeability of the coat lower the available oxygen to below the necessary amount for germination?

Table 5.2. Some Treatments That Remove Coat-Imposed Dormancy[a]

Species	Tissue removal	Puncturing coat	High oxygen
Acer pseudoplatanus	+++ (testa)	+++	–
Avena fatua	+++ (hull)	–	++
Betula pubescens	+++ (pericarp, testa, endosperm)	+	++
Hordeum spp.	+++ (hull)	++	++
Oryza sativa	+++ (hull)	++	++
Phacelia tanacetifolia	+++ (endosperm)	–	++
Triticum spp.	+++ (pericarp, testa)	+	+++
Xanthium pennsylvanicum	+++ (testa)	–	+++

[a]+++, Strong dormancy-relieving effect; ++, moderate effect; +, slight effect; –, no effect.

It has been found in several cases that the permeability of the seed coat to oxygen is less than that of water of equivalent thickness. In *Sinapis arvensis*, for example, it is lower by a factor of about 10^4 and in *Xanthium pennsylvanicum* by about 10^2. The reasons for these differences are not clear in all instances, but one possibility is that there is resistance to the entry of oxygen offered by the layer of mucilage around many seeds (*Sinapis*, for example). Another way in which the coats act is by consuming oxygen themselves. A good example of this is the dispersal unit of sugar beet (*Beta vulgaris*) in which the ovary cap (the operculum) is the major barrier, for when this is removed germination can occur. Though dead, the operculum when wetted in air consumes relatively large quantities of gas, presumably oxygen (Fig. 5.8). This is probably caused by the enzymatic oxidation of various chemical constituents, an occurrence that is known in the testa of apple and other seeds, where various phenolic compounds (e.g., phloridzin and chlorogenic acid) are implicated. The dormancy-imposing hull of rice (*Oryza sativa*) and the glumellae of barley (*Hordeum vulgare*) also consume oxygen. As dormancy slowly diminishes during dry storage of the grains, so the oxygen-consuming capacity of the hull decreases; this correlation supports the possibility that the hull imposes dormancy because it deprives the embryo of oxygen. Also correlated with the extent of dormancy in different cultivars of rice is the activity of peroxidase in the hull—an enzyme that may form part of an oxygen-consuming complex.

The answer to the first of our three questions therefore supports the possibility that the coat prevents germination by limiting the embryo's oxygen supply. But when we come to answer the second question about how much oxygen an embryo needs, the situation becomes less clear. In barley grains, 40–50% of the total oxygen uptake is accounted for by the activity of the glumellae, which impose dormancy, yet this does not reduce the energy charge (essentially, ATP production) of the embryo. Hence, sufficient oxygen enters to satisfy the demand in ATP generation. It is known that embryos of several

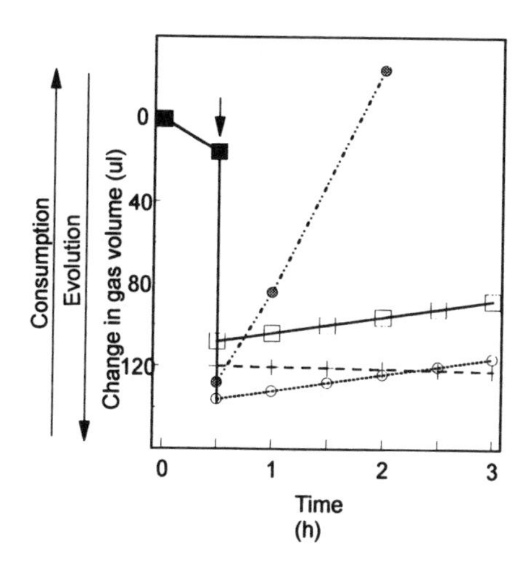

Figure 5.8. Gas uptake by components of the dispersal units of sugar beet (*Beta vulgaris*). Naked seeds (○), opercula (●), testae (+), and pericarps (□) were placed separately in Warburg flasks, wetted (arrow), and the subsequent gas exchange measured. There is initially a gas evolution (usually occurring when dry seed parts are moistened) followed by gas, probably oxygen, uptake. Redrawn from Coumans *et al.* (1976).

species are satisfied by extremely low partial pressures of oxygen. Isolated embryos of *Betula pubescens* (birch), *Sinapis arvensis*, and wheat can germinate even in nitrogen! Furthermore, the oxygen consumption of embryos of both *Sinapis arvensis* and *Xanthium pennsylvanicum* is considerably less than that permitted by the amount of oxygen that can diffuse through the testa, as we can see from the results shown in Fig. 5.9. It is not the case, therefore, in answer to our third question, that the respiratory demand for oxygen by the embryo exceeds the possible supply. Taking the answers to our three questions into account, then, we are bound to conclude that coats do not seem to impose dormancy on the enclosed embryo simply by restricting the amount of oxygen available for respiration.

But we are still faced with the findings that both high partial pressures of oxygen and facilitated access to air (by pricking or scratching the seed coats) cause intact, dormant seeds to germinate. If the oxygen is not needed to support respiration, what *is* it for? Inhibitors have been invoked, in some cases, to answer this. In *Xanthium*, for example, growth inhibitors have been found in the embryo which, in the intact seed, are oxidized under high oxygen concentrations but not in air; hence, the effect of exposing intact seeds to oxygen is to remove these inhibitors. The inhibitors (which have not been identified chemically) can also diffuse out of the isolated embryo when it is set on a moist substratum, but they do not cross the seed coat of an intact seed. It is arguable, therefore, that in this case the major beneficial effect of removing the coat is to permit the escape of inhibitors and not to make more oxygen available. There is a twofold effect of the intact coat: primarily, it causes the retention of inhibitor, and secondarily, because it does act as a barrier to oxygen, it prevents the entry of sufficient

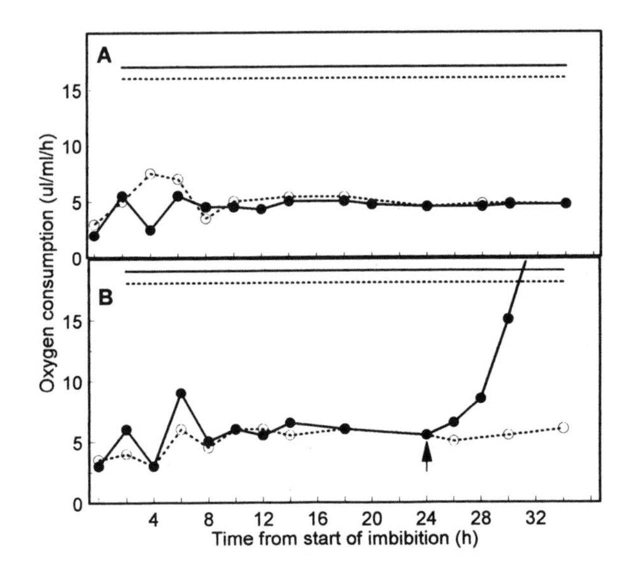

Figure 5.9. Permeability of the coat and oxygen consumption in *Xanthium pennsylvanicum*. The burr of *Xanthium* has two seeds, an upper one and a lower one. Newly ripened seeds of both kinds are dormant (embryo dormancy). A period of afterripening causes loss of embryo dormancy in both kinds, but the upper seed retains a coat-imposed dormancy. In (A), the permeability of the coats and the oxygen uptake of both kinds of newly ripened seed are shown. In (B), these features in afterripened seed are plotted. ——, Lower seed coat permeability to oxygen; - - - -, upper seed coat permeability to oxygen; O, upper seed oxygen uptake; ●, lower seed oxygen uptake. Arrow: radicle emergence from nondormant (afterripened), lower seed. Note the following: (1) Permeabilities of upper and lower seed coats, of both dormant and nondormant seeds [compare (A) and (B)] ,are identical and higher than that needed to satisfy the oxygen uptake by the whole seed (i.e., presumably by the embryos inside). (2) Oxygen uptake by embryos of nondormant, afterripened seed is identical to that of dormant (coat-imposed) seeds, as shown in (B). As expected, however, oxygen uptake increases when the radicle emerges from the nondormant seed. After Porter and Wareing (1974).

oxygen from air to support the oxidation of the inhibitor. Inhibitors may be involved in a different way in *Sinapis arvensis*. Here, there is evidence that these substances are actually formed under conditions of low oxygen tensions; thus, the effect of the coat is to lower the amount of oxygen reaching the embryo, and this, though it is still adequate to support respiration, encourages the production of inhibitor. Of course, the process occurs in isolated embryos held on moist filter paper under very low oxygen tensions, but in this situation the inhibitor can probably diffuse out. So here, too, there is a dual role for the coat, involving effects on gaseous exchange and on the exit of inhibitor.

Effects of oxygen and the seed coat in *Xanthium* may also be connected with the production and action of ethylene. Ethylene can break dormancy in some species of seed, including *Xanthium*, and can, in fact, be produced by the embryo itself. But relatively high oxygen tensions are needed for the operation of the so-called aerobic ethylene-producing system, a requirement that is not satisfied in the intact seed. Hence, the embryo may not be able to produce enough ethylene to relieve its dormancy. Which of the two sets of interpretation of coat and oxygen effects in *Xanthium* is more plausible (i.e., the "inhibitor hypothesis" or the "ethylene hypothesis") is not easy to judge at this time.

5.2.5.4. Prevention of Exit of Inhibitors

We have already seen that the testa of *Xanthium* stops the loss of inhibitors from the embryo. Inhibitors of different chemical classes have been found in seeds of many species, some in the embryo and some in the coat itself (Table 5.3). If these inhibitors are retained by the imbibed seed, instead of being lost to the external medium, germination of the embryo may be blocked. We must be clear, though, that the discovery of an inhibitor in a seed does not necessarily mean that it functions in the dormancy mechanism. Important considerations that help us decide whether it does or not are: (1) Where is the inhibitor located? Is it available to the embryonic axis, the region whose germination is arrested? (2) Even if it is present in the axis itself, is it sequestered in cell compartments or is it freely distributed? (3) Is it present at concentrations likely to be effective in maintaining dormancy? (4) Does it indeed have activity against the seed from which it is extracted or is it simply active (e.g., in preventing growth) in some general bioassay system such as the oat or wheat coleoptile? Unfortunately, in no case do we know the answers to all or even most of these questions. We can nevertheless assess whether the seed coat stops the escape of inhibitor by collecting diffusion products from intact and isolated embryos. The latter approach has been taken in the case of *Xanthium*, where it seems clear that the inhibitor readily diffuses out of the isolated embryo but not from the intact seed. In *Avena fatua* the hull is the part of the coat that imposes dormancy, so the naked caryopsis germinates when set on moist filter paper. If only part of the hull is removed from one side, to leave the caryopsis in a boat-shaped portion, germination occurs only if the caryopsis is in contact with the wet paper and not if the unit is placed caryopsis uppermost with the hull in contact. Water uptake is unaffected by the intervening hull, and so it seems that outward movement of substances from the caryopsis might be impeded by the hull. Inhibitors in the caryopsis do not, in fact, leave the intact dispersal unit, but they do move out of a naked caryopsis. To test the effect of stopping outward diffusion in both *Xanthium* and *Avena fatua* the simple device has been used of allowing the isolated embryo or caryopsis to imbibe in a humid atmosphere, not in contact with liquid water: this does not reduce the total water uptake. Dormancy is retained after such treatment, whereas it is lost on moist filter paper. The interpretation is that an inhibitor cannot diffuse out in the humid atmosphere because no liquid is present.

5.2.5.5. Supply of Inhibitors to the Embryo

Examples of coats with inhibitors are given in Table 5.3. We should again note that their presence does not mean that they function in dormancy. In some cases, however, the coats do seem to have a powerful inhibitory effect, of a chemical rather than physical origin. Embryos of *Rosa* germinate when placed on a moist substratum, but if pieces of pericarp and testa are also present, many of the embryos fail to germinate; and dormancy is almost complete if each embryo is actually covered by a piece of pericarp. The inhibitor ABA is present in these coat tissues. In several cases, where repeated washing (leaching) of the seeds relieves dormancy, inhibitors are known to be removed.

5.2.6. Coat-Imposed Dormancy—A Summary

We have seen that there are several possible constraints imposed by the structures enclosing the embryo. In several species, e.g., in some of the Leguminosae, interference with water uptake is possibly the only factor involved. But in many species, the picture seems quite complex and it is not unlikely that more than one factor operates simultaneously. For example, it is clear that the coat

Table 5.3. Some Seeds Containing Germination Inhibitors

Species	Location	Inhibitor
Acer negundo	Pericarp	ABA
Avena fatua	Not determined	ABA
		Short-chain fatty acids
Beta vulgaris	Pericarp	Phenolic acids
		cis-Cyclohexene-1,2-dicarboximide
		Inorganic ions
Corylus avellana	Testa, embryo	ABA
Eleagnus angustifolia	Pericarp, testa	Coumarin
	Embryo	Coumarin
Fraxinus americana	Pericarp	ABA
	Embryo	ABA
Medicago sativa	Endosperm	ABA
Prunus domestica	Embryo	ABA
Rosa canina	Pericarp, testa	ABA
Taxus baccata	Embryo	ABA
Triticum spp.	Pericarp/testa	Catechin, tannins

must offer some mechanical resistance and that to penetrate it the embryo must exert a certain minimum force, or thrust, or must degrade the restraining tissue by enzymes. The ability to do one or both of these may be curtailed by the presence of inhibitors either held in the embryo by the impermeability of the coat or supplied to the embryo by the external layers. To overcome the inhibition, adequate oxygen must be available; this could be secured by elevating the external oxygen tensions or by puncturing the seed coat. Hence, in this scenario we can see how several of the possible constraints might act together to maintain the intact seed in a dormant state.

5.3. EMBRYONIC INADEQUACY—THE CAUSES

We now have to consider why the embryo in the intact seed cannot overcome the constraints to which it is subjected—the intrinsic ones and those originating in the coat. One possibility we will discuss is that the embryo is metabolically deficient in some way.

5.3.1. Metabolism of Dormant Seeds

One problem here is that we have not yet been able to identify any particular aspect of metabolism that is peculiar to a germinable seed. Is there, for example, some metabolic pathway that occurs especially in a germinable seed and are there some special "germination enzymes" and "germination biochemical processes"? Since we do not know the answers to these questions, we cannot begin to investigate the fine differences between dormant and germinable seeds, and we must therefore deal only with generalized aspects of metabolism.

An approach frequently taken is to compare the metabolism of dormant seeds with that of afterripened, i.e., nondormant, seeds; where possible we will make such comparisons here. To ensure that the comparison is meaningful we must not contrast a dormant seed with one that has germinated, so in this discussion we will deal only with metabolism occurring in the early phases after imbibition, before radicle emergence from the nondormant seed. Beginning with respiration, dormant seeds are capable of carrying out this activity to the same extent as nondormant material. Dormant and afterripened seeds of *Xanthium* and *Avena fatua*, for example, show equivalent oxygen consumptions up to the time of emergence of the radicle from the germinable seed (Table 5.4). In some cases (e.g., barley and rice) dormant grains exhibit even higher oxygen consumption than the nondormant ones, but in these instances part of the consumption could be taking place in extraembryonic tissues. And hand in hand with the equal oxygen utilization by dormant and nondormant seeds, the ATP amounts in both are comparable.

Notwithstanding these observations, several aspects of respiratory metabolism have received much attention with regard to their possible participation in dormancy. Especially prominent is the pentose phosphate pathway which has been suggested to play a key role; this is discussed later in the context of breakage of dormancy by chemicals and other factors. Quantitative and qualitative differences have been recorded in some hexose and triose phosphates between dormant and nondormant seeds. Higher concentrations of fructose 2,6-bisphosphate are achieved in nondormant than in dormant grains of *Avena sativa* during the first few hours of imbibition. This is a consequence of the higher amounts of phosphoenolpyruvate and glycerol 3-phosphate in dormant grains, two compounds which inhibit phosphofructokinase, the enzyme phosphorylating fructose 6-phosphate to fructose 2,6-bisphosphate. One important action of fructose 2,6-bisphosphate is to regulate gluconeogenesis through its inhibitory effect on fructose 1,6-bisphosphatase, but the possible significance of this finding with regard to dormancy is obscure. Changes in the concentration of this hexose phosphate have been reported only in one other species (*Oryza* spp.), so it is not known if the effect is widespread.

Another area of metabolism that may be particularly important for germination is nucleic acid and protein synthesis, for it is plausible that herein lie the key differences between embryos in dormant and nondormant seeds. As we indicated previously, perhaps dormant seeds cannot make the right enzymes, because of an inability to transcribe certain genes. Again, however, dormant seeds seem to suffer no general deficiency in this regard. They are able to incorporate radioactive amino acids into protein at a rate no less than that of their fully germinable counterparts, and, similarly, RNA synthesis apparently proceeds normally in embryos of dormant seeds, from which active polysome preparations can also be made. A note of warning should be sounded here. The techniques used to study these metabolic processes have only covered overall macromolecular synthesis and have not yet been able to resolve fine differences that might arise, say, from the lack of synthesis of a small number of RNA species and proteins. With the techniques now available to approach these problems, it is likely that we will see some important information

Table 5.4. Gas Exchange in Dormant and Nondormant *Avena fatua*[a]

	O_2 uptake (μl/h per 10 embryos)	CO_2 evolution (μl/h per 10 embryos)
Dormant	12.07 ± 0.55	11.74 ± 0.63
Nondormant	12.16 ± 0.41	11.59 ± 1.14

[a]Determined during the first 10 h after the start of imbibition. After Simmonds and Simpson (1971).

emerging, to enable us to answer the question: are there special germination macromolecules that an embryo in a dormant seed cannot make, or, alternatively, are there special dormancy macromolecules? This is discussed in the next section and in Section 5.5.1.8.

5.3.2. Gene Expression and Dormancy

Since dormancy involves an inability of embryos to complete the germination process it would seem logical to suppose that they lack some essential component(s) of the germination mechanism. The component may be a protein, the product of expression of a gene. No protein has been found that is produced by nondormant embryos and not by dormant ones; and, of course, no gene has been found that is differentially expressed in the appropriate manner.

Recent work suggests, however, the reverse situation may exist—that genes are expressed specifically in dormant embryos! The first indication of this came from the isolation of several cDNA clones which showed that expression of certain genes was maintained in embryos of dormant, but not nondormant, wheat grains. A cDNA clone has also been obtained using mRNA from hydrated, dormant embryos of *Bromus secalinus* (a grass) which is absent from their nondormant counterparts hydrated for the same period. Using this cDNA as a probe it appears that the matching mRNA is initially present in both dormant and nondormant embryos and increases with time in the former but not in the latter, which germinate (Fig. 5.10). It may be, therefore, that dormancy requires the expression of a gene whose product actually prevents the completion of germination. The sequence of the "dormancy cDNA" of *Bromus* yields no clues as to the identity of the gene product. Although the expression of the gene is enhanced by ABA (Fig. 5.10A), the cDNA sequence bears no resemblance to other well-known ABA-induced proteins such as the LEAs. The wheat "dormancy genes" are also ABA-responsive. We should note that "dormancy mRNA" is present in dry embryos of *Bromus* whether dormant or not, so the gene must be expressed during seed development and/or maturation. Since the nondormant embryos fail to produce new mRNA when they are rehydrated, one can infer that dormancy breakage has suppressed the potential for expression of the gene. It remains to be seen if these findings have wide significance.

5.3.3. Membranes and Dormancy

The state of cell membranes in relation to dormancy has received special attention. In some species, a striking feature of seed dormancy is its temperature dependence, i.e., so-called relative dormancy, where the dormant condition is

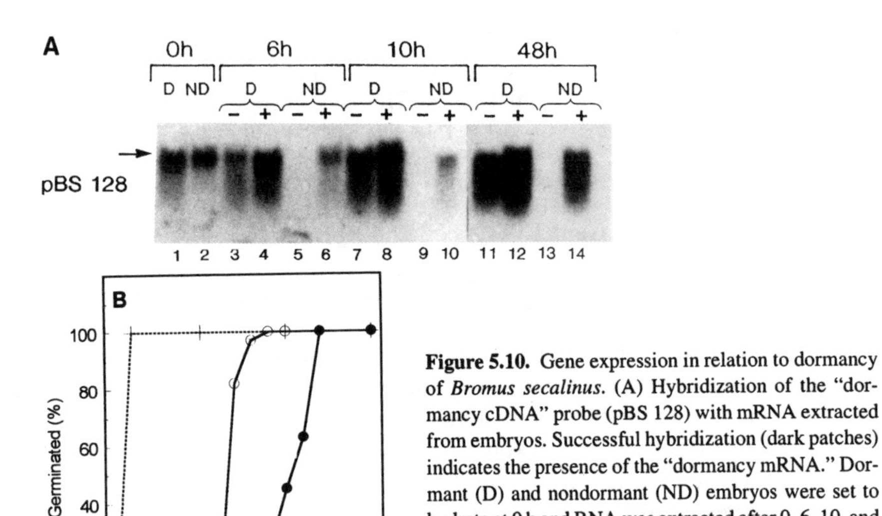

Figure 5.10. Gene expression in relation to dormancy of *Bromus secalinus.* (A) Hybridization of the "dormancy cDNA" probe (pBS 128) with mRNA extracted from embryos. Successful hybridization (dark patches) indicates the presence of the "dormancy mRNA." Dormant (D) and nondormant (ND) embryos were set to hydrate at 0 h and RNA was extracted after 0, 6, 10, and 48 h. Embryos were treated with (+) or without (–) 50 μM ABA before extraction. (B) For reference purposes the time courses of germination are shown of isolated nondormant (- - - -) and dormant (O) embryos and whole dormant seeds (●). After Goldmark *et al.* (1992).

exhibited only at certain temperatures (Fig. 5.11). The noticeable characteristic of this is the abruptness with which the dormancy response to temperature occurs, suggestive of a sudden change in the cells. Events that are known to occur in this abrupt manner in response to temperature are phase transitions of cell membranes; these undergo a sudden change from a crystalline or gel phase at lower temperatures to a fluid, liquid-crystalline phase at higher temperatures. Many properties of the membrane (e.g., control of solute passage, activity of bound enzymes) are altered when this happens, possibly with profound effects on the physiology of the cells, which could interfere with processes necessary for germination. Evidence that the temperature dependence for dormancy may in reality be connected with membrane changes comes from two observations: (1) Leakage of amino acids from certain seeds occurs at the same temperature at which dormancy is expressed (Fig. 5.12); (2) a crude membrane fraction from temperature-sensitive seeds shows phase transitions, when tested by certain biophysical techniques (fluorescent probes), at the same temperature at which dormancy is expressed. Further support for a special role of cell membranes in the dormancy mechanism comes from the ability of certain chemicals to release seeds from dormancy. These chemicals are anesthetics—ethyl ether, chloroform, acetone, and ethanol—which are known to act on cells by entering membranes,

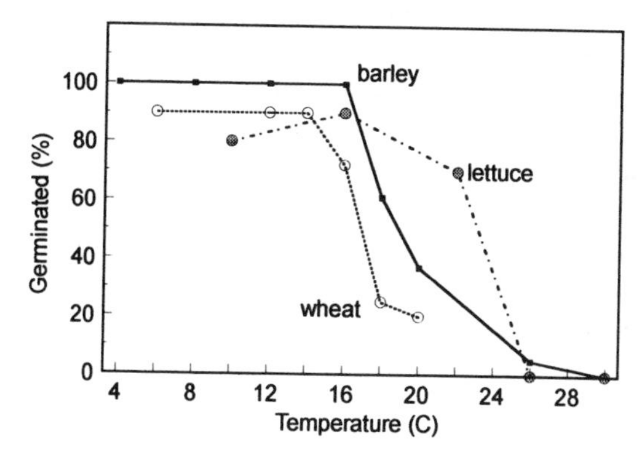

Figure 5.11. Relative dormancy in three species. In all three species there is no dormancy in seeds held at temperatures below approximately 15°C, i.e., almost all of the seeds germinate at these low temperatures. Dormancy is expressed as temperatures rise above approximately 15°C and is present in almost all of the barley grains and lettuce (cv. Grand Rapids) seeds at 25°C and in most of the wheat grains at 20°C. Data for barley after Roberts and Smith (1977) and for wheat and lettuce by M. Black.

thus altering relationships between membrane components. It is of great interest that these chemicals, especially ethanol, are remarkably effective in terminating dormancy in seeds of several grass species, e.g., *Panicum* spp. and *Digitaria* spp.; propanol has a similar effect on lettuce seeds. It should be noted, however, that there is evidence that ethanol might have effects other than on membranes, for in some cases it must be metabolized by alcohol dehydrogenase. This qualification notwithstanding, we are justified in suspecting that there is particular participation of membranes in dormancy, and the evidence just outlined points to several lines of investigation worthy of further exploration.

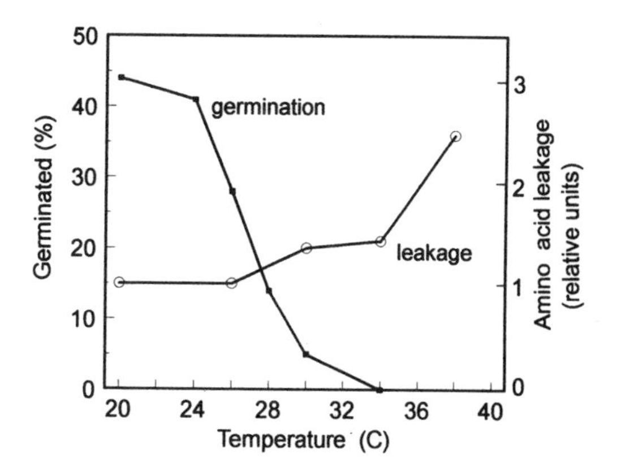

Figure 5.12. Temperature dependence of dormancy and amino acid leakage. In this batch of lettuce seeds (cv. Grand Rapids) there is a sharp drop in the percentage of germinated seeds in darkness (i.e., a rise in dormancy) beginning at 25–26°C. At these temperatures, amino acid leakage from the seed shows a sharp rise suggesting that a change in permeability of the plasma membrane of at least some embryo cells has occurred. Adapted from Hendricks and Taylorson (1979).

The reader will have concluded that we cannot identify with certainty any area of metabolism or biochemistry of the dormant seed that accounts for the inadequacy of the embryo and its failure to overcome the constraints imposed on it. There is, however, an approach to the problem that we have not yet covered fully in our discussion: this is to elucidate the biochemical changes in the embryo that occur when dormancy is broken. We have already touched on this subject very briefly with respect to afterripening, but later we will examine it in more detail in connection with the action of chilling, light, and certain chemicals in the termination of dormancy (Section 5.5.1.9).

5.4. DEVELOPMENT OF DORMANCY

A seed may become dormant while still on the parent plant, or after dispersal, to have primary or secondary dormancy (Section 5.2.2). Sometimes a seed may pass from primary dormancy into secondary dormancy, a transition we can recognize because different factors are required to break the two kinds of dormancy. There is no evidence at present to suggest that the mechanisms of primary and secondary dormancy differ in any fundamental way.

5.4.1. When Does Primary Dormancy Occur?

The stage in the seed's development and maturation at which primary dormancy sets in varies from species to species. In some, such as certain embryo-dormant strains of *Avena fatua*, it is a very early event, detectable 10 days after fertilization, when the embryo is isolated and placed on a liquid medium. Other cereals—wheat is an example—also display at an early age what seems to be a kind of dormancy. When young (e.g., 20 days after fertilization) grains are tested, they are incapable of germination unless the pericarp is removed or pricked, or if they are treated with high oxygen concentrations—typical treatments to break dormancy. Thus, it seems that these grains have a coat-imposed dormancy at quite an early age. On the other hand, dormancy may be delayed until a seed is almost fully mature. This happens in *Sida spinosa* and *Medicago lupulina* (Fig. 5.13), for example, and in both cases, changes in the seed coat (apparently precipitated by drying in *Sida*) seem to be responsible for initiating the dormant state. Because dormancy in *Medicago* is such a late event, seeds still on the parent plant frequently germinate when the plants fall and the fruiting heads touch the damp earth.

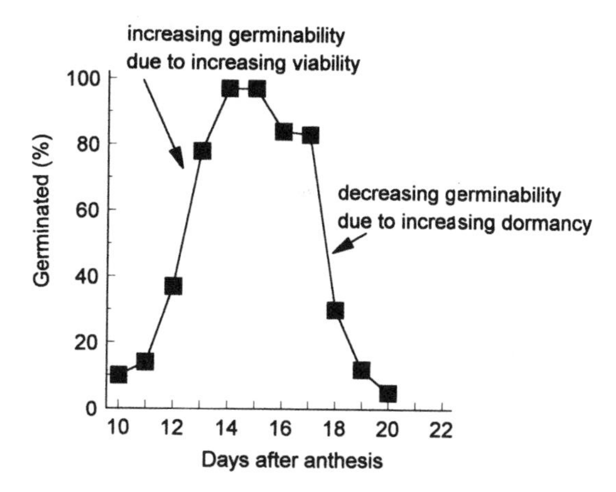

Figure 5.13. Onset of dormancy in *Medicago lupulina*. Germinability of seeds was determined at different ages (i.e., days after anthesis–dehiscence of the anther, which is very closely followed by pollination and fertilization). After approximately 15 days, germinability declines owing to the development of dormancy, which is complete in 20-day-old seeds. Adapted from Sidhu and Cavers (1977).

5.4.2. Genetic Control of Dormancy

Although environmental factors and correlative effects are important, the entry into dormancy is also under genetic control.

Pure lines of certain species have been isolated that show contrasting degrees of dormancy. There is a wide range of dormancy encountered in natural populations of wild oat (*Avena fatua*), from types which have a short-lived dormancy to those whose dormancy is very prolonged. From these, deep embryo dormancy has been inbred into one line. Pure lines have also been isolated of two groups whose grains respond to temperature in distinct ways, one showing little or no dormancy at temperatures in the range 4–32°C and the second exhibiting dormancy at intermediate temperatures. And in lettuce, different lines of the cultivar Grand Rapids exist, some showing the classical dormancy symptoms and others with no dormancy at all. We should recall, in this context, that dormancy can depend on an interaction between coat and embryo, and therefore three genetically distinct components are involved—the diploid embryo, with maternal and paternal genes, the diploid testa, pericarp, and hull, of maternal constitution only, and the triploid endosperm, bearing two sets of maternal genes. It may well be, then, that the genetics of coat-imposed dormancy are rather complex.

5.4.3. Correlative Effects in Dormancy

Though dormancy must have a genetic basis, there is obviously a great deal of plasticity in phenotypic expression, as determined by correlative phenomena within the plant, and by the environment. The phenomenon of polymorphism indicates the existence of the former, since genetically identical seeds or dispersal units might differ in their dormancy characteristics, depending on their position on the plant. In some species there are clearly influences of the seeds on each other. Spikelets of *Avena ludoviciana*, for example, produce a small distal caryopsis with a fairly deep dormancy and a large, proximal one which is less dormant. But if the latter is removed during grain development, the remaining distal caryopsis is much less dormant at maturity. This suggests that caryopses may compete for factors whose supply can determine the degree of dormancy. These factors might include hormones, but no information is available as to how these are distributed to different seeds in an inflorescence or to different inflorescences. However, the hormonal status of a seed, as established during its development and maturation, can certainly influence the degree of dormancy. For example, if gibberellic acid is fed to plants of *Avena fatua* and *A. ludoviciana* while the grains are developing, the mature grains have no dormancy, though the unfed, control grains are fairly deeply dormant. This artificial elevation of the growth regulator supply to the seeds might mirror the situation that could obtain in nature, where variations in hormonal flux from the parent plant to the seeds could be brought about by changes in the plant's nutrition and by environmental factors.

5.4.4. Dormancy Induction by Abscisic Acid

The growth regulator (or hormone) that has received considerable attention with respect to the onset of dormancy is ABA. This substance is found developing seeds of many species, in many cases even when the mature seed is nondormant, such as the pea. So the presence of the inhibitor does not necessarily lead to the inception of dormancy, though this could be because that ABA content in many cases falls considerably just before full maturity is reached. It has been claimed that in rice, however, the ABA is retained into grain maturity by those cultivars with dormancy, but not by those which are nondormant when mature. This seems to be an isolated example, however, and we often find no clear correlation between the retention of ABA into maturity and the depth of dormancy: the nondormant Great Lakes cultivar of lettuce, for example, contains relatively high amounts of ABA! The temporal relationship between entry into primary dormancy and the ABA content has been followed in detail in several species, including peach and apple (embryo dormancy), and we will use the latter to illustrate the findings (Fig. 5.14). No consistent relationship between the development of embryo dormancy and the ABA concentration in the embryonic axis is discernible. A rise in free ABA in the cotyledons does occur, however, at about the time when the embryos begin to move into dormancy. It is possible that this increase in ABA is responsible for the commencement of dormancy of the axis (recall the role of the cotyledons in embryo dormancy), but one would have to suggest that ABA is working through an intermediary, since there are no indications here that the ABA content of the cotyledons is matched by a rise in the axis. It is important to remember the evidence that in embryo dormancy the steady production of ABA by synthesis (*Helianthus annuus*) or release (apple) (Section 5.2.4) appears to be responsible for the maintenance of dormancy. And so, in these cases, dormancy is not induced by a short-term event during development.

We should be aware, also, that in other cases, say of embryos in coat-imposed dormancy, an involvement of ABA in dormancy inception does not necessarily require a continuous correlation between inhibitor content and the extent of dormancy. It is conceivable, for example, that once the inhibitor has reached a certain threshold level, events are slowly set in train that lead to dormancy. Alternatively, sensitivity to ABA may change and only when the embryo becomes responsive does dormancy begin. Supporting the latter possibility is the finding, not in seeds but in the aquatic plant *Spirodela*, that induction of dormant buds (turions) by added ABA can occur only during a fairly short time period, when the plant is sensitive to the inhibitor.

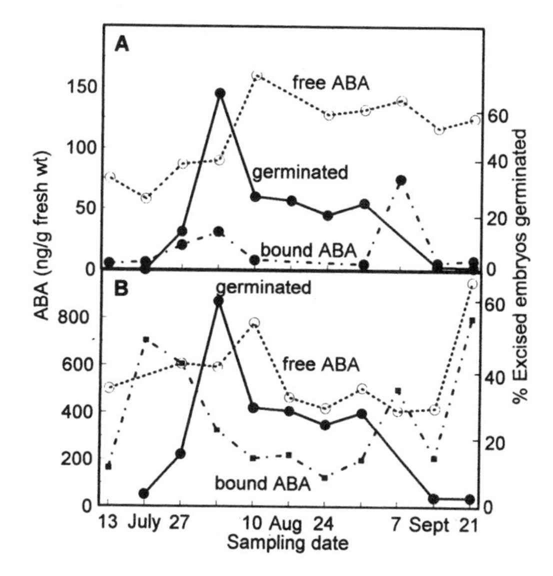

Figure 5.14. Embryo dormancy and concentration of ABA in developing apple embryos. ABA (free, and bound to glucose) in the cotyledons (A) and embryonic axes (B) was determined at stages of embryo development during July, August, and September. Germinability of the embryos, isolated from the seeds, was also determined. Note that germinability declines (owing to the onset of dormancy) after August 3. After Balboa-Zavala and Dennis (1977).

Clarification of some of these points comes from work with mutants, especially of *Arabidopsis thaliana*. Here, dormancy (coat-imposed) in the wild type sets in about half way through seed development,but in the ABA-deficient mutants it does not appear at all. The ABA content of wild-type seeds rises during their development to reach a peak approximately at the mid-point after which it declines, so that mature seeds have hardly any. Mutant seeds contain very little ABA at all times (Fig. 5.15). It appears, therefore, that dormancy is initiated by ABA during seed development but that the regulator's continued presence is not required. Genetic crosses between the wild type (*Aba*) and the mutant (*aba*) followed by backcrossing the heterozygote F_1 (which can make ABA) with the mutant produces offspring, some of which are *aba/aba* embryos lying in *Aba/aba* maternal tissue while other embryos are *Aba/aba*. Dormancy is instituted in the developing seeds only if the embryo itself has the *Aba* allele. Hence, induction

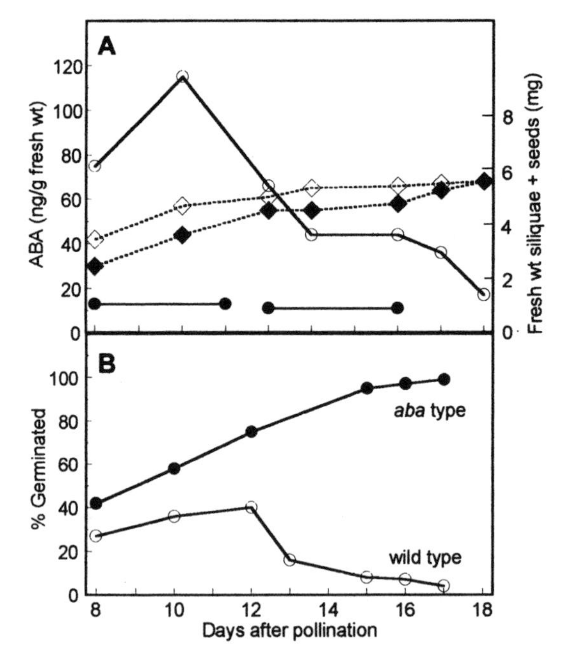

Figure 5.15. Absence of dormancy inception in an ABA-deficient mutant of *Arabidopsis thaliana*. (A) ABA contents (circles) of seeds and fresh weights (diamonds) of seed-containing fruits at different times after pollination of wild-type *Arabidopsis* (open symbols) and ABA-deficient mutants (*aba*) (closed symbols). The horizontal lines indicate the time over which samples were taken. (B) Germination after 14 days of *aba* seeds and wild-type seeds tested at different developmental ages. Note the lower germination percentages achieved by wild-type seeds and the entry into full dormancy by approximately 16 days. Adapted from Karssen *et al.* (1983).

of dormancy is by embryonic, not maternal, ABA. A similar situation exists in tomato, again elucidated by the use of a mutant, the ABA-deficient *sitiens*. Consistent with this finding is the observation that seeds in mutant plants cannot be made dormant by application of ABA. Other *Arabidopsis* mutants—the ABA-insensitive (*abi*) and particularly the recombinants *abi/aba*—also produce nondormant seeds; indeed, they may even begin to germinate while still in the fruit.

Interference with ABA production by biochemical means also modifies the onset of dormancy. In *Helianthus annuus*, which has embryo dormancy, developing seeds treated with fluridone (which inhibits ABA synthesis) fail to become dormant. The initiation and maintenance of embryo dormancy in this species apparently requires a continuous input of ABA.

There is evidence that sensitivity to ABA can determine whether or not dormancy becomes expressed. This is illustrated by two cultivars of wheat whose seeds have more or less coat-imposed dormancy. The concentration of ABA is similar in embryos of both kinds but those embryos isolated from the more dormant seeds are much more sensitive to applied ABA (which inhibits germination) than those from the less dormant seeds. It has been suggested, therefore, that the high sensitivity to ABA of embryos of one cultivar might account for the dormancy of the seed.

The question now arises as to the mechanism by which ABA might initiate dormancy. The nondormant seeds of the ABA-deficient and ABA-insensitive mutants of *Arabidopsis*, and the various recombinants, display several kinds of biochemical differences from the wild type, such as failure to make certain reserve proteins, lack of certain oligosaccharides, etc., but none of these differences is likely to account for the absence of dormancy. ABA is a potent regulator of expression of several genes, especially some LEAs and storage proteins, but there is no reason yet to implicate these in the dormancy syndrome. The ABA-sensitive embryos of the dormant wheat cultivar discussed above produce dehydrins in response to ABA, but again these are unlikely to be involved in dormancy. There may, of course, be other ABA-regulated "dormancy genes" which await discovery. The gene in *Bromus secalinus* (Section 5.3.2) may be one such kind which gives us a tantalizing glimpse of the possibilities.

Clues as to the possible site of action of ABA in the induction of dormancy come from findings made with mutant tomato. Even after dormancy has been terminated, differences in the elongation capacity are evident between radicles of the never-dormant *sitiens* and those of the previously dormant wild type. These differences are displayed in response to osmotica which are more inhibitory to radicle extension of the wild type (Fig. 5.16). This outcome could reflect a persistent effect of ABA on the capacity for radicle elongation, i.e., to generate enough force to overcome the osmotic restraint. One possibility is that ABA has a long-lasting effect on wall stretchability (Section 5.5.1.10). In tomato seeds, the resistance of the endosperm impedes radicle growth so this too may be a contributory factor. In this case, the ability of the radicle to soften the endosperm cell walls may be involved (Sections 4.2 and 5.5.1.8).

5.4.5. The Environment in Dormancy Inception

Although there is no certainty as to the nature of the internal controls that force a seed into dormancy, there is no doubt that environmental factors play an important part. The effect of these factors partly explains why dormancy varies with provenance. Certain species used commonly by researchers on seed physiology (e.g., the Grand Rapids cultivar of lettuce) are notorious in this respect, and one cannot assume that a particular batch will have any dormancy at all: this depends on where it was grown and what environmental factors were operative at the time. What are these factors which can have such a profound effect on the development of dormancy? The edaphic factor is likely to be important, but relatively little information is available about this. Much more is known about the effects of temperature and of light—its quantity, daily distribution, and spectral quality.

A theory of dormancy inception has been put forward by Vegis which attributes potent effects to relatively high temperatures and partial anaerobiosis—the latter enforced by the tissues covering the embryo. As a result, the seed moves in time through predormancy, dormancy, and then after maturity into postdormancy. Each phase is characterized by a response of the seed to temperature: in the first, as the seed approaches dormancy, there is a narrowing of the temperature range over which germination can occur until in full dormancy, it might not occur at any temperature; finally, the temperature range widens as full dormancy declines during the postdormant phase. Although the theory is attractive in some respects, especially in that it draws attention to

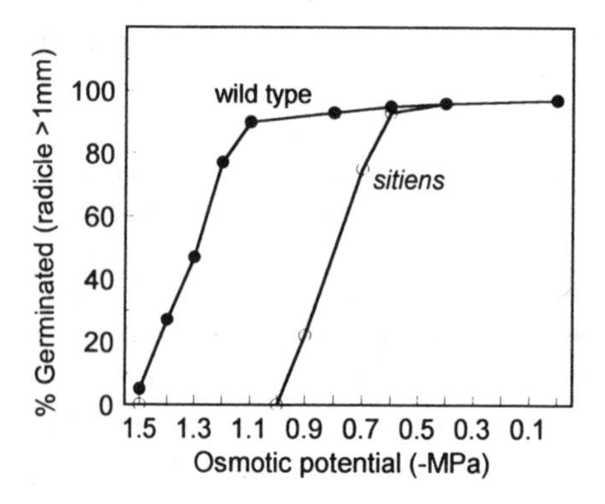

Figure 5.16. Effect of osmotic potential on germination of wild type and ABA-deficient *sitiens* mutant of tomato. Dormancy of wild type was removed by afterripening. Seeds of wild type (O) and *sitiens* mutant (●) were incubated on osmotica (solutions of polyethylene glycol) of different strengths. Germination (radicle > 1 mm) was recorded after 12 days. Note that *sitiens* radicles elongate in osmotic potentials as low as −1.2 MPa which is completely inhibitory to wild-type radicles. After Groot and Karssen (1992).

changes in response to temperatures as dormancy lessens, parts of it seem to have some inherent difficulties. There are cases, for example, where development of dormancy is promoted not by high temperatures, but by relatively low ones, as in *Rosa* spp. and in grasses such as *Avena fatua* and the cereals, wheat, and barley. It is hard to understand, also, how polymorphism can be accommodated within the theory, or how various day-length effects might operate. There are, nevertheless, some cases in which dormancy is induced by the elevated temperatures. *Syringa vulgaris*, for example, can be made to produce dormant seeds by holding the plant at relatively high temperatures (18–24°C) during the last week of seed maturation, a treatment that appears to make the endosperm (the tissue imposing dormancy) tougher.

Photoperiodic effects in the inception of seed dormancy are well known in several species, and *Chenopodium* spp. provide good examples. *C. album* (Fig. 5.17) has deeply dormant seeds when the fecund plants are held under long days, but nondormant seeds under short days. Not only is the dormancy pattern affected by day-length but so is the structure of the seeds, for those maturing in long days are smaller and thicker coated than those in short days. Dormancy induced by photoperiod is not always associated with coat thickness, however, since a short-term dormancy of seeds with thin coats is brought about by long photoperiods given for just a few days after the end of flowering. An effect of photoperiodic conditions on coat structure is also seen in seeds of several other species. When grown in long days (the last 8 days of maturation are most important), seeds of *Ononis sicula* are relatively large with thick, yellow coats, are poorly permeable to water, and show low germinability. After short days, they are, in contrast, smaller, green, and thin coated and, because they readily take up water, most of them germinate.

Light, quite apart from the photoperiod, has an important role in dormancy induction in several species. Seeds of *Arabidopsis thaliana* maturing

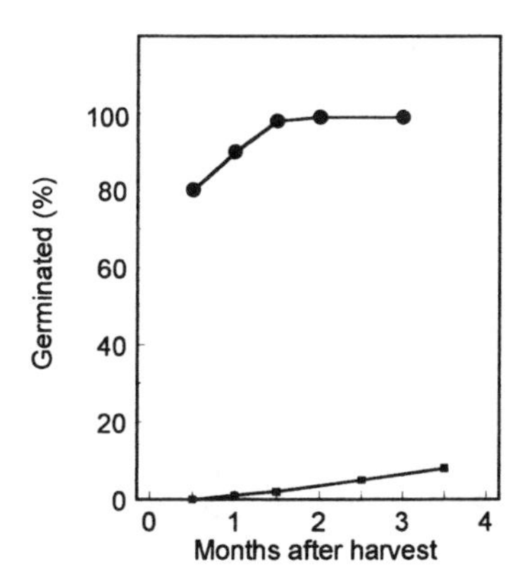

Figure 5.17. Germination of *Chenopodium album* seeds that developed under different photoperiods. *Chenopodium album* plants with developing seeds were held under short days or long days. After harvesting, germination was tested at intervals. Note that seeds developing and maturing under short days (●) have a high germination percentage whereas those from long-day-treated plants (■) have a dormancy that lasts for at least 3.5 months. All germination tests were carried out in darkness. Adapted from Karssen (1970).

in white fluorescent light have little dormancy when harvested, whereas those which have experienced incandescent light remain deeply dormant for at least several months (Fig. 5.18). The explanation of this effect is that white fluorescent light is relatively rich in the red wavelengths whereas incandescent light has a relatively high component of far-red light. Under the former illumination conditions there is a higher concentration of the active Pfr form of phytochrome (Section 5.5.1.4) set up in seeds than there is under the latter type of light. Seeds with a high Pfr content can germinate in darkness while those with a low Pfr concentration remain dormant (see Section 5.5.1.4f for further discussion). Treatment with far-red light alone, of course, has the same effect: cucumber seeds (*Cucumis sativus*) are made dormant when fruits are irradiated with far-red light. It appears that this kind of phenomenon also occurs in nature, where the source of far-red light is light filtered through green tissues (Section 6.4). Chlorophyll absorbs red light (peak ca. 660 nm) but not wavelengths longer than about 710 nm; hence, the transmitted light, being rich in the far-red component, serves to lower the amount of Pfr. Seeds of many species mature and dry while the surrounding tissues (fruit or seed coat) are still green, and thus the embryos are in a far-red-rich environment: such seeds are dormant, whereas those whose covering tissues have little chlorophyll are nondormant when mature (Fig. 5.19).

5.4.6. Development of Hard Coats

We have already mentioned that photoperiodic conditions can affect seed coat thickness and color. The hardness of seed coats also depends on the rate and

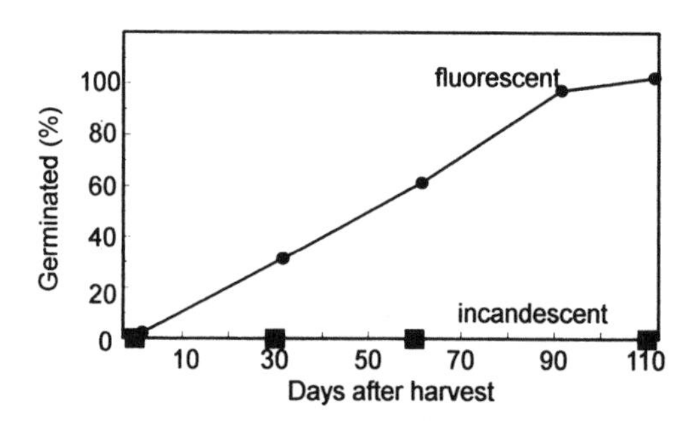

Figure 5.18. Light quality during seed maturation and subsequent dormancy. Seed-bearing plants of *Arabidopsis thaliana* were kept in white incandescent or white fluorescent light during seed maturation. Seed germination was subsequently tested in darkness. Seeds maturing in white fluorescent light (●) have little dormancy compared with those maturing in incandescent light, which are deeply dormant for more than 100 days (■). Adapted from Hayes and Klein (1974).

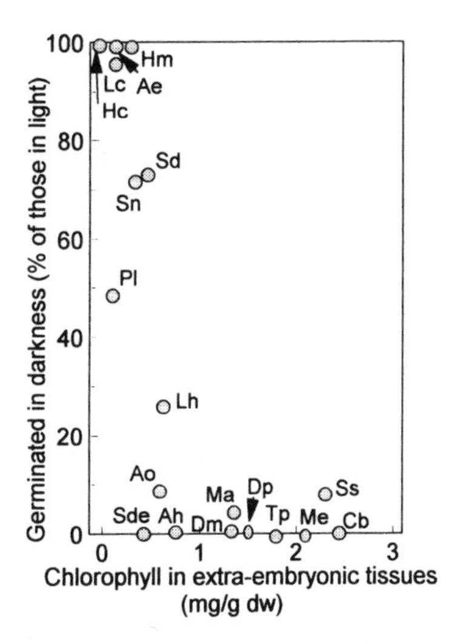

Figure 5.19. Extraembryonic chlorophyll content and subsequent seed dormancy. Chlorophyll content of the tissues surrounding the maturing embryos was determined at the midpoint of seed drying. Mature seeds were collected and their germination in the light and dark was tested. Germination in the dark as a percentage of that in the light is shown for each species. Note that all of the species that have little or no requirement for light (i.e., no light-sensitive dormancy)—Hc, Ae, Hm, Lc, Sn, Sd—have a low chlorophyll content in the extraembryonic tissues. Of the light-requiring seeds, the majority (Dm, Ma, Dp, Tp, Me, Ss, Cb) have moderate to high extraembryonic chlorophyll. Ae, *Arrhenatherum eliatus;* Ah, *Arabis hirsuta;* Ao, *Anthoxanthum odoratum*; Cb, *Capsella bursa-pastoris;* Dm, *Draba muralis;* Dp, *Digitalis purpurea;* Hc, *Helianthemum chamaecistus;* Hm, *Hordeum murinum;* Lc, *Lotus corniculatus;* Lh, *Leontodon hispidus;* Ma, *Myosotis arvensis;* Me, *Millium effusum;* Pl, *Plantago lanceolata;* Sd, *Silene dioica;* Sde, *Sieglingia decumbens;* Sn, *Silene nutans;* Ss, *Senecio squalidus;* Tp, *Tragopogon pratensis.* After Cresswell and Grime (1981).

degree of drying, and while drying in air is proceeding, phenolic substances are oxidized to dark-colored compounds that may contribute to coat impermeability. In some leguminous seeds, drying is controlled by the hilum. When the relative humidity is low, the cells bordering the hilar fissure shrink owing to loss of water, thus opening the tissue and allowing even more drying to occur. Under damp conditions, however, the cells expand to close the fissure. Drying of the seed is thus regulated according to the prevailing environmental conditions.

5.4.7. Secondary Dormancy

Secondary dormancy develops in already dispersed, mature seeds in response to unfavorable germination conditions (see also Section 6.8). Some examples are given in Table 5.5. The mechanism of induction is unknown but inhibitors have been invoked, though there is no evidence that ABA is involved. Indeed, even seeds of ABA-deficient mutants of *Arabidopsis thaliana* can enter secondary dormancy in response to high temperature. In general, secondary dormancy is characterized by a loss of sensitivity to dormancy-breaking factors such as light or nitrate. One concept to explain this is based on the high-temperature induced loss of key receptors for nitrate or Pfr perhaps in the plasma membrane. Proteinaceous receptors might be temperature labile or altered chemically say by phosphorylation. This speculative scenario awaits experimental confirmation.

5.5. THE EXTERNAL CONTROLS

We will divide our discussion of the external factors controlling germination into two parts. First, we will see how external factors are involved in the termination of dormancy, and second, we will examine the ways in which germination of a nondormant seed is regulated by the environment.

5.5.1. The Release from Dormancy

In the following account the termination of dormancy by several factors will be considered. For the purposes of this discussion we must treat these factors

Table 5.5. Factors Inducing Secondary Dormancy

Inducing factor	Example
Anaerobic conditions	*Xanthium pennsylvanicum*
Darkness (skotodormancy)	*Lactuca sativa*
	Lamium amplexicaule
	Phleum pratense
Prolonged white light (photodormancy)	*Lactuca sativa*
	Nemophila insignis
	Phacelia tanacetifolia
Prolonged far-red light (photodormancy)	*Arabis hirsuta*
	Amaranthus caudatus
	Lactuca sativa
Temperatures above maximum for germination	*Ambrosia trifida*
	Avena sativa
	Chenopodium bonus-henricus
	Taraxacum megalorhizon
Temperatures below minimum for germination	*Phacelia dubia*
	Taraxacum megalorhizon
	Torilis japonica
Water stress	*Lactuca sativa*

separately and examine some of the characteristics of their effects. It must be appreciated, however, that in nature, a seed is not subject to the influence of just one factor but to several simultaneously. Most species of seed are affected by more than one factor. For example, dormancy of many light-requiring seeds is also broken in darkness by chilling, by alternating temperatures, by exposure to nitrate in the soil, and by afterripening in the "dry" or near-dry state. The termination of dormancy in the field is not likely to be the result of just one factor in the seed's environment but will be influenced by several, and in some circumstances, seeds of the same species might have their dormancy ended by different agencies. Moreover, loss of dormancy may take place gradually. This is seen very well in those cases of temperature-dependent dormancy, i.e., relative dormancy (Section 5.2.2), where a widening range of temperatures at which germination can occur develops as dormancy is relieved. With these qualifications in mind we can now consider the factors that operate in dormancy breakage.

5.5.1.1. Afterripening

Dormant seeds, when "dry," slowly lose their dormancy by the process of afterripening, perhaps requiring as little time as a few weeks (e.g., barley) or as long as 60 months (e.g., *Rumex crispus*). "Dry" seeds, we should note, can have up to 18–20% water content. In ways that are not all understood, dry seeds change, and these changes somehow lead to the termination of dormancy. The efficacy of afterripening depends on the environmental conditions—moisture, temperature, and oxygen. Since afterripening generally occurs in dry seeds, or more accurately in seeds below a certain water content, it may be prevented in the presence of water. Indeed, at intermediate moisture contents not only might afterripening fail but the seed may also lose viability, and at higher water contents secondary dormancy may set in. It seems, nevertheless, that a minimum water content is required for afterripening, and if seeds become too dry (e.g., 5% water content), the process is delayed. The rate of afterripening depends on the temperature, as is shown in Table 5.6, a property that is sometimes exploited to accelerate the loss of dormancy in agriculturally important species such as barley and wheat. Afterripening is delayed when the oxygen tension is low and accelerated when it is high. But here we mean concentrations of oxygen of near zero to 100%, a range that dry seeds in nature would never encounter, although it might be employed experimentally in the laboratory.

5.5.1.2. Low Temperatures (Chilling)

A high proportion of species—probably the majority of nontropical species—can be released from dormancy when, in the hydrated condition, they

Table 5.6. Effect of Temperature on Afterripening Rate in Rice (*Oryza sativa*)[a]

Temperature (°C)	Time for 50% loss of dormancy (days)
27	50
32	30
37	15
42	8
47	5

[a]Based on data in Roberts (1965).

experience relatively low temperatures, generally in the range 1–10°C, but in some cases as high as 15°C (see Table 5.7). The operation of this kind of control in nature is obvious: dormancy of the hydrated seed is slowly broken over the winter and, as mentioned before, this is presumably a means of preventing germination until after the winter has passed (see also Section 6.5). Chilling of seeds to break dormancy is a long-standing practice in horticulture and forestry and is generally referred to as stratification, because the seeds are sometimes arranged in layers (i.e., stratified) in moist substrata.

Chilling is effective in seeds with embryo, coat-imposed, primary, relative, and secondary dormancy. We can recognize, in apple for example, how the different components of dormancy are differentially broken by chilling. Figure 5.20 shows that chilling for 60 days suffices to remove the embryo dormancy; the presence of the endosperm increases the required time to about 80 days, whereas the whole seed needs even longer than this. In general, woody species require fairly extensive treatment times, sometimes as much as 180 days (*Crataegus mollis*), but usually 60–90 days is satisfactory. In contrast, dormancy in some herbaceous species is broken by just a few days of low temperature (7 and 14, respectively, in *Poa annua* and *Delphinium ambiguum*) and by 12 h in wheat!

Recorded optimum temperatures are generally in the region of 5°C, but this figure may be misleading if the situation exemplified by *Rumex obtusifolius* is widespread. Provided the seeds are illuminated to prevent the onset of secondary dormancy, they are released from dormancy almost as effectively by 1.5, 10, and 15°C, at least with 2-week treatment times. As the treatment time lengthens, secondary dormancy sets in, the temperatures of 10 and 15°C cease to be as efficacious, and it then appears that 1.5°C is the most suitable temperature (Fig. 5.21). Temperatures as high as 10°C also serve to promote dormancy breakage in lettuce and wheat, a point to which we will return when we consider the chilling mechanism. To return briefly to secondary dormancy and chilling: the onset of this type of dormancy could explain why an interruption of chilling

Table 5.7. Termination of Dormancy by Various Factors

Factor				
Afterripening	Alternating temperatures	Chilling	Light	Species
+		+		*Acer pseudoplatanus*
+		+		*Ambrosia trifida*
+	+		+	*Agrostis tenuis*
+		+		*Avena fatua*
+		+	+	*Betula pubescens*
+			+	*Chenopodium album*
+		+		*Corylus avellana*
+		+		*Delphinium ambiguum*
+		+		*Hordeum* spp.
+			+	*Kalanchoë blossfeldiana*
+		+	+	*Lactuca sativa* (some cvs.)
	+	+	+	*Lepidium virginicum*
	+		+	*Lythrum salicaria*
	+		+	*Nicotiana tabacum*
		+	+	*Pinus sylvestris*
+		+		*Prunus domestica*
		+	+	*Poa annua*
+		+	(+)?	*Pyrus malus*
+	+	+	+	*Rumex obtusifolius*
+		+	+	*Senecio vulgaris*
+		+		*Triticum aestivum*

[a]+, Effective in dormancy breaking.

by high temperatures seems to cancel out the previous low-temperature experience and why the seed then has to undergo the full chilling treatment, without interruption, once more, i.e., short periods of chilling are not cumulative. Secondary dormancy might set in during the high-temperature interpolation.

What changes does chilling bring about in the seed to terminate its dormancy? The first problem we have in answering this question is to understand how the low temperature is sensed. Several possibilities exist. First, low temperatures might act simply to lower the rate of enzymatic reactions taking place in the seed. It would be expected that all of the metabolic processes are retarded by the chilling treatment, but it is not easy to see how this would have a positive effect in removing dormancy unless some of the affected processes are ones stopping the germination mechanism; i.e., they could be inhibitory, dormancy-imposing processes: it is then conceivable that if these are arrested, certain germination steps could therefore proceed, albeit slowly. On the other hand, the low temperature might have differential effects, perhaps because of differences in the activation energies of separate reactions. This would lead to some processes being less affected than others by the chilling treatment, and in this way

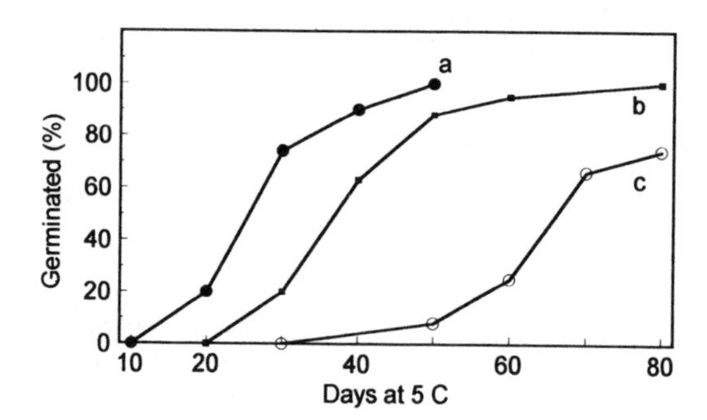

Figure 5.20. Termination of apple seed dormancy by chilling. Imbibed seeds were kept at 5°C and periodically removed for testing. The percentage germination of the following was determined: (a) isolated embryos; (b) seeds with testa removed but endosperm intact; (c) intact, whole seeds. Note that embryo dormancy is terminated by 40–50 days chilling, endosperm-imposed dormancy by 70–80 days chilling, and whole-seed dormancy by more than 80 days chilling. After Visser (1956).

the germination reactions could be favored. Other effects of low temperatures might include differential changes in enzyme concentration or in enzyme production. It must be admitted, however, that there is no convincing experimental evidence that any of these possible sensing mechanisms actually occurs.

One particular feature of the chilling syndrome nevertheless gives us a clue and suggests exciting avenues for exploration. From some cases that have been studied in detail, we see that termination of dormancy can be achieved by temperatures as high as 12–15°C, as in wheat and *Rumex obtusifolius*. And in lettuce, where chilling enhances the dormancy-breaking effect of light, temperatures up to about 10°C suffice. When these temperatures are exceeded, there is a fairly abrupt decline in effectiveness, seen very strikingly in *Rumex*, and rather less so in lettuce (Fig. 5.22). If we look at this in another way, it seems that as the temperature is decreased, a value is reached when temperatures sharply

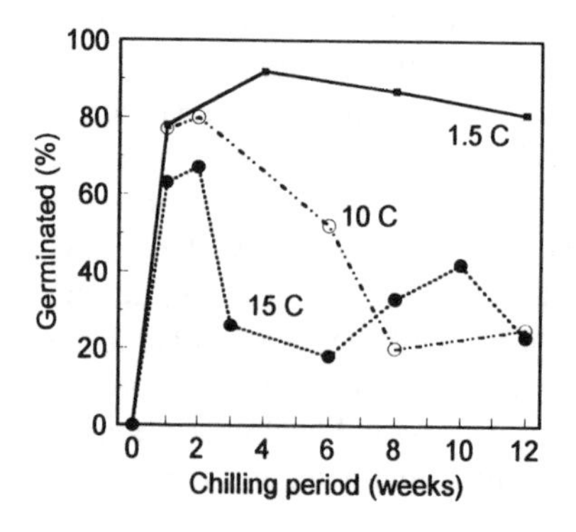

Figure 5.21. Temperature–time relationships for the termination of dormancy in *Rumex obtusifolius*. Imbibed seeds were held in the light at three temperatures for times up to 12 weeks. They were then transferred to 25°C for 4 weeks, after which the percentage of germinated seeds was determined. After Totterdell and Roberts (1979).

become efficacious in chilling, i.e., somewhere between 20 and 15°C in *Rumex*, about 12°C in wheat, and about 13°C in lettuce. Now we have already mentioned an event in cells that occurs at certain critical temperatures—the phase transition of cell membranes (Section 5.3.3). The characteristics of the temperature dependence for chilling would be consistent with the occurrence of such transitions. One might speculate, therefore, that a membrane could be the sensor for the low-temperature breakage of dormancy, and at a certain critical temperature (let us say around 15°C in the case of chilling), the membrane becomes crystalline (a gel), a change that would interfere with its functioning. Homeostatic mechanisms involving membranes are well known in many organisms—bacteria, algae, protozoa, and higher animals and plants—which, in response to temperatures lower than a critical value, adjust the composition of the membrane to maintain a certain degree of fluidity. This acclimation—sometimes called homeoviscous adaptation—involves alterations in membrane components, partly by increasing the proportion of unsaturated fatty acids in the phospholipids. A membrane (or membranes) could thus be altered in response to the low temperature, and such a change might be an essential prelude to the dormancy-breaking mechanism, which then proceeds when the seeds later experience higher temperatures.

5.5.1.3. Other Effects of Temperature on Dormancy

In the field, dormant seeds are commonly subjected to fluctuating temperatures, for example, low night temperatures and high daytime temperatures. Such temperature fluctuations, or temperature alternations, are frequently effective in dormancy breakage, in cases such as *Bidens tripartitus, Nicotiana tabacum,* and *Rumex* spp., which all have coat-imposed dormancy. The effect of alternating temperatures on *Rumex obtusifolius* is

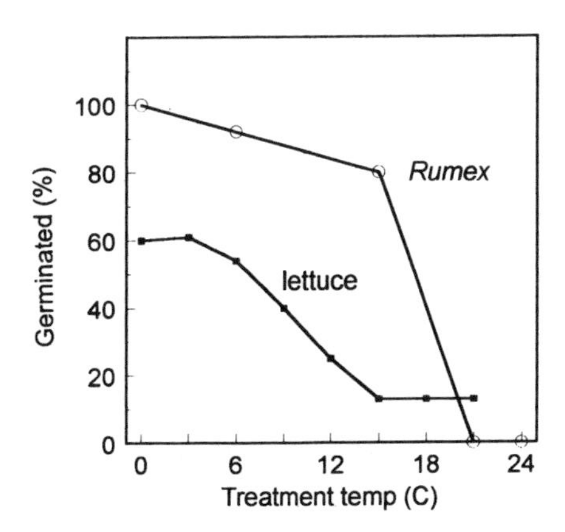

Figure 5.22. Temperature requirements for termination of dormancy. Imbibed seeds of *Rumex obtusifolius* were held at different temperatures for 4 weeks in dim, diffuse light, after which they were transferred to 25°C for germination. Based on data from Totterdell and Roberts (1979). Imbibed lettuce seeds (cv. Grand Rapids) were held at different temperatures for 6 h, then given far-red light for a few minutes, to generate low amounts of Pfr (see Section 5.5.1.4). They were then transferred to 20°C in darkness for 48 h after which germinated seeds were counted. After Van Der Woude and Toole (1980).

shown in Fig. 5.23. At constant temperatures, the seeds remain dormant, as shown by the valley running across the diagonal of the figure. Dormancy is broken when different temperatures are combined, with the maximum effect at the greatest temperature differentials (i.e., amplitudes). So to be effective, the temperature alternation must have a certain minimum amplitude, which in some species need be only a few degrees. Other parameters are also important. Without going into detail, we can recognize that the temperatures of the pair must be above and below certain values, that the duration of exposure to each is important, and that the number of cycles of fluctuating temperatures can be decisive.

In a few species, relatively high temperatures can break, or assist in breaking, dormancy. Seeds of *Hyacinthoides non-scripta* require several weeks at 26–31°C followed by a germination phase at 11°C. Several species that are chilling sensitive need a period at relatively high temperature before the cold. *Fraxinus* spp., for example, require a few weeks at about 20°C prior to chilling at 1–7°C. Softening of the seed coat of woody species might occur at the higher temperature.

5.5.1.4. Light

Light is an extremely important factor for releasing seeds from dormancy (see examples in Table 5.7). Almost all light-requiring seeds have coat-imposed dormancy. Seeds of many species are affected by exposure to white light for just a few minutes or seconds (e.g., lettuce), whereas others require intermittent illumination (e.g., *Kalanchoë blossfeldiana*). Photoperiodic effects also exist, so

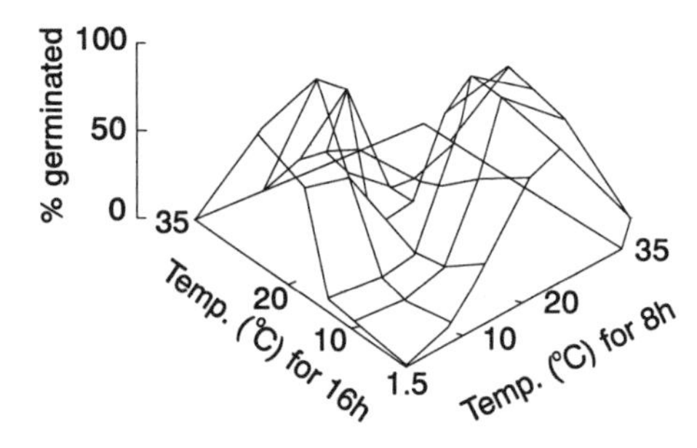

Figure 5.23. Breaking dormancy in *Rumex obtusifolius* by alternating temperatures. Imbibed seeds were held in darkness for 28 days at different temperature combinations—16 h at one temperature followed by 8 h at another. Germinated seeds were counted after 28 days. Note that high germination percentages (i.e., termination of dormancy) occur in the following temperature combinations: (a) 16 h at 25–35°C, 8 h at 1.5–15°C; (b) 8 h at 25–35°C, 16 h at 1.5–20°C. After Totterdell and Roberts (1980).

that some species require exposure to long days and others to short days (Table 5.8). The light requirement frequently depends on the temperature. Grand Rapids lettuce, for example, generally is dormant in darkness only above about 23°C, so below this value seeds germinate without illumination. Seeds of some species of *Betula*, on the other hand, are dormant in darkness at lower temperatures (e.g., 15°C) but not at 25–30°C. Sensitivity to light in many species is enhanced by chilling, but in some (e.g., *Betula maximowicziana*) light terminates dormancy only in seeds that have previously been chilled. Various temperature alternations and temperature shifts also interact with light, as we shall see later in this section.

5.5.1.4a. Action Spectra and Phytochrome. In nature, white light (i.e., sunlight) breaks dormancy, but we know that the wavelengths in the orange/red region of the spectrum are most effective. In 1954, a detailed action spectrum for the breaking of dormancy in the Grand Rapids cultivar of lettuce, obtained by Borthwick, Hendricks, and their colleagues, revealed that major activity is at 660 nm (Fig. 5.24A). Prior to this finding, inhibitory parts of the spectrum were also known, an especially potent waveband being far-red light, i.e., wavelengths longer than about 700 nm; the action spectrum showed that 730 nm is the wavelength of maximum activity (Fig. 5.24B). At about the same time as the action spectrum was discovered, Borthwick, Hendricks, Parker, E. H. Toole, and V. K. Toole showed that the red and far-red light are mutually antagonistic. This was done by exposing lettuce seeds to a sequence of red and far-red irradiations: only when the last exposure in the sequence was to red light was dormancy terminated (see Table 5.9). This established the fact of photoreversibility, i.e., the two wavelengths 660 and

Table 5.8. Illumination Conditions Required for the Breaking of Dormancy

Illumination conditions	Examples
Seconds or minutes	*Agrostis tenuis*
	Chenopodium album
	Lactuca sativa cv. Grand Rapids
	Nicotiana tabacum
Several hours (or intermittent)	*Hyptis suaveolens*
	Lythrum salicaria
Days (or intermittent)	*Epilobium cephalostigma*
	Kalanchoë blossfeldiana
Long days	*Begonia evansiana*
	Betula pubescens (at 15°)
	Chenopodium botrys (at 30°)
Short days	*Chenopodium botrys* (>20°)
	Tsuga canadensis
	Betula pubescens (>15°)

730 nm are able to reverse each other's effect. Light is, of course, absorbed by molecules of a pigment, and the one participating in the breaking of dormancy and in other photoresponses came to be called *phytochrome*. Phytochrome therefore exists in two forms. One is present in unirradiated, dormant seeds; it absorbs red light (peak at 660 nm) and is therefore designated as Pr. This form of the phytochrome obviously cannot break dormancy (if it could, the seeds would not be dormant!), but when activated by 660-nm light, it is changed into an active, dormancy-breaking form. But this active form absorbs far-red light (730 nm)—hence it is designated as Pfr—and having done so it reverts to Pr as follows:

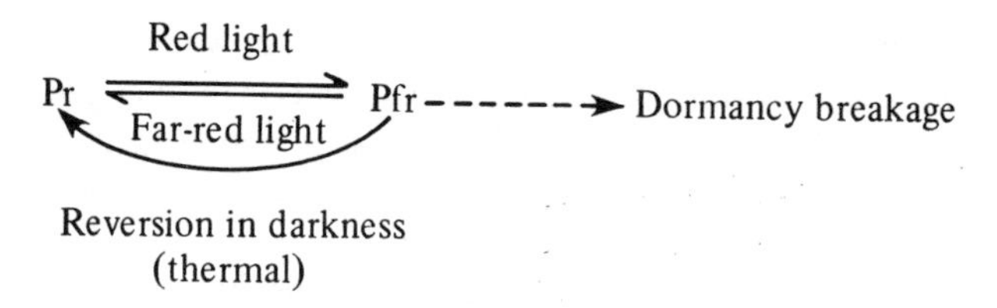

Ignoring for the moment the dark-reversion path and destruction (we will consider these soon), it is easy to see how the reversibility of dormancy breakage works: if Pfr is left in the seed at the end of the radiation sequence, dormancy is terminated, but if Pr is left, dormancy is retained. Note that photoreversion only stops the breaking of dormancy if Pfr has been given insufficient time in which to act. If dosage with far-red light is delayed for a few hours, Pfr can then operate,

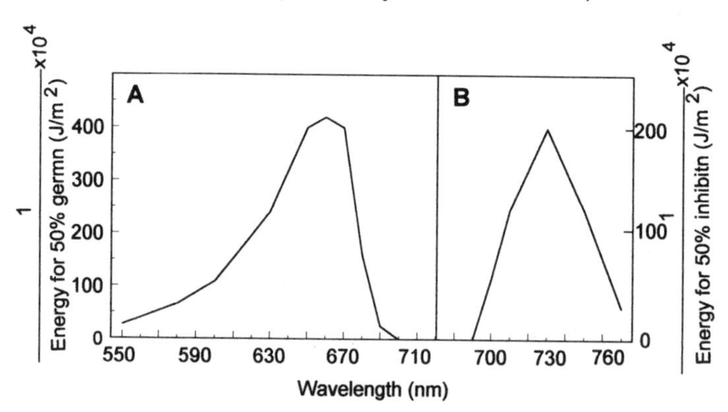

Figure 5.24. Action spectra of light effects on seed dormancy in lettuce cv. Grand Rapids. In (A), seeds were imbibed for 16 h in darkness and then irradiated with various energy levels of light at wavelengths between 560 and 690 nm. They were then returned to darkness and the percentage of germinated seeds was later counted. The energy at each wavelength required to produce 50% germinated seeds (i.e., to break dormancy in 50% of the seeds) was determined. In (B), energies at different wavelengths of far-red light effective in reversing the red-light termination of dormancy were determined. Redrawn from Borthwick *et al.* (1954).

Table 5.9. Phytochrome Photoreversibility and the Breaking of Dormancy[a]

Irradiation sequence	Germinated (%)
None (darkness)	4
R	98
FR	3
R, FR	2
R, FR, R	97
R, FR, R, FR	0
R, FR, R, FR, R	95

[a]Seeds of the Grand Rapids cultivar of lettuce were imbibed in darkness and then exposed to red light (640–680 nm) (R) for 1.5 min and far-red light (> 710 nm) (FR) for 4 min in the sequence shown. After irradiation, they were returned to darkness for 24 h before germinated seeds were counted.

and even if photoreversion occurs later, dormancy nevertheless has been terminated. The time needed for Pfr to act, after which photoreversion is inconsequential, is called the *escape time* (Fig. 5.35).

The energies needed to carry out these photoconversions are relatively small. A saturating dose of red light in lettuce seeds is about 10 Jm^{-2}—an amount given by about 0.2 s of direct, summer sunlight. Up to saturation value, the effect is directly proportional to the total amount of energy, irrespective of how the energy is delivered, i.e., by a lower fluence rate (irradiance) for a longer time, or by a higher fluence rate for a shorter time. As long as the products, fluence rate × time, are equal, the same effect (e.g., percentage of seeds breaking dormancy) is achieved. This is to say that the phytochrome system shows *reciprocity*. In a more complex response type, e.g., where repeated doses of red light are required, reciprocity is shown for each exposure, and not for the total amount of light involved. The dosage of far-red light needed for photoreversion is higher; in lettuce, for example, approximately 600 Jm^{-2} of light at 730 nm causes about 50% reversion of the effect of saturating red light. In the laboratory, higher doses of far-red are usually secured by higher fluence rates and/or longer irradiation times.

Turning back to the scheme for pigment photoconversion given previously, we must now explain the dark-reversion component. This was discovered when it was found that lettuce seeds, when transferred to a relatively high temperature for a few hours immediately after exposure to red light, fail to germinate (Fig. 5.25): they come to require a second dose of red light. Thus, the high temperature causes a slow loss of active phytochrome by a reversion of Pfr to Pr—so-called *dark reversion*. Loss of Pfr might also occur by destruction—Pfr is labile and has a half-life of about 1.25 h.

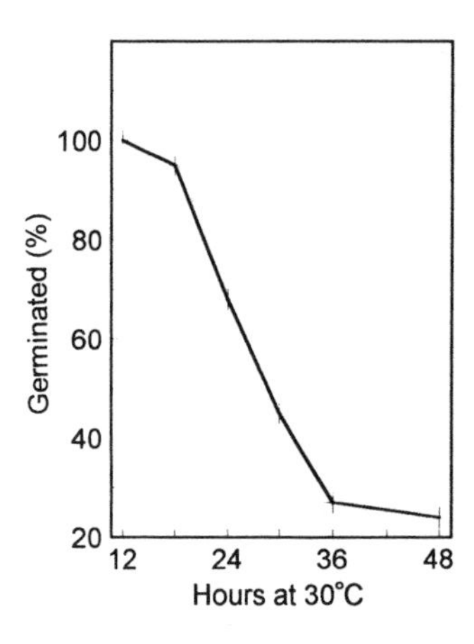

Figure 5.25. Inhibitory effect of high postillumination temperature. Lettuce seeds (cv. Grand Rapids) were exposed to a saturating red irradiation at 20°C and then kept in darkness at 30°C for periods up to 48 h. They were then returned to 20°C, still in darkness. The percentage of seeds that later germinated was determined. Note that more than 18 h at 30°C causes some reversion of the red-light potentiation of germination. After Borthwick *et al.* (1954).

Our discussion so far has centered on the photoconversions of phytochrome by red and far-red light. In nature, seeds are exposed to mixtures of wavelengths such as that which exists in sunlight. Conversions of phytochrome can occur at any wavelength or mixtures thereof—in both directions—to generate a certain proportion of Pfr. And it is the proportion of Pfr which determines dormancy breakage. How these proportions are formed is explained in the next section.

5.5.1.4b. Photoequilibria. The properties of phytochrome that we have so far discussed—the Pr and Pfr forms, their peak absorptions at 660 and 730 nm, respectively, the photoreversibility, and the dark reversion—were all suspected simply on the basis of experimental work on the physiology of light action in lettuce seed dormancy. Complete confirmation of these points came when phytochrome was isolated from plants—not from seeds, but from dark-grown (i.e., nongreen) tissues such as oat or rye coleoptiles. Phytochrome is a blue chromoprotein, the chromophore being an open-chain tetrapyrrole not unlike the phycocyanins of the blue-green algae. In solution, it shows photoreversibility with red and far-red light, and the absorption spectrum of the two forms can be determined (Fig. 5.26A). Peak absorption of Pr occurs at 660 nm and of Pfr at 730 nm—just as predicted from the experiments with lettuce seeds! An important point to note is the overlap in absorbance of the Pr and Pfr. Because both forms absorb over the spectrum from about 300 to about 730 nm, irradiation with monochromatic light in this range sets up an equilibrium mixture of Pr and Pfr—the photoequilibrium or photostationary state, Pfr/P_{total} or φ (Fig. 5.26B). At 660 nm, for example, Pr is photoconverted to Pfr, but Pfr also absorbs at this wavelength and hence some Pfr molecules are phototransformed back to Pr. A mixture of approximately 80% Pfr and 20% Pr is thus established (Pfr/P_{total}, φ = approximately 0.8). Even at 730 nm, where Pfr absorbs most strongly, there is also some absorption by Pr; here the φ value is 0.02, i.e., 2% Pfr. Inspection of the absorption spectrum shows that no waveband region can produce 100% Pfr; on the other hand, photoreversion to give almost 100% Pr can be achieved by irradiation at 740–800 nm. It will be appreciated that irradiation with mixed wavelengths also establishes a certain φ value. In midday sunlight, φ is about 0.55, in white incandescent light it is about 0.45, and in white fluorescent light about 0.65. Now the effectiveness of phytochrome in terminating dormancy is determined by the φ value set up in the seeds, the required value depending on the species, as can be seen in Table 5.10. Lettuce is clearly satisfied by the φ value brought about by sunlight, and *Wittrockia superba* can even be stimulated to germinate by broad-band far-red light ($\varphi > 0.02$).

What is the function of this apparently complex photoreceptive system? The importance of control by the φ value lies in the fact that light-requiring seeds can use phytochrome to detect different light qualities. In nature, the quality varies mainly when the proportions of red and far-red light change. This happens to some extent according to the time of day, but more importantly it occurs when light is transmitted through green leaves, whose chlorophyll absorbs red light but allows far-red light to pass, or through the soil which also transmits far-red light. Thus, a seed under a leaf canopy is in light rich in far-red, which sets up a φ value too low to satisfy most light-requiring seeds. In a sense, phytochrome is the device used by the seed to detect where it is, especially in relation to other plants. Phytochrome is also involved in the photoinhibition of germination, when light of rather high fluence rates given for relatively long periods of time prevents even nondormant seeds from germinating. We will return to these aspects of the ecophysiology of dormancy in Chapter 6.

5.5.1.4c. Where Is Phytochrome Located? Selective irradiation (using narrow beams of light) of different parts of light-requiring seeds—the radicle/hypocotyl and the cotyledons—shows that dormancy is terminated only when the former, i.e., the embryonic axis, is illuminated. This suggests that phytochrome is located there, and indeed, the technique of *in vivo* spectrophotometry, by which phytochrome can be measured in intact plant tissues, confirms its presence in the axis and its very low concentration in the cotyledons.

5.5.1.4d. Photoconversions of Phytochrome. To appreciate fully the role of phytochrome in dormancy and germination we should understand something of the conversion pathway of the pigment. Photoconversion does not occur directly between Pr and Pfr but through several intermediates, as shown in Fig. 5.27. Some of the steps can occur only in highly hydrated seeds, whereas others can proceed in seeds at a very low moisture content. Thus, "dry" seeds have some light sensitivity, but all of the photoconversions cannot take place. While the seed is developing on the plant, phytochrome conversions proceed normally, until the

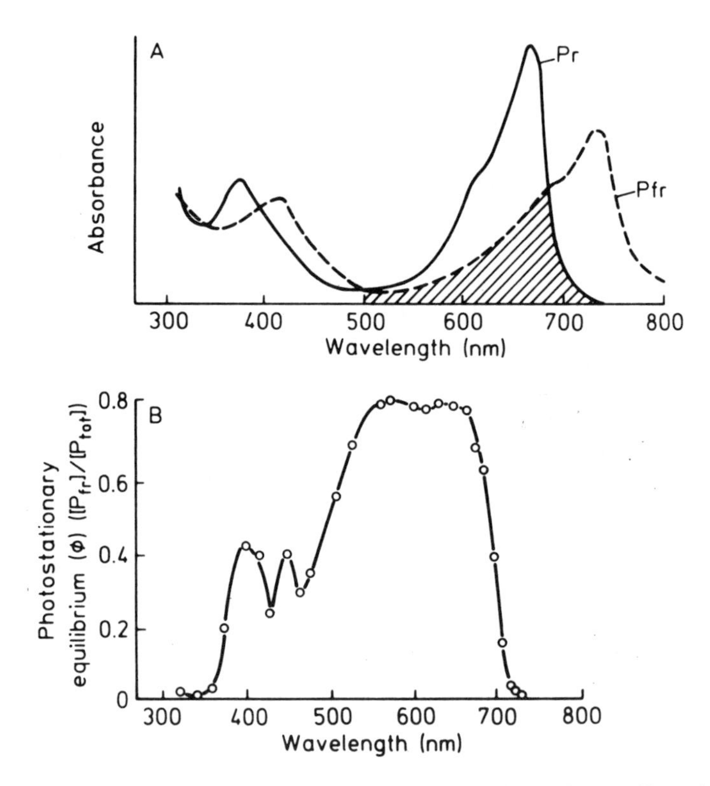

Figure 5.26. Absorption spectra and photostationary equilibria of phytochrome. Absorption spectra were determined using solutions of Pr and Pfr (a). Photostationary equilibria were measured in hypocotyls of dark-grown *Sinapis alba* (white mustard) seedlings (B). A, after Hartmann (1966); B, after Hanke *et al.* (1969).

maturing seed begins to dry. Some of the intermediates (meta Rb and possibly meta Fa) are then trapped, so the mature, dry seed has consigned to it some Pfr, meta Rb and meta Fa, and some Pr, in proportions depending on the quality of light reaching the embryo and on the relative sensitivity of intermediate steps to different hydration levels. On rehydration, meta Rb can pass to Pfr in darkness, and if there is much Pfr already present, the total might be sufficient to promote germination. Such a seed would be nondormant. But if sufficient Pfr, or intermediates that can produce it, is not present in the seed, it matures into a dormant, light-requiring seed.

5.5.1.4e. Phytochrome—The Sensor. The formation of Pfr alone does not break dormancy. Rather, Pfr sets in train some events that culminate in the termination of dormancy, as manifest by the completion of germination by the seed.

Table 5.10. Photoequilibrium Values of Phytochrome (Pfr/P_{total} φ) Required for Dormancy-Breakage[a]

Species	Pfr/P_{total} (φ)
Amaranthus retroflexus	0.001
Amaranthus caudatus	0.02
Wittrockia superba	0.02
Sinapis arvensis	0.05
Cucumis sativus	0.1–0.15
Chenopodium album	0.3
Lactuca sativa	0.59

[a]Values given are sufficient to break dormancy in most seeds of the population.

Stated another way, we can suggest that there is some transduction mechanism that intervenes between Pfr formation and the final display—the germinated seed. Let us now consider what the earliest events in this transduction process might be.

There is good evidence from several physiological systems other than seeds that phytochrome acts on cell membranes. Fast changes in the flux of ions and water occur in some cells, for example. Several pieces of evidence, mainly relating to effects of temperature, suggest that in seeds also there may be an association of Pfr with membranes. Many dormant seeds—*Amaranthus retroflexus* is a good example—are stimulated to germinate when they experience a shift in temperature, perhaps for only 1–2 h. In *Amaranthus*, Pfr cannot act to break dormancy if the temperature is too high, say 40°C. But a drop in temperature to 26°C for about 60 min *after* (but *not* before) Pfr is generated by light enables Pfr to operate, even if the seeds are transferred back to 40°C, so that dormancy is terminated. When this is investigated carefully, it becomes clear that there is a critical temperature below which Pfr action is initiated—it is almost exactly 32°C (Fig. 5.28). So if the seed falls below 32°C for a relatively short time, the Pfr that is present is able to initiate events leading to dormancy breakage, irrespective of later temperatures. The critical nature of the temperature dependence is highly suggestive of a cell membrane transition, and the implication is that a membrane must be in the correct condition for Pfr to interact with it; in this case, it seems that the membrane must be in the gel state. Crude membrane fractions extracted from the seed show temperature-controlled transitions at this temperature. The enhancement of Pfr action by the chilling of lettuce seeds prior to illumination might also depend on membrane effects. Here, too, it seems that there may be a critical temperature (Fig. 5.22), though the effect is not as sharp as in *Amaranthus*. As well as chilling, other treatments (some anesthetics, gibberellins, nitrate, and other chemicals) sensitize seeds to Pfr, when effects of very low fluence rates become apparent. This enhanced sensi-

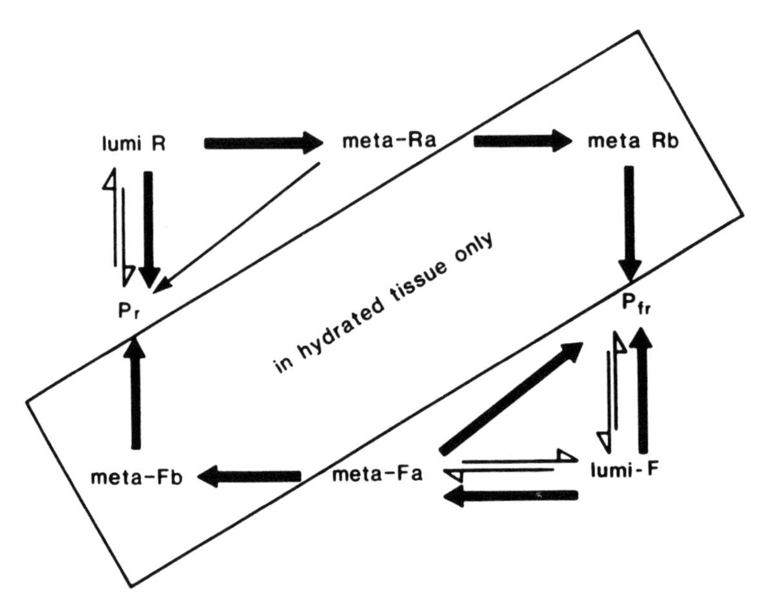

Figure 5.27. Transformation of phytochrome. Conversion of Pr to Pfr and Pfr to Pr occurs in each case through three intermediates, involving light-driven steps (thin arrows) and nonphotochemical steps (thick arrows). Adapted from Kendrick and Spruit (1977).

tivity is interpreted on the basis of the interaction of Pfr (or a Pfr:Pr dimer) with membranes.

This is attractive, preliminary evidence suggesting that newly formed Pfr in seeds associates with a cell membrane. The process clearly occurs fairly quickly, in 60 min at most, and we can envisage it as the primary action of Pfr. Once the association with the membrane has taken place, certain membrane-dependent events might follow, such as alterations in ion movement or possibly activation of membrane-bound enzymes; unfortunately, there is no information to tell us what events do occur and the matter therefore remains speculative. Presumably, however, these events eventually allow processes to occur that terminate in emergence of the radicle from the seed, which occurs by cell elongation. Hence, the ultimate effect of phytochrome may be to control the capacity for cell enlargement, a point which is discussed further in Section 5.5.1.10.

Before ending this section reference should be made to aspects of phytochrome action in other plant parts which may have some relevance to seeds. One

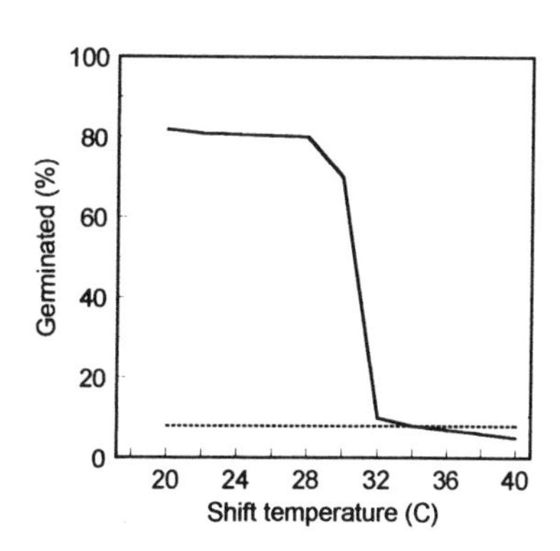

Figure 5.28. Effect of a temperature shift on phytochrome action in *Amaranthus retroflexus*. Imbibed seeds were held in darkness at 40°C except for a shift to lower temperatures for 2 h. Using far-red light, a small amount of Pfr was generated before (——) or after (- - - -) the shift. Note that a temperature of 32°C or below for 2 h in the presence of Pfr breaks the dormancy. Seeds not given the temperature shift (i.e., at 40°C throughout) do not respond to Pfr. Adapted from Hendricks and Taylorson (1978).

effect which is now receiving clarification by the use of molecular techniques is the involvement of phytochrome in the regulation of gene expression. Several genes, for enzymes and other proteins, are regulated by phytochrome and this could occur in seeds; it needs to be investigated. Another important action of phytochrome involves calcium, which participates as a second messenger in several regulatory systems in plant cells. Interactions between Ca^{2+} and phytochrome occur in the stimulation of fern spore germination by light. Similar interactions might occur in seeds but, again, it is a field of investigation that needs more attention.

5.5.1.4f. New Light on Phytochrome. Studies of the molecular biology of phytochrome show that there is more than one species of the pigment in plant tissues. The major difference between them lies in the apoprotein part of the phytochrome molecule which in each case is immunologically distinct. Two kinds that have received the most attention are type I, or phytochrome A, and type II, or phytochrome B, which are encoded by different genes, *phy A* and *phy B*, respectively. Phytochrome A accumulates in dark-grown seedlings while B, a light-stable form, is found in nonetiolated, green plants; it is now clear that the two types have different roles in photomorphogenesis. The parts played by the two phytochromes in seed physiology have not been established but it is sometimes assumed that phytochrome A will be formed in seeds hydrating in darkness—the type occurring in etiolated seedlings.

The cDNAs encoding the apoproteins of phytochromes A and B have been prepared from various sources and in some cases the genes themselves have been isolated. Several phytochrome mutants are known, in *Arabidopsis*, tomato, and cucumber. Transgenic and mutant technology now makes it possible to explore phytochrome participation in seeds, giving new insights into the different roles of the pigment types. The *hy3* mutant of *Arabidopsis*, for example, produces seeds that germinate poorly in darkness, indicating that little active Pfr is present in the mature seed as a legacy of seed development. This mutant is low in phytochrome B so it appears that it is this type of phytochrome Pfr which normally (i.e., in the wild type) would be consigned to the seed during maturation on the parent plant, to enable it to germinate in darkness when mature. This possibility is consistent with the fact that phytochrome B, unlike A, is stable in the light; of course, developing seeds are illuminated daily. Interestingly, germination of mature *hy3* seed is stimulated by light, so presumably phytochrome B is not needed for the light-promoted breaking of dormancy. Phytochrome A may be the important species here, but this has not been shown experimentally.

We saw in Section 5.4.5 that the photoenvironment experienced by maturing seeds influences their subsequent dormancy. Particularly relevant is that light comparatively rich in far-red induces the production of dormant *Arabidopsis* seeds. Transgenic seeds of this species containing oat phytochrome B are much less sensitive to high far-red applied during their maturation than are those of the wild type. An explanation is that the overexpression of phytochrome B in the transgenic leads to a high *concentration* of Pfr (even at a low φ value) in the maturing seeds—sufficient to promote their germination on rehydration.

The use of seeds of phytochrome transgenics is beginning to provide information on other aspects of seed photophysiology discussed in Section 5.5.2.1. There is no doubt that this approach, together with the availability of various mutants, will greatly elucidate the parts played by the different phytochromes in seed development, dormancy, and germination.

5.5.1.5. Seeds with Impermeable Coats

The environment is important in softening hard coats that are impermeable to water. Microbial attack is thought to be important, as well as abrasion by soil particles. Of particular interest is the effect of high temperatures on some leguminous seeds. During exposure to heat, cracks appear in the seed coat of some species, especially in the region of the micropyle. A spectacular response is seen in seeds such as *Albizzia lophantha,* in which the strophiolar plug is audibly ejected from the seed as high temperature is reached, leaving a strophiolar crater through which water can enter. Such effects of high temperature are thought to be important in pyric species whose seedlings emerge as a consequence of forest fires.

5.5.1.6. Breaking of Dormancy by Chemicals

A selected list of chemicals that can break dormancy is given in Table 5.11. Only a few of these are likely to be encountered by seeds in their natural environment, but they are nevertheless of great interest because they may help us understand the mechanism of dormancy breakage. The possible action of many of these substances is discussed elsewhere in this chapter—anesthetics in relation to membranes (Section 5.3.2), respiratory inhibitors, nitrate, nitrite, methylene blue in relation to the pentose pathway, and oxygen in connection with various oxidation processes (Section 5.5.1.9). Something needs to be said about the growth regulators, however, since their effects have suggested that their naturally occurring counterparts participate in the dormancy mechanism. The growth regulators, gibberellin (usually gibberellic acid GA_3, GA_4, and GA_7), cytokinin (usually kinetin, benzyladenine), and ethylene, variously affect seed dormancy (Table 5.12). Seeds that normally require chilling or light or afterripening exhibit striking responses to these substances, which often also accelerate germination of nondormant seeds. It should be said, however, that the success of these regulators is mixed, and many species of seed do not respond at all.

The growth regulators with the widest spectrum of activity are the gibberellins. As an example, we can quote their effect on dormant lettuce seeds, which respond to a concentration of around 10^{-4} M GA_3 or 10^{-5} M GA_{4+7}. Cytokinins are less widely effective, and even when they do act, they often induce abnormal germination; in lettuce, for example, the cotyledons tend to emerge from the seed before the radicle. Ethylene is also only limitedly effective, and many species remain unaffected. Frequently, the growth regulators interact with other factors or among themselves. Kinetin, for example, promotes normal germination of dormant lettuce in combination with low levels of light, and ethylene stimulates *Chenopodium album* most effectively in the presence of light and gibberellin. Numerous studies have demonstrated interactions among the growth regulators and with other promoters, such as nitrate.

5.5.1.7. The Mechanism of Dormancy Release

What changes are brought about in dormant seeds by the action of the various factors we have just considered? Are the changes different, according to the particular factor, or do they all have the same fundamental effect? In short, is there a universal dormancy-breaking mechanism? We should be clear at the outset that whether we are dealing with embryo dormancy or coat-imposed dormancy, it is the embryo that is affected by the active factor—and possibly only the axis, since its growth is the culmination of the germination process. The embryo nevertheless might bring about changes in the restraining tissues and these have been found in some species just before radicle protrusion occurs (Section 5.5.1.8). Permeability of the coats to oxygen and carbon dioxide does not change. Germination of seeds with hard, water-impermeable coats is permitted when the coats crack or become gradually softened in the earth, but it is debatable whether this is dormancy breakage in the strict sense.

Table 5.11. Some Chemicals That Break Seed Dormancy

Class	Example
Respiratory inhibitors	
Cyanide	*Lactuca sativa*
Azide	*Hordeum distichum*
Iodoacetate	*Hordeum distichum*
Dinitrophenol	*Lactuca sativa*
Sulfhydryl compounds	
Dithiothreitol	*Hordeum distichum*
2-Mercaptoethanol	*Hordeum distichum*
Oxidants	
Hypochlorite	*Avena fatua*
Oxygen	*Xanthium pennsylvanicum*
Nitrogenous compounds	
Nitrate	*Lactuca sativa*
Nitrite	*Hordeum distichum*
Thiourea	*Lactuca sativa*
Growth regulators	
Gibberellins	*Lactuca sativa*
Cytokinins	*Lactuca sativa*
Ethylene	*Chenopodium album*
Various	
Ethanol	*Panicum capillare*
Methylene blue	*Hordeum distichum*
Ethyl ether	*Panicum capillare*
Fusicoccin	*Lactuca sativa*

A concept that has attracted a great deal of attention is the hormonal theory of dormancy. This attributes the control of dormancy to various growth regulators or hormones—inhibitors, such as ABA, and promoters, such as gibberellins, cytokinins, and ethylene. According to the theory, dormancy is maintained (and possible even induced) by inhibitors such as ABA, and it can end only when the inhibitor is removed or when promoters overcome it. The theory owes its inception primarily to known effects of applied growth regulators on dormancy, some of which, as we saw earlier, cause a dormant seed to germinate whereas others (e.g., ABA) inhibit germination of a nondormant seed. A second concept is that important metabolic changes occur as a consequence of the action of the dormancy-breaking factor. One such change is thought to involve the synthesis of RNA and protein, and another, the operation of the pentose phosphate pathway. We will discuss these, and others, in the following sections.

5.5.1.8. Hormones in Dormancy Breaking

Dormant seed of some species have appreciable concentrations of ABA and at one time it was thought that chilling acts by causing a drop in ABA content.

Table 5.12. Breaking of Dormancy by Some Growth Regulators

Species	Factor breaking dormancy	Effect of growth regulator[a]		
		GA[b]	CK[c]	Ethylene
Apium graveolens	Light	+	–	+
Avena fatua	Afterripening, chilling	+	–	+
Corylus avellana	Chilling	+	–	–
Hordeum distichum	Afterripening, chilling	+	–	+
Lactuca sativa	Light	+	±	–
Lamium amplexicaule	Light	+	nr	nr
Pyrus malus	Chilling	–	+	nr
Ruellia humilis	Chilling	+	nr	nr
Xanthium pennsylvanicum	Afterripening	+	–	+

[a]+, Regulator effective; –, regulator ineffective; ±, moderately effective; nr, not recorded.
[b]GA, gibberellin (usually gibberellic acid GA_3, or GA_4, GA_7).
[c]CK, cytokinin (usually kinetin, benzyladenine).

Although the amount of ABA in seeds does indeed decrease dramatically during the chilling treatment, it also does so when seeds are held imbibed at nonchilling temperatures, so the effect is not specific. Another finding that makes it certain that a decline in ABA alone is insufficient to break dormancy is that this decline occurs long before the effective duration of cold has been experienced: in apple seeds, for example, free ABA is no longer present after 3 weeks of chilling, when the seeds are still dormant! Although it can be concluded that a reduction in ABA cannot on its own account for dormancy breakage, the loss of the inhibitor, nevertheless, might be essential to it, but we do not know if this is really the case or not. Changes in sensitivity to ABA might also be part of dormancy breakage. This has been reported to occur in one variety of apple in which ABA contents of axes in chilled seeds increase in the later stages of stratification yet the seeds are stimulated to germinate, indicating that they now possess little sensitivity to ABA. We must remember, also, that not all dormant seeds have ABA so in some cases it may not function at all in dormancy breakage. Other hormonal changes are also implicated in the termination of dormancy. Transient increases in various gibberellins and cytokinins take place in seeds of several species during chilling, but the timing of these events does not seem to be related in any clear way to the course of dormancy breakage, unless a sequential appearance of the promotive hormones is necessary.

One case is known—hazel (*Corylus avellana*)—where chilling has a marked effect on the capacity for GA biosynthesis. GA production takes place when the chilled seed subsequently experiences higher temperatures and leads to substantial increases in GA_1 and GA_9 (Table 5.13). Significantly, the application of 2-chloroethyltrimethyl ammonium chloride (a substance that blocks GA biosynthesis) to seeds just after the cold treatment stops germination, and so it seems that the chilling-induced build-up of GA is essential for dormancy breakage to culminate in germination. Dormant, nonchilled hazel seeds germinate after treatment with exogenous GA_3; presumably, the endogenous GA which appears after chilling is also effective in promoting germination—certainly the concentration formed is high enough to be physiologically active. The termination of hazel seed dormancy by chilling therefore involves an enhancement of the capacity subsequently to make GA, and not an increase in the promoter during the chilling treatment itself. The reason for the enhancement is unknown. It is important to discover if seeds of other species behave in the same way as hazel, but this has not been reported.

Transient increases in GA and cytokinins have been found to occur shortly after illumination of some seeds, such as pine (*Pinus sylvestris*) and *Rumex* spp.; and in lettuce, fairly high increases in GA_9 content have been reported. Whether the hormones play any role in dormancy breakage is unclear, however. Some doubt arises from the fact that the increases occur quickly, well within the escape time for Pfr (Section 5.5.1.4a). This means that even though the hormones may be formed, further Pfr action is nevertheless still needed for dormancy to be relieved, suggesting that termination of dormancy cannot depend solely on the Pfr-induced formation of endogenous gibberellin or cytokinin.

There is circumstantial evidence that connects the action of light on dormancy with both gibberellin synthesis and sensitivity to the hormone. Dormancy in *Sisymbrium officinale* is broken and germination is stimulated by treating seeds with light and nitrate together, but their effect is blocked by an inhibitor of GA biosynthesis, tetcyclasis. This indicates that completion of germination (radicle emergence) depends on the biosynthesis of GA, and it may be that this synthesis is provoked by the joint action of light and nitrate. In the absence of these two factors, germination can be promoted by applied GA, whose efficacy is enhanced by a relatively short exposure to red light. This suggests that light (presumably Pfr) increases sensitivity to GA. Breaking of dormancy and the completion of germination in this species therefore involves the production of GA and sensitization to it. A similar interpretation applies to seeds of *Arabidopsis thaliana*. Seeds of the GA-deficient mutant, *ga1*, cannot germinate unless supplied with GA, and they do not respond to light in the absence of GA, unlike wild-type seeds which do. But *ga1* seeds are sensitive to light in the presence of GA; in fact, light increases their sensitivity to the regulator. Again, these findings are evidence that GA biosynthesis (where possible) and light-induced sensitization to the hormone are parts of the dormancy-breakage (germination) behavior.

Prechilling of *Arabidopsis* seeds also increases the response to GA, of both wild-type and *ga1* seeds (Fig. 5.29). That even mutant seeds are affected by this dormancy-breaking treatment, though they cannot produce GA, supports the

Table 5.13. Enhancement by Chilling of the Gibberellin-Biosynthetic Capacity in *Corylus avellana* Embryos[a]

Treatment	GA content (nmol/seed)	
	GA_1	GA_9
None	1.02	< 0.01
42 days at 5°C	0.12	< 0.01
42 days at 5°C, then 8 days at 20°C	4.92	3.06

[a]Based on Williams *et al.* (1974).

view that the relief of dormancy in the strict sense does not involve the synthesis of GA.

Further contributions to our understanding of the role of GA come from studies on the *gib1* mutant of tomato. Seeds of this GA-deficient mutant, like its *Arabidopsis* counterpart, cannot germinate unless they are supplied with GA. Wild-type tomato seeds germinate after a period of afterripening (to break dormancy), though applied GA can substitute for this. It is clear, therefore, that endogenous GA must play an important part in germination of the wild type. Dormancy in tomato is coat-imposed and it is lost when a portion of the coat (testa plus endosperm) surrounding the radicle is removed. This operation also allows mutant seeds to germinate, so coat removal alleviates the inability to make GA and it has the same beneficial effect as does the application of GA. When wild-type seeds germinate, the part of the endosperm restraining the radicle becomes degraded by hydrolytic enzymes, allowing the radicle to protrude. It is this process that requires GA—to induce the production of the hydrolytic enzyme, endo-β-mannanase, that degrades the cell-wall mannans. This enzyme is formed in the endosperm in response to GA from the radicle in the wild type, or supplied externally in the case of the mutant (see Fig. 4.11).

Dormant tomato seeds do not germinate, partly because the embryo does not produce GA. As we have seen (Section 5.4.4) the dormancy is induced during seed development by ABA. It may be that one long-lasting effect of ABA is to curtail the embryo's ability to make GA or to interfere at some other level in GA-controlled endosperm degradation. It is important to note that regulation of endosperm endo-β-mannanase production is not the sole action of GA. Even when the coat is removed, radicle extension in the wild type is more rapid than in the mutant, indicating that cell elongation is also affected by the regulator. A model incorporating the above points is given in Fig. 4.12, and in the section on radicle elongation (Section 4.2).

As a result of these studies with mutants it now seems certain that GA must be produced for germination to be completed—and dormancy could involve the regulation of GA production, or sensitivity to the hormone. However, because

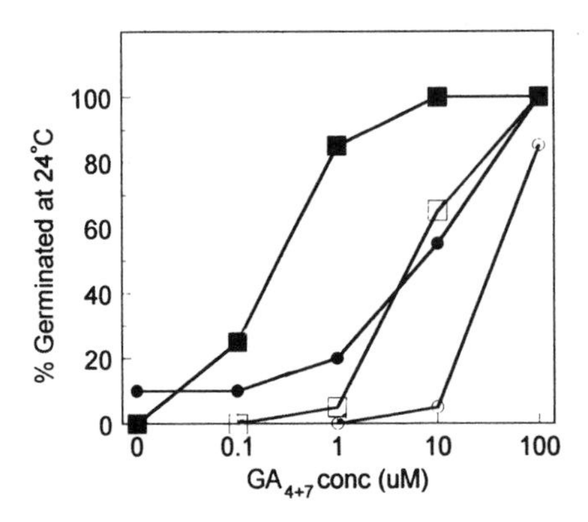

Figure 5.29. Response of mutant and wild-type *Arabidopsis thaliana* to a dormancy-breaking treatment. Seeds of the wild type (○,●) and the *ga1* mutant (□,■) were placed on solutions of GA_{4+7}, with (●,■) or without (○,□) a prior chilling treatment to lessen dormancy. After Hilhorst and Karssen (1992).

these concepts are derived from studies on a very limited number of species, they now must be extended to test the general applicability of the findings.

Ethylene is another hormone that might function in the dormancy-breaking mechanism. The production of ethylene by *Xanthium* and peanut embryos is thought to be important for the loss of dormancy. Imbibed afterripened peanut seeds, for example, eventually evolve more ethylene than dormant seeds, though it is difficult to decide whether the substance is produced simply as a result of germination or is really the cause of it. Evidence from work on apple seeds stands against the likelihood that ethylene is needed for dormancy breakage or germination. The stimulation or inhibition of ethylene production by these seeds has no effect on dormancy or on their capacity for germination. On the other hand, ethylene appears to be a very important factor in the germination of seeds of certain parasitic angiosperms, *Orobanche ramosa* and *Striga* spp. Germination of so-called "conditioned" seeds of *Striga hermontica*, for example, is promoted by ethylene or by aminocyclopropane carboxylic acid (ACC), a precursor of ethylene. Seeds of parasites normally are stimulated to germinate by exudates from the roots of the host. Ethylene substitutes for this exudate. In fact, the exudate causes production of ethylene by the seed whose germination is thereby promoted.

5.5.1.9. Dormancy Breakage and Metabolism

Metabolic activity that has been investigated in relation to the breaking of dormancy by light, chilling, afterripening, and growth regulators such as gibberellic acid includes nucleic acid and protein synthesis, and aspects of respiratory metabolism—gas exchange, ATP production, the pentose phosphate and reverse glycolytic pathways, and the so-called alternative (i.e., cyanide-insensitive) pathway. The rationale is that dormant seeds may be metabolically deficient in some respect and that in the breaking of dormancy the deficiencies are made good.

There is no evidence that synthesis of RNA or protein is promoted by light or chilling. Dormant, light-requiring seeds are able to carry out macromolecular syntheses in darkness, and abundant polysomes are produced; there is some slight evidence, however, that polysome numbers in lettuce and pine spp. increase after irradiation, but the effect awaits confirmation. These observations, however, do not rule out the possibility that the termination of dormancy involves the production of discrete RNA (and hence protein) types—the techniques that have been used to date would certainly fail to detect small quantitative or qualitative changes. Application of GA to hazel seeds causes an increase in total RNA of the embryonic axes, but the effect, prior to radicle emergence itself, is small. RNA synthesis, as detected by the incorporation of radioactive precursors, apparently is promoted, and on this basis it has been suggested that the growth regulator increases DNA template availability and RNA polymerase activity, i.e., GA derepresses certain genes. However, there is only limited evidence that RNA synthesis is enhanced, and no evidence that there is synthesis of mRNA(s) for proteins essential for germination. Some stimulation by GA of polysome formation and protein synthesis occurs in lettuce and charlock, but whether this event is related to germination is unknown. There is no evidence that dormancy-breaking treatments enhance the expression of any genes. On the other hand, GA does antagonize the effect of ABA in promoting expression of the putative dormancy-specific genes in *Bromus secalinus* (Section 5.3.2). Similarly, cytokinins increase RNA and protein synthesis, especially when they overcome the inhibitory effects of ABA.

Respiration, as measured by gaseous exchange, on the whole seems unresponsive to light or chilling. There are no convincing changes in oxygen consumption or production of ATP in illuminated or chilled seeds. Chilling seems to affect phosphate metabolism in embryonic axes of cherry. Here, organic phosphates (nucleotides) accumulate during cold stratification, as opposed to inorganic phosphates during the warmth. It is possible that in this species the change in phosphate metabolism brought about by cold treatment makes more precursors available for nucleic acid synthesis, but this has not been shown to be the case.

The concept that the pentose phosphate pathway (PPP) plays a unique role in dormancy breakage arises largely from studies of the effect of certain chemicals on dormancy. Dormancy of several different kinds of seeds, including lettuce, rice, and barley, is broken by application of inhibitors of respiration. Active substances include those which inhibit terminal oxidation and the tricarboxylic acid cycle in the mitochondria (e.g., cyanide and malonate) and some which inhibit glycolysis (e.g., fluoride). Electron acceptors such as nitrate, nitrite, and methylene blue can also break dormancy. High oxygen concentrations are also active. An explanation for the effects of these substances is as follows: The consumption of oxygen by conventional respiration is blocked by the inhibitors, and oxygen then becomes available for other processes; alternatively, the oxygen could be made available by elevating the external oxygen concentration. The PPP has been suggested to be the important "other" process, and it may require oxygen for the oxidation of reduced NADP (NADPH) which the pathway generates. Oxidation of NADPH can also be brought about by the electron acceptors methylene blue, nitrate, and nitrite, in which case oxygen is not needed (Fig. 5.30).

Although the special significance and function of the PPP has never been explained, the hypothesis, nevertheless, has been thoroughly investigated in several ways. First, the contribution of the pathway to glucose oxidation has been measured in several kinds of dormant seeds and in seeds whose dormancy is being broken. The results are mixed, some of them showing a marked participation of the PPP during dormancy breakage, whereas the others are less convincing. Moreover, some criticisms have been leveled at the techniques used in these determinations, and it has been suggested that they might give misleading results. A second approach is to examine the activity of two key enzymes in the pathway, glucose–6-phosphate dehydrogenase and 6-phosphogluconate dehydrogenase. In wild oats and barley, the extracted enzymes from nondormant grains are no more active than those from dormant material (Fig. 5.31). It could be argued, however, that *in vivo* activity is more important. Some enhancement of activity seems to occur in hazel cotyledons during chilling, which might be significant since the cotyledons are involved in the regulation of embryo dormancy. Third, amounts of reduced and oxidized NADP and NAD have been studied, but these are completely unrelated to the dormancy status of seeds. Finally, although the suggested role for oxygen is for the oxidation of NADPH, such an oxidation system seems to be unknown in plants. The evidence for the participation of the PPP in dormancy breakage is therefore, at best, equivocal. Information is accumulating in the research literature that casts doubt on the validity of the hypothesis that the termination of dormancy requires a switch in respiratory metabolism toward increased PPP activity.

Notwithstanding the doubt surrounding the PPP, an "alternative" respiratory pathway (one that is insensitive to cyanide and therefore does not use the normal terminal oxidation mechanism; see Section 4.4) has been suggested to be important in the ending of dormancy. Little is known about this pathway, though, and its special role in dormancy breakage is not clear. In the case of *Xanthium*, however, it seems that chilling temperatures enhance subsequent electron flow through the cyanide-insensitive pathway, an effect that appears to increase ATP concentrations and the total adenylate pool (but not the energy charge). It has been suggested that such changes promote subsequent germination, but for reasons that remain to be elucidated.

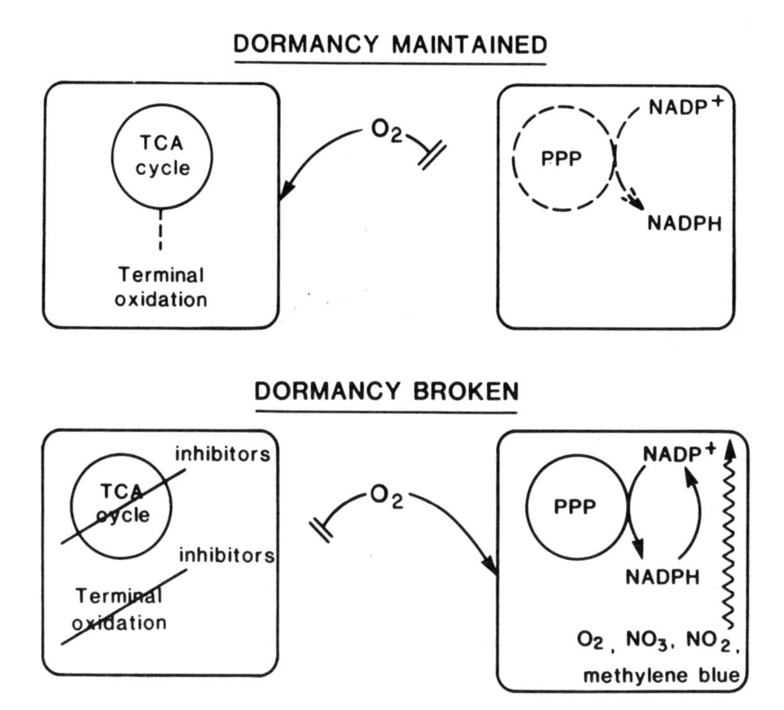

Figure 5.30. The pentose phosphate pathway and dormancy breakage. See text for explanation.

Before closing this section, mention should be made of fructose 2,6-bisphosphate which can be produced from the glycolytic intermediate fructose 6-phosphate. We saw in Section 5.3.1 that nondormant oat seeds develop relatively high amounts of this compound during early imbibition. It is also produced in the first few hours of dormancy breakage of red rice seeds induced by certain chemicals. The significance of this change is yet to be explained.

5.5.1.10. Dormancy Breakage and Cell Enlargement

The release from dormancy, in the end, means that the embryo is enabled to develop enough growth thrust to overcome the constraints on it. Measurements on lettuce embryos demonstrate that one dormancy-breaking factor—light—has this effect. The lettuce is a seed in which dormancy is coat-imposed (actually the endosperm is the effective tissue). Therefore, isolated embryos have no apparent dormancy and germinate in darkness. But their dormant behavior is restored if they are placed in solutions of high osmotic strength (mannitol or polyethylene

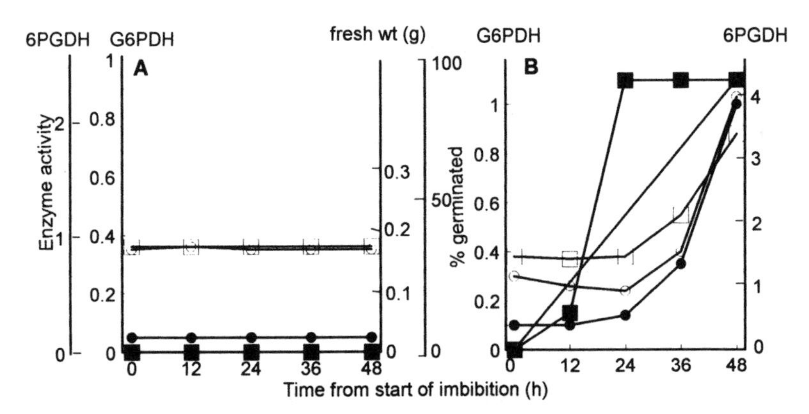

Figure 5.31. Enzymes of the pentose phosphate pathway and dormancy in wild oats (*Avena fatua*). For this experiment, two strains of wild oats were used: a highly dormant line (A) and a nondormant line (B). In both, the activity was determined of two enzymes in the pentose phosphate pathway, glucose–6-phosphate dehydrogenase (G6PDH) and 6-phosphogluconate dehydrogenase (6PGDH). Units of enzyme activity refer to the amount of reduced nicotinamide dinucleotide phosphate produced in the reaction, measured in absorbance units. Activity was determined at time intervals over 48 h, during which the percentage of germinated seeds and the change in fresh weight in grams (a measure of seedling growth) were followed. □, 6PGDH; ○, G6PDH; ■, % germinated seeds; ●, fresh weight. Note the absence of significant differences in enzyme activity between nondormant and dormant grains until after germination of the former has occurred. This shows that dormancy is not associated with lower enzyme activity. Adapted from Upadhyaya *et al.* (1981).

glycol), when they become light requiring once more. Embryos that have received no light require a concentration equivalent to 0.1 molal mannitol to make them dormant (i.e., to prevent their germination and growth). When embryos are illuminated, the necessary concentration to stop growth is about 0.35 molal mannitol (Fig. 5.32). Now, the arrest of embryo growth occurs when the external osmotic potential is lower than the embryo's water potential. Clearly, since a higher osmotic strength (i.e., lower osmotic potential) is needed to inhibit growth of irradiated embryos, these embryos must have a lower water potential than the unirradiated ones; in short, light has brought about a lowering of embryonic water potential. It follows, then, that embryos exposed to light are more able to take up water, i.e., to generate a growth thrust. Is this ultimately the effect of light in dormancy breakage? And if so, how is it achieved? The water potential of the embryo (ψ_{emb}) is comprised of its pressure or turgor potential (ψ_p) and its osmotic potential (ψ_π) (remember that ψ_π is a negative quantity):

$$\psi_{emb} = \psi_p + \psi_\pi$$

Thus, a low ψ_{emb} could be achieved by decreasing ψ_p or by making ψ_π more negative. There is no evidence for a substantial change in ψ_π as a result of illumination, and so a fall in ψ_p must occur. Now this means that the resistance of the cell walls is lowered, i.e., they become softer. Softening of cell walls is known to occur during growth of plant cells, so perhaps it is not surprising that it seems to take place in the lettuce embryo capable of growth. But the implication is that here, light (i.e., Pfr) ultimately is responsible. Cell wall softening is thought to occur as a consequence of the secretion of hydrogen ions by the protoplast into the wall. From measurements of the effects of red-light-promoted lettuce embryos on the acidity of the external liquid, it does seem that Pfr can stimulate the hydrogen-ion-secreting system (Fig. 5.33). We might conjecture, therefore, that whatever the preceding effects of Pfr are in the dormancy-terminating mechanism, the final result might be the activation of a proton pump, leading to softer cell walls and thereby the capacity for radicle extension. That chemical stimulation of germination might act via activation of a proton pump is less clear-cut, and some reservations are considered in Section 4.2. A point that may be relevant in this content is that when ABA inhibits germination i.e., reduces the growth capacity of embryos it apparently does so by action on the cell walls. There is good evidence, for example in *Brassica napus*, that the yield threshold is increased and the extensibility of radicle cell walls is reduced under the influence of added ABA. Thus, an inhibitor which has the semblance of a dormancy inducer acts on the cells in an opposite direction to a dormancy-breaking factor.

5.5.2. Environmental Control of Germination

Several factors in the environment—water, oxygen, light, temperature, and chemicals—determine whether or not germination occurs and the rate at which it does so. These controls apply to seeds whose dormancy has been broken, as well as to those which never had any dormancy. As far as the dormant seed is concerned, controlling factors that act over a fairly long duration might have effects on dormancy and germination which merge, in some cases antagonistically. For example, exposure of a dormant seed to an extended period of cold may serve to break its dormancy, but it will not germinate because the temperature is too low, and it could even be forced into a secondary dormancy. Similarly, a light-requiring seed under long-term illumination with white light could initially have its dormancy broken by the light but then have its radicle extension inhibited by the same irradiation, and again, it too might enter secondary dormancy.

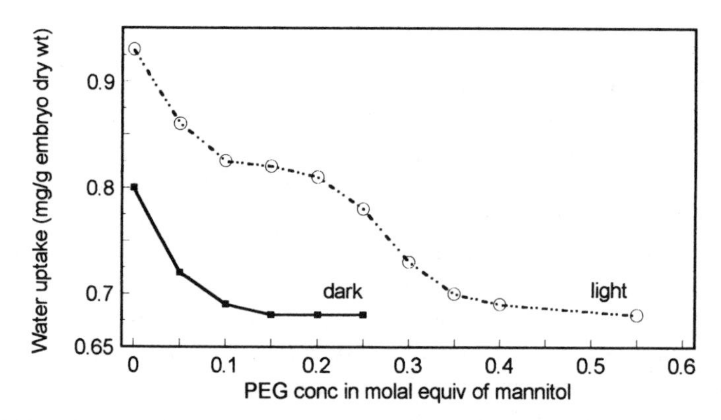

Figure 5.32. Effect of light on the water potential of lettuce embryos. Lettuce embryos were isolated from red-light-treated (-‧‧‧-) and dark-treated (——) seeds (cv. Grand Rapids). They were then placed in darkness on polyethylene glycol solutions (PEG) of different osmotic strengths, expressed as molal equivalents of mannitol. Water uptake after 15–16 h was determined by measuring the change in weight. Water uptake in dark-treated embryos is stopped by concentrations higher than 0.1 molal equivalent, whereas in light-treated embryos, water uptake continues in solutions up to approximately 0.35 molal equivalent. Adapted from Nabors and Lang (1971).

We will not, in this section, consider environmentally limited water, which is covered briefly in Section 4.1, and some effects of chemicals have been mentioned earlier. Attention will be confined to light and temperature.

5.5.2.1. *Effects of Light*

Seed germination in many species is inhibited by continuous white light. Well-known examples are *Nemophila insignis, Phacelia tanacetifolia, Amaranthus caudatus,* and several cultivars of lettuce. Such seeds are normally dark germinators: they are frequently referred to as light-inhibited or negatively photoblastic seeds. But even some whose dormancy is broken by light can also be inhibited by prolonged exposure, especially if the fluence rate is high. Dual effects of this kind are neatly illustrated by dormant *Oryzopsis miliaceae*, which germinates in response to a few minutes of light but does not do so under continuous exposure (Fig. 5.34). Even the classical light-requiring seed—Grand Rapids lettuce—behaves like this under certain circumstances.

We have emphasized that white light inhibits when exposure is prolonged; hence, the phenomenon is time dependent. In some cases, however, inhibition is brought about by intermittent light of a few hours each day. Since the proportion of inhibited seeds increases with the duration of each light period, a quasiphotoperiodic effect results, when most of the seeds can germinate under short days

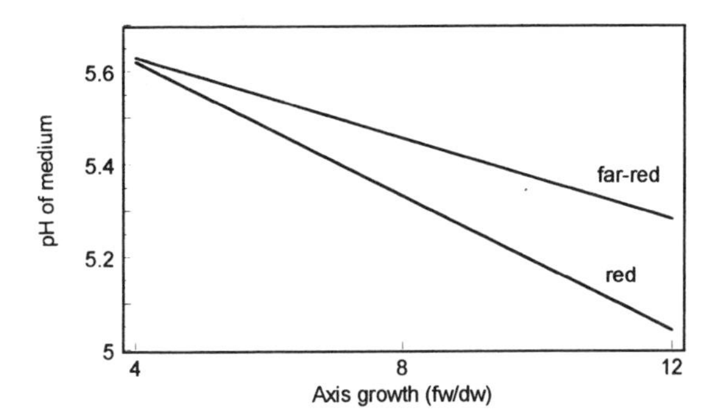

Figure 5.33. Effect of Pfr on hydrogen ion secretion. Embryos of lettuce (cv. Grand Rapids) were isolated from red-light-treated and far-red-treated seeds and placed in a medium of 1 mM phosphate buffer for 24 h. Over this period, the growth of the embryonic axes and the pH of the external medium were determined. Note the greater acidification of the medium by red-light-treated (i.e., containing Pfr) embryos. After Carpita *et al.* (1979).

but few can do so in long days of, say, about 20 h light per day. *Nemophila insignis* shows this kind of behavior. Photoinhibition is also fluence-rate dependent, the degree of inhibition generally increasing linearly with the logarithm of the fluence rate. There is no reciprocity, which distinguishes the response from the phytochrome effects discussed earlier (Section 5.5.1.4a). The degree of inhibition also may depend on temperature, but no general rule can be formulated. *Nemophila insignis*, for example, at moderate fluence rates is inhibited only at temperatures above 21°C; *Amaranthus* spp., however, are photoinhibited only at lower temperatures! It is interesting that for photoinhibition to occur, the embryo must be under constraint. If the embryo of *Nemophila insignis* is isolated from the enclosing tissues, or even if just the radicle tip is freed, light ceases to inhibit. On the other hand, application of a constraint to a seed not normally affected by light can make it inhibitable. This happens in cucumber and radish seeds when they are placed on osmotica (about 0.5 M mannitol): only a few of them complete germination in continuous light whereas all do so in darkness, although on water they are almost light insensitive.

It seems that if a seed is to be inhibited it must receive light at a critical stage in its germination. The most sensitive time for a population of *Nemophila* is 40–60 h after the start of imbibition, which is the period when the radicles would otherwise start to emerge. Hence, seeds are susceptible to photoinhibition just prior to and during the time when the radicle cells are about to elongate: the inhibitory effect of light is therefore probably on cell elongation.

Photoinhibition of seed germination by prolonged illumination is a manifestation of the high-irradiance reaction (HIR) of photomorphogenesis. It has received most experimental attention in seeds in connection with the action of extended periods of far-red light, which act after the escape time has passed and therefore do not simply reverse Pfr (Fig. 5.35A). The effectiveness of the inhibitory far-red light depends on its duration and fluence rate (Fig. 5.35A,B). It is not appropriate here to enter into a detailed discussion of the mechanism of the HIR, but it will suffice to say that the prolonged far-red effect undoubtedly involves the operation of phytochrome in a special way. Most importantly, it requires a high cycling rate of the pigment, i.e., a high rate of the photoconversions Pr ⇌ Pfr. This gives us a clue, for high cycling rates also occur under white light, the rates obviously being dependent on the fluence rate. This, then, accounts partly for the inhibitory effect of white light, though the mechanism is unknown. But white light also inhibits because of its blue component. Prolonged, pure blue light—again it is fluence-rate dependent—inhibits germination, the sensitive stage again being around the time of incipient radicle emergence. Part of the effect of blue light may be the result of Pr ⇌ Pfr cycling, since phytochrome absorbs in the blue region of the spectrum (Fig. 5.26), but probably a substantial part operates through another pigment system—the special blue-light receptor. Phytochrome cycling and this receptor, therefore, together account for photoinhibition by white light. A few words are appropriate about the type of phytochrome involved in the HIR. Evidence is now emerging, based on studies of mutants, that this is phytochrome A. Seeds of the *aurea* mutant of tomato (this lacks phytochrome A), for example, are insensitive to prolonged far-red.

5.5.2.2. *Inhibition by Short Periods of Far-Red Light*

Many dark-germinating seeds are inhibited by far-red light given for just a few minutes or intermittently. This happens in seeds of certain cultivars of

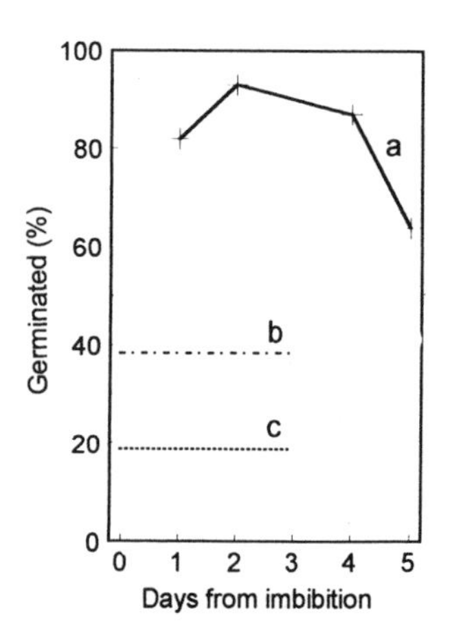

Figure 5.34. Dual effects of light on seeds of *Oryzopsis miliaceae*. Imbibed seeds were held in darkness (b) or continuous white fluorescent light (c) or given a few minutes of such light at different times (a). The percentage of germinated seeds was determined at 8 days after the start of the experiment. After Koller and Negbi (1959).

tomato and lettuce and in cucumber, for example. In these cases, far-red light is not acting through the HIR but through phytochrome, in the low-energy mode, which is why inhibition can be reversed by just a few minutes of red light. The short periods of far-red light therefore convert these nondormant seeds into dormant, light-requiring ones. The explanation for the phenomenon is that the dark-germinating seeds initially have no dormancy because they mature on the parent plant with sufficient Pfr already in them, or because they contain phytochrome intermediates (such as meta Rb, Section 5.5.1.4d) that generate Pfr when the seed is hydrated. Those seeds which have Pfr, but no intermediates, are inhibited by a single dose of far-red light (e.g., some cultivars of tomato). In those with intermediates which slowly generate Pfr, intermittent far-red light is required to drive the developing Pfr back to Pr (e.g., some cultivars of lettuce). This process might happen in the field when a seed that contains Pfr hydrates in light rich in far-red such as under leaf canopies.

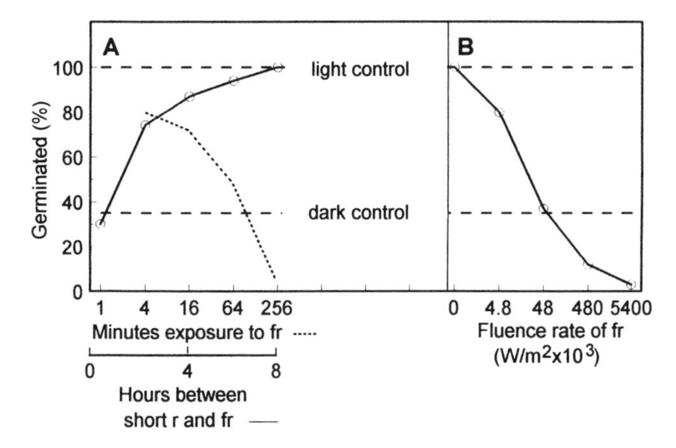

Figure 5.35. Effects of far-red light on lettuce seeds (cv. Grand Rapids). (A) Imbibed seeds were exposed to 1 min red light; this caused 100% of the seeds to germinate (- - - -, light control). Other seeds were exposed to 1 min red light and then, at different intervals, to 4 min far-red light. The solid line shows the effect of these treatments on the subsequent germination. This curve shows the escape time, i.e., the time taken for the dormancy-breaking process to escape from the effects of reversal of Pfr. Reversal of Pfr after 8 h has no effect on subsequent germinability, and reversal after 4 h has only a relatively small effect. Seeds were also exposed to 1 min red light and then allowed to spend 4 h in darkness, after which they were irradiated with different durations of far-red light (····). Even though seeds have almost completely escaped from the effect of a short far-red exposure, they can nevertheless be inhibited by prolonged exposure (e.g., completely inhibited by 256 min far-red light). (B) Seeds were given the following treatment: 1 min red light, 4 h darkness, 256 min far-red light at different fluence rates. Note the increasing inhibition as the fluence rate rises. Adapted from Mohr and Appuhn (1963).

5.5.2.3. *Effects of Temperature*

It is worth emphasizing that when considering the effects of temperature on germination we must be careful to exclude dormancy. For example, when seeds with relative dormancy are tested at different temperatures, germination is found to occur only over a certain range (see Fig. 5.11). But this is not strictly the temperature range for germination—it is the temperature range over which there is no dormancy. When dormancy of such seeds is removed, germination then occurs over a substantially wider range of temperatures (Fig. 5.36). Similarly, dormant seeds with a requirement for chilling may actually succeed in germinating at the same low temperature needed for the termination of dormancy, albeit sluggishly; they cannot germinate at higher temperatures (since they are dormant) so they appear to have a very narrow temperature range for germination. In this section, then, we omit complications related to dormancy and consider only the control by temperature of germination in nondormant seeds.

Temperature affects both the capacity for germination and the rate of germination. Seeds have the capacity to germinate over a defined range, characteristic for each species (Fig. 5.37); hence, there are clear minimum and maximum temperatures for germination, and between them a broad range over which germination of all seeds can be attained. Species can have widely different temperature minima and maxima; a few examples are shown in Table 5.14. It is sometimes useful for experimental and descriptive purposes not to state maximum or minimum temperatures, but to define the temperature at which 50% germination occurs. These are referred to as the GT_{50}, or the temperature cutoff points. Interestingly, the GT_{50} of several species is affected by growth regulators. The cytokinin, kinetin, raises the upper GT_{50} for lettuce seeds in the light from 31°C to 40°C, for example.

Although all seeds of a species may germinate over a fairly wide range of temperatures, the time needed for the maximum germination percentage to be reached varies with the temperature; that is to say, the rate of germination is temperature dependent, the maximum rate occurring over a range of just 1–2°C—in the example shown, at about 30°C (Fig. 5.38). Because of the high rate of germination at certain temperatures, a seed population whose germination total is determined early in a test appears to have a narrow optimum temperature (Fig. 5.39), but this, of course, is only the optimum after a limited time period. The Q_{10} for germination can be estimated from germination rates. The rate curves generally consist of two linear portions (Fig. 5.39), but the Q_{10} values on both the rising and falling parts of the curve are not equal. In *Dolichos*, for example, as temperatures increase, the Q_{10} between 10 and 20°C is approximately 5, while that between 15 and 25°C is 2.5. Because of the differences in rate, seeds in a population germinate at different times. A few succeed in

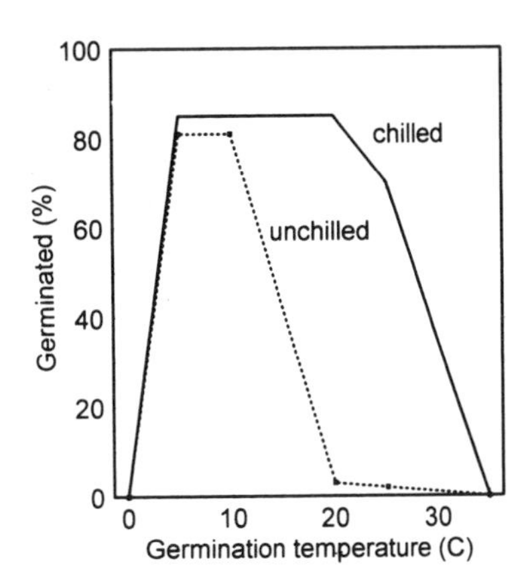

Figure 5.36. Temperature effects on dormancy and germination. Unchilled seeds of *Delphinium ambiguum* were set to germinate at different temperatures. A similar experiment was carried out with seeds that had been chilled at 6°C for 2 weeks to remove dormancy. All of the unchilled seeds are dormant at 20°C and above. The chilled seeds have no dormancy and their temperature range for germination extends to well above 30°C. Data adapted from Ezumah (Ph.D. thesis, 1980).

germinating quickly even at low temperatures, but the majority need more time. Thus, to a certain extent, this heterogeneous response to temperature acts like dormancy, by distributing germination in time.

Rate data for germination can be treated mathematically in the same manner as an enzyme reaction, and activation energies can be calculated. The values obtained are those characteristic of protein denaturation, and so it is suggested that as the temperature rises, changes in protein conformation occur which actually promote the germination process, but further conformational changes occur which are deleterious to it as temperatures become too high. Because of the lack of agreement between effects of temperature on respiration

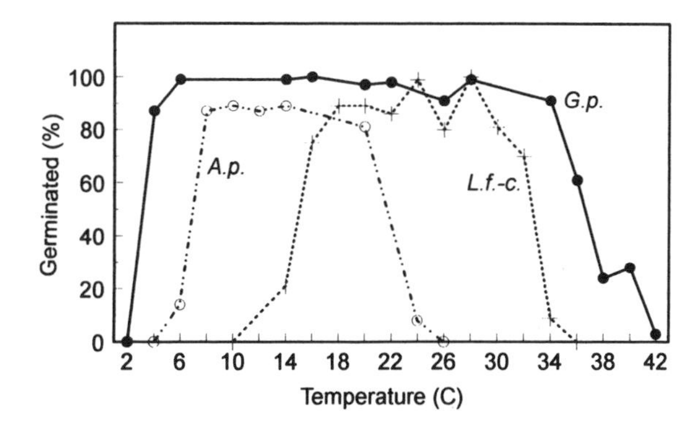

Figure 5.37. Temperature ranges for germination. Nondormant seeds of three different species were set to germinate over the range 1–42°C. Note that each species has a distinct range, which differs as far as the maximum and minimum temperatures for germination are concerned. ——, *Gypsophila perfoliata;* -·-·-, *Allium porrum;* - - - -, *Lychnis flox-cuculi.* Adapted from Thompson (1973).

Table 5.14. Temperature Maximum and Minimum for Seed Germination of Some Species

Species	Minimum (°C)	Maximum (°C)
Allium porrum	7	23
Apium graveolens (cv. Golden self-blanching)	10	15
Brassica oleracea (var. gemmifera)	4	42
Dolichos biflorus	6	42
Gypsophila perfoliata	2	40
Lychnis flos-cuculi	9	35
Lycopersicon esculentum cv. Top cross	12	36
Silene gallica	2	32

and on germination, an increase in respiratory metabolism is not thought to account for the increase in germination rate as the temperature rises to the optimum. There are indications, however, that changes in membrane states could be involved.

5.5.2.4. Water Stress

Water stress can reduce both the rate and percentage of germination. The range of response among species is wide, from the very sensitive (e.g., soybean) to the resistant (e.g., pearl millet) (Table 5.15). Resistant seeds may have an ecological advantage in that they can establish plants in areas in which drought-sensitive seeds cannot do so.

Water stress is sometimes used in the laboratory, and to some extent in horticulture, to delay germination. Seeds treated with osmotica carry out many of the germination processes, but radicle growth cannot occur; when introduced to water, these seeds, nearly all of which are at the same germination stage, show rapid, almost synchronous radicle emergence (Fig. 5.40). Advantages of this treatment—called osmotic priming—are that treated seeds take less time to complete germination, and they produce a crop of uniform age.

5.5.2.5. Oxygen and Carbon Dioxide

Seed germination in the majority of species is delayed by oxygen concentrations less than in air. A few exceptions occur, and seeds of certain aquatic plants (e.g., *Typha latifolia*) are actually inhibited by air. High carbon dioxide concentrations (e.g., 4%) can prevent germination of several species (e.g., *Capsella bursa-pastoris*). In the terrestrial habitat, concentrations of oxygen low enough to affect germination adversely are rarely found. The air in soil, for example, contains about 19% oxygen and no more than 1% carbon dioxide; the latter is not likely to inhibit germination.

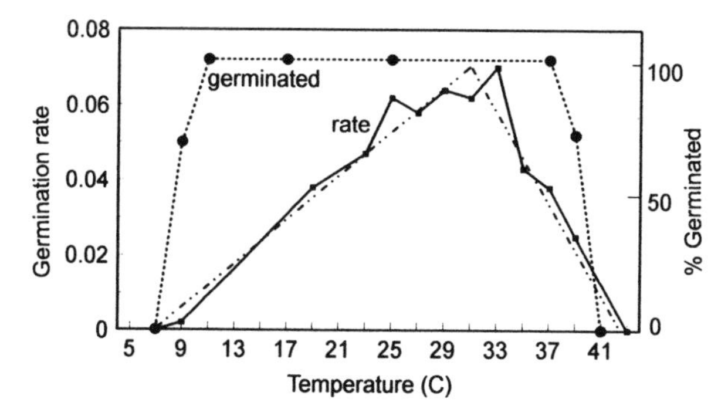

Figure 5.38. Temperature and the rate of germination. Seeds of *Dolichos biflorus* were allowed to germinate at different temperatures. Percentages of germinated seeds were recorded (- - - -) and the rate of germination (1/hours taken to reach maximum percentage) was determined (——). The regression line for the germination rate is shown (-···-). Adapted from Labouriau and Pacheco (1979).

5.6. CONTROL OF GERMINATION—A SUMMARY

The ramifications of the effects of factors and the processes involved in the control of germination are extensive, and in the foregoing account we have considered the subject from several different viewpoints. In this summary, we will give a general overview which, we hope, will enable the reader to set the many individual items that we have discussed into a broad perspective.

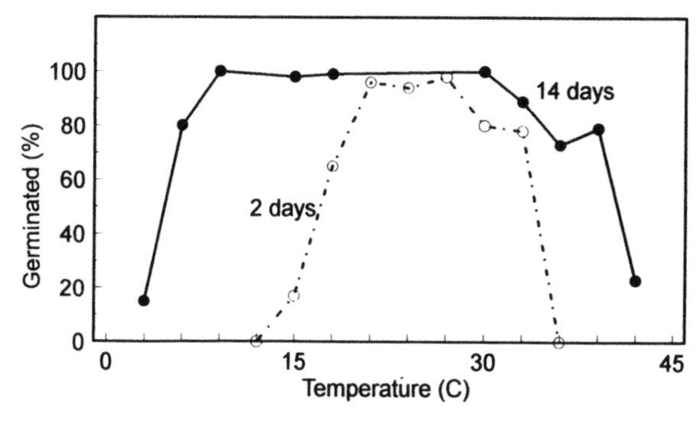

Figure 5.39. Time and the temperature optima for germination. Brussels sprout seeds (*Brassica oleracea* var. *gemmifera*) were set to germinate over the temperature range 4–42°C. Germination percentages were recorded after 2 and 14 days. The highest rates of germination are at temperatures around 25°C, but all the seeds are capable of germinating at temperatures from about 10°C to about 35°C. Adapted from Thompson (1973).

We have divided the controls of germination into two groups: the internal and the external ones. The internal controls—within the seed itself—are responsible for the syndrome of dormancy. Dormancy essentially is caused by a block to the completion of germination because of an inadequacy within the embryo. In some cases, the inadequacy is exacerbated by the structures enclosing the embryo and it might be revealed predominantly when these structures are present (coat-imposed dormancy). In other cases, the deficiency of the embryo is so profound that dormancy is manifest even in the absence of other tissues (embryo dormancy).

Central problems regarding dormancy are to identify the block and to determine where in the germination process it lies. Dormancy may be the inhibition of some key events in germination that could occur at any point up to the time of radicle emergence, even, for example, the interference with incipient cell elongation itself. No particular aspect of metabolism has been found that accounts for the failure of the dormant seed to complete germination but it may be too soon to rule out the possibility of a metabolic deficiency. Evidence is now being produced, however, that there are certain genes whose expression is specifically connected with dormancy. So far, genes which might qualify for the appellation "dormancy genes" have been suggested to occur in seeds of a few grasses (including the cereal, wheat). This is an intriguing development which is sure to attract much research attention in the next few years.

There are strong indications that ABA participates in dormancy. Studies of mutants (e.g., *Arabidopsis*, tomato) furnish compelling evidence that the hormone operates in developing seeds to initiate dormancy, which in these two cases is coat imposed. Experiments with inhibitors of ABA synthesis support the view that the onset and maintenance of embryo dormancy requires the action of ABA, and other approaches lead to the conclusion that in cases of embryo dormancy a continued input of ABA seems to be required. These findings raise the possibility that in

Table 5.15. Effect of Water Stress on Germinability and Subsequent Radicle Growth

Species	Radicle length (mm)[a]			
	0 MPa	−0.3 MPa	−0.6 MPa	−1 MPa
Dandelion	13	15	6	0
Bitter sneezeweed	4	1	1	0
Hemp sesbania	10	7	0	0
Jimson weed	33	22	13	0
Soybean	68	12	0	0
Prickly sida	32	38	33	16
Pearl millet	124	125	94	102

[a]Germinated for 96 h at 29°C.

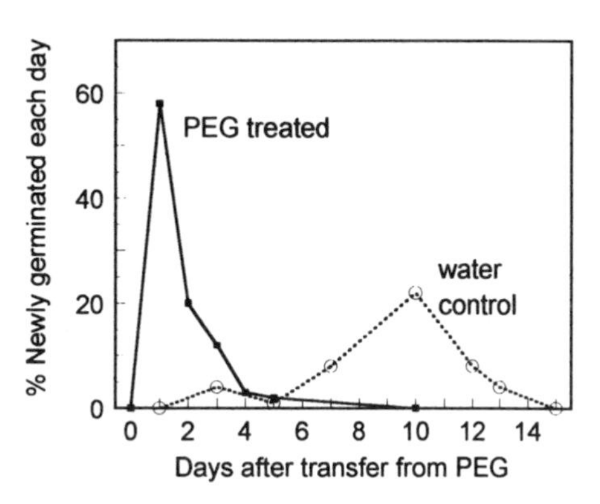

Figure 5.40. Synchronizing radicle emergence by osmotic treatments. Onion seeds were treated with a solution of polyethylene glycol (PEG) at –1 MPa for 23 days at 10°C. They were then transferred to water for germination (——). Controls (i.e., seeds not given a prior PEG treatment) are shown (- - - -). Note that nearly 80% of the PEG-treated seeds completed germination in the first 2 days on water. After Heydecker *et al.* (1973).

coat-imposed dormancy, where the dormancy of the embryo is relatively shallow, the initiation of dormancy is a "one-off" event occurring over a fairly short period in seed development. Embryo dormancy, on the other hand, may have the same causal factor—ABA—but is more profound because the factor operates continuously. If ABA is indeed involved we will need to know how it operates; in this connection it may be that the "dormancy genes" are ABA-responsive. It should be noted that it is not necessary to postulate an exclusive role for ABA in the establishment of dormancy. For example, in those cases in which the photo-environment is important, the development of dormancy can be attributed to effects on phytochrome, in which the seed matures with insufficient Pfr or intermediates leading to it. It remains possible, nevertheless, that the Pfr is required in order to nullify inhibitory events set in train by ABA.

To be released from dormancy, the seed must experience the influence of certain factors—or sufficient time must elapse with the seed in a "dry" state to allow dormancy gradually to wane. These factors are environmental—light, temperature, and, in some cases, certain chemicals such as nitrate. The subtleties of operation and interaction among these factors are extremely important in the natural situation and account for the complex ecophysiological behavior displayed by seeds. The sensor for light is phytochrome, but that (or those) for temperature is still a matter for speculation: cell membranes and/or their constituents may be involved. Membranes may also have a role in connection with phytochrome, but an understanding of the details of the interaction and what the eventual consequences might be has yet to emerge. The biochemical or molecular basis for the breaking of dormancy requires elucidation. There is evidence that gibberellins are involved—not their biosynthesis during dormancy breakage, but their appearance just before the completion of germination, and also the sensitivity of the seed to them. Herein might lie the actions of afterripening, Pfr, and low temperature.

Seeds that have lost their dormancy, or were never dormant, are still sensitive to environmental factors which now influence germination itself. Light, again, is important because of its inhibitory effect, probably by acting on the very last stage, the initiation of radicle elongation itself. Temperature is profoundly important, each seed population having a characteristic range over which germination can take place. This factor determines how many seeds in a population can complete germination, and the rate at which they do so. Unfortunately, the explanation for the basic action of temperature on germination is elusive.

Although much has been learned about the control of germination, with some highly significant recent advances in our knowledge, this aspect of seed physiology continues to pose many of the challenging questions on the subject.

USEFUL LITERATURE REFERENCES

SECTION 5.2

Ballard, L. A. T., 1973, *Seed Sci. Technol.* **1:**285–303 (seed coat effects).

Barton, L. V., 1965, in: *Encyclopedia of Plant Physiology,* Volume 15/2 (W. Ruhland, ed.), Springer, Berlin, pp. 909–924 (general review on dormancy types).

Bewley, J. D., and Black, M., 1982, *Physiology and Biochemistry of Seeds*, Volume 2, Springer-Verlag, Berlin (all aspects of dormancy).

Cavers, P. B., and Harper, J. L., 1966, *J. Ecol.* **54:**367–382 (polymorphism in *Rumex*).

Côme, D., and Thevenot, C., 1982, in: *The Physiology and Biochemistry of Seed Development, Dormancy and Germination* (A. A. Khan, ed.), Elsevier, Amsterdam, pp. 271–298 (embryo dormancy).

Coumans, M., Côme, D., and Gaspar, T., 1976, *Bot. Gaz.* **137:**274–278 (oxygen consumption by beet seed coats).

Datta, S. C., Evenari, M., and Gutterman, Y., 1970, *Isr. J. Bot.* **19:**463–483 (polymorphism in *Aegilops*).

Edwards, M. M., 1969, *J. Exp. Bot.* **20:**876–894 (oxygen and coat effects in *Sinapis arvensis*).

Esashi, Y., and Leopold, A. C., 1968, *Plant Physiol.* **43:**871–876 (coat strength and embryo thrust).

Hamly, D. H., 1932, *Bot. Gaz.* **93:**345–375 (seed coat structure).

Lenoir, C., Corbineau, F., and Côme, D., 1986, *Physiol. Plant.* **68:**301–307 (oxygen uptake by enclosing tissues of barley seed).

Le Page–Degivry, M.-T., 1973, *Biol. Plant.* **15:**264–269 (ABA and embryo dormancy).

Le Page–Degivry, M.-T., and Garello, G., 1992, *Plant Physiol.* **88:**1386–1390 (ABA and induction of dormancy).

McKee, G. W., Pfeiffer, R. A., and Mohsenin, N. N., 1977, *Agron. J.* **69:**53–58 (seed coat impermeability to water).

Porter, N. G., and Wareing, P. F., 1974, *J. Exp. Bot.* **25:**583–594 (coat permeability to oxygen).

Rolston, M. P., 1978, *Bot. Rev.* **44:**365–396 (coat impermeability to water).

Thevenot, C., and Côme, D., 1973, *C. R. Acad. Sci. Ser. D* **277:**1873–1876 (cotyledons and embryo dormancy).

SECTION 5.3

Bewley, J. D., and Black, M., 1982, *Physiology and Biochemistry of Seeds,* Volume 2, Springer-Verlag, Berlin (metabolism, control of dormancy).

Goldmark, P. J., Curry, J., Morris, C. G., and Walker-Simmons, M. K., 1992, *Plant Mol. Biol.* **19:**433–441 ("dormancy genes" in *Bromus*).

Hendricks, S. B., and Taylorson, R. B., 1979, *Proc. Natl. Acad. Sci. USA* **76:**778–781 (membranes, dormancy, and germination).

Larondelle, Y., Corbineau, F., Dethier, M., Côme, D., and Hers, H. G., 1987, *Eur. J. Biochem.* **166:**605–610 (fructose 2,6-bisphosphate and oat dormancy).

Morris, C. F., Armstrong, R. J., Goldmark, P. J., and Walker-Simmons, M. K., 1991, *Plant Physiol.* **95:**814–821 ("dormancy genes" in wheat).

Roberts, E. H., and Smith, R. D., 1977, in: *The Physiology and Biochemistry of Seed Dormancy and Germination* (A. A. Khan, ed.), North-Holland, Amsterdam, pp. 385–411 (pentose phosphate pathway and germination).

Ross, J.D., 1984, in: *Seed Physiology, Volume 2. Germination and Reserve Mobilization* (D. R. Murray, ed.), Academic Press, New York, pp. 45–75 (metabolic aspects of dormancy).

Simmonds, J. A., and Simpson, G. M., 1971, *Can. J. Bot.* **49:**1833–1840 (respiration of dormant wild oats).

SECTION 5.4

Balboa-Zavala, O., and Dennis, F. G., 1977, *J. Am. Soc. Hortic. Sci.* **102:**633–637 (ABA and the onset of dormancy in apple).

Cresswell, E. G., and Grime, J. P., 1981, *Nature* **291:**583–585 (green enclosing tissues and the onset of dormancy).

Groot, S. P. C., and Karssen, C. M., 1992, *Plant Physiol.* **99:**952–958 (dormancy and germination in ABA-deficient tomato mutants).

Gutterman, Y., 1980/81, *Isr. J. Bot.* **29:**105–117 (onset of dormancy).

Hayes, R. G., and Klein, W. H., 1974, *Plant Cell Physiol.* **15:**643–663 (spectral quality of light and onset of dormancy).

Karssen, C. M., 1970, *Acta Bot. Neerl.* **19:**81–94 (photoperiodic induction of dormancy in *Chenopodium*).

Karssen, C. M., 1980/81, *Isr. J. Bot.* **29:**45–64 (secondary dormancy).

Karssen, C. M., Brinkhorst-van der Swan, D. L. C., Breekland, A. E., and Koornneef, M., 1983, *Planta* **157:**158–165 (ABA mutants and *Arabidopsis* dormancy).

Le Page–Degivry, M. T., and Garello, G., 1992, *Plant Physiol.* **98:**1386–1390 (ABA and induction of dormancy).

Sawhney, R., and Naylor, J. M., 1979, *Can. J. Bot.* **57:**59–63 (genetic aspects of dormancy in *Avena fatua*).

Sidhu, S. S., and Cavers, P. B., 1977, *Bot. Gaz.* **138:**174–182 (onset of dormancy in *Medicago*).

SECTION 5.5

Bartley, M. R., and Frankland, B., *Nature* **300:**750–752 (high-irradiance inhibition of germination).

Bewley, J. D., 1979, in: *The Plant Seed* (I. Rubenstein, R. L. Phillips, C. B. Green, and B. E. Gengenbach, eds.), Academic Press, New York, pp. 219–239 (hormonal and chemical effects on dormancy, phytochrome).

Black, M., 1980/81, *Isr. J. Bot.* **29:**181–192 (hormones and dormancy).

Borthwick, H. A., Hendricks, S. B., Toole, E. H., and Toole, V. K., 1954, *Bot. Gaz.* **115:**205–225 (action spectrum for breaking of dormancy in lettuce).

Carpita, N. C., Nabors, M. W., Ross, C. W., and Petretic, N. L., 1979, *Planta* **144:**225–233 (phytochrome and cell elongation).

Esashi, Y., Ishihara, N., Saijoh, K., and Saitoh, M., 1983, *Plant Cell Environ.* **6:**47–54 (cyanide-resistant respiration and dormancy).

Hanke, J., Hartmann, K. M., and Mohr, H., 1969, *Planta* **86:**235–241 (phytochrome photoequilibria).

Hartmann, K. M., 1966, *Photochem. Photobiol.* **5:**349–354 (absorption spectrum of phytochrome).

Hendricks, S. B., and Taylorson, R. B., 1978, *Plant Physiol.* **61:**17–19 (temperature shifts and phytochrome).

Heydecker, W., Higgins, J., and Gulliver, R. L., 1973, *Nature* **246:**42–44 (advancing germination with PEG).

Hilhorst, H. W. M., and Karssen, C. M., 1988, *Plant Physiol.* **86:**591–597 (light, nitrate, and gibberellin—effects on germination).

Hilhorst, H. W. M., and Karssen, C. M., 1992, *Plant Growth Regul.* **11:**225–238 (ABA, GA in dormancy and germination: mutants).

Kendrick, R. E., and Frankland, B., 1982, *Phytochrome and Plant Growth*, Arnold, London (a general account of phytochrome).

Koller, D., and Negbi, M., 1959, *Ecology* **40:**20–36 (dual effect of light on germination).

Logan, D.C., and Stewart, G. R., 1992, *Seed Sci. Res.* **2:**179–190 (germination of seeds of parasites).

McCormac, A. C., Smith, H., and Whitelam, G. C., 1993, *Planta* **191:**386–393 (germination of phytochrome transgenics).

Mohr, H., and Appuhn, U., 1963, *Planta* **60:**274–288 (high-irradiance far-red effects).

Nabors, M. W., and Lang, A., 1971, *Planta* **101:**1–25 (light and embryo water relations).

Roberts, E. H., 1965, *J. Exp. Bot.* **16:**341–349 (temperature and afterripening).

Roberts, E. H., and Smith, R. D., 1977, in: *The Physiology and Biochemistry of Seed Dormancy and Germination* (A. A. Khan, ed.), North-Holland, Amsterdam, pp. 385–411 (pentose phosphate pathway and germination).

Schopfer, P., and Plachy, C., 1985, *Plant Physiol.* **77:**676–686 (ABA effects on wall extensibility).

Taylorson, R. B., 1982, in: *The Physiology and Biochemistry of Seed Development, Dormancy and Germination* (A. A. Khan, ed.), Elsevier, Amsterdam, pp. 323–346 (light and other factors in seed germination).

Thomas, T. H., 1977, in: *The Physiology and Biochemistry of Seed Dormancy and Germination* (A. A. Khan, ed.), North-Holland, Amsterdam, pp. 111–114 (growth regulators and germination).

Thompson, P. A., 1973, in: *Seed Ecology* (W. Heydecker, ed.), Butterworths, London, pp. 31–58 (temperature and germination).

Totterdell, S., and Roberts, E. H., 1979, *Plant Cell Environ.* **2:**131–137 (chilling of *Rumex*).

Totterdell, S., and Roberts, E. H., 1980, *Plant Cell Environ.* **3:**3–12 (alternating temperatures and germination).

Upadhyaya, M. K., Simpson, G. M., and Naylor, J. M., 1981, *Can. J. Bot.* **59:**1640–1646 (pentose phosphate pathway enzymes and dormancy).

Van Der Woude, W. J., and Toole, V. K., 1980, *Plant Physiol.* **66:**220–224 (chilling and phytochrome action).

Vincent, E. M., and Roberts, E. H., 1977, *Seed Sci. Technol.* **5:**659–670 (interacting factors and dormancy).

Visser, T., 1956, *Proc. K. Ned. Akad. Wet. C* **59:**314–324 (chilling and apple seed dormancy).

Walker-Simmons, M. K., 1987, *Plant Physiol.* **84:**61–66 (sensitivity to ABA and dormancy in wheat).
Williams, P. M., Bradbeer, J. W., Gaskin, P., and MacMillan, J., 1974, *Planta* **117:**101–108 (chilling and gibberellins in hazel).

Chapter 6

Some Ecophysiological Aspects of Germination

6.1. INTRODUCTION

We pointed out in the previous chapter that the biological importance of dormancy must be seen in relation to the ecology of germination—when and where a seed germinates. Interactions between the dormancy-releasing agents—light, temperature, and afterripening—and the sensitivity of germination to light, temperature, and, say, water stress are responsible for determining if a seed will germinate in a particular situation and season. The germination and dormancy mechanisms are therefore of great adaptive importance in ensuring that seedling emergence occurs at the most advantageous time and place. In this chapter we will consider some examples to illustrate the ecological significance of these control processes.

6.2. SEED BURIAL

6.2.1. The Seed Bank

A great quantity of seeds is buried in the soil—the seed bank. One determination for the top layer of cultivated land put the number of buried seeds therein at tens of millions per hectare. The distinction has been drawn between transient and persistent seed banks, the former consisting of seeds that mostly are viable in the soil for no more than 1 year, and the latter having a significant proportion of the seeds remaining viable for many years. In Fig. 6.1, which illustrates the different types of seed bank, we can see the seasonal acquisition of germinability which gives each group of plants its characteristic emergence pattern. These seasonal patterns are largely controlled by the seeds' response to environmental factors, particularly light, temperature, moisture, and various chemicals. Also involved are seasonal environmental factors that govern the onset of dormancy. It is worth noting, however, that the environment may not have exclusive control, for in a limited number of species (e.g., *Mesembryanthe-*

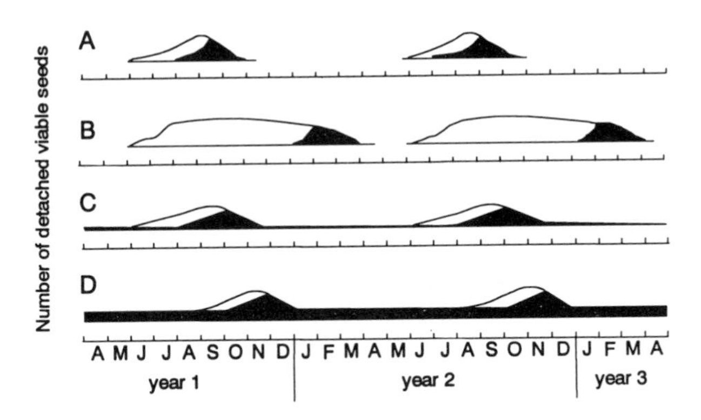

Figure 6.1. Seed banks common in soils of temperate regions. ■, Seeds that can germinate immediately when tested in the laboratory (i.e., nondormant seeds). □, Viable seeds that are dormant. (A) Seeds of annual and perennial grasses of dry or disturbed habitats (e.g., *Lolium perenne*). (B) Seeds that germinate in early spring, colonizing gaps in the vegetation. They include annual and perennial herbs, trees, and shrubs (e.g., *Impatiens glandulifera, Acer pseudoplatanus*). (C) Seeds of winter annuals, which germinate mainly in the fall. A small, persistent seed bank of nondormant seeds is maintained (e.g., *Erophila verna*). (D) Seeds of annual and perennial herbs and shrubs, which have large, persistent seed banks (e.g., *Stellaria media*). After Grime (1979).

mum spp.) the seeds seem to have an inbuilt rhythm of germinability that governs seedling emergence.

6.2.2. Light and Seed Burial

Clearly, germination of buried seeds is held in check, sometimes for many years, perhaps until some disturbance occurs. If cultivated soil is turned over, or if natural disturbance takes place as in a river bank, many of the seeds germinate and a flush of seedlings results. One investigation reported that weeds such as *Sinapis arvensis, Polygonum aviculare, Veronica persica,* and others increased from about 35 to 780 seedlings per square meter after the ground was dug. A factor that has major responsibility for this effect is light, for when steps are taken to exclude light when the earth is disturbed, very few seedlings appear; and as little as 0.5 s exposure can cause many seeds to germinate and produce seedlings. But many of the species that appear as a consequence of soil disturbance and the accompanying illumination do not disperse light-requiring seeds. And so the seeds of these species (such as *Heracleum sphondylium, Senecio vulgaris,* and *Spergula arvensis*) must develop a light requirement after their dispersal, i.e., during burial. This is probably because burial induces secondary dormancy by mechanisms that are still unexplained. Light sensitivity can also be lost during burial, as in *Barbarea vulgaris* and *Chenopodium bonus-henricus*. Seeds of both of these species are light-requiring when freshly dispersed, become resistant to light after a period of burial, and then acquire light sensitivity once more. Throughout this sequence all of the seeds are completely dormant in darkness (Fig. 6.2). The development of a type of secondary dormancy called skotodormancy could account for the loss of photosensitivity; this is a phenomenon that occurs in light-requiring seeds when they are held in darkness for extended periods of time. Why photosensitivity is regained is unclear, but it might be a consequence of chilling; in Chapter 5 we mentioned that prior treatment with low temperatures can confer photosensitivity on seeds of several species. The onset of secondary dormancy in buried seeds can be an important contribution to the germination strategy of a species. We will discuss some examples later in our considerations of the role of temperature in the temporal distribution of germination.

Burial is also important in protecting seeds from the inhibitory effects of light (Section 5.5.2.1). For example, several cultivars of lettuce seeds are unable to germinate in full sunlight but can do so when buried. It has been observed that the light-inhibited *Nemophila menziesii* germinates in cracks in the earth where the seeds are shaded from direct sunlight, whereas seeds a few centimeters distant, exposed on the soil surface, are inhibited.

The transmittance of light in soil is clearly of great importance in light-controlled dormancy and germination. How effective is the soil as a light filter and how deep must seeds be buried to escape from the effects of light? Less than 2% of the light passes through 2 mm of sand, and the light that is transmitted consists only of wavelengths longer than about 700 nm (Fig. 6.3). A wide part of the spectrum can pass through clay loams, but if the grain size is small (when they will be closely packed), a layer 1.1 mm thick is virtually opaque. Soil moisture content affects transmission, increasing it somewhat in the clay loam but decreasing it in larger-grained sand. Lettuce seeds remain dormant under 6 mm of sand even when the surface is illuminated, but beneath a thinner layer (2 mm), about 24% of the seeds eventually germinate when given 1 h of illumina-

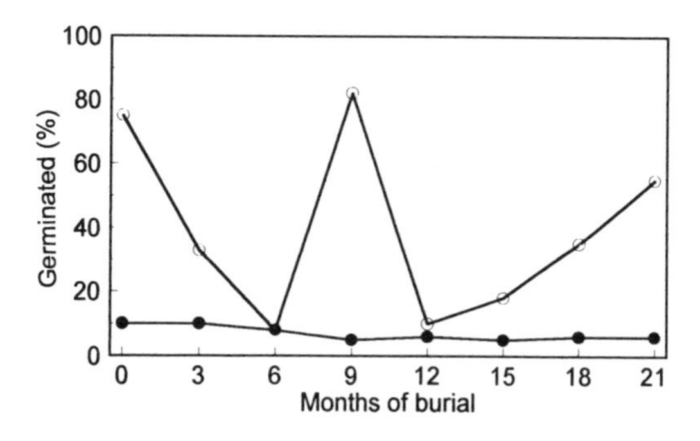

Figure 6.2. Changes in light sensitivity of buried seeds of *Barbarea vulgaris*. Seeds were buried at a depth of 2.5 cm and samples were taken periodically for germination tests in darkness (●) or after red light (○). After Taylorson (1972).

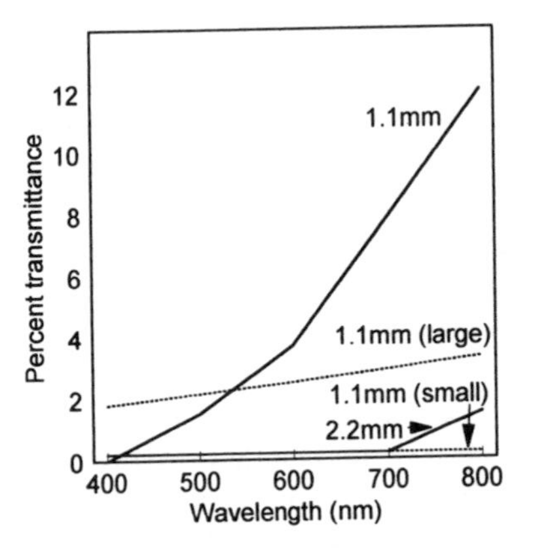

Figure 6.3. Transmission of light through soils. Two soils were used, sand (——) and a clay loam (- - - -), at thicknesses indicated on the curves and, in the case of clay, at large (approximately 0.9 mm) and small (approximately 0.45 mm) grain size. Adapted from Woolley and Stoller (1978).

tion, and 72% are promoted by 10 h of light. These values are a good illustration of the importance of burial depth to a light-requiring seed. Inhibition of germination by other soils may require greater depths, however. About 8 mm of loam is needed to completely suppress germination of *Plantago major*, whereas light-promoted germination of *Digitalis purpurea* still occurs under 10 mm of sand (Fig. 6.4). The quality of transmitted light is, of course, as important as the quantity. Since the longer wavelengths penetrate earth more easily, a buried seed experiences a relatively high proportion of far-red light (i.e., > 700 nm), which obviously affects the phytochrome photoequilibrium (φ) (Section 5.5.1.4b). Light passing through 1 mm of dry, sandy soil sets up a photoequilibrium of about 0.45, which should be sufficient to break dormancy in a high percentage of light-requiring seeds. That it does not necessarily do so may be because at the

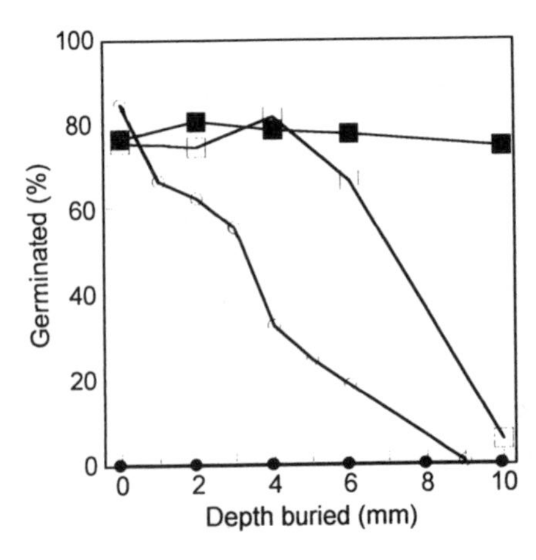

Figure 6.4. Effect of burial on germination. *Plantago major* seeds were buried at different depths in wet loam held in the light (○) or dark (●). Seeds of *Digitalis purpurea* (■) and *Cecropia obtusifolia* (□) were buried at different depths in wet sand exposed to light. Germination of the last two species in darkness was almost zero. Adapted from Frankland and Poo (1980) and Bliss and Smith (1985).

very low fluence rates in the soil the rate at which Pfr is formed is extremely low, and thermal reversion may succeed in removing a large proportion of the Pfr. The germination of dormant, light-requiring seeds of some species is therefore restricted to the uppermost layers of soil where the light stimulus for dormancy breakage can operate. This is extremely important for small seeds with limited food reserves, because if germination is completed at too great a depth the seedling's reserves may be exhausted before it is able to reach daylight and begin to photosynthesize. Most species of light-requiring seeds are, in fact, small (< 1 mg seed weight). It must be remembered that light requirements of seeds are much modified by factors such as temperature and nitrate which clearly will affect seed behavior in the field.

6.3. GERMINATION IN DIRECT SUNLIGHT

Germination in many species is inhibited when seeds are on the soil surface. Often, the relative dryness of the soil accounts for this. In one well-known case, the oxygen concentration of the air is inhibitory: this happens in *Scirpus juncoides*, a weed of rice paddies. Seeds of a large number of species, however, are affected by high fluence rates (photon flux densities) of light on the soil surface, which probably act through the HIR (Section 5.5.2.1). The negatively photoblastic seeds fall in this group (e.g., *Amaranthus caudatus, Nemophila insignis, Oldenlandia corymbosa*) and also seeds which are promoted by low fluence rates, such as *Urtica urens* and *Sinapis arvensis*. It has been argued that such sensitivity is a mechanism for discouraging germination under high solar radiation when the seedling would be subjected to harsh, drying conditions. Interestingly, photoinhibition is much enhanced by, or may only be manifest under a degree of water stress (see Section 6.7).

6.4. GERMINATION UNDER LEAF SHADE

Chloroplasts absorb light most strongly at approximately 675 nm and allow wavelengths longer than 720 nm to be transmitted completely. Because of this, sunlight passing through green leaves has a very low red/far-red ratio (Fig. 6.5). The phytochrome photoequilibrium established by such light can be as low as 0.15, if the light has passed through a sufficient number of leaves. Most light-requiring, dormant seeds will not germinate under such conditions. Indeed, if they are exposed to this light for many hours over several days they will probably be forced into secondary dormancy. Nondormant seeds can also be inhibited by canopy light (Table 6.1) for two reasons: first, the Pfr that some contain is reversed to Pr, and second, if the far-red-rich light is at a high enough

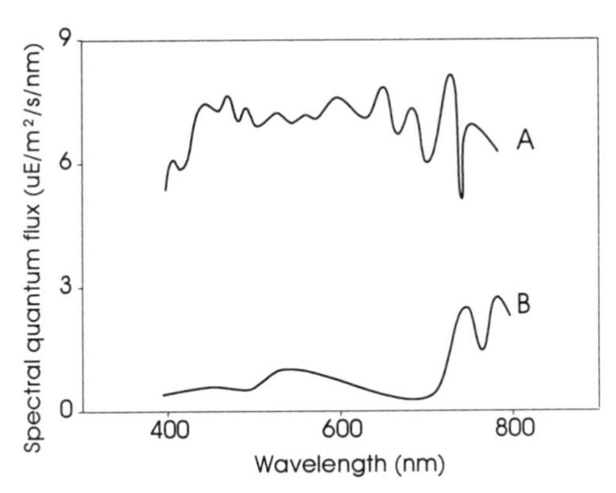

Figure 6.5. Spectral energy distribution of sunlight and vegetational shade light. (A) Direct sunlight at 11.00 h. (B) Shade light (under sugar beet plants). The ratio of red to far-red light (660/730 nm) in A is 1.18 and in B is 0.12. After Holmes and McCartney (1975).

fluence rate, it will inhibit through the HIR (Section 5.5.2.1). These seeds, therefore, become dormant as a result of irradiation with canopy light and will germinate later only when a dormancy-releasing factor, such as chilling or light, has been experienced. Seeds of *Bidens pilosa*, for example, when forced into dormancy by canopy light, become light-requiring and subsequently germinate when exposed to light of suitable quality to establish a relatively high photoequilibrium value of phytochrome; direct sunlight would, of course, suffice. This mechanism accounts for the paucity of seedlings on forest floors and the flush that follows the appearance of a gap in the leaf canopy, brought about when individual trees die and fall or when tree clearance occurs. That is, the seed bank in forest soils may be rich but germination is held in check, to a large extent by the far-red environment. One example will illustrate this. In the forests of Brazil, seeds of *Cecropia glaziovi* are distributed by bats and other small mammals; yet seedlings are not found there. These emerge only after tree clearance. In the field, vegetational shade is provided by grasses and herbs, of course, as well as by trees. Plants of *Arenaria* spp., *Veronica* spp., and *Cerastium* spp. establish themselves on vegetation-free ant mounds but not in the surrounding pasture. Seeds of these species are inhibited by light filtered through green leaves, which probably explains their difficulty in germinating among grasses and herbs: they germinate well on bare earth, in other words, in sites where the seedlings will

Table 6.1. Some Nondormant Seeds Whose Germination Is Inhibited by Light Filtered through Green Leaves

Achillea millefolium	*Cucumis sativus*	*Ruellia tuberosa*
Amaranthus caudatus	*Lactuca sativa*	*Rumex acetosa*
Bidens pilosa	*Leontodon hispidus*	*Tragopogon orientalis*
Bromus tectorum	*Melandrium album*	*Veronica arvensis*

face less competition with other plants for light for photosynthesis. Some species are, of course, able to colonize shaded sites. It is known that seeds of some of these species, e.g., *Centaurium erythrea*, are certainly less sensitive to far-red canopy light.

As mentioned, the spectral energy distribution of canopy light is determined by the number of leaves through which the light passes. And hence the effects of germination will vary according to the density of leaf cover. As far as we know, no investigations on this have been carried out in field situations, but it has been simulated under experimental conditions using seeds of *Plantago major*. When placed under increasing leaf cover, provided by *Sinapis alba* plants, germination of these light-requiring seeds is increasingly inhibited as the cover becomes more dense (measured as the leaf area index) and the phytochrome photoequilibrium beneath the leaves decreases (Fig. 6.6). Variations in the density of a leaf canopy occur in the field, of course, as buds break and leaves expand and later senesce and fall. Thus, light-sensitive seeds among vegetation are exposed to changing light environments throughout the growing season, with concomitant effects on their dormancy breakage and germination. We must appreciate, however, that other environmental factors, such as temperature, are also influential, so that even when shade ends, seed germination will not necessarily proceed.

6.5. TEMPERATURE

Temperature acts to regulate germination in the field in three ways: (1) by determining the capacity and rate of germination, (2) by removing primary and/or secondary dormancy, and (3) by inducing secondary dormancy. Unfortunately, it is not always possible to distinguish from the available experimental

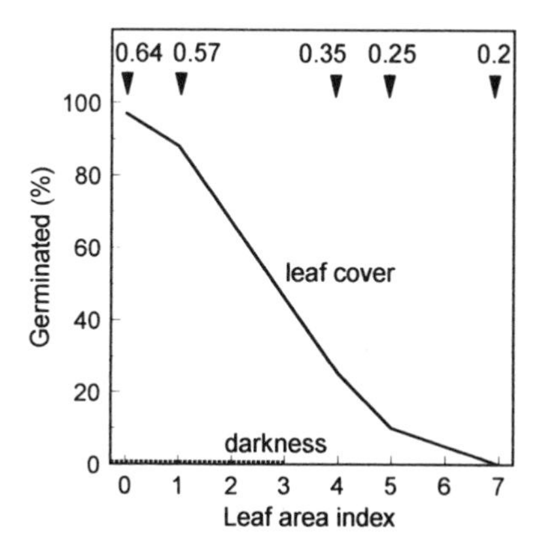

Figure 6.6. Germination of light-requiring seeds under leaf cover. Seeds of *Plantago major* were sown beneath plants of *Sinapis alba* (white mustard) at different densities, as indicated by the leaf area index (i.e., unit leaf area per unit land area). P_{fr}/P_{total} (φ) values (at soil level) at different densities are indicated at the top of the graph. - - - -, Germination in darkness. From data in Frankland and Poo (1980).

work which of these is involved, but where we are able to do so we will make the distinction. The reader should refer also to Sections 6.8 and 6.9.

In Chapter 5, we saw how species have a particular temperature range for germination, and how temperature also determines germination rate. This sensitivity to temperature limits germination of some species to particular times of the year. *Amaranthus retroflexus* prefers relatively high temperatures for germination (> 25°C), whereas *Chenopodium album* and *Ambrosia artemisiifolia* are both satisfied by lower temperatures (> 10°C and > 15°C, respectively). Of these three, then, *Amaranthus* germinates only in late spring and early summer in Kentucky, but the other two can germinate in early spring. The *Ambrosia* seeds that do not succeed in germinating in early spring enter secondary dormancy and must await chilling to be released from this. The combination of a low-temperature requirement for germination and the induced secondary dormancy thus serves to restrict *Ambrosia* germination to early spring. *Nemophila menziesii* is another interesting example that illustrates the ecological importance of the germination temperature. Although these seeds are light-inhibited dark germinators, they are nevertheless incapable of germinating even in darkness at temperatures above a certain value, and freshly mature seeds are inhibited by a lower temperature than are older seeds (Fig. 6.7). In California, the seeds are dispersed in springtime (the plant is a winter annual) when the mean daily temperature is about 15°C, and so, even though moisture is available, they do not germinate. To do so, the seeds must await the passage of time, which widens their temperature tolerance, and they can germinate in September. This strategy prevents germination in spring, after which the young seedlings would be exposed to the hot, dry summer, and restricts it to late summer, just prior to the wet fall and winter. Summer and winter annuals of the flora of the Mojave and Colorado deserts also use their germination temperature requirement to achieve seasonal emergence, the winter annuals germinating only at lower temperature (e.g., 10°C) and the summer ones only at higher values (e.g., 26–30°C).

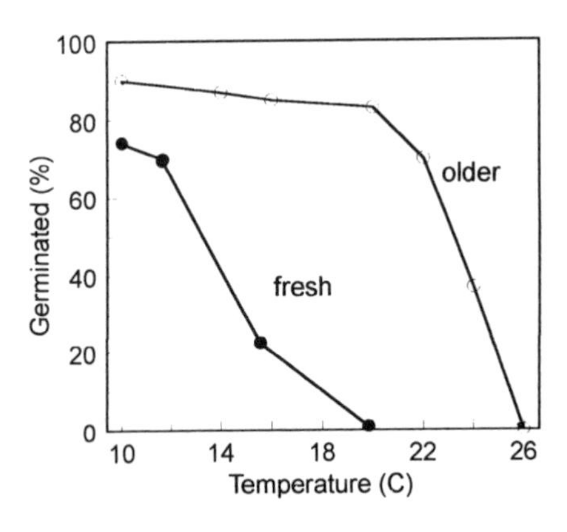

Figure 6.7. Sensitivity to temperature of germination in *Nemophila menziesii*. The percentage of seeds germinating at different temperatures is shown for freshly mature seeds (●) and seeds several months old (O). After Cruden (1974).

Seeds whose dormancy is broken by chilling germinate when temperatures begin to rise in early spring, which has dual advantages. First, seedling emergence immediately before the inclement winter is prevented. Second, seedlings can establish themselves at a favorable time when there is no shading by leaves. Plants inhabiting meadows, for example, benefit from this mechanism because vegetation density in such habitats is lowest in early spring. Species inhabiting deciduous woodlands, including the nascent tree seedlings themselves, also derive benefit from germinating before the leaf canopy appears.

The bluebell (*Hyacinthoides non-scripta*), a herbaceous plant of woodlands, uses another strategy involving temperature to ensure germination before the leaves are out. Dormancy in these seeds is best terminated by relatively high temperatures (26–31°C) for a few weeks. Germination cannot occur immediately but it does take place at an optimum of approximately 11°C. Many of the seeds shed in midsummer are warmed up in the soil and germinate later, in the fall. Seedlings in southern English woodlands therefore emerge in late fall, overwinter, and establish vigorous growth in spring before the development of the leaf canopy.

Afterripening, as a dormancy-relieving mechanism, operates under field conditions, and here, too, the importance of temperature can be discerned. For example, dormant seeds of the winter annuals *Sedum pulchellum* and *Valerianella umbilicata* require a fairly high temperature for afterripening, so that after being shed in winter they must experience the hot, dry summer before losing their dormancy; germination is therefore restricted to the fall and avoids the unfavorable midsummer droughts.

Dormancy of many species in the field is broken by alternating temperatures. As far as buried seeds are concerned, it seems that the vegetation cover is important, for it can influence the amplitude of the temperature fluctuation. This effect is seen rather well in an experiment in which the diurnal temperature fluctuation in soil under different sizes of gaps in the overlying vegetation was determined, together with the germination of buried *Holcus lanatus* seeds (Fig. 6.8). Gap size clearly influences the amplitude of the temperature fluctuation, which in turn appears to correlate with the germination percentages. The temperature amplitude that a soil can achieve after vegetation clearance varies according to the depth at which it is measured (Fig. 6.9). It has been suggested that the dampening of the amplitude can be used by dormant seeds to sense the depth at which they are buried. Clearance of vegetation can therefore have important effects in addition to an alteration of the light environment in the top layer of the soil, and thus buried seeds can use a mechanism other than the phytochrome system to given them positional information.

Temperature is important in the breakdown of the hard, waterproof coats present in seeds of many species (see Section 5.2.5.1). Heat and fluctuating

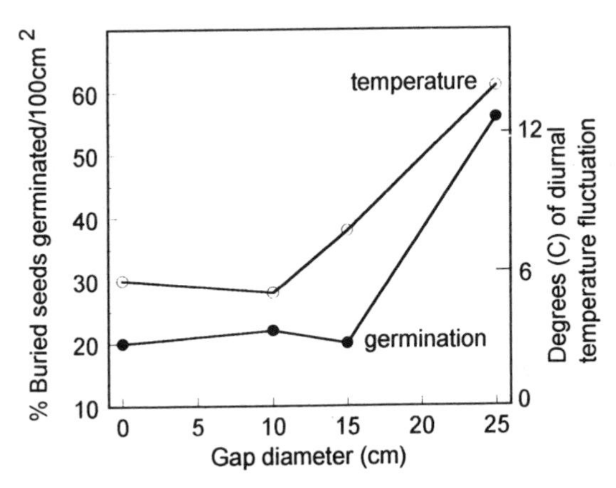

Figure 6.8. Effect of vegetation gap size on soil temperature fluctuation and seed germination. Gaps of different sizes were made in the grass of a pasture in northern England. The mean daily temperature fluctuation at a depth of 1 cm during May was recorded (O), together with the percentage of germination (●) of a naturally buried seed population of *Holcus lanatus*. After Grime (1979).

temperatures are especially important in hard seeds of tropical, subtropical, and Mediterranean climates. High solar radiation, which heats the dry seeds, followed by low night temperatures act to crack the hard coats. Under some conditions, fire is equally effective in certain species.

6.6. WATER

Seeds of many species are nondormant when shed from the parent plant and can germinate over a wide temperature range. It seems that in these cases the major, perhaps the only, determinant of germination is the availability of water. Clearly, if the seeds are nondormant and germinate when sufficient moisture is present, there is no reservoir of seeds left in the soil, i.e., there is no persistent seed bank (A in Fig. 6.1). Seeds of this type are characteristic of the common grasses of European meadows and pastures, such as *Festuca* spp., *Lolium* spp., *Bromus* spp., and *Dactylis glomerata*.

It will be appreciated that in desert habitats germination should be rigorously linked to water supply if the newly produced seedlings are going to survive. This means that seedlings should not emerge when there is just enough water to support germination but rather when the soil has enough water in it to sustain subsequent plant growth. Interesting findings on the germination physiology of certain desert plants suggest how this might occur. Many seed

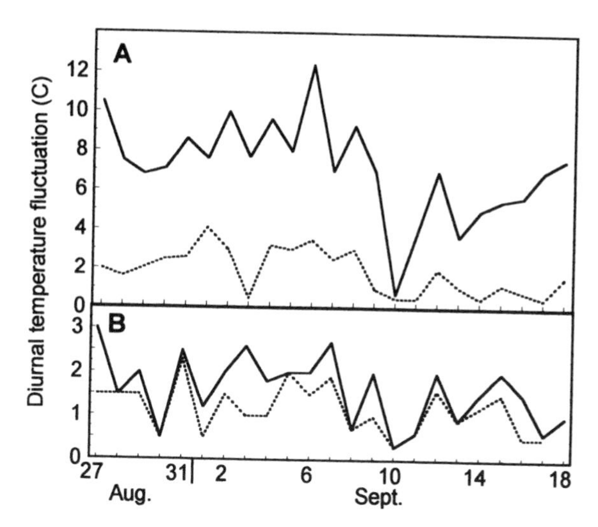

Figure 6.9. The effect of vegetation clearance on soil temperatures. The daily temperature fluctuation was recorded in soil at a depth of 1 cm (A) and 5 cm (B). The amplitude of fluctuation in gaps of 20-cm diameter cleared of overlying grass (——) and in uncleared soil (- - - -) is shown. After Grime (1979).

species in desert soils germinate only after heavy rain (e.g., 12–15 mm), not after light rain, even when the latter is composed of several, intermittent falls. Moreover, the precipitation (say 15 mm) is more effective when spread over 10 h than when received over 1 h. The explanation for this is that the seeds (e.g., of *Euphorbia* spp. and *Pectis papposa*) contain water-soluble germination inhibitors, which must be leached out before germination can commence. Thus, the seeds have a device that acts like a rain gauge and ensures that germination takes place only when the earth has received sufficient rain to support the establishment and growth of the seedling. Another way by which sufficient rainfall can be detected is through the seeds' sensitivity to salt solutions. Germination of *Baeria chrysostoma* is reported to be inhibited by a solution having an osmotic potential of approximately –0.1 MPa, and it therefore occurs only when the salt in the top layer of earth has been diluted to an osmotic potential above this value. Desert seeds with very hard coats, such as *Cercidium aculeatum*, can germinate only after the seeds have been carried along by heavy rain in the company of abrasive sand and small stones, which scarify the testa. This species, in company with several other desert inhabitants, is found only in the former courses of such torrents.

6.7. INTERACTIONS

From time to time we have indicated that various factors operate together to control germination in the field. Perhaps a few more specific examples would be appropriate at this juncture. Chilling, light, and alternating temperatures appear to interact in *Aster pilosus*, for example, where at maturity seed dormancy can be broken only by light combined with high, alternating temperature. This cannot occur when the seeds are dispersed in the fall because the temperatures are not high enough. The temperature requirement is lowered by chilling, however, so that the seeds can then subsequently germinate in the alternating temperatures of the fairly cool spring of Kentucky. Germination before the very cold winter is thus prevented, and in the spring a vegetative rosette can be established before the summer drought begins. Temperature, light, and available water all regulate germination in *Nemophila menziesii*. We can see from Fig. 6.10 that high temperature and light, which are both inhibitory, would permit some germination as long as water stress is not serious. Thus, some *Nemophila* seeds might germinate in summer and fall in years when water is adequate, but not otherwise.

6.8. SECONDARY DORMANCY AND SEASONAL GERMINATION

Secondary dormancy is of great importance in germination ecophysiology, since it develops in many species exposed to unfavorable conditions. The residual content of dormant seeds in the persistent seed bank (Fig. 6.1) includes seeds that failed to germinate and entered secondary dormancy. Some of these seeds succeed in germinating at a later stage, and those which do not, enter dormancy once more. This partially accounts for the rhythm of withdrawal from the seed bank indicated in Fig. 6.1. Seeds of winter and summer annuals are forced into dormancy by different factors. Winter annual seeds, such as those of

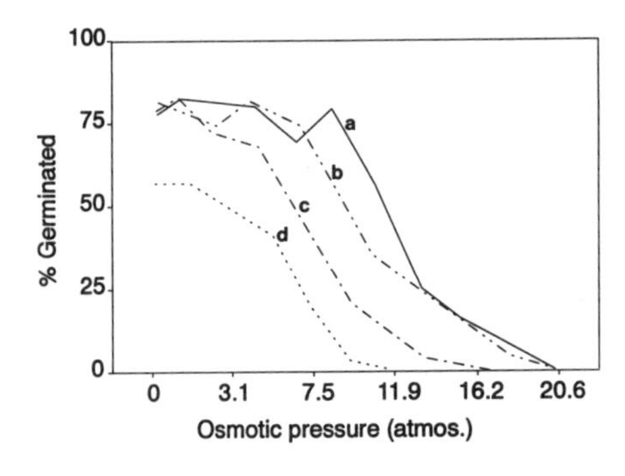

Figure 6.10. Interacting factors controlling germination of *Nemophila menziesii*. Seeds were held on solutions at various osmotic concentrations (a) in darkness, (b) in darkness at temperatures at incipient inhibition (4 h at 23°C, 20 h at 10°C), (c) in the light (4 h per day) or (d) in the light plus inhibitory temperature. Note that 10 atm is approximately 1 MPa, and osmotic "pressure" (potential) is a negative value. Maximum germination percentages are shown. Adapted from Cruden (1974).

Veronica hederofolia and *Phacelia dubia*, become dormant because of the low winter temperatures they experience after dispersal. They slowly emerge from dormancy over the summer, partly by afterripening, and germinate in the fall. Any seed that does not germinate goes back into secondary dormancy and follows the sequence once more. The rhythm of their germination (Fig. 6.11), therefore, shows peaks in the fall seasons. The summer annuals (e.g., *Polygonum persicaria*), on the other hand, develop secondary dormancy in response to the high temperatures of summer. It is broken by chilling, at least in many of the seeds, so that germination occurs in springtime (Fig. 6.11). Again, the unsuccessful seeds are forced back into secondary dormancy, and reenter the bank of dormant seeds in the soil. Some of them germinate after the following winter. Their germination rhythm therefore peaks in springtime (Fig. 6.11), so the seeds of the winter and summer annuals are out of phase.

A contribution to the establishment of such patterns is also made by the primary, relative dormancy of the seeds. That of the summer annuals prevents germination in temperatures that are too low (i.e., the seeds are dormant) whereas that of the winter annuals arrests germination (because of dormancy) in temperatures that are too high. Hence, seeds of the summer annual, released from the parent plant in the fall, cannot germinate because the temperature is not high enough; dormancy remains until it is broken by the winter chilling, and germi-

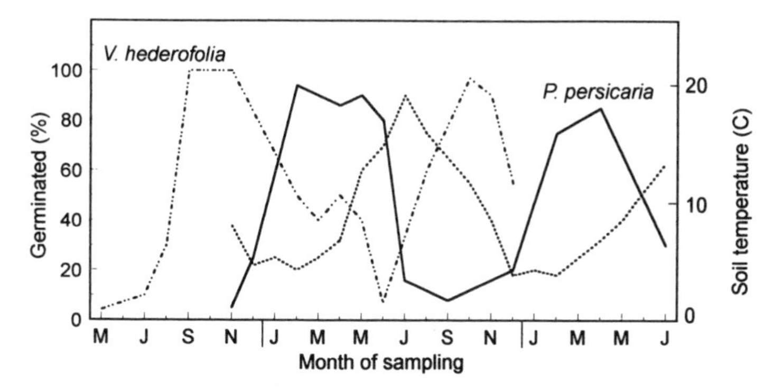

Figure 6.11. Patterns of seed dormancy during burial. The patterns of germination and dormancy of *Veronica hederofolia* and *Polygonum persicaria* are shown. Seeds of both species were buried in sandy loam at 5 cm *(Veronica)* or 10 cm *(Polygonum)*. Germination at alternating temperatures in the light was tested at intervals. The dotted line shows the average morning soil temperature at the site of burial of *Polygonum*. N.B. *Veronica hederofolia* is a winter annual; *Polygonum persicaria*, a summer annual. Note that the two species enter a deep secondary dormancy at different times of the year, in response to low and high temperatures, respectively. After Roberts and Lockett (1978) and Karssen (1980/81).

nation then occurs in early spring. Those individuals whose dormancy is not relieved may enter secondary dormancy and must await the next round of dormancy breakage. The opposite pertains in the winter annuals whose seeds, ripening in spring and early summer, do not germinate when released because the temperature is not sufficiently low. Their dormancy is lost by afterripening over the summer and germination occurs in the fall.

In these scenarios seed responses to temperature play the major role. Evidence is now available showing that other factors participate in the regulation of seasonal dormancy and germination patterns. In *Sisymbrium officinale*, for example, changes in sensitivity to light and nitrate accompany the seasonal rhythm of dormancy—sensitivity to both factors is low at times of deepest dormancy. Temperature still plays a dominant role, however, as these alterations in sensitivity are provoked by the dormancy-inducing and -relieving temperature changes. These examples are a striking illustration of how the germination physiology has important ecological, adaptive significance.

6.9. GERMINATION, PLANT DISTRIBUTION, AND PLANT ORIGIN

In Chapter 5, some examples were given of species differences in the temperature requirements for germination. Such differences are important in determining the distribution of plants, for they obviously limit germination to regions that have suitable temperatures. It follows, also, that indigenous species of a particular region show characteristic temperature requirements, since they are adapted to the temperature conditions prevailing in the environment. To illustrate this we can take one family, the Caryophyllaceae, and see how a few members differ. One way of doing this is to define the "germination character" of a species by determining the time taken at each temperature to reach 50% of maximum germination. The curve so derived therefore expresses rate data and also indicates the cutoff temperatures for germination. Germination characters of three caryophyllaceous species are shown in Fig. 6.12, each typifying geo-

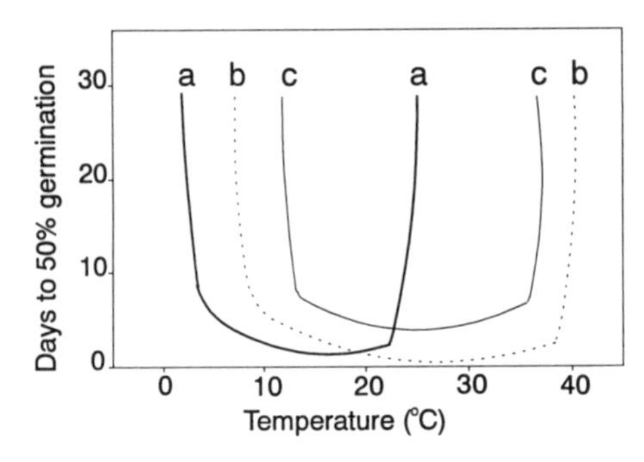

Figure 6.12. Germination "character" curves. Nondormant seeds of three species were held at different temperatures, and the number of days taken to reach 50% germinated seeds was determined. a, *Silene echinata* (Mediterranean); b, *Petrorhagia prolifera* (continental grassland); c, *Silene dioica* (European woodland). Adapted from Thompson (1973b).

graphical origin. The species from the continental grassland (steppe) *(Petrorhagia prolifera)* completes germination quickly at the favorable temperatures (12–40°C), with high minimum (ca. 8°C) and maximum (ca. 42°C) temperatures. The Mediterranean species (*Silene echinata*) also completes germination fairly rapidly at median temperatures but, in contrast to the grassland type, has rather low minimum (< 5°C) and maximum (ca. 25°C) requirements. Finally, the European woodland species *(Silene dioica)* is relatively slow to complete germination at median temperatures, with a high minimum (10–15°C) and moderate maximum (ca. 35°C) temperature requirement. The Mediterranean character, belonging to the winter annuals, has been interpreted as favoring fall germination of the shed seed, in anticipation of the winter growing season. Seeds of the European woodland species, with a median temperature range, if shed in summer, would complete germination at once, whereas fall-shed seed would have to await the following spring. Here, germination only in spring/summer is encouraged. The grassland species have an opportunist character and are able to germinate over a wide range of temperatures; they would germinate when the seeds are shed in mid- to late summer. Of course, the same species may be found in a wide variety of climatic regions: here, the germination behavior may differ according to provenance. This is seen in the case of *Tsuga canadensis* where distinct ecological types can be recognized—the seed coming from northerly latitudes (Quebec) showing temperature requirements distinguishable from those from southerly regions (Tennessee) (Fig. 6.13). The germination characters of representatives of single species from different regions are very revealing (Fig. 6.14). Seeds of *Silene vulgaris* from three origins show the typical germination characters of the regional types (Mediterranean, continental grassland, and European woodland—cf. Fig. 6.12), so clearly adaptation to the local conditions occurs within a species. On the other hand, some species have remained remarkably stable as far as their germination pattern is concerned, even in widely distributed populations. Figure 6.15 shows germination charac-

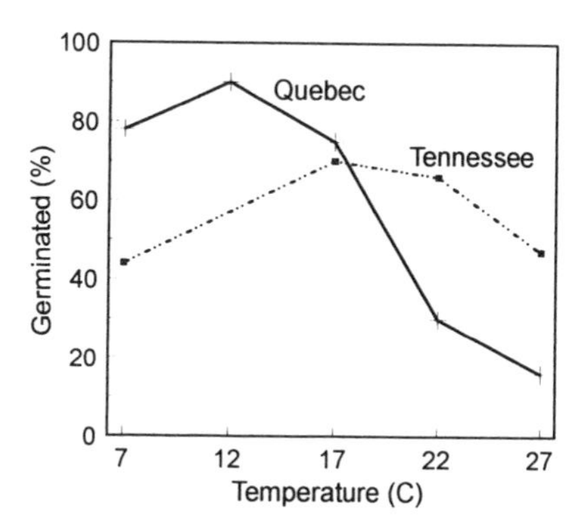

Figure 6.13. Germination of *Tsuga canadensis* seeds of two provenances. Seeds collected from Quebec, Canada, and eastern Tennessee, USA, were cold stratified to remove dormancy. Germination tests were carried out over a range of temperatures: maximum germination percentages are shown. Adapted from Stearns and Olson (1958).

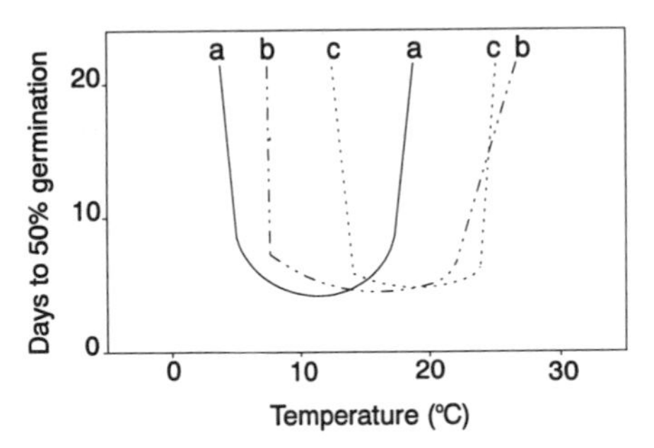

Figure 6.14. Germination "character" curves of *Silene vulgaris* seeds from different regions. Seeds of *Silene vulgaris* from three geographical localities were tested at a range of temperatures. The number of days for the germinated seeds to reach 50% was determined at each temperature. a, Seeds from Portugal; b, seeds from Czechoslovakia; c, seeds from England. After Thompson (1973a).

ters of seeds of *Agrostemma githago* (the weed corncockle) from four widely separated origins in Europe. Despite the geographical origins the characters are strikingly close and bear the greatest resemblance to the Mediterranean Caryophyllaceae shown in Fig. 6.12 (i.e., *Silene echinata*). It would seem, therefore, that *A. githago* has a Mediterranean origin and has retained this character in its spread through Europe, which accompanied the cultivation of wheat, i.e., over the last 3500–4500 years. Seeds of this species are difficult to separate from wheat grains, and so they have been sown and harvested along with the crop, generally in spring and late summer, respectively. Thus, although their germination is adapted to a Mediterranean climate, the seeds would not have been subjected to the same selective pressures that a feral population would experience. Interestingly, there is evidence of the decline of the species in Europe in the last 100 years, probably as seed-cleaning methods improved.

Finally, we should note that the germination character can give a clue as to the geographical origin of crop plants. When we examine the germination characters of lettuce and carrot seeds (Fig. 6.16), we see similarities to the Mediterranean and continental types, respectively (see Fig. 6.12), which suggest their origins. Throughout a long history of cultivation, the characters of these

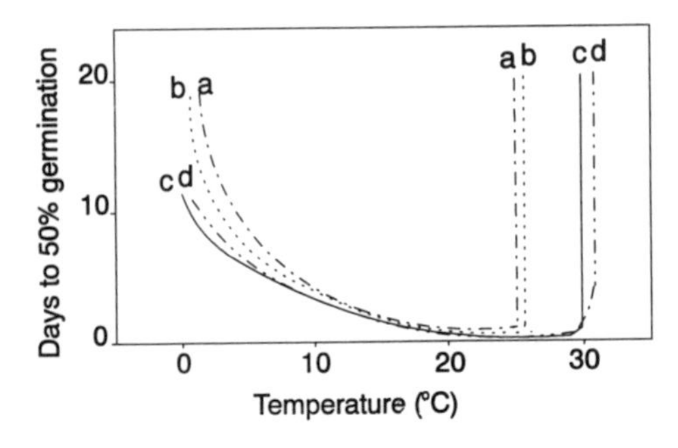

Figure 6.15. Germination "character" curves for *Agrostemma githago* seeds from different geographical localities: a, from Switzerland; b, from Poland; c, from Greece; d, from Germany. Adapted from Thompson (1973a).

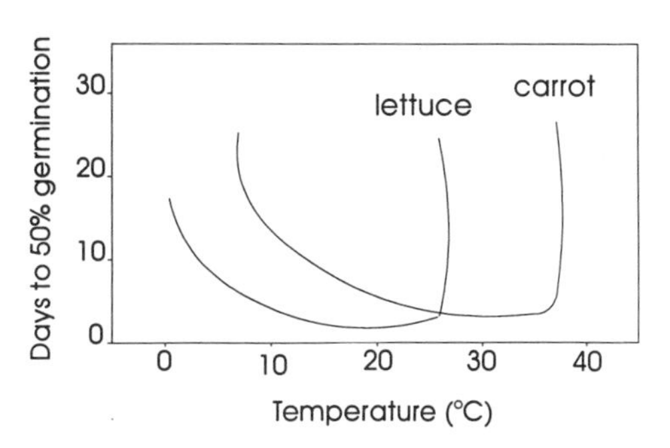

Figure 6.16. Germination "character" curves of lettuce and carrot seeds. Note the similarity of the curve for lettuce to the Mediterranean type (curve a), and that for carrot to the continental type (curve b) in Fig. 6.12. After Thompson (1973a).

species must have remained unchanged, though they do, of course, limit the climatic range over which cultivation can be practiced.

6.10. CHEMICALS IN THE NATURAL ENVIRONMENT

Seed germination can be affected by inorganic and organic chemicals that are present in the soil. The most important inorganic compound is undoubtedly nitrate, followed by ammonium ions. Both of these stimulate germination of many species, often interacting with light and temperature. Whether or not the nitrate in the soil is active depends on several factors, especially soil water content—the nitrate obviously must be dissolved and in a diffusible form. Germination is stimulated by concentrations up to about 50 mM—commonly reached in soils. Effective concentrations within seeds extend from about 0.1 to about 10 μmol/g seed (Fig. 6.17). Seeds of *Sisymbrium officinale* and *Chenopodium album* can accumulate nitrate to concentrations in this range; and indeed, without its internal nitrate the *Sisymbrium* seed cannot respond to light.

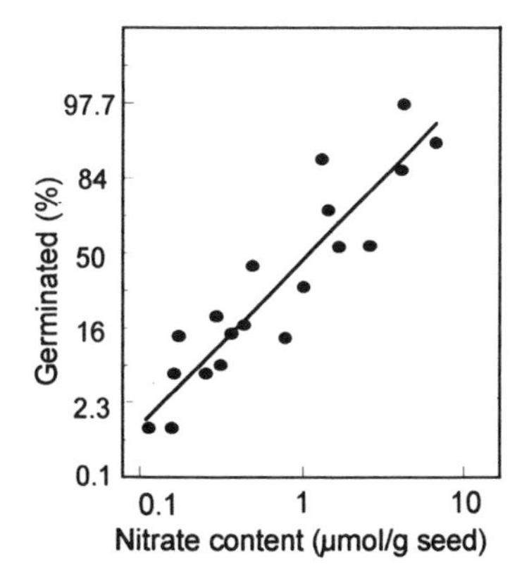

Figure 6.17. Effect of nitrate content on germination of *Sisymbrium officinale*. Seeds were taken from 20 lots, derived from plants grown in liquid culture or in the field. Seeds were treated with red light and germinated at 24°C. Nitrate content is plotted on a log scale and germination percentage on a probit scale. Note the poor germination at low nitrate content. After Karssen and Hilhorst (1992).

Another important factor is soil salinity. This is often inhibitory, and for seeds of certain desert species the salinity must be lowered by heavy rainfall before germination proceeds. This may be a rain gauge to measure whether sufficient water has fallen for successful seedling establishment. On the other hand, seeds of some species (e.g., *Atriplex halinus*) are stimulated to germinate by low concentrations of salts.

Various organic compounds produced by plants can affect seed germination. Such allelopathic chemicals, which are almost always inhibitory, are thought to be important factors in several ecological situations, for example in the succession of species that colonize abandoned agricultural land as it reverts to prairie. Pioneer weeds (e.g., *Chenopodium album*, *Sorghum halepense*) inhibit each other's germination by chemicals released as their vegetation decays. Seeds of a succeeding species, the grass *Aristida oligantha*, are not inhibited by the same chemicals.

The best known allelopathic promoters are the root exudates of the hosts (e.g., *Sorghum bicolor*) of parasitic angiosperms (e.g., *Striga asiatica*). There is good evidence that these induce the seed of the parasite to produce ethylene which then promotes germination. The dose–response relationships with regard to germination are such that the parasite's seeds are stimulated to germinate only in close proximity to the host.

An important constituent of the soil atmosphere is ethylene. Seeds of several weed species (e.g., *Sinapis arvensis*) are thought to be promoted by soil ethylene. In several cases (e.g., *Chenopodium album*, *Portulaca oleracea*) there are interactions of the gas with soil nitrate.

USEFUL LITERATURE REFERENCES

SECTION 6.2

Bewley, J.D., and Black, M., 1982, *Physiology and Biochemistry of Seeds*, Volume 2, Springer, Berlin, Chapter 6 (ecophysiological aspects of seeds).

Bliss, D., and Smith, H., 1985, *Plant Cell Environ.* **8:**475–483 (light and buried seeds).

Frankland, B., and Poo, W. K., 1980, in: *Photoreceptors and Plant Development* (J. DeGreef, ed.), Proc. Ann. Eur. Symp. Plant Photomorphogenesis, University Press, Antwerp, pp. 357–366 (light, canopy, and seed germination).

Grime, J. P., 1979, *Plant Strategies and Vegetation Processes*, Wiley, New York (seed banks, dormancy in the field, temperature effects).

Harper, J. L., 1977, *Population Biology of Plants*, Academic Press, New York, Chapters 2–5 (seed ecology).

Pons, T. L., 1992, in: *Seeds: The Ecology of Regeneration in Plant Communities* (M. Fenner, ed.), CAB International, Wallingford, pp. 259–284 (seed responses to light).

Taylorson, R. B., 1972, *Weed Sci.* **20:**417–422 (light sensitivity of buried seeds).

Wesson, G., and Wareing, P. F., 1969, *J. Exp. Bot.* **20:**414–425 (buried seeds and light).

Woolley, J. T., and Stoller, E.W., 1978, *Plant Physiol.* **61:** 597–600 (light transmittance of soil).

SECTION 6.3

Bliss, D., and Smith, H., 1985, *Plant Cell Environ.* **8:**475–483 (light-inhibited seeds).
Corbineau, F., and Côme, D., 1982, *Plant Physiol.* **70:**1518–1520 (light inhibition in *Oldenlandia corymbosa*).
Ellis, R. H., Hong, T. D., and Roberts, E. H., 1989, *J. Exp. Bot.* **40:**13–22 (effects of high photon flux densities on germination).

SECTION 6.4

Frankland, B., 1981, in: *Plants and the Daylight Spectrum* (H. Smith, ed.), Academic Press, New York, pp. 187–204 (shade and germination).
Frankland, B., and Poo, W. K., 1980, in: *Photoreceptors and Plant Development* (J. DeGreef, ed.), Proc. Ann. Eur. Symp. Plant Photomorphogenesis, University Press, Antwerp, pp. 357–366 (shade, light, and seed burial).
Górski, T., Górska, K., and Rybicki, J., 1978, *Flora* **167:**289–299 (leaf canopy light effects on different species).
Holmes, M. G., and McCartney, H. A., 1975; in: *Light and Plant Development* (H. Smith, ed.), Butterworths, London, pp. 446–476 (shade light quality).
Vazquez-Yanes, C., and Orozco-Segovia, A., 1990, *Oecologia* **83:**171–175 (leaf canopy effect on tropical species).

SECTIONS 6.5 AND 6.7

Baskin, J. M., and Baskin, C. C., 1977, *Oecologia* **30:**377–382 (temperature in germination ecology).
Cruden, R. W., 1974, *Ecology* **55:**1295–1305 (ecophysiology of *Nemophila* germination).
Grime, J. P., 1979, *Plant Strategies and Vegetation Processes,* Wiley, New York (dormancy in the field and environmental effects).
Karssen, C. M., 1980/81, *Isr. J. Bot.* **29:**65–73 (dormancy and seed burial).
Probert, R. J., 1992, in: *Seeds: The Ecology of Regeneration in Plant Communities* (M. Fenner, ed.), CAB International, Wallingford, pp. 285–325 (temperature and germination ecophysiology).

SECTION 6.8

Baskin, J. M., and Baskin, C. C., 1980, *Ecology* **61:**475–480 (temperature and secondary dormancy).
Bouwmeester, H. J., and Karssen, C. M., 1993, *New Phytol.* **124:**179–191 (detailed analysis of seasonal dormancy in *Sisymbrium officinale*).
Derkx, M. P. M., and Karssen, C. M., 1993, *Plant Cell Environ.* **16:**469–479 (rhythms of sensitivity to light and nitrate).
Karssen, C. M., 1980/81, *Isr. J. Bot.* **29:**45–64 (secondary dormancy and the environment).
Karssen, C. M., Derkx, M. P. M., and Post, B. J., 1988, *Weed Res.* **28:**449–457 ("seasonal" variation in dormancy in a condensed temperature cycle).
Roberts, A., and Lockett, P. M., 1978, *Weed Res.* **18:**41–48 (seed dormancy and burial).

SECTION 6.9

Karssen, C. M., and Hilhorst, H. W. M., 1992, in: *Seeds: The Ecology of Regeneration in Plant Communities* (M. Fenner, ed.), CAB International, Wallingford, pp. 327–348 (chemical environment and seed germination).

Stearns, F., and Olson, J., 1958, *Am. J. Bot.* **45:**53–58 (seed provenance and germination).

Thompson, P. A., 1973a, in: *Seed Ecology* (W. Heydecker, ed.), Butterworths, London, pp. 31–58 (geographical adaptation of seeds).

Thompson, P. A., 1973b, *Ann. Bot.* **37:**133–154 (temperature and germination character).

Chapter 7

Mobilization of Stored Seed Reserves

The major mobilization of stored reserves in the storage organs commences after radicle elongation, i.e., it is a postgerminative event. In the growing regions (i.e., axis) some mobilization can occur before germination is completed; here the reserves are generally present in minor amounts, although the products of their hydrolysis might be important for early seedling establishment.

As the high-molecular-weight reserves contained within the storage organs of the seed are mobilized, they are converted into forms that are readily transportable to the sites where they are required (usually the most rapidly metabolizing and growing organs) for the support of energy-producing and synthetic events. Reliance on the stored reserves diminishes as the seedling emerges above the soil and becomes photosynthetically active (i.e., autotrophic). For the purpose of clarity we have divided this chapter into sections, each of which covers the mobilization of one major type of storage reserve. It must be remembered, however, that storage organs usually contain substantial quantities of two or more major reserves (Table 1.2) and that hydrolysis and utilization of these can occur concurrently.

7.1. STORED CARBOHYDRATE CATABOLISM

Starch is the most common reserve carbohydrate in seeds, and hence most attention will be paid to this. Other stored carbohydrates of note are the hemicelluloses (such as galactomannans—see Chapter 1), which are to be found particularly as cell wall components of the endosperm of various legumes.

7.1.1. Pathways of Starch Catabolism

There are two catabolic pathways of starch: one hydrolytic and the other phosphorolytic.

The amylose and amylopectin in the native starch grain are first hydrolyzed by α-amylase, which breaks the $\alpha(1 \rightarrow 4)$ glycosidic links between the glucose residues randomly throughout the chains. The released oligosaccharides are further hydrolyzed by α-amylase (or with the cooperation of α-glucosidase; see below) until glucose and maltose are produced:

$$\text{Amylose} \xrightarrow{\alpha\text{-Amylase}} \text{Glucose + maltose}$$

Multiple forms of this enzyme occur in germinated seeds of many species. Germinated wheat, for example, contains over 20 α-amylase isoenzymes which fall into two groups separated by isoelectric focusing on the basis of their electrical charge at particular pHs. But α-amylase cannot hydrolyze the α(1→6) branch points of amylopectin, and hence highly branched cores of glucose units, called limit dextrins, are produced:

$$\text{Amylopectin} \xrightarrow{\alpha\text{-Amylase}} \text{Glucose + maltose + limit dextrin}$$

The small branches must be released by enzymes specific for the α(1→6) link (debranching enzyme) before being hydrolyzed to the monomer. Two enzymes have been described, which have been called limit dextrinase and the R-enzyme (or sometimes isoamylase and the R-enzyme). The former was considered to be specific for the small limit-dextrins, and unable to release branches from amylopectin, which R-enzyme could do, but this strict distinction may not be justified.

$$\text{Limit dextrin} \xrightarrow[\text{(Limit dextrinase)}]{\substack{\text{Isoamylase} \\ \text{Debranching (R-)enzyme}}} \text{Glucose oligomers} \xrightarrow{\alpha\text{-Amylase}} \text{Maltose + glucose}$$

$$\text{Maltose} \xrightarrow{\alpha\text{-Glucosidase}} \text{Glucose}$$

Another amylase, β-amylase, cannot hydrolyze native starch grains; rather it cleaves away successive maltose units from the nonreducing end of large oligomers released by prior α-amylolytic attack. Again, amylopectin cannot be completely hydrolyzed, and the involvement of a debranching enzyme is essential:

$$\text{Amylose} \xrightarrow{\beta\text{-Amylase}} \text{Maltose}$$

$$\text{Amylopectin} \xrightarrow{\beta\text{-Amylase}} \text{Maltose + limit dextrin}$$

The disaccharide maltose, produced by α- and β-amylase action, is converted by α-glucosidase to two glucose molecules. This enzyme can also cleave glucose from low-molecular-weight maltosaccharides:

$$\text{Maltose} \xrightarrow[\text{(Maltase)}]{\alpha\text{-Glucosidase}} \text{Glucose}$$

Starch phosphorylase releases glucose-l-phosphate by incorporating a phosphate moiety, rather than water, across the $\alpha(1\rightarrow4)$ linkage between the penultimate and last glucose at the nonreducing end of the polysaccharide chain. Complete phosphorolysis of amylose is theoretically possible, and amylopectin can be degraded to within two or three glucose residues of an $\alpha(1\rightarrow6)$ branch linkage. The enzyme cannot attack starch granules, which first must be partly degraded by other enzymes.

$$\text{Amylose/amylopectin} + P_i \xrightarrow{\text{Starch phosphorylase}} \text{Glucose-1-}P + \text{limit dextrin}$$

The relative activities of the two amylases and phosphorylase vary between different seed species: β-amylase tends to be more active in germinated rice than in some other cereals, and although phosphorylase activity is low or negligible in cereals, it is appreciable in legumes. Examples are given in Table 7.1.

7.1.2. Synthesis of Sucrose

The products of starch (and triacylglycerol) catabolism eventually are transported as sucrose into the root and shoot of the developing seedlings. Glucose-1-phosphate (Glc-l-*P*) released by phosphorolysis can be used directly as a substrate for sucrose synthesis, but glucose released by amylolysis first must be phosphorylated to glucose-6-phosphate (Glc-6-*P*) and then isomerized to Glc-l-*P*. This combines with a uridine nucleotide (UTP) to yield the nucleotide sugar uridine diphosphoglucose (UDPGlc), which in turn transfers glucose to free fructose or to fructose-6-phosphate:

$$\text{Glc-1-}P + \text{UTP} \xrightarrow{\text{UDPGlc pyrophosphorylase}} \text{UDPGlc} + PP_i \text{ (pyrophosphate)}$$

$$\text{UDPGlc} + \text{fructose} \xrightleftharpoons{\text{Sucrose synthetase}} \text{Sucrose} + \text{UDP}$$

$$\text{UDPGlc} + \text{fructose-6-}P \xrightleftharpoons{\text{Sucrose-6-}P\text{ synthetase}} \text{Sucrose-6-}P\text{+ UDP}$$

It is generally accepted that this latter reaction is the predominant, if not the only, one involved in sucrose synthesis, whereas the sucrose synthetase is

Table 7.1. Activities of Starch-Hydrolyzing Enzymes during the Period of Rapid Starch Breakdown in Two Cereals and a Legume[a]

Enzyme	Barley	Rice	Pea
α-Amylase	34.4	31.8	19
β-Amylase	11.4	120	Very low
Starch phosphorylase	Negligible	0.09	14.6

[a]The values are in milligrams starch hydrolyzed per hour per seed. Based on Ap Rees (1974).

important for sucrose breakdown. The phosphate moiety is cleaved from sucrose-6-*P* by sucrose phosphatase. In the seedling tissues, sucrose can be hydrolyzed to free glucose and fructose by a β-fructofuranosidase (e.g., sucrase, invertase), or converted to UDPGlc and fructose by sucrose synthetase.

7.2. MOBILIZATION OF STORED CARBOHYDRATE RESERVES IN CEREALS

7.2.1. The Embryo Reserves

Although the endosperm is the major source of carbohydrate reserves in cereals, some low-molecular-weight sugars are stored in much lower quantities within the embryo. These probably provide an early source of respirable substrate during germination and early seedling growth, until the hydrolytic products of starch are made available from the endosperm. Sucrose and raffinose (galactosyl sucrose) are present in the ratio of about 3:1 in dry grains of Proctor barley, and both decline to imperceptible levels within the first 1–2 days from the start of imbibition. A replenishment of carbohydrate from the endosperm, as sucrose, begins on the third day.

7.2.2. The Endosperm Reserves

Taking barley as an example, the first change during starch hydrolysis that the endosperm appears to undergo is the digestion of the intermediate (or depleted) layer adjacent to, and in between, the cells of the absorptive epithelium of the scutellum (Fig. 7.1C). The enzymes thought to be responsible for digestion of the cell walls of the intermediate layer are endo-β-glucanases, a class of enzymes capable of degrading hemicelluloses [glucans containing β(1→3) and β(1→4) links]. In some cereal grains (e.g., those of the oat family, but not barley or wheat), the scutellum and its epithelial cells then elongate (Fig. 7.2), thus

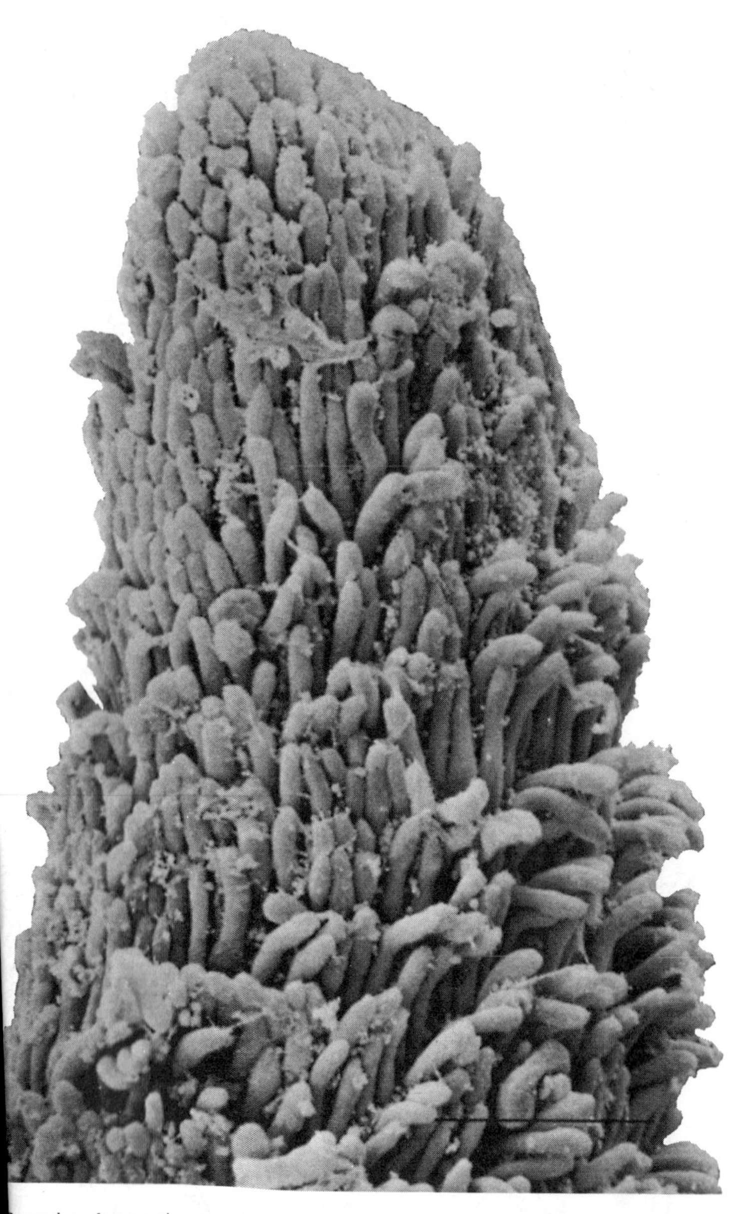

Scanning electron micrograph of the epithelial cells of the scutellum of germinated *Avena* the extent of their elongation. Bar is 0.1 mm. The scutellum itself grows through the perm and by 8 days from imbibition may extend from the proximal (embryo) end to the urtesy of J. Sargent and M. Negbi.

presenting a much increased surface area for absorption of the hydrolytic products of the endosperm reserves.

The starch granules in the endosperm are degraded to glucose by the action of four enzymes: α-amylase, β-amylase, the debranching enzyme(s), and α-glucosidase. Accompanying the mobilization of starch, carbohydrates of the endosperm cell walls are also hydrolyzed, which promotes easier access by the four starch-degrading enzymes.

The initial production of α-amylase occurs in the region of the scutellum (Fig. 7.1C) in some species of grain, perhaps in the epithelial layer of this organ (e.g., rice), or the entire scutellum (e.g., sorghum), and in others in the few aleurone cells that penetrate the peripheral regions of the scutellum (e.g., barley). In barley, the α-amylase is then released into the starchy endosperm and diffuses away from the scutellum (Fig. 7.3A–C). Later the enzyme is synthesized within the aleurone layer (Figs. 7.1, 7.3D) and is secreted inward to complete the hydrolysis of starch reserves. Although α-amylase from the scutellum may be important at early stages of mobilization, in barley, wheat, rye, maize, and oat, it is likely that later most hydrolysis is effected by isoenzymes from the aleurone layer. In rice, however, the major source of α-amylase following germination is the scutellum. Utilization of endosperm reserves in some cereals (or cultivars) appears to be controlled by the embryo and mediated by the production of the plant hormone gibberellin (GA) (see Chapter 8 for details). GA is synthesized by the embryo during and after germination and is released through the scutellum into the starchy endosperm. It is thought that the hormone diffuses to the aleurone layer and there sets off a sequence of events that culminate in the synthesis and release of α-amylase. In wheat, one group of α-amylase isoenzymes released from the scutellum is also under hormonal control, but it is not clear if there is a similar situation in other cereals.

β-Amylase is involved also in the digestion of cereal starch, but in most cases this enzyme is not *de novo* synthesized, nor is it released from either the scutellum or the aleurone layer. Instead it is present in the starchy endosperm of the mature, ungerminated grains, such as barley (Fig. 7.3E) where, presumably, it becomes active (or activated) after initial digestion of the starch grains by α-amylase. The enzyme exists in free and bound forms, and in wheat kernels a large proportion is present in the inactive, latent form bound by disulfide linkages to protein (glutenin) bodies, or their remnants, in the starchy endosperm. Extracts of the enzyme can be activated through release from its bound form, either by digestion of the glutenin with proteolytic enzymes or by disrupting the binding linkages with sulfhydryl-containing compounds. In the grain, activation of β-amylase occurs as a result of proteolytic attack by proteinases produced by the scutellum and aleurone layer. There is evidence in some cases that proteinase synthesis is regulated by GA. In rice grains there is initially some *de novo* synthesis of β-amylase in the scutellum and release therefrom. Later there is activation of a latent form of this enzyme, which is bound to starch grains, within the endosperm. Some doubt about the significance of β-amylase in starch degradation arises, however, from studies of mutants of rye and barley that lack the enzyme. Notwithstanding this deficiency, starch mobilization is unhindered and postgerminative seedling establishment is normal. It has been suggested that the relatively high content of β-amylase protein in some cereals is indicative of a storage function rather than an essential enzymatic role.

Complete amylolysis of the amylose component of starch can be achieved by α- and β-amylases, but digestion of amylopectin ceases with the production of limit dextrins. Little is known about the site and mode of production of the debranching enzymes that are needed for further degradation, but in barley it appears that at least the limit dextrinase is synthesized *de novo* in the aleurone layer at the time of starch breakdown, possibly induced by GA. A portion of the enzyme is converted to a bound form and later is slowly released. In rice grains, on the other hand, debranching enzyme is synthesized in an active form during the ripening stage of development, is present in an inactive and insoluble form in the dry seed, and is subsequently solubilized and activated after germination.

The major product of α- and β-amylolysis is maltose which is hydrolyzed by α-glucosidase (maltase), an enzyme present in low amounts in the embryo and aleurone layer of ungerminated barley. The enzyme increases in activity at both sites after germination although only that released from the aleurone layer is involved in the mobilization of endosperm reserves. That in the axis participates in local sugar metabolism.

A notable variation from this general pattern of events is found in sorghum grains. Here, α- and β-amylase and α-glucosidase are present in the dry grain, sequestered in organelles (lytic bodies) along with other hydrolytic enzymes, e.g., proteinase, phosphatase, and nuclease. Presumably, these enzymes are released at the appropriate time after germination to effect hydrolysis of the stored reserves.

Enzymatic solubilization of the endosperm cell walls accompanies digestion of the starch reserves in many cereals (sorghum is one exception). The cell walls of most cereals contain mainly arabinoxylan and a mixed-linkage (1→3, 1→4)-β-glucan: there is relatively little cellulose or pectin, although the latter is present in rice, together with glycoprotein. Because the walls are relatively simple in composition, their mobilization may be effected more easily. Cell wall dissolution may precede starch hydrolysis in any given cell, and thus the

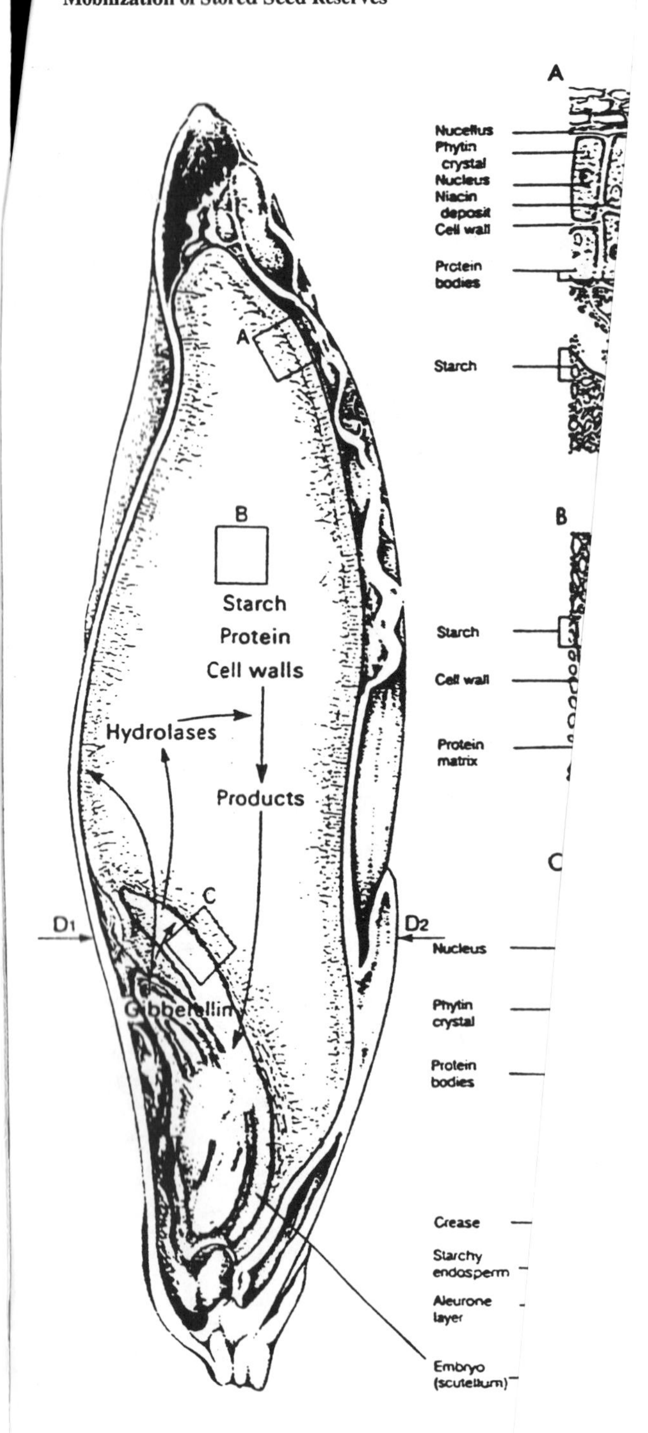

Figure 7.1. Anatomy of a barley grain. A longitudinal section of the grain is on the left. Parts of it (blocks A, B, C) are magnified on the right. Bottom right is a transverse section across $D_1 \rightarrow D_2$. From Jones and Jacobsen (1991); originally drawn by S. I. Wong, and supplied by G. Fulcher, Plant Research, Agriculture Canada.

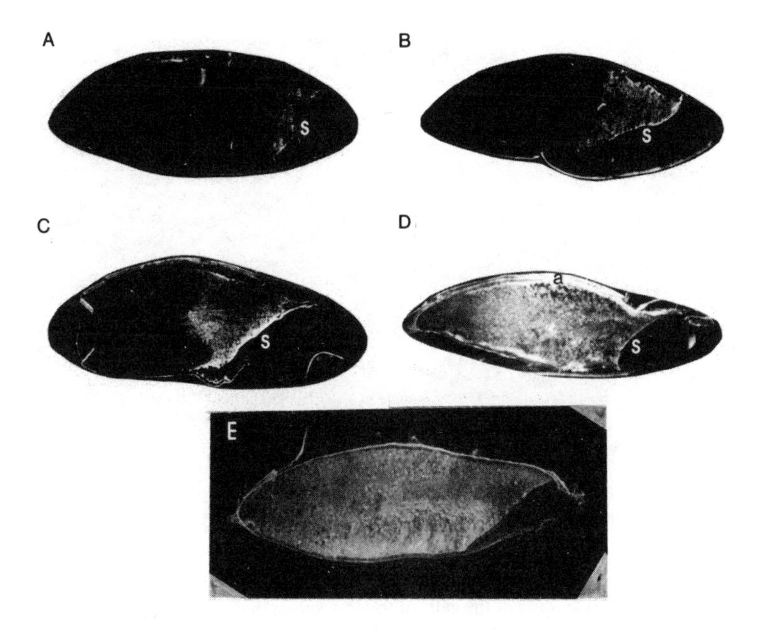

Figure 7.3. (A–D) Immunofluorescent localization of α-amylase in barley grains at various times after the start of imbibition. (A) After 30 h (fluorescence near scutellum). (B) After 54 h (fluorescence spreading from scutellum). (C) After 78 h (fluorescence spreading from scutellum; little near aleurone layer). (D) After 7 days (fluorescence throughout endosperm; more near aleurone layer). Rabbit antibodies specific for α-amylase were made (monospecific antibodies). Grains were germinated and at various times thereafter fixed and sectioned. The antibodies were then added, which tightly bound to the α-amylase in the sections. After this, a fluorescent probe attached to a rabbit antiserum protein was introduced to the sections (this bound specifically to the rabbit antibody–α-amylase complex) and was detected and photographed using a fluorescence microscope. (E) Immunofluorescent localization of β-amylase in mature dry barley grains using monospecific β-amylase antibodies. This enzyme is found throughout the starchy endosperm of the dry grain. The fluorescence in the outer coat region on all sections is an artifact of the technique. S, scutellum; a, aleurone layer. Photographs kindly provided by G. C. Gibbons. Details in Gibbons (1979).

synthesis and release of several arabinoxylanases (pentosanases) and glucanases [especially (1→3, 1→4)-β-glucanase] may, in turn, precede those of α-amylase. In barley there are two isoenzymes of the glucanase, one released from the scutellum and one from the aleurone layer. In wheat, release of pentosanases from this latter tissue results first in the hydrolysis of endosperm walls in the subaleurone layer, and then a wave of degradation progresses toward the center of the endosperm (Fig. 7.4), the loss of cell walls accompanying an increase in the formation of water-soluble arabinoxylans and small glucans.

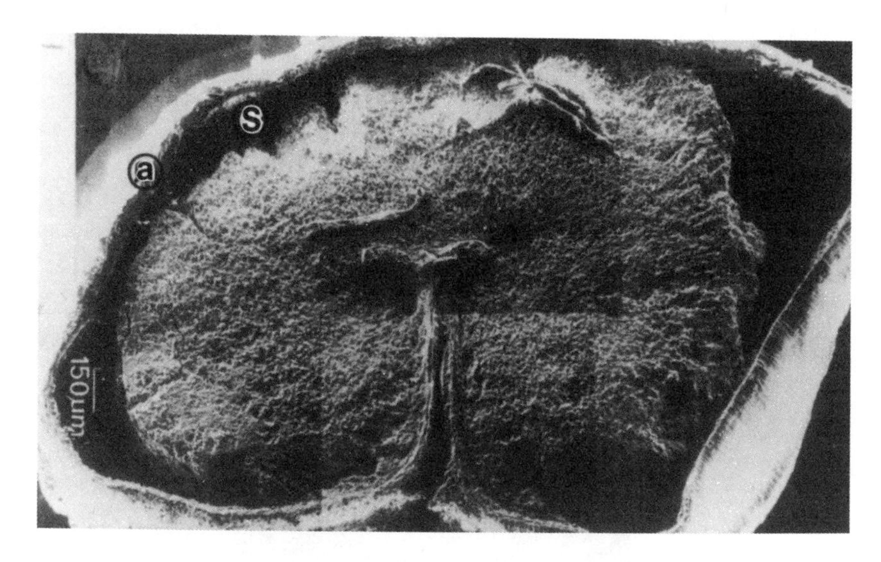

Figure 7.4. Scanning electron micrograph of a median cross section of a wheat grain after 2 days from imbibition to show dissolution of the cells of the endosperm in the subaleurone region (s). The aleurone layer (a) remains adhered to the grain coat. From Fincher and Stone (1974).

7.2.3. The Fate of the Products of Starch Hydrolysis

The growing embryonic axis and the endosperm are separated by the shield-shaped scutellum (Fig. 7.1)—the modified single cotyledon of the Gramineae. The products of starch digestion (mostly glucose and maltose) are absorbed by the scutellum, which in some cases increases in surface area (Fig. 7.2), and are converted there to sucrose. This is transported to the axis (perhaps through the phloem of the vascular system connecting this with the scutellum) and is utilized by the growing root and shoot tissues. Glucose is absorbed by the scutellum both passively and by active transport; maltose is probably hydrolyzed very quickly by α-glucosidase contained within the free space of the scutellar cells. Some of the absorbed glucose is converted to fructose-6-phosphate, but neither of these sugars accumulates in the scutellum, for they are quickly converted to sucrose, via sucrose-6-phosphate.

For a generalized summary of events associated with endosperm reserve hydrolysis, based on the observations for certain cultivars of barley grains, see Table 7.2.

Table 7.2. Probable Sequence of Events Involved in the Mobilization of Stored Starch Reserves in the Cereal Endosperm[a]

1. Completion of germination
2. Release (probably after synthesis) of gibberellins from embryo via scutellum
3. Dissolution of intermediate layer and expansion of epithelial layer cells
4. Release of α-amylase and cell-wall-hydrolyzing enzymes (after *de novo* synthesis) from scutellum into endosperm. Start of endosperm cell wall dissolution and starch hydrolysis
5. Release of α-amylase, α-glucosidase, debranching enzymes, and cell-wall-hydrolyzing enzymes (after *de novo* synthesis) from aleurone layer into endosperm. Grand phase of endosperm hydrolysis
6. Activation of β-amylase (concurrent with 4 and 5)
7. Absorption of glucose by scutellum. Synthesis of sucrose therein for transport to growing axis (concurrent with 4–6)

[a]Note that this scheme is a general guide to events. Variations will occur between grains of different cereal species and cultivars, and timing of the events will vary depending on the conditions (e.g., temperature, moisture content) of germination and growth.

7.3. MOBILIZATION OF STORED CARBOHYDRATE RESERVES IN LEGUMES

Legume seeds have been divided into two groups on the basis of the location of their major stored reserves. The nonendospermic legumes are those in which the endosperm is absorbed during seed development, the cotyledons then becoming the major storage organs (e.g., pea, dwarf and broad bean). Endospermic legumes, which are members of the tribe Trifolieae, retain their endosperm until maturity. In some seeds (e.g., fenugreek, carob, honey locust, and guar) it is the major storage organ, although in others (e.g., soybean) it is only residual and the cotyledons store most of the reserves. Fewer studies have been conducted on the mobilization of carbohydrates in legumes than on this process in cereals; nevertheless, a general pattern of events has emerged.

7.3.1. Nonendospermic Legumes

Early seedling growth in peas draws on reserves of carbohydrate, protein, and triacylglycerol within the radicle itself, and sucrose, raffinose, and stachyose serve as the primary sources of respirable substrate. Both α- and β-amylase are present in the axes of ungerminated pea seeds, which also contain starch: β-amylase activity increases appreciably after germination, in association with the growth of the epicotyl. After these early events have passed, the further development of the root and shoot depends on a supply of hydrolytic products from the cotyledons. Their stored carbohydrate (and other reserves) are hydrolyzed and transported to the growing axis.

Hydrolysis of reserves in intact legume cotyledons commences after emergence and elongation of the radicle. The depletion of starchy reserves from the cotyledons of peas is biphasic (Fig. 7.5A), an initial slow rate that lasts for 5–6 days being followed by a phase of rapid decline. The free sugars and dextrins released by starch hydrolysis are rapidly translocated to the growing axis; they do not accumulate to any extent within the cotyledons (Fig. 7.5A). The initial slow degradation of starch might be by phosphorolysis since starch phosphorylase is the first enzyme to increase in activity in the cotyledons (Fig. 7.5B). The more rapid degradation at 8 or 9 days probably requires amylolysis and is coincident with an increase in α- and β-amylase activity. It is not known what proportion of the increase in phosphorolytic and amylolytic activity is related to *de novo* synthesis of enzymes, and what proportion is related to activation of latent forms. Debranching enzyme (limit dextrinase) activity is high in dry pea seeds and remains so in the rehydrated seed for several days before declining. This decline occurs prior to the major increase in amylolytic activity, and since

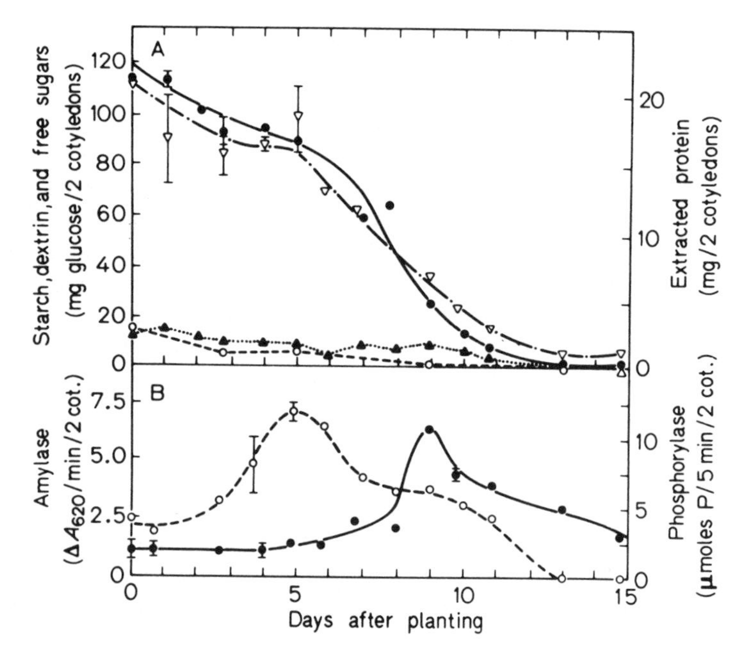

Figure 7.5. (A) Changes in the amount of starch and dextrin (●), dextrin (○), free sugars (▲), and extracted protein (▽) in the cotyledons of Alaska pea. (B) Changes in starch phosphorylase (○) and amylase (●) enzymes in Early Alaska pea. After Juliano and Varner (1969).

the limit dextrinase can only debranch products of amylolysis, its role in the overall scheme of starch breakdown is enigmatic. Suffice it to say, at the present time, that debranching must accompany amylolysis, but the appropriate enzyme remains to be identified. In another legume, green gram (*Phaseolus vidissimus*), α- and β-amylases are found in the cotyledons (at a ratio of 20:1) at the time of starch hydrolysis, and α-glucosidase is present too.

The immediate fate of the products of phosphorolysis and amylolysis is virtually unknown, although eventually they must be made available to the growing axis. Starch phosphorylase activity in the cotyledons of peas produces Glc-l-*P*; this reacts with UTP to form UDPGlc, which is then converted to sucrose by sucrose phosphate synthetase and sucrose phosphatase (see Section 7.1.2). Enzymes for the hydrolysis and transformation of sucrose are absent from the cotyledons but are present in the axis, to which this sugar is presumably transported. Since little α-glucosidase activity has been detected in the cotyledons, maltose might be a major transported form of sugar released by amylolysis, although some glucose will be produced directly by this process and also translocated to the axis. Only low amounts of α-glucosidase have been detected in the axis, but it is presumed that this is where maltose is hydrolyzed.

7.3.2. Endospermic Legumes

In many species of the Trifolieae a well-developed endosperm, comprised largely of the storage carbohydrate galactomannan, lies between the seed coat and the cotyledons. In fenugreek (*Trigonella foenum-graecum*) extensive deposition of this polymer on the inside of the primary walls during seed development results in the gradual occlusion of the living contents until in the mature seeds the cells are dead (Section 2.3.2). The outermost layer of the endosperm is the aleurone layer which, unlike the rest of the endosperm, is made up of living cells devoid of galactomannan reserves (Fig. 7.6A). In some endospermic legumes, e.g., carob, the endosperm cell walls do not completely occlude the cytoplasm, and all cells have living contents.

Under suitable conditions of light and moisture, isolated embryos of fenugreek germinate and grow as well as the embryos of intact seeds. Support for this growth comes from the hydrolytic products of proteins and lipids stored in the cotyledons. Hence, the endosperm may not be an essential food source for the embryo, although it is difficult to extrapolate from experiments using isolated embryos, germinated and grown under ideal laboratory conditions, to the situation in the natural environment. It is clear, however, that the endosperm has another function besides that of a storage organ. The high affinity of galactomannans for water (when imbibed, many become mucilaginous) allows the endosperm to regulate the water balance of the embryo during germination; this

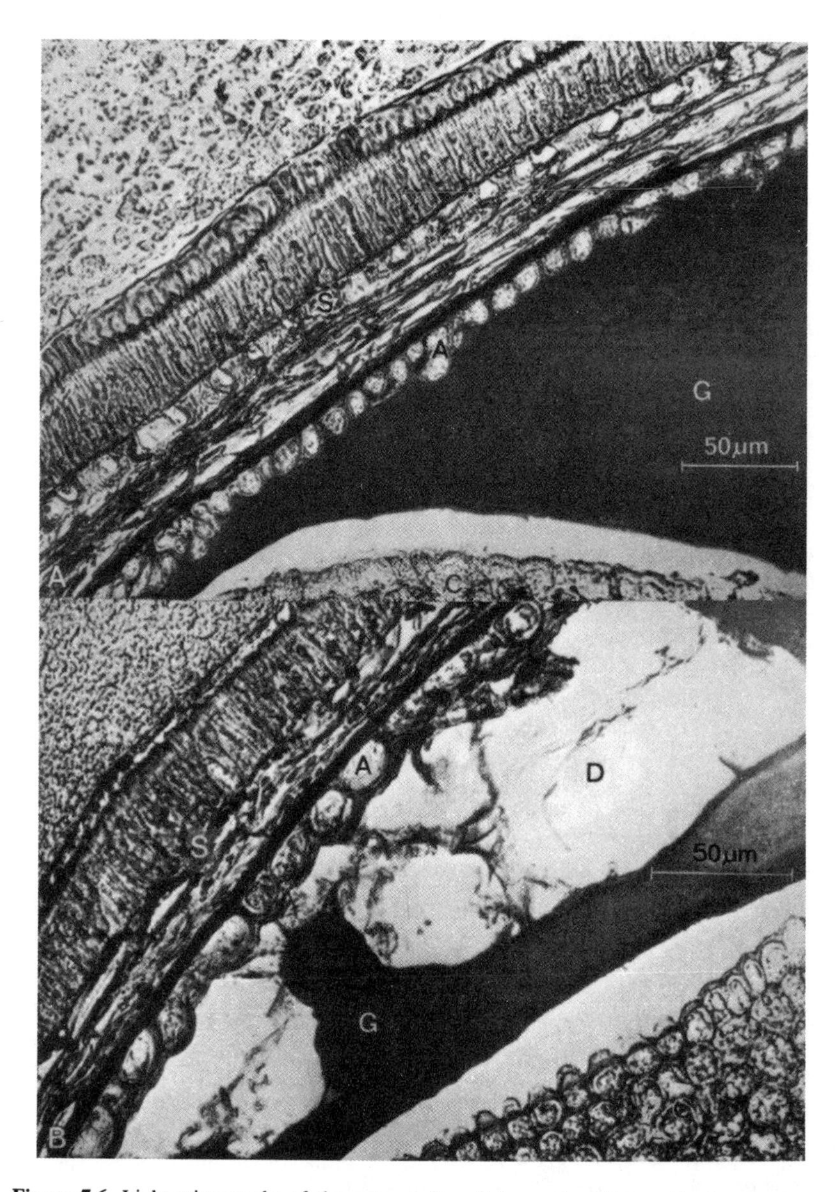

Figure 7.6. Light micrographs of the outer region of the seed of *Trigonella foenum-graecum* (fenugreek). (A) Before mobilization of the seed reserves. The three-layered seed coat (S), a small part of the cotyledon (C), and the endosperm layer (A and G) are shown. The aleurone layer (A) is the outer cell layer of the endosperm, the rest (G) being comprised of large cells with thin primary walls to the inside of which is deposited dark-staining galactomannan that appears to completely fill the cell. (B) During galactomannan breakdown the reserves in the endosperm (G) are being dissolved. The dissolution zone (D) begins at the aleurone layer (A) and spreads toward the cotyledons (C). (C) The endosperm (E) is depleted and only a remnant remains between the seed coat (S) and the cotyledon (C). The aleurone layer has disintegrated. From Reid (1971).

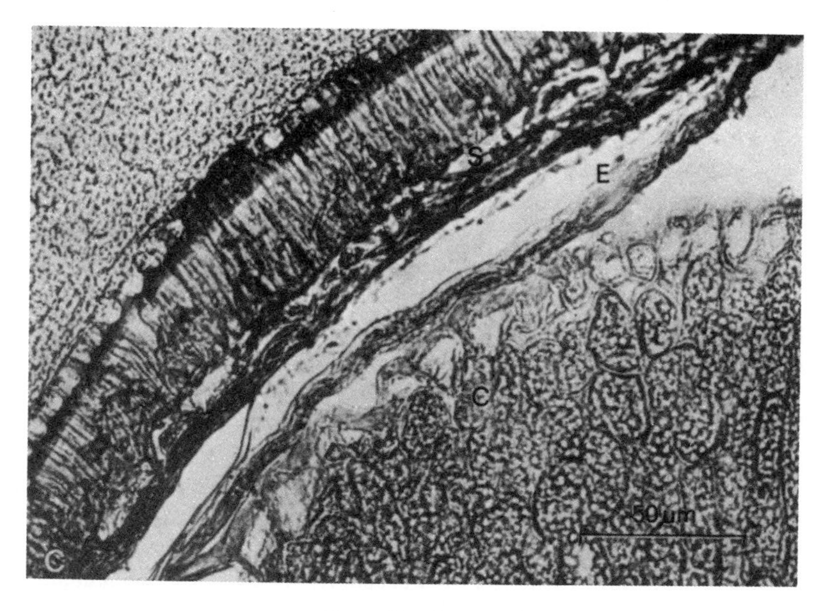

Figure 7.6. (*Continued*)

may be important to plants in their native habitat, since many members of the tribe Trifolieae appear to have their origins in the dry regions of the eastern Mediterranean. Perhaps it is safe to assume that, under natural conditions, the endosperm has a dual role: to regulate water balance during germination, and to serve as a substrate reserve for the developing seedling after germination.

As in the cereal grains and seeds of nonendospermic legumes, the raffinose-series oligosaccharides (especially raffinose and stachyose) in the embryonic tissues of the fenugreek seed are the first carbohydrates to be utilized, by hydrolytic cleavage of the α-galactosidic link to yield sucrose and galactose. This commences soon after imbibition. After emergence of the radicle, the galactomannan in the endosperm begins to be mobilized. There is a wave of hydrolysis commencing close to the aleurone layer and moving toward the cotyledons (Fig. 7.6B) until the reserves are depleted (Fig. 7.6C). This reflects the synthesis and release from this layer of at least three critical enzymes: α-galactosidase, β-mannosidase (exo-β-mannanase), and endo-β-mannanase. α-Galactosidase is an exopolysaccharidase that cleaves the α-(1→6) link between the unit galactose side chains and the mannose backbone.

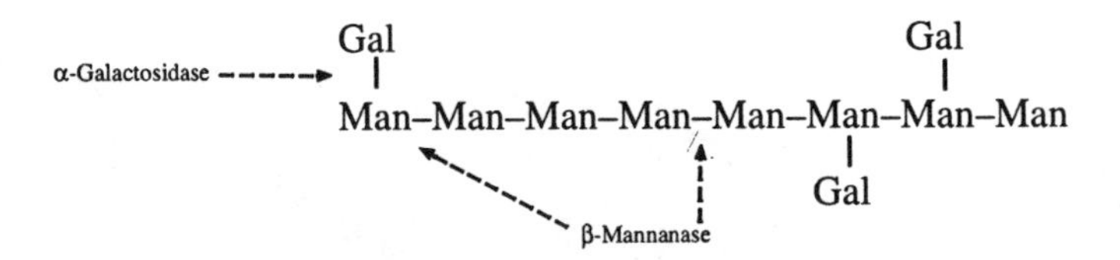

Endo-β-mannanase is an endoenzyme that hydrolyzes oligomers of mannose (tetramers or larger) to mannobiose or mannotriose, and β-mannosidase then converts these residues to mannose. This latter enzyme might also act as an exomannopolysaccharidase and hydrolyze single mannose residues from the oligomannan chain. Mannan breakdown by phosphorolysis appears not to occur. A similar sequence of events occurs in other endospermic legume seeds.

The released galactose and mannose are absorbed by the cotyledons, the former by passive diffusion, but the latter requires active uptake utilizing a carrier-specific component. Neither sugar accumulates in the cotyledons, but instead they are metabolized further, perhaps by initially being phosphorylated (to Gal-1-*P* and Man-6-*P*). If not used directly for energy metabolism, they are transformed to sucrose and then to starch, which is mobilized when the sucrose content of the cotyledons falls after its transport to the axis. A small amount of starch builds up in the cotyledons in the absence of the endosperm, but the massive accumulation of starch which occurs only in its presence (Fig. 7.7A) shows that the products of galactomannan breakdown are made available for the major starch synthesis. This sequestering of sugars as a large polymer is a convenient strategy for the removal of potentially osmotically damaging monomers, and for the retention of useful metabolites. Not surprisingly, an increase in α-amylase activity within the cotyledons coincides with starch hydrolysis (Fig. 7.7B). A summary of the events involved in galactomannan breakdown in endospermic legumes is shown in Fig. 7.8. It should be noted, however, that not

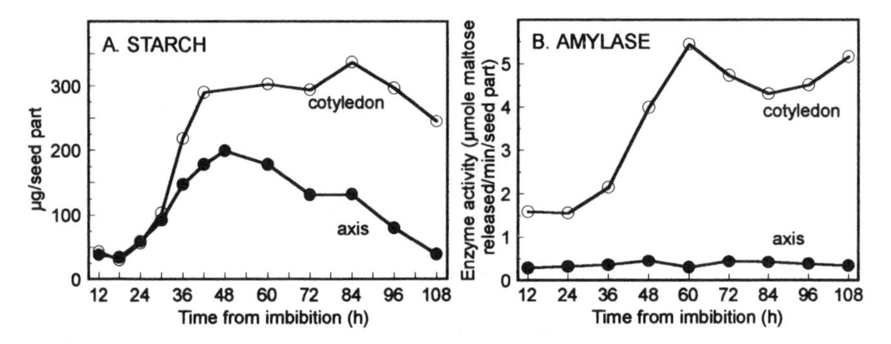

Figure 7.7. (A) Transitory starch accumulation and (B) soluble amylase activity in the cotyledons and axes of intact germinated fenugreek seeds. By L. M. A. Dirk. See also Bewley *et al.* (1993).

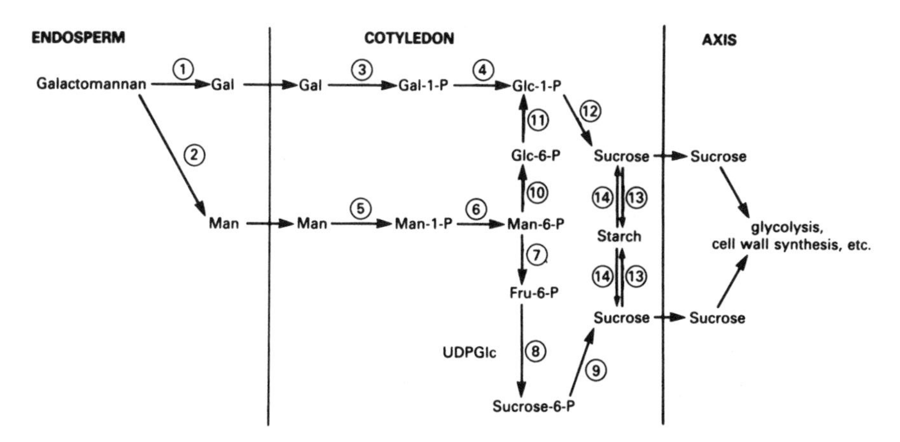

Figure 7.8. Flow diagram to illustrate the potential fate of products of galactomannan mobilization in endospermic legumes. Enzymes: (1) α-galactosidase; (2) endo-β-mannanase and β-mannosidase (exo-β-mannanase); (3) galactokinase; (4) hexose phosphate uridyl transferase (a group of three enzymes that convert Gal-1-*P* + UTP → UDPGal → UDPGlc → Glc-1-*P* + UTP); (5) mannokinase; (6) phosphomannomutase; (7) phosphomannoisomerase; (8) sucrose-6-*P* synthetase; (9) sucrose phosphatase; (10) C_2 epimerase; (11) phosphoglucomutase; (12) sucrose synthetase or sucrose-6-*P* synthetase (see Section 7.1.2); (13) see Section 2.3.1; (14) see Sections 7.1.1 and 7.1.2. Gal, galactose; Man, mannose; Glc; glucose; Fru, fructose.

all of the enzymes required for the conversion of mannose to sucrose and starch have been located within the cotyledons, although it is reasonable to assume they are there.

How mobilization of reserves within the fenugreek endosperm is controlled by the embryo is not clear, but it is assumed that at least some temporal control exists because the hydrolytic enzymes do not increase in activity, and are not released from the aleurone layer, until after germination is completed. Abscisic acid (ABA) is present in the endosperm, and if it is leached out from isolated endosperms, then endo-β-mannanase activity increases. There is no requirement for a stimulatory hormone such as GA. Application of ABA to leached endosperms prevents any increase in enzyme activity. Diffusible saponin-like substances are also present in the fenugreek endosperm, and these strongly inhibit any increase in α-galactosidase activity. These, and perhaps ABA, may play a role in limiting galactomannan hydrolysis until germination is completed, although it remains unknown how their inhibitory effects are negated, and whether some role is played by the germinated embryo.

The similarities between the mobilization of galactomannans in fenugreek and of starch in cereals are quite striking, although the endospermic legume is unique in that the cotyledons do not contain starch until it is synthesized therein as a temporary reserve after the germination is completed. In both cases, reserve hydrolysis is by enzymes released from a living, peripheral aleurone layer, and the products are absorbed and modified by the cotyledons (the scutellum being a reduced cotyledon in cereals) before being passed to the growing axis. In a sense, therefore, more contrast is shown between the patterns of hydrolysis in the endospermic and nonendospermic leguminous seed than between endospermic legumes and cereals, although the mode of regulation of hydrolytic enzyme production appears to be distinctly different.

7.4. HEMICELLULOSE-CONTAINING SEEDS OTHER THAN LEGUMES

A number of nonleguminous plants also store mannans, although few have received much attention as far as mobilization of their reserves is concerned. Hydrolysis of polysaccharides in the endosperm (89% mannose deposited in the secondary walls) of date palm (*Phoenix dactylifera*) occurs when a haustorial projection from the seedling grows into it. This results in preformed hydrolytic enzymes being released from protein bodies in the endosperm, and these come into contact with the wall following loss of membrane integrity. The galactomannan is converted to its constituent monomers, which are absorbed by the haustorium and transported to the growing axis; there they are converted to sucrose.

Mobilization of galactomannans from the cell walls of the lettuce seed endosperm commences when endo-β-mannanase activity increases within the endosperm itself, immediately after germination is completed. α-Galactosidase is present as a constitutive enzyme within the endosperm. The products of hydrolysis diffuse to the cotyledons, and small oligomannans are cleaved further by a β-mannosidase located in the cell walls of the cotyledons; the resultant mannose residues are taken up by the cotyledon cells (see also Fig. 8.19). The breakdown of galactomannans within the endosperm of tomato also requires the synthesis of endo-β-mannanase. The importance of this in relation to cell-wall weakening and the completion of germination was discussed in Section 4.2.

Different patterns of mannan hydrolysis between different nonleguminous seeds are to be expected. Since the deposition of mannans within the cell walls of the endosperm and perisperm of hard seeds like date, coffee, and ivory nut results in the destruction of the cell cytoplasm during seed development, a source of enzymes other than the endosperm itself is required. On the other hand, the lettuce endosperm is still comprised of living cells at maturity, and mobilization of their walls is essentially by autolysis.

The cell walls of the cotyledons of nasturtium (*Tropaeolum majus*) contain "amyloids" or galactoxyloglucans, which account for about one-third by weight of the reserves. These are mobilized following germination as the hydrolytic activities of endo-β-glucanase, β-galactosidase, and α-xylosidase increase. β-Glucosidase is also involved, but this enzyme is present in the dry seed and does not increase as the cell wall is degraded.

7.5. STORED TRIACYLGLYCEROL CATABOLISM

7.5.1. General Catabolism

As outlined in Chapter 1, triacylglycerols are the major storage lipids in seeds. Their initial hydrolysis (lipolysis) is by lipases, enzymes that catalyze the three-stage hydrolytic cleavage of the fatty acid ester bonds in triacylglycerols (TAGs), ultimately to yield glycerol and free fatty acids (FFA):

$$\text{Triacylglycerol} \longrightarrow \text{Diacylglycerol + FFA} \longrightarrow \text{Monoacylglycerol + 2FFA} \longrightarrow \text{Glycerol + 3FFA}$$

Glycerol enters the glycolytic pathway after its phosphorylation and oxidation to the triose phosphates (dihydroxyacetone phosphate ⇌ glyceraldehyde-3-phosphate), which are then condensed by aldolase in the reversal of glycolysis to yield hexose units (Fig. 7.9, step 21). Alternatively, the triose phosphates may be converted to pyruvate and then oxidized through the citric acid cycle (Fig. 7.9).

The FFA released by lipase may be utilized in oxidation reactions to yield compounds containing fewer carbon atoms. The predominant oxidation pathway is β-oxidation, in which the fatty acid is first "activated" in a reaction requiring ATP and coenzyme A, and then, by a series of reactions involving the sequential removal of two carbon atoms, this "active fatty acid" is broken down to acetyl-CoA (Fig. 7.9). Saturated fatty acids with an even number of carbon atoms yield only acetyl-CoA. Chains containing an odd number of carbon atoms, if completely degraded by β-oxidation, will yield the two-carbon acetyl moieties (acetyl-CoA) and one three-carbon propionyl moiety (propionyl-CoA, CH_3CH_2CO-S-CoA). This, in turn, can be degraded in a multistep process to acetyl-CoA. The acetyl moiety may be completely oxidized in the citric acid cycle to CO_2 and H_2O or utilized initially via the glyoxylate cycle for carbohydrate synthesis. This latter process is the most important during seedling establishment, i.e., following germination.

α-Oxidation of fatty acids is a process that involves the sequential removal of one carbon at a time from FFA of chain lengths C_{13}–C_{18}. Although it is unlikely that complete oxidation of fatty acids occurs by this pathway, it could serve to shorten odd-chain fatty acids to even lengths and thus facilitate their complete degradation to acetyl-CoA by β-oxidation. Alternatively, it could convert even-length FFA to odd-length FFA, resulting in increased propionic acid production by β-oxidation; this compound is an important precursor of coenzyme A. At present, however, it remains to be demonstrated clearly that α-oxidation occurs in the same organelle (the glyoxysome) as β-oxidation.

The oxidation of unsaturated fatty acids (e.g., oleic acid, 18:1) is by the same general pathways, although some extra steps are required. The double bonds of naturally occurring unsaturates are in the *cis* configuration, but for step 4 (Fig. 7.9) of β-oxidation to be effected, they must be in the *trans* position. Hence a *cis,trans*-isomerase (enoyl-CoA isomerase) is necessary to convert the fatty acid to its oxidizable form:

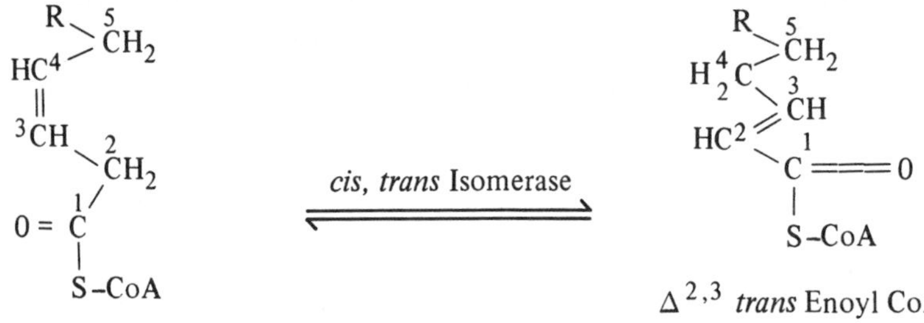

$$R-{}^{5}CH_2-{}^{4}CH={}^{3}CH-{}^{2}CH_2-{}^{1}C(=O)-S\text{-}CoA \xrightleftharpoons{\textit{cis, trans}\ \text{Isomerase}} R-{}^{5}CH_2-{}^{4}CH_2-{}^{3}CH={}^{2}CH-{}^{1}C(=O)-S\text{-}CoA$$

$\Delta^{3,4}$ *cis* Enoyl CoA $\quad\quad$ $\Delta^{2,3}$ *trans* Enoyl CoA

Figure 7.9. Pathways of triacylglycerol catabolism and hexose assimilation. Enzymes: (1) lipases; (2) fatty acid thiokinase; (3) acyl-CoA dehydrogenase; (4) enoyl-CoA hydratase (crotonase); (5) β-hydroxyacyl-CoA dehydrogenase; (6) β-ketoacyl thiolase; (7) citrate synthetase; (8) aconitase; (9) isocitrate lyase; (10) malate synthetase; (11) malate dehydrogenase; (12) catalase; (13) succinate dehydrogenase; (14) fumarase; (15) malate dehydrogenase; (16) phosphoenolpyruvate carboxykinase; (17) enolase; (18) phosphoglycerate mutase; (19) phosphoglycerate kinase; (20) glyceraldehyde-3-phosphate dehydrogenase; (21) aldolase; (22) fructose-1,6-bisphosphatase; (23) phosphohexoisomerase; (24) phosphoglucomutase; (25) UDPGlc pyrophosphorylase; (26) sucrose synthetase or sucrose-6-P synthetase and sucrose phosphatase. (i) Glycerol kinase; (ii) α-glycerol phosphate oxidoreductase. Substrates: TAG, triacylglycerol; MAG, monoacylglycerol; Gly, glycerol; FFA, free fatty acid; PEP, phosphoenolpyruvate; 2PGA, 2-phosphoglyceric acid; 3PGA, 3-phosphoglyceric acid; DPGA, 1,3-diphosphoglyceric acid; G3P, glyceraldehyde-3-phosphate; FruDP, fructose-1,6-bisphosphate; Fru-6-P, fructose-6-phosphate; Glc-6-P, glucose-6-phosphate; Glc-1-P, glucose-1-phosphate; UDPGlc, uridine diphosphoglucose; α-Gly P, α-glycerol phosphate; DHAP, dihydroxyacetone phosphate. Coenzymes and energy suppliers: FAD(H), flavin adenine dinucleotide (reduced); NAD(H), nicotinamide adenine dinucleotide (reduced); GTP, guanosine triphosphate; ATP, adenosine triphosphate; UTP, uridine triphosphate; GDP, guanosine diphosphate; ADP, adenosine diphosphate; AMP, adenosine monophosphate; CoA, coenzyme A.

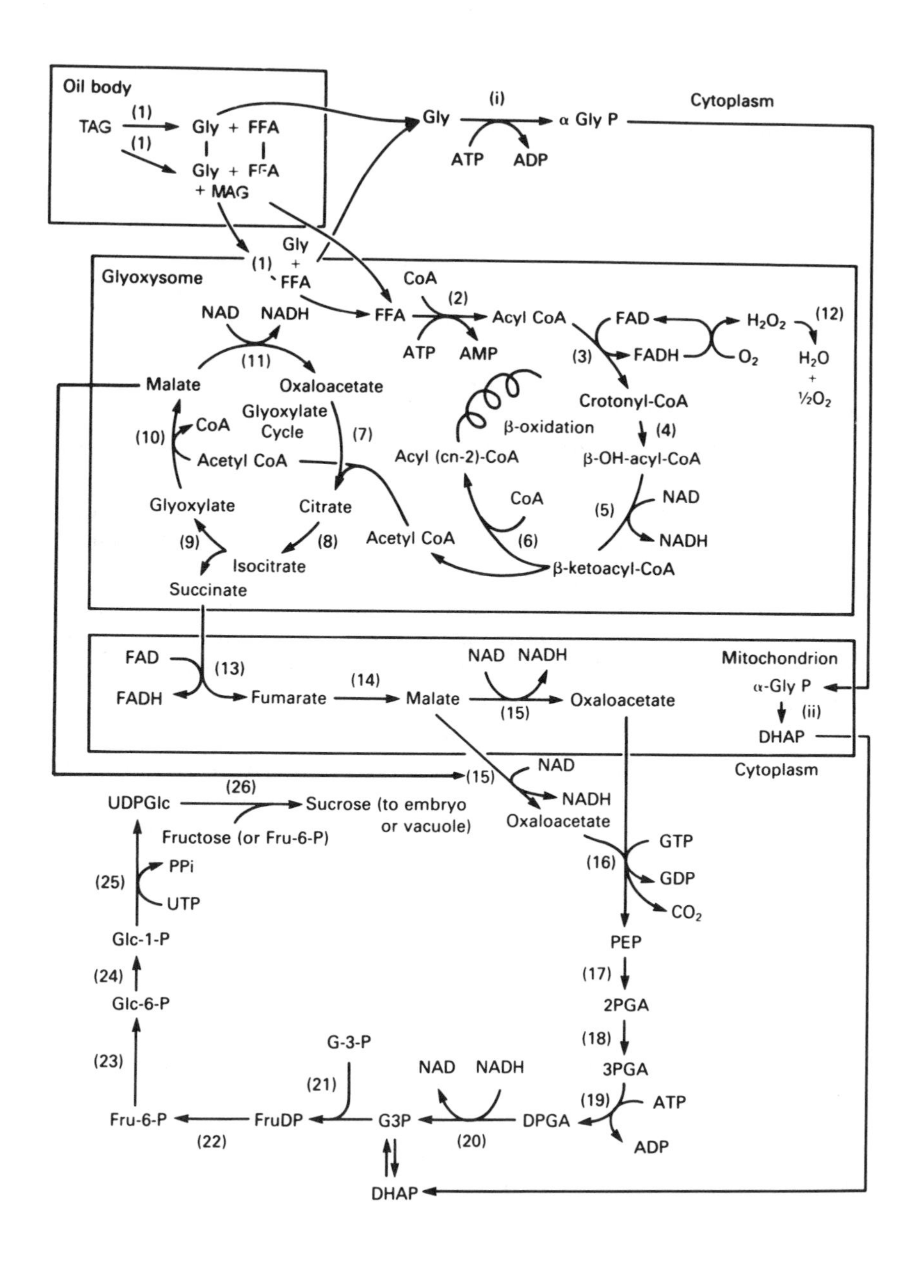
Oil body
TAG
(1)
Gly + FFA
Gly + FFA + MAG
Gly + FFA
Gly
(i)
α Gly P
ATP ADP
Cytoplasm
Glyoxysome
CoA
(2)
FFA
Acyl CoA
ATP AMP
FAD
FADH
(3)
H_2O_2
(12)
O_2
H_2O + $½O_2$
NAD NADH
(11)
Malate
Oxaloacetate
Glyoxylate Cycle
(7)
(10)
CoA
Acetyl CoA
Crotonyl-CoA
(4)
β-oxidation
β-OH-acyl-CoA
Acyl (cn-2)-CoA
NAD
(5)
NADH
Glyoxylate
Citrate
CoA
(6)
(9)
(8)
Acetyl CoA
Isocitrate
β-ketoacyl-CoA
Succinate
FAD
FADH
(13)
Fumarate
(14)
Malate
(15)
Oxaloacetate
Mitochondrion
α-Gly P
(ii)
DHAP
Cytoplasm
(26)
UDPGlc
Sucrose (to embryo or vacuole)
Fructose (or Fru-6-P)
Oxaloacetate
GTP
(16)
GDP
CO_2
PPi
(25)
UTP
Glc-1-P
PEP
(24)
(17)
Glc-6-P
2PGA
G-3-P
(18)
(23)
3PGA
(21)
(19)
ATP
ADP
Fru-6-P
FruDP
G3P
DPGA
(22)
(20)
DHAP

which is the normal substrate for the next enzyme in the β-oxidation pathway, enoyl-CoA hydratase (4). Polyunsaturated fatty acids containing two or more double bonds (e.g., linoleic acid, 18:2; linolenic acid, 18:3) cannot be degraded simply by β-oxidation either, but the appropriate enzymes (2,3 enoyl-CoA isomerase, 3-OH acyl-CoA epimerase, and 2,4 dienoyl-CoA reductase) which are required for the continuation of β-oxidation are present within the glyoxysome.

Directly coupled to the β-oxidation pathway is the glyoxylate cycle, which takes the acetyl-CoA and, in a series of enzymatic reactions, links this to the glycolytic pathway, which then operates to produce hexose (Fig. 7.9). The key enzymes for forging this link are malate synthetase and isocitrate lyase. Acetyl-CoA is first converted to citrate (in the same manner as initiates its entry into the citric acid cycle: Fig. 7.9, step 7), then to isocitrate, but the decarboxylating steps in the citric acid cycle between isocitrate and succinate are avoided by the action of isocitrate lyase, which cleaves isocitrate directly to succinate and glyoxylate. Another acetyl-CoA is incorporated into the cycle (step 10) and is condensed with glyoxylate by malate synthetase to yield malate. With each turn of the cycle one molecule of succinate is released (step 9) and is converted to oxaloacetate by citric acid enzymes in the mitochondria (steps 13–15), and then into the glycolysis pathway as phosphoenolpyruvate (step 16).

Catabolism of triacylglycerol reserves within the storage tissues of germinated seeds involves three distinct organelles found within oil-containing cells. These are: (1) the TAG-storing oil body, (2) the glyoxysome, and (3) the mitochondrion (Fig. 7.10), and they function as follows: (1) hydrolysis of TAGs to FFA and glycerol commences within the oil bodies; (2) within the glyoxysome the fatty acids are oxidized, and synthesis of succinate occurs via the glyoxylate cycle; (3) the succinate is converted to malate or oxaloacetate within the mitochondrion. The malate or oxaloacetate is processed further in the cytosol to yield sucrose. We will consider these stages in more detail, and since many of the "classical" studies on TAG mobilization in seeds have been carried out on the endosperm of castor bean (*Ricinus communis*), work on this tissue features prominently in the following account.

7.5.2. Mobilization of Triacylglycerols from Oil Bodies

In high-oil seeds, such as peanut (*Arachis hypogaea*) and castor bean, the oil bodies change little in the general appearance during mobilization; they gradually decrease in size as the reserves are depleted.

Since storage TAGs are present within discrete oil bodies, it is not an unreasonable expectation that the enzymes responsible for their hydrolysis should be intimately associated with these structures. However, in peanut and

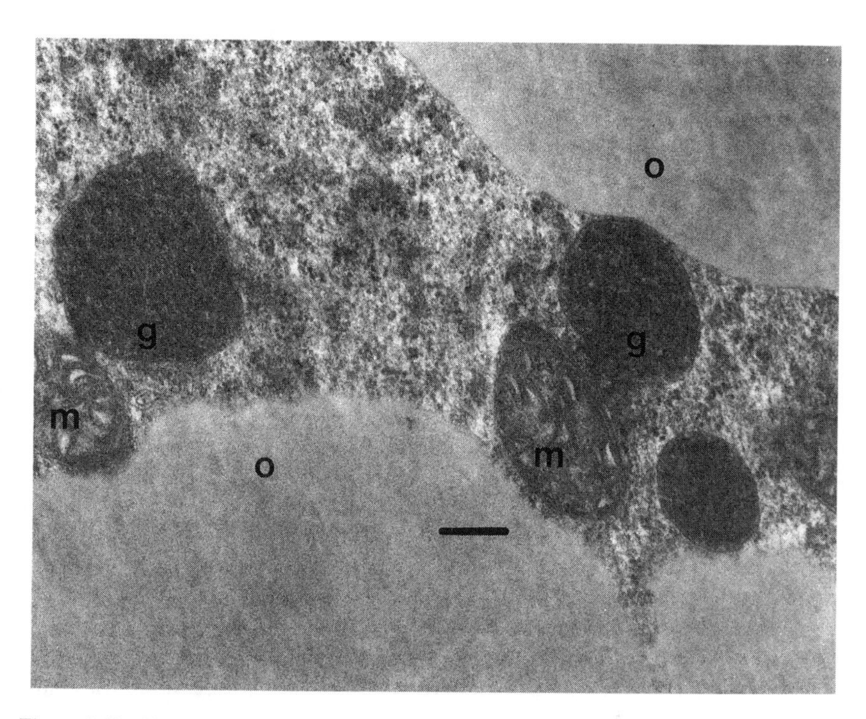

Figure 7.10. Electron micrograph of the cells of the castor bean endosperm during the period of mobilization of the lipid reserves. Note the close association of the oil bodies, glyoxysomes, and mitochondria. o, Oil body; g, glyoxysome; m, mitochondrion. Bar = 0.2 μm. Courtesy of J .S. Greenwood.

soybean, for example, there is very little lipase activity in the oil bodies, and most is associated with the glyoxysomes. It remains to be explained how, in these seeds, the TAGs are initially mobilized, and how the FFA, or monoacylglycerols reach the required location, the glyoxysome, for further conversion. In cotton cotyledons, it is likely that a lipase is synthesized in the cytoplasm, and then imported into the oil body. Lipases become associated with oil bodies of castor bean, maize, and rapeseed also, but only in castor bean is the enzyme present in the dry seed; in other species enzyme activity increases after germination. Even in castor bean, however, the acid lipase present in the dry seed is probably not involved in lipolysis; its activity declines on imbibition (Fig. 7.11), and the pH of the cell is too high (> pH 5.5) for the enzyme to function effectively. A second lipase, with optimum activity at neutral pHs, increases within the oil bodies after 2 days from the start of imbibition at the time when TAG mobilization commences (Fig. 7.11). Complete mobilization of the TAGs takes about 4 days. By

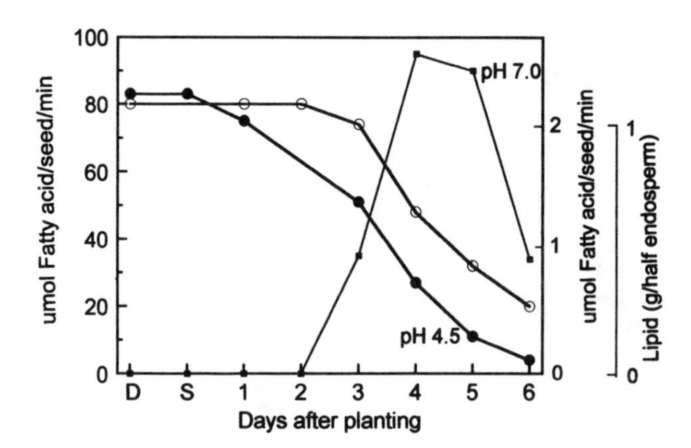

Figure 7.11. Change in lipolytic activity of oil bodies on their endogenous substrate during the growth of the castor bean seedling. Activities at pH 4.5 (●) and 7.0 (■) are given. (O) Decline in stored lipid content. D, dry seed; S, 1-day-soaked seed; the numbers refer to days after planting. After Hills and Beevers (1987) and Muto and Beevers (1974).

then the endosperm is liquified and the assimilates have been absorbed by the expanding cotyledons attached to the growing root–shoot axis. In castor bean, and other oil-storing seeds, a lipase associated with the membrane of the glyoxysome might play a role in converting monoacylglycerols, which diffuse from the oil bodies, to FFA and glycerol. Compared with other processes in the mobilization of TAGs, lipases have been studied very little, and in the seeds of only a few species.

7.5.3. The Fate of Glycerol and Fatty Acids

The products of lipolysis, glycerol and FFA, usually are rapidly metabolized. In persistent storage organs some of the released glycerol may be reutilized for TAG synthesis, some is respired, and the rest is converted to sucrose for transport to the growing axis. As might be expected, in nonpersistent organs the reincorporation of glycerol into TAGs is very limited: most of the glycerol is converted to sucrose for export, with some being lost in respiratory processes. The conversion to sucrose requires first that the glycerol be phosphorylated by glycerol kinase in the cytosol, to give α-glycerol phosphate. This is then oxidized in the mitochondria by cytochrome-linked α-glycerol phosphate oxidoreductase, to yield dihydroxyacetone phosphate (Fig. 7.9, step ii). Finally, this is released into the cytosol for conversion to hexose.

Although seeds of most dicots and cereals do not accumulate FFA during lipolysis, the storage tissues of a few species do. For example, in the germinated seeds of the West African oil palm (*Elaeis guineensis*) a specialized structure, the haustorium, invades the endosperm, and through its vascular system the products of TAG catabolism are transported to the growing root and shoot. During oil degradation there is a buildup of FFA in the endosperm, and those that are not respired are absorbed directly by the haustorium without prior conversion to sucrose. FFA accumulate also in the haustorium itself and there might be reconverted to TAG for temporary storage until required by the growing axis.

7.5.4. Role of the Glyoxysome, Mitochondrion, and Cytosol in Gluconeogenesis

The degradation of saturated triacylglycerols and their conversion to carbohydrate is summarized in Fig. 7.9. The oil bodies, glyoxysomes, mitochondria, and cytosol are all involved, and the enzymes for each step in this process have been located in their appropriate position in the cell. In view of the cooperative role of the three organelles, it is not surprising that they are found in juxtaposition within the cell (Fig. 7.10). The pathways involved in lipolysis, fatty acid oxidation, and gluconeogenesis have been outlined in Section 7.5.1 and will not be considered in detail again; only a few essential features of the pathways will be mentioned here. FFA produced by lipolysis enter the glyoxysome and are activated with CoA therein. Their passage through the β-oxidation pathway and glyoxylate cycle is completed within this organelle too. There are no enzymes present in the matrix of the glyoxysome for the reoxidation of NADH, an event that is essential for the sustained supply of NAD to accept electrons during β-oxidation and operation of the glyoxylate cycle within this organelle (Fig. 7.9, steps 5 and 11). The acyl-CoA dehydrogenase of β-oxidation avoids this by transferring electrons to O_2 (step 3). Glyoxysomal membranes contain NADH:ferricyanide reductase and NADH:cytochrome *c* reductase activities. These may be responsible for the transport of reducing equivalents, produced during reoxidation of NADH within the organelle, across the glyoxysomal membranes to external electron acceptors (yet unidentified) which would mediate the transport of electrons between the glyoxysomes and the outer mitochondrial membrane, where the reoxidation cycle is completed. An alternative shuttle system, whereby reducing equivalents are transferred between the glyoxysomes and mitochondria using malate and aspartate as intermediates, has been proposed, but this may not function efficiently enough to regenerate NAD at the required rate.

The involvement of the mitochondrion is essential for further utilization of succinate released from the glyoxylate cycle by isocitrate lyase (Fig. 7.9, step 9). Either malate or oxaloacetate (steps 14 and 15) is transferred to the cytosol from the mitochondrion, and after conversion to phosphoenolpyruvate (PEP) (step 16) the glycolytic pathway operates in reverse to produce hexose. It has been suggested that the reaction sequence from PEP to sucrose (steps 17–26) occurs within the plastids, from which enzymes of the glycolytic pathway have been isolated. However, the activities of several of these plastid enzymes seem to be too low to account for the *in vivo* rate of gluconeogenesis, whereas those of all glycolytic pathway enzymes in the cytosol are quite adequate.

One aspect of triacylglycerol degradation that the scheme in Fig. 7.9 does not take into account is the fact that storage oils and fats are usually unsaturated and, in the case of ricinoleic acid in castor bean, also hydroxylated. Completed oxidation of unsaturated and polyunsaturated fatty acids requires their modification by processes other than β-oxidation (Section 7.5.1): the appropriate enzymes (e.g., *cis,trans*-isomerase, epimerase, and hydratase) are present in the glyoxysomes. For β-oxidation of ricinoleic acid (12-OH 18:1), the C_8-intermediate (2-OH 8:0) fatty acid requires conversion by an α-hydroxy acid oxidase and oxidative decarboxylation to circumvent the metabolic barrier caused by the hydroxyl group. The heptanoyl-CoA so formed can be catabolized further by β-oxidation.

7.5.5. Glyoxysome Biosynthesis and Degradation

The formation of glyoxysomes and the synthesis of their constituent enzymes have been investigated quite intensively since the appearance of the first edition of this book. This has resulted in some major changes in our understanding of these events.

Regardless of the species or tissue in which they are found, mature glyoxysomes (which are a class of peroxisomes prevalent in the storage tissues of oil seeds) are remarkably conservative in both size and composition. They have an equilibrium density of 1.25 g/cm^3, are surrounded by a single unit membrane, and, with the exception of the membrane-associated alkaline lipase, have very similar specific activities of individual enzymes. As noted previously, when triacylglycerol degradation is occurring most actively, glyoxysomes are frequently found in juxtaposition to oil bodies and mitochondria (Fig. 7.10). We will now follow the synthesis and fate of the glyoxysome in two types of storage organ: (1) that which disintegrates as the major stored reserves are mobilized (exemplified by the castor bean endosperm) and (2) that which persists after reserve mobilization and eventually becomes photosynthetic, using as examples those seeds which store TAGs in their cotyledons, and which exhibit epigeal growth. A major difference between glyoxysomes in the castor bean endosperm, and those in the cotyledons of other oil seeds, is that these organelles are formed during the middle to late stages of development in the latter, and are present in the dry seeds. This appears not to be the situation in castor bean, where the glyoxysomes are formed only during postgerminative oil mobilization.

1. *Cotyledons.* Glyoxysomes have been identified in the cotyledons of mature dry seeds of several species. In general, they contain some of the enzymes that are expected to be present within this organelle, but by no means the full complement. Following germination the glyoxysomes enlarge considerably in volume (sevenfold in cotton seedlings by 48 h after imbibition). Many of the glyoxysomal enzymes, including isocitrate lyase, increase greatly in activity as the proteins accumulate in the matrix of this organelle. To accommodate the increase in size of the glyoxysomes, a considerable amount of membrane lipid and protein must be available for membrane expansion. Glyoxysome membranes contain about 50% of their phospholipids as phosphatidylcholine (PC) and 10% as phosphatidylethanolamine (PE), but the enzymes which synthesize these components are absent from the organelle. Hence, they must be imported from elsewhere within the cell. While the primary site of phospholipid-synthesizing enzymes (especially for PC) in the cell is the endoplasmic reticulum (ER), there is no evidence for the direct transfer of membrane lipids from the ER to the enlarging glyoxysomes. Rather, PC and PE are transferred from the ER to developing mitochondria; the depleting oil bodies are the source of membrane lipids for the glyoxysomes (Fig. 7.12). Both PC and nonpolar lipids are transferred from the oil bodies to the glyoxysome membrane; the PC is incorporated into the expanding membrane, but the nonpolar lipids are hydrolyzed, presumably by a lipase associated with the bounding membrane. The resultant FFA may be incorporated, in part, into the glyoxysome membrane which contains about 10% by weight of nonpolar lipids, 60% of which are FFA. Proteins must also be incorporated into the expanding membranes of the glyoxysome, but the site of their synthesis is not known. The enzyme complement of the glyoxysomes is greatly increased as they expand. Most studies have been carried out on isocitrate lyase (ICL) and malate synthetase (MS), but it is generally assumed that most or all of the enzymes are synthesized on free polysomes within the cytosol, from newly formed transcripts, and then inserted into the expanding glyoxysomes. Since these organelles are present in the developing seed, albeit in a small, immature form, it is not surprising that the mRNAs for MS are present, and expressed at this time. In contrast, glyoxysomes do not contain ICL while the seed is undergoing development, and the mRNA for this enzyme is virtually absent (Fig. 7.13). Following germination, when the full complement of enzymes is inserted into the glyoxysomes, there is a substantial increase in mRNAs for both ICL and MS, and in these proteins, within the cotyledons (Fig. 7.13). The

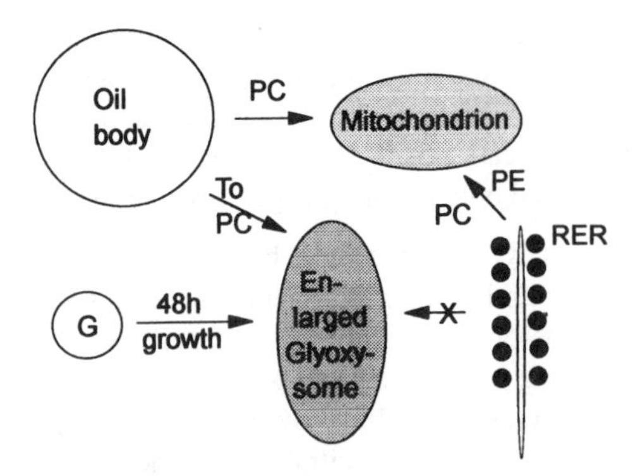

Figure 7.12. Proposed scheme to show the intracellular trafficking of lipids in cotyledons of cotton seedlings during glyoxysome enlargement and storage oil mobilization. Preexisting glyoxysomes in the dry seed enlarge greatly during postgerminative growth (48 h), and the phospholipid [phosphatidylcholine (PC)] and nonpolar lipids (To) are supplied by the mobilization of the oil body contents. While phospholipids [PC and phosphatidylethanolamine (PE)] are not transferred from the endoplasmic reticulum (ER) to the glyoxysome, this is a source for the developing mitochondria, as are the oil bodies. After Chapman and Trelease (1991). Reproduced by copyright permission of Rockefeller University Press.

expression of the genes for these two enzymes is coordinately regulated following germination, and the amount of transcript (mRNA) produced is generally proportional to the amount of enzymes present in the cotyledons. This is indicative that, to a large extent at least, the appearance of the enzymes is transcriptionally regulated. Interestingly, and in contrast, during seed development the gene for MS is activated, but that for ICL only to a very small extent. Presumably, then, the signal(s) which coordinately upregulates both genes during seedling growth is different from that present during seed development, which only upregulates the MS gene. How the glyoxysomes are formed in the cotyledons during development is not known. At present it is assumed that they are formed in the same way as castor bean endosperm glyoxysomes (see next subsection), by the budding off of segments of the ER. But, while glyoxysomes are differentiated during development, they do not become this specialized organelle until after seed maturation and subsequent germination.

Following their synthesis on free polysomes in the cytoplasm, the glyoxysome matrix enzymes are targeted to, and enter, this organelle posttranslationally. In general, uptake into the glyoxysome is not accompanied by proteolytic processing of the matrix proteins, although there are exceptions, e.g., malate dehydrogenase, which has an amino-terminal signal (transit) peptide. How the matrix proteins are specifically targeted to the glyoxysomes is still under investigation, but a tripeptide at the carboxy-terminus of some imported proteins appears to be a sufficient targeting signal.

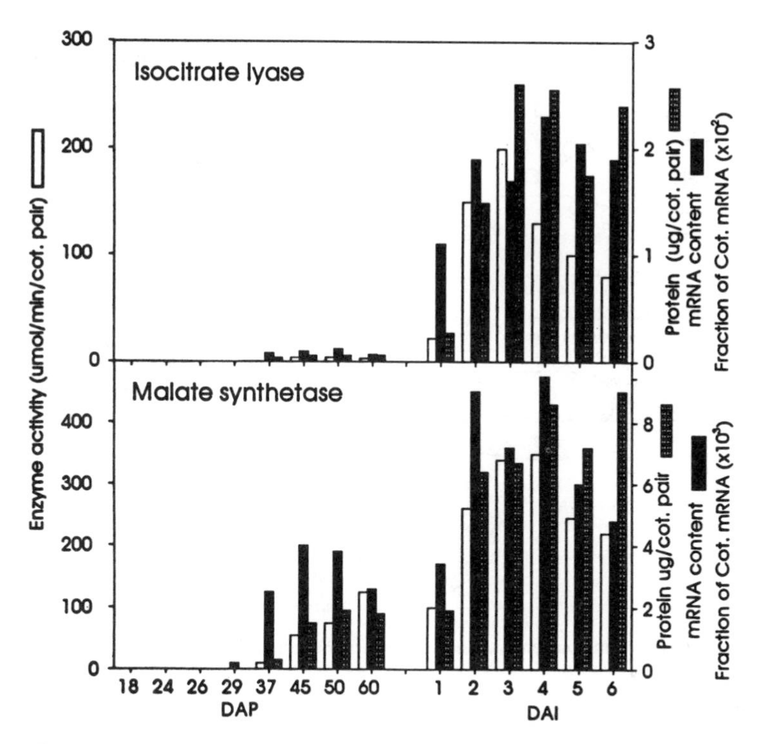

Figure 7.13. Expression of isocitrate lyase and malate synthetase genes during development and following germination of rapeseed (*Brassica napus*). Enzymes activities, total protein content, and mRNA amounts for the enzymes are shown at various days after pollination (DAP) or days after imbibition (DAI) of the mature dry seed. After Olsen and Harada (1991).

2. *Endosperm*. Glyoxysomes are absent from the mature dry endosperm of castor bean seeds, and hence must both be formed and furnished with the appropriate complement of enzymes by the time TAG mobilization commences following germination. The glyoxysomes are formed *de novo* by the budding off of segments from the ER (Fig. 7.14) which, therefore, contributes membrane lipid and (glyco)proteins directly. These glyoxysomes do not expand, for they are released from the ER at their mature size. The appropriate enzymes are synthesized *de novo* on cytoplasmic polysomes, from newly transcribed mRNAs, and are inserted into the matrix of the glyoxysomes both during their formation on the ER and after their release therefrom. Some of the matrix proteins are glycoproteins, and these must be routed from the cytoplasmic polysomes through the ER for glycosylation before insertion into the organelle. At one time it was thought that some enzymes were actually synthesized on the rough endoplasmic

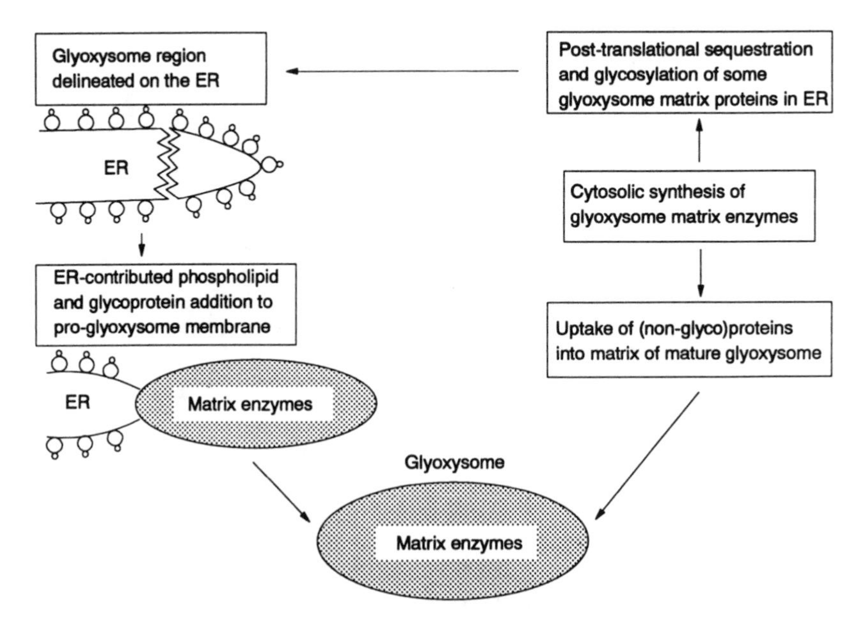

Figure 7.14. Glyoxysome biogenesis in the endosperm of germinated castor bean seeds. Based on Gonzalez (1986).

reticulum (RER), inserted into the ER lumen, and transported vectorially to the newly forming glyoxysomes; this is now known not to be the case. To ensure correct targeting of the matrix enzymes to the glyoxysomes only, and not to the ER from which they are derived, the organelles must presumably incorporate specific receptors within their membrane to act as target sites.

7.5.6. Utilization of the Products of Triacylglycerol Catabolism

Although the products of TAG hydrolysis, FFA and glycerol, may be used for the resynthesis of oils and membrane lipids (particularly in persistent cotyledons), they are to a large extent converted to hexose, and finally to sucrose, by a sequence of reactions outlined in Fig. 7.9. Castor bean endosperms contain high amounts of sucrose-6-*P* synthetase, sucrose phosphatase, and also sucrose synthetase (see Section 7.1.2 for their role in sucrose synthesis). Here the major product of TAG mobilization is sucrose, which is taken up by active transport into the cotyledons. More than 80% of this sucrose is redistributed to the growing axis. If the embryonic axis is removed, there is temporary storage of sucrose in the endosperm, in vacuoles that develop as the storage products are degraded. Sucrose uptake by the cotyledons is thereby drastically reduced. Thus, as far as the growing seedling is concerned, removal of the sink (axis) alters replenishment at the source (cotyledons).

The cotyledons of some seeds (e.g., pumpkin, watermelon, sunflower) can utilize acetyl-CoA arising from β-oxidation of fatty acids for amino acid synthesis via partial reactions of the glyoxylate and citric acid cycle. The usual products are glycine, serine, glutamic acid, glutamine, and γ-aminobutyric acid.

The cotyledons of those species whose mode of seedling growth is epigeal turn green as they emerge from the soil into the light. During greening there is a gradual loss of glyoxysomes and an increase in a similarly sized microbody called the peroxisome. It is now generally accepted that the glyoxysome persists as a peroxisome, and fatty-acid-catabolizing enzymes such as isocitrate lyase, malate synthetase, and β-oxidation pathway enzymes are destroyed, while others, e.g., catalase and malate dehydrogenase, are partially or wholly retained. Controlled removal of certain enzymes might involve a special class of inactivating proteins, perhaps proteinases with specificity for particular glyoxysomal enzymes. This mechanism would ensure the destruction of some proteins and conservation of others within the same organelle. Other enzymes become inserted after synthesis on cytoplasmic polysomes.

7.6. STORED PROTEIN CATABOLISM

7.6.1. General Catabolism

Hydrolysis of storage protein (polypeptides) to their constituent amino acids requires a class of enzymes called proteinases, some of which effect total hydrolysis whereas others produce small polypeptides that must be degraded further by peptidases. The proteinases can be categorized in relation to the manner in which they hydrolyze their substrates as follows:

1. Endopeptidases: these cleave internal peptide bonds to yield smaller polypeptides.
2. Aminopeptidases: these sequentially cleave the terminal amino acid from the free amino end of the polypeptide chain.
3. Carboxypeptidases: as (2), but single amino acids are sequentially hydrolyzed from the carboxyl end of the chain.

Both (2) and (3) are exopeptidases.

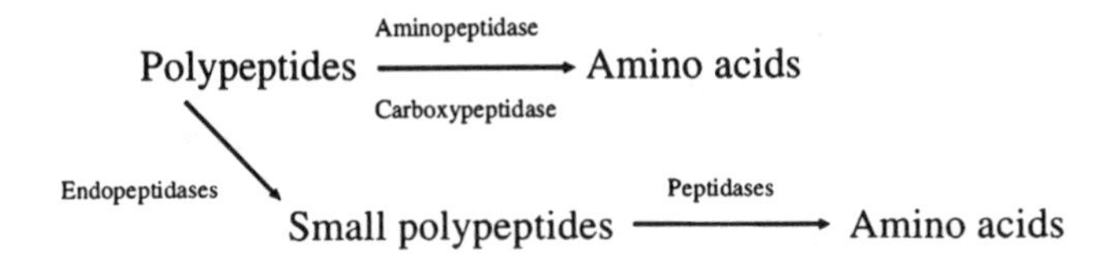

The liberated amino acids may be reutilized for protein synthesis or be deaminated to provide carbon skeletons for respiratory oxidation. Ammonia is produced by deamination, but this is prevented from reaching toxic concentrations by fixation into glutamine and asparagine, two commonly translocated forms of amino acid.

7.6.2. Protein Mobilization in Cereals

Reserve proteins are stored in two separate regions of the cereal grain: in the aleurone grains of the aleurone layer (up to 30% of total), and in the protein bodies (sometimes disrupted, e.g., wheat, barley) of the starchy endosperm. A minor amount of storage protein is present in the scutellum and the axis, and this may be hydrolyzed to provide amino acids for the growing axis prior to mobilization of the major endosperm reserves.

There appear to be three major sites of proteolytic activity in the germinated cereal grain:

1. Aleurone layer. Proteinases are produced in the aleurone layer, in some species or cultivars under the control of gibberellin (see Section 8.1.5), and these may be synthesized *de novo*. One or more of these enzymes hydrolyzes the proteins within the aleurone grains, and the resultant amino acids are utilized for the synthesis of proteins (e.g., α-amylase) in this tissue.

2. Starchy endosperm. Proteinases effective in breaking down the major protein reserves within the starchy endosperm may come from two sources: the aleurone layer or the starchy endosperm itself. In barley, for example, endopeptidases released from the aleurone layer may play a central role in endosperm protein hydrolysis. But there are also preformed proteinases within the endosperm of dry grains and these also may be involved in protein mobilization. Besides their action in degrading reserve proteins, proteinases in the starchy endosperm also serve to release and activate bound enzymes (e.g., β-amylase), and to aid in dissolution of the cell walls by hydrolyzing links between glucan and protein components.

3. Axis and scutellum. Peptidase activity in the scutellum is responsible for the hydrolysis of peptides being taken up by this structure from the endosperm. Proteolytic activity is also present in the axis for hydrolysis of the small amount of reserve proteins stored therein. In addition, there are several pro-

teolytic enzymes within the embryo that are involved with the normal turnover of proteins associated with growth.

Recently, there has been an increase in interest in proteolytic enzymes in seeds, particularly in cereal grains, and several have now been well-characterized. It is evident, however, that many proteinases and peptidases are present during reserve mobilization, and the precise location of synthesis of all of them, and their cooperative role in protein breakdown remain to be elucidated. The complexity of proteolytic enzyme activities will be illustrated by germinated maize kernels, but similar multiple- and sequential-enzyme activities have been reported for other cereal grains during storage-protein mobilization.

In maize kernels at least 15 different endopeptidase activities are detectable during the first 6 days after the start of imbibition. Four groups of enzymes have been identified, based on the time of their appearance (Fig. 7.15). Group I is present in the dry seed; it contains two metallo-endopeptidases which decline in activity soon after imbibition. They appear not to be involved in the initial mobilization of zein, the major storage protein in maize. Group II endopeptidases

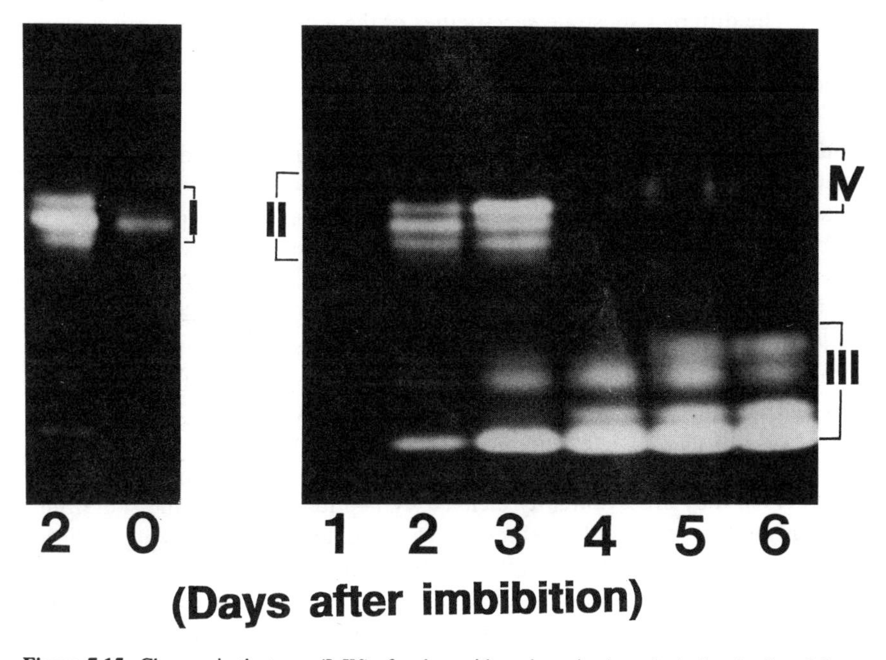

Figure 7.15. Changes in the types (I–IV) of endopeptidases in maize kernels during the first 6 days after imbibition. An activity gel, using gelatin as substrate, shows where the enzymes are located following their separation on a native polyacrylamide gel. The light areas show where digestion of the substrate has occurred. Provided by W. Mitsuhashi and A. Oaks; see also their 1994 publication.

increase following germination and reach peak activity after 3 days. These are SH-(cysteine)endopeptidases and have a high affinity for γ-zein, the form of this storage protein which is located peripherally in the protein bodies, and thus is the first to be subjected to proteolysis. Group III enzymes achieve maximum activity after 5 days and are mostly SH-endopeptidases which cleave α-zein, the form located internally within the protein body. Group IV enzymes increase in activity only after day 3, and their specificity is for α-zein. They are unable to hydrolyze γ-zein, but by the time they are present in the endosperm this form of zein is likely to have been completely mobilized. The site of synthesis of the endopeptidases which hydrolyze zein is either the scutellum or the aleurone layer, but it is not known which enzymes are capable of mobilizing the storage proteins within the latter tissue. In addition to these groups of endopeptidases, it is likely that there are also several amino- and carboxy-peptidases involved in completing proteolysis, as well as the peptidases.

A general pattern of hydrolysis in both cereals and dicots seems to be emerging in that metallo-endopeptidases are present first, then a series of cysteine-endopeptidases, followed by the terminal-acting (amino- and carboxy-) peptidases and the enzymes which hydrolyze the resultant peptides, the peptidases. The different substrate specificities of the enzymes, as they arise, could account for the order in which storage proteins and their component forms are mobilized.

An interesting observation is that the gliadins and glutelins in the wheat endosperm are reduced by an NADP-dependent thioredoxin reductase prior to their proteolysis. This appears to increase their susceptibility to proteolytic enzymes. A similar situation occurs in barley. The storage proteins in the imbibed endosperm are not contained within discrete bodies, and thus are susceptible to cytoplasmically located enzymes. Whether a similar reducing system can operate in cereal endosperms in which the storage proteins are sequestered in protein bodies (e.g., maize) remains to be determined.

Proteolytic activity within the starchy endosperm results in the production of amino acids, dipeptides, and a number of small oligopeptides. These soluble products are rapidly absorbed by the embryo, via the scutellum. Although this uptake of peptides does not appear to involve or require their hydrolysis, they are eventually cleaved by peptidases within the scutellum, and only free amino acids accumulate to any extent in the growing embryo. Active, i.e., energy-dependent, uptake mechanisms within the scutellum can distinguish between peptides and amino acids. Several uptake systems have been identified in cereal grain scutella. In those of wheat and barley there are at least four systems for amino acids alone: two nonspecific amino acid systems, one specific for proline, and another specific for basic amino acids. Maize and rice also possess multiple uptake systems, but with some differences in specificity. The efficiency with which scutella take up certain

Table 7.3. The Relative Uptake of Several Peptides by Scutella Isolated from Cereal Grains, in Relation to Uptake of Glutamine, over a 20-Min Incubation Period[a]

	Barley	Wheat	Rice	Maize
		Ratio of total uptake		
Gly-Sar/Gln	0.20	0.25	0.27	0.35
Ala-Ala/Gln	0.55	1.10	0.77	1.54
Ala-Ala-Ala/Gln	0.31	0.84	0.74	1.32

[a]The uptake was measured of di- and tripeptides of Ala (alanine), the dipeptide Gly-Sar (glycine–sarcosine, which is taken up similarly to glycine dipeptide), and the amino acid Gln (glutamine). After Salmenkallio and Sopanen (1989).

peptides, compared with amino acids, also varies between species. For example, the relative rates of uptake of Gly-Sar in relation to Gln into the scutella of barley, wheat, rice, and maize are about the same, whereas the uptake of Ala di- and tripeptides is proportionately greater in relation to Gln in those of wheat and maize (Table 7.3).

The protein components of the peptide and amino acid carriers, which are incorporated into the outer (plasma) membrane of scutellar cells, are synthesized *de novo* about the time that protein mobilization commences within the endosperm. In barley, the scutellum has little capacity to transport a glycine–phenylalanine peptide, nor glycine, at 12 h after the start of imbibition, but this capacity develops by 15 h, and increases thereafter (Fig. 7.16). This increase in the protein components of the carrier system is accompanied by an increased capacity to take up the products of protein mobilization; the scutellum of dry and early-imbibed grains take up peptides and amino acids only poorly. Studies to identify and characterize the membrane-associated carrier proteins are under way.

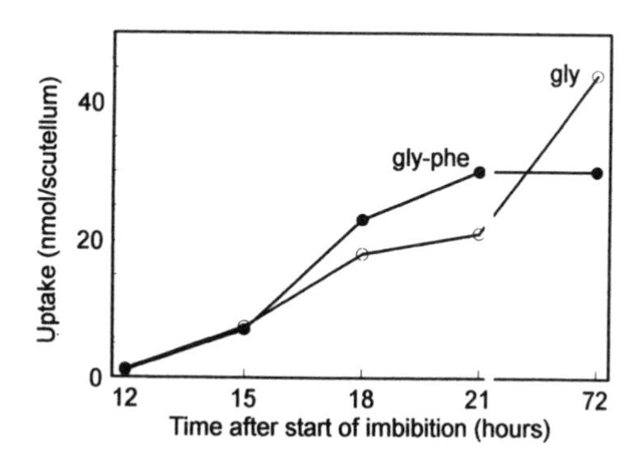

Figure 7.16. Development of the capacity for Gly-Phe (●) and Gly (O) transport into the scutellum of barley embryos following imbibition. Increased transport is related to the synthesis of protein components of the carriers in the plasmalemma. After Walker-Smith and Payne (1985).

7.6.3. Protein Mobilization in Dicots

While proteinases are present in the storage tissues of dry seeds, these are not regarded as being important in the initial mobilization of storage proteins, even though they may be located within the protein body itself. An exception may be the buckwheat seed metalloproteinase (see next section). The general pattern of proteolysis outlined below has been found consistently in several leguminous and nonleguminous dicots, although the "model" tissues on which it is based are the cotyledons of pumpkin (*Cucurbita moschata*), vetch (*Vicia sativa*), and soybean (*Glycine max*).

The first group of proteinases to increase in activity following germination, as a result of *de novo* synthesis, have been called "proteinase(s) A," which are usually SH-dependent acid endopeptidases. These enzymes act on the insoluble native 11 S (legumin) and 7 S (vicilin) proteins, resulting in the following consecutive series of events (Fig 7.17):

1. Short-chain peptides are cleaved from the native proteins, which results in a marked increase in their susceptibility to proteinases.
2. Further proteolysis leads to an increase in their negative charge, perhaps resulting from the removal of short basic peptides.
3. An increased solubilization of the proteins resulting from the initial proteolysis.

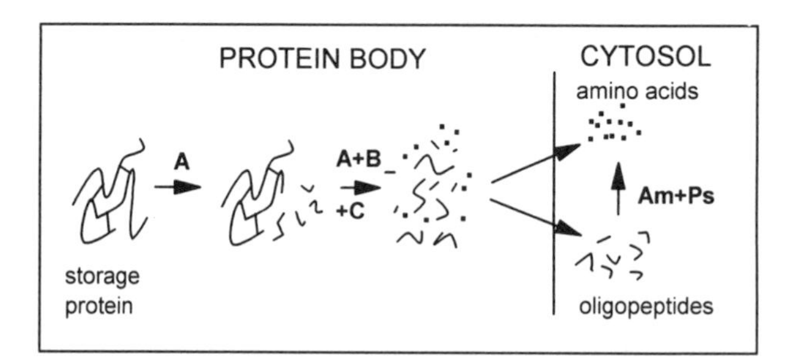

Figure 7.17. A generalized pathway for the mobilization of storage proteins in the protein bodies of dicot seeds. The native storage protein, here a disulfide-linked insoluble legumin, is initially trimmed by proteinase A (endopeptidase) activity (A) to render it more soluble as small oligopeptides are released. Further proteinase A activity, and that of proteinase B endopeptidases (A and B) result in hydrolysis of the protein to amino acids and small peptides which are transported into the cytosol. Hydrolysis of the polypeptides in the protein bodies is aided by carboxypeptidases (C). The released oligopeptides are degraded further by aminopeptidases (Am) and peptidases (di- and tripeptidases, Ps) within the cytosol to yield amino acids. Based on Wilson (1986).

4. A stepwise decrease in the size of the acidic subunits of the 11 S protein, which are cleaved into polypeptides of intermediate size (10–18 kDa, from the native 20–40 kDa). There is no concomitant hydrolysis of the basic subunits; this occurs later. The size of the 7 S subunits is also decreased at an early stage of proteolysis.

Proteinase B activity (Fig. 7.17), again related to endopeptidase(s), accomplishes extensive degradation of the proteins modified by proteinase A. It is inactive against native storage proteins, and the hydrolytic activity of B is enhanced by the more prolonged activity of A. Proteinase B enzymes increase as a result of their *de novo* synthesis. Carboxypeptidases in the protein body also contribute to hydrolysis of the products of proteinase A activity, as new carboxy-terminal sequences arise because of fragmentation of the polypeptides of the 7 S and 11 S proteins. The product of their activity is free amino acids. Some carboxypeptidases are present in the protein bodies of dry seeds, in some species, and become active as the substrate is made available. Others are synthesized *de novo* on cytoplasmic polysomes and inserted into the protein body.

Protein bodies contain only endopeptidases and carboxypeptidases, and these cannot effect the complete hydrolysis of storage proteins to amino acids. Small oligopeptides (along with the amino acids released by carboxypeptidase activity) are transported into the cytosol, where they are further hydrolyzed by aminopeptidases and peptidases. Whereas the acidic environment within the protein bodies favors the activity of carboxypeptidases, which have pH optima of 5 to 6, the more alkaline environment within the cytosol is closer to the pH optima of aminopeptidases (pH 6.5–8). Aminopeptidases have been reported in dry seeds and they retain their activity until the time of reserve hydrolysis; more rarely, increases in activity have been reported in some seeds.

It will be appreciated that different seeds contain several to many enzymes with protein- and peptide-hydrolyzing abilities, and that their sequential activities will result in the degradation of the stored proteins. Specific enzymes will initiate the mobilization of the proteins, and the timing and extent of mobilization of a particular protein can be fine-tuned by regulating the appearance and quantity of these initiating proteinases. Even different polypeptide components of the storage proteins are hydrolysed to different extents at different times during mobilization. An appreciable number of proteolytic enzymes has now been purified and characterized, which will lead to further work on the identification of their genes, the regulation of their synthesis, their targeting to protein bodies, and any posttranslational processing which is required for their activation.

The cellular changes that precede and accompany proteolysis have been studied most thoroughly in the cotyledons of mung beans (*Vigna radiata*). Here, the major storage protein is vicilin, which comprises some 70–80% of the total, and the enzyme responsible for its hydrolysis is an endopeptidase, vicilin peptidohydrolase; with some participation by a carboxypeptidase. The ER is the site of vicilin peptidohydrolase synthesis, and associated with the synthesis of this enzyme the ER undergoes various modifications. Dry and early-imbibed cotyledons contain tubular ER, which is dismantled some 12–14 h after the start of imbibition. Although, overall, there is a net loss of membrane, at the same time there is a proliferation of a new type of ER, with ribosomes attached and with obvious cisternae. The vicilin peptidohydrolase is synthesized *de novo* on polysomes attached to this new ER, and the enzyme is inserted into the ER lumen and packaged into vesicles. These are transported to the protein bodies, and degradation of the vicilin therein commences only after the peptidohydrolase has been inserted into these organelles. Initially the vicilin is cleaved from 50- to 63-kDa components to 20- to 30-kDa components, and these are then hydrolyzed more slowly.

Eventually the protein bodies come to contain hydrolytic enzymes other than the peptidohydrolase. During protein hydrolysis a *de novo*-synthesized ribonuclease is inserted into the protein body (after synthesis on ER and transport via vesicles). These organelles have other hydrolase activities also: α-mannosidase and a glucosaminidase (for the hydrolysis of mannose and glucosamine residues from vicilin, a glycoprotein), acid phosphatase, phosphodiesterase, and phospholipase D (for digestion of membrane phospholipids). Of these, α-mannosidase is known to be synthesized during development of jack bean (*Canavalia ensiformis*) on the ER and is transported to the protein body, via the Golgi complex, in the same manner as the storage proteins which are accumulating concurrently. The precursor form of the enzyme is posttranslationally cleaved within the protein bodies, and is present in a potentially active form in the dry seed.

As protein digestion proceeds, the emptying storage vesicles fuse to form a large vacuole containing an array of hydrolytic enzymes, analogous in function to the lysosome. Digestion of cell contents by the enzymes of the vacuole is achieved when vesicles are internalized by an autophagic process in which a portion of the cytoplasm is engulfed and sealed off by the protein body membrane (Fig. 7.18). Hence, disintegration of the cotyledons can be achieved by cellular autolysis. Figure 7.19 summarizes the sequence of events in mung bean cotyledons associated with protein, and ultimately cell, hydrolysis. It is likely that this scheme of events has features that are to be found in many dicotyledon storage organs.

7.6.4. Proteinase Inhibitors

Within both monocot and dicot seeds are proteins that specifically inhibit the action of proteinases in animals and, to a lesser extent, in plants. The function

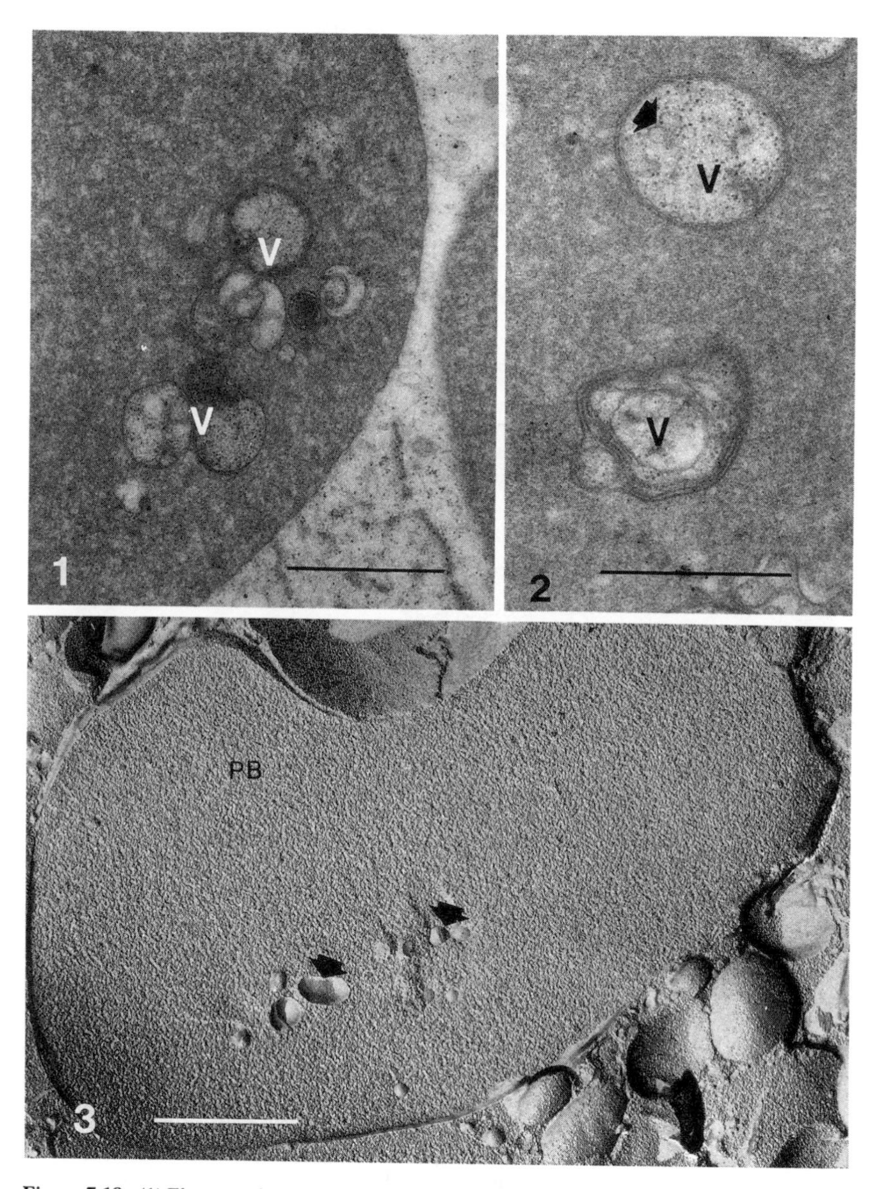

Figure 7.18. (1) Electron micrograph of autophagic vesicles (V) in a protein body of a cotyledon cell of a 3-day-old seedling of mung bean. The vesicles contain sequestered membrane elements and numerous ribosomes. (2) Portion of protein body with two autophagic vesicles (V). The top vesicle contains free ribosomes and membranous vesicles (arrow). The bottom vesicle contains several sequestered membranes. (3) Freeze-fracture replica of a protein body (PB) from a 3-day-old seedling cotyledon cell showing small vesicles (arrows). Bars = 1 μm. Photographs courtesy of M. J. Chrispeels. See also Chrispeels and Jones (1980/81) and Herman *et al.* (1981).

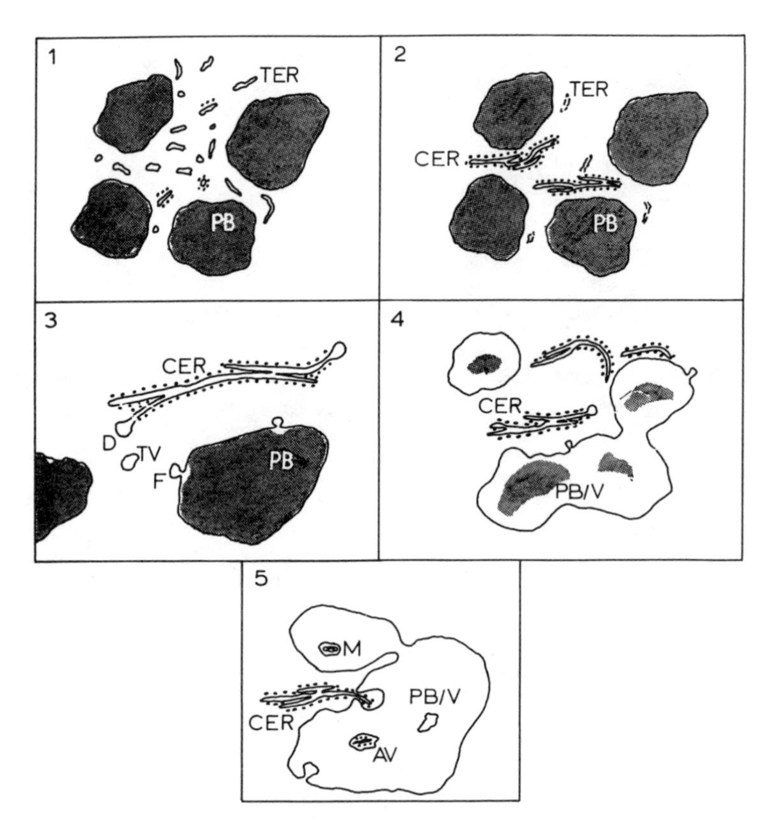

Figure 7.19. Part of a cell from a cotyledon of mung bean illustrating the changes undergone by the protein body and ER during reserve hydrolysis and cell autolysis. (1) Dry state and up to 12 h from the start of imbibition. Intact protein bodies and tubular endoplasmic reticulum (TER) with few ribosomes attached. (2) Starting 12–24 h from imbibition. Dismantling of TER and synthesis of cisternal endoplasmic reticulum (CER) with ribosomes attached. No change to protein bodies. (3) Three to five days from start of imbibition. (Expanded scale to show one protein body.) Vicilin peptidohydrolase synthesized on polysomes attached to newly formed CER and inserted into CER lumen. Dilations (D) of the cisternae form containing the enzymes; these break off as transport vesicles (TV) and carry the peptidohydrolase to the protein bodies, with which they fuse (F). The proteinase commences hydrolysis of the vicilin. Other enzymes, e.g., ribonuclease, start to be inserted into the protein body. (4) As proteins are hydrolyzed the protein bodies coalesce to form large vacuoles (PB/V) and other hydrolytic enzymes are inserted. (5) Autophagic vacuoles (AV) form, engulfing cell contents such as the CER and mitochondria (M). More protein bodies fuse to form a large central vacuole, with autolytic enzymes. Note that this is an illustrative scheme of events and is not drawn to scale. Moreover, organelles and cell structures other than the ER and protein bodies are deliberately omitted for clarity. Based on the studies of Chrispeels and co-workers.

of these inhibitors is incompletely understood, but it could include one or more of the following:

1. Storage. Proteinase (trypsin) inhibitors constitute some 5–10% of the water-soluble proteins of some cereal grains, but in legumes the amounts are much smaller, varying from 0.25 to 3.6 g/kg seed.

2. Control of endogenous proteins. It is now generally accepted that in most seeds it is not the role of the inhibitors to control the activity of endogenous proteinases. In some instances, the evidence is clearly against this. Vicilin peptidohydrolase activity in mung bean cotyledons increases as the concentration of two inhibitors declines (Fig. 7.20). But these two phenomena are not causally related, for fractionation of the cells shows that the peptidohydrolase is associated exclusively with the protein bodies, and the inhibitors are only in the cytosol. Perhaps the inhibitors function to protect the cytoplasm should the proteinase-containing bodies accidentally rupture. In contrast, a metalloproteinase exists in the protein bodies of dry buckwheat (*Fagopyrum esculentum*) seeds, complexed with an inhibitor of its activity. This proteinase–inhibitor complex can be disrupted *in vitro* by divalent cations, allowing enzyme activity. It is suggested that in the initial stages of storage protein hydrolysis in the

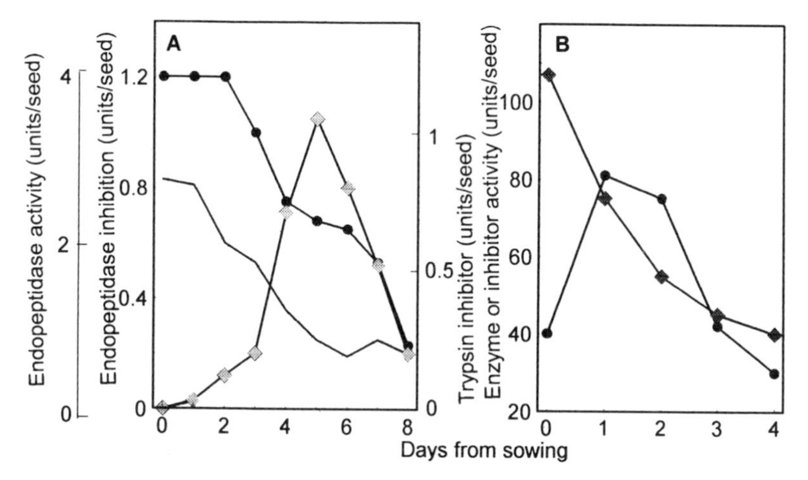

Figure 7.20. Relationship between the presence of proteinase inhibitor and the activity of proteinases of germinated mung bean and buckwheat seeds. (A) Time course of increase of endopeptidase (vicilin peptidohydrolase activity) (♦) in the protein bodies and decline of two proteinase inhibitors, large endopeptidase inhibitor (O) and trypsin inhibitor (●) in the cytosol of the cotyledons of mung bean. (B) Changes in activity of metalloproteinase (●) and its inhibitor (♦) located in protein bodies during the growth of buckwheat seedlings. A, after Baumgartner and Chrispeels (1976); B, after Elpidina *et al.* (1991).

germinated seed, divalent cations are released from phytin as this is mobilized (Section 7.7.2). These cations disrupt the enzyme–inhibitor complex, and metalloproteinase activity increases and initiates the mobilization of the major 13 S globulin storage protein (Fig. 7.20B).

3. Protection or dissuasion. Some proteinase inhibitors can inhibit proteolytic digestive enzymes of invading insects or the secreted proteinases of invading microorganisms. Some inhibitors are bifunctional, and inhibit both proteinases and α-amylases.

7.6.5. Utilization of the Liberated Amino Acids

The major transported forms of amino acids from the storage organs into and throughout the growing seedlings are the amides, i.e., asparagine and glutamine. Hence, the amino acids liberated from storage proteins must be further metabolized, including the conversion of amino nitrogen to amido nitrogen. The extent to which the carbon skeletons of liberated amino acids undergo interconversions can be seen by comparing the amino acid composition of the cotyledon exudate of mung beans (i.e., the translocated amino acids) with that of the major storage protein, vicilin (Table 7.4). Of particular interest is the increase in aspartyl amino acids in the exudate, with the predominance of asparagine as the transported amino acid being evident. The synthesis of asparagine involves the donation of an amino group from the glutamine in a reaction catalyzed by the ATP-dependent enzyme asparagine synthetase (AS) as follows:

$$\text{Aspartate} + \text{glutamine} + \text{ATP} \xrightarrow{\text{Asparagine synthetase}} \text{Asparagine} + \text{glutamate} + \text{AMP} + PP_i$$

Glutamine itself is formed from glutamate as follows:

$$\text{Glutamate} + NH_3 + \text{ATP} \xrightarrow{\text{Glutamine synthetase}} \text{Glutamine} + \text{ADP} + P_i$$

and glutamine can donate its amino group to form glutamate from α-ketoglutarate using the enzyme GOGAT (glutamate synthetase):

$$\text{Glutamine} + \alpha\text{-ketoglutarate} + \text{NAD(P)H} \xrightarrow{\text{GOGAT}} 2\ \text{Glutamate} + \text{NAD(P)}$$

Alternatively, glutamate:NAD(P) oxidoreductase (deaminating), also known as glutamate dehydrogenase (GDH), can add ammonia to α-ketoglutarate to yield glutamate in the presence of NAD(P)H.

The various fates of the amino acids released from storage proteins in relation to their transport and subsequent utilization in the growing seedling are detailed in Fig. 7.21.

In cotyledons of various legume seeds, not surprisingly, there is an increase in the activity of the aforementioned enzymes involved in glutamate, glutamine, and asparagine synthesis at a time when the major protein reserves are being hydrolyzed. Hence, as amino acids are being liberated from the stored form they undergo the appropriate conversions to the readily transportable asparagine. GDH, glutamine synthetase, and AS are present only in low amounts in the cotyledons of mature dry cotton seed (*Gossypium hirsutum*) (Fig. 7.22) but increase appreciably during the first 2 days from the start of imbibition (perhaps by *de novo* synthesis), peaking approximately at the time of, or just prior to the commencement of, protein mobilization. In this seed, as in the legumes, the major transport form of amino acid is asparagine. The pattern of amino acid

Table 7.4. Comparison of the Amino Acid Composition of the Cotyledon Exudate from Cotyledons after 4–5 Days from the Start of Imbibition, and the Composition of the Major Storage Glycoprotein Fraction (Vicilin)[a]

	Amino acid composition (mole%)	
	Cotyledon exudate	Vicilin
Asp	4.2	13.4[b]
Asn	31.3	
Thr	4	3
Ser	3.8	7.3
Glu	3.8	
Gln	2.0	19.9[b]
Pro	5.9	2.9
Gly	0.2	5.4
Ala	2.1	5.6
Val	8.9	6.6
Ile	4.8	4.5
Leu	6.7	9.4
Tyr	2	1.9
Phe	5.7	6.1
Lys	4.8	6
His	4.6	2
Arg	4.8	5.5

[a]Data taken from Ericson and Chrispeels (1973) and Kern and Chrispeels (1978).
[b]Amino acid and amide not determined separately in the vicilin fraction.

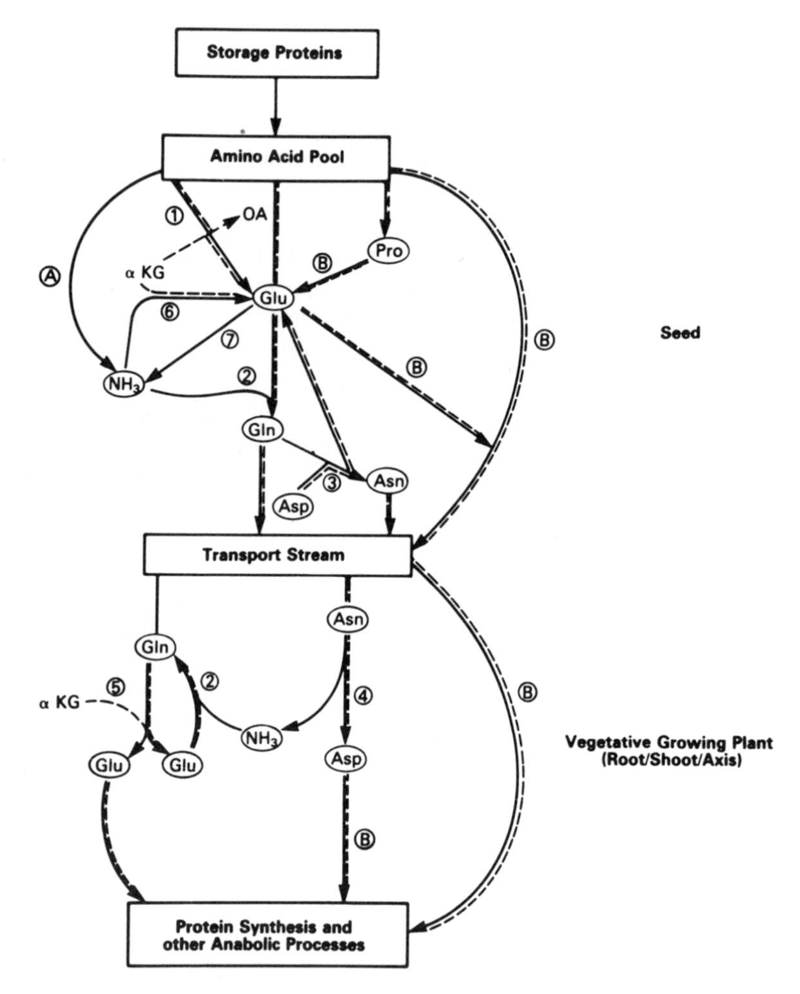

Figure 7.21. The fate of amino acids liberated by storage protein hydrolysis, with emphasis on the fact that glutamine and asparagine are the major transport amino acids. Enzymes: (1) aminotransferase; (2) glutamine synthetase; (3) asparagine synthetase; (4) asparaginase; (5) GOGAT; (6) glutamate dehydrogenase; (7) deaminase. Reactions: (A) specific deaminations; (B) direct interconversions of amino acid skeletons, or direct transfer without interconversion. Compounds: Glu, glutamic acid; Gln, glutamine; Asp, aspartic acid; Asn, asparagine; NH_3, ammonia; Pro, proline (high in amino acid pool of cereals when storage prolamins are broken down); OA, oxaloacetic acid; α KG, α-ketoglutaric acid. Solid lines show the path of N, and dashed lines the path of C. Based on Miflin *et al.* (1981) and P. J. Lea (personal communication).

metabolism in pea (*Pisum sativum*) cotyledons is unusual in that a major transported form of amino acid, and the one that accumulates in the cotyledon, is homoserine. Glutamine is another transport form of amino acid, but not asparagine to any great extent.

In castor bean seeds the site of protein storage is the endosperm, and the predominant form of transported nitrogen from this region is glutamine. Some of the amino acids released from the storage protein by hydrolysis, e.g., aspartate, glutamate, alanine, glycine, and serine, can be converted to sucrose and transported as the sugar. The amide nitrogen derived from the deamination of these gluconeogenic amino acids is probably used in the production of glutamine. By comparison, amino acids that are not gluconeogenic are probably transported unchanged to the growing seedling; some might undergo modifications of their carbon skeleton to form glutamate.

The major transported amino acid from the endosperm of castor bean, glutamine, is taken up into the cotyledons by an active process, even against a concentration gradient. In species where the cotyledons are the storage organs, these are connected to the growing regions by a network of vascular bundles, and the products of hydrolysis must be translocated to the axis largely within the phloem. Loading of amino acids into the translocation stream might be aided in some species, e.g., broad bean (*Vicia faba*), by the presence of transfer cells that

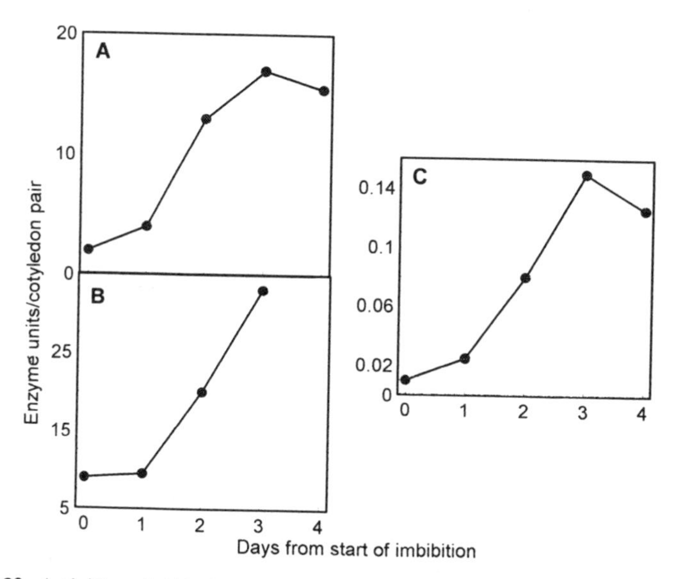

Figure 7.22. Activities of (A) glutamate dehydrogenase, (B) glutamine synthetase, and (C) asparagine synthetase in the cotyledons of cotton. After Dilworth and Dure (1978).

border the xylem and phloem; these are specialized cells with an increased surface area to aid the transport of solutes over short distances. In other species, e.g., mung bean (*Vigna radiata*), there are no transfer cells, but the parenchyma cells adjacent to the phloem have extensive evaginations of the plasmalemma to form fine tubules (plasmalemmasomes): these also serve to increase the surface area for transport.

7.7. STORED PHOSPHATE CATABOLISM

7.7.1. General Catabolism

Phytic acid (*myo*-inositol hexaphosphate) is the major phosphate reserve in many seeds, and since in its storage form it is a mixed salt with such elements as K^+, Mg^{2+}, and Ca^{2+} (and as such is called phytin or phytate), it is also a major source of these macronutrient elements in the seed.

Phytase hydrolyzes the phytin to release phosphate, its associated cations, and *myo*-inositol. Breakdown of the phytin is rapid and complete, for *myo*-inositol phosphate esters with fewer than six phosphate groups do not accumulate within seeds. The released *myo*-inositol may be used by the growing seedling for cell-wall synthesis, since this compound is a known precursor of pentosyl and uronosyl sugar units normally associated with pectin and certain other cell wall polysaccharides.

Lipid phosphate, protein phosphate, and nucleic acid phosphate occur in smaller amounts in seeds. Phospholipids and phosphoproteins are probably dephosphorylated during their hydrolysis (acid phosphatases may play a role here); the free phosphate is translocated to the growing axis, but the lipid and protein moieties are catabolized *in situ* as outlined in previous sections.

7.7.2. Phosphate Catabolism in Seeds

Mobilization of phosphate-containing compounds has been studied in most detail in oat (*Avena sativa*) grains. In Fig. 7.23, the changes in various phosphate fractions over the first 8 days from the start of imbibition are shown. Essentially the results indicate that the most abundant source of phosphate in the dry grain is phytin (Fig. 7.23, acid soluble-P) in the endosperm (actually the aleurone layer; see later), which accounts for over 50% of the total seed phosphorus. Phytin, along with other phosphate-containing fractions, declines in the endosperm after about 2 days, accompanied by a rise in various phosphate fractions within the growing axis (Fig. 7.23). This is suggestive of a release of phosphate from its storage form in the endosperm, its transport to the axis, and its reutilization there for the synthesis of essential phosphate-containing compounds. Phytin does not accumulate in the axis at any time.

The largest store of phytin in the cereal grain is within the aleurone layer, where it is present as globoids in the aleurone grains. No phytin is associated with the protein bodies of the starchy endosperm, although a little is present in protein bodies in the scutellum. Generally, phytase activity is quite high in the mature dry cereal grain and appears to be closely associated with the aleurone grains. The enzyme increases in activity after germination, although it has been calculated that sufficient enzyme is present in the dry aleurone layer to hydrolyze all of the stored phytate; hence, the significance of the increase is not known.

Changes in the phosphate fractions in the cotyledons and axes of the legumes exhibit a pattern similar to that shown already for cereals. That is, there is a decline in total phosphate (comprised largely of phytin, phospholipids, phosphoproteins, and nucleic acids) in the cotyledons after germination, coincident with an increase in phytase and acid phosphatase activity. Phytase activity

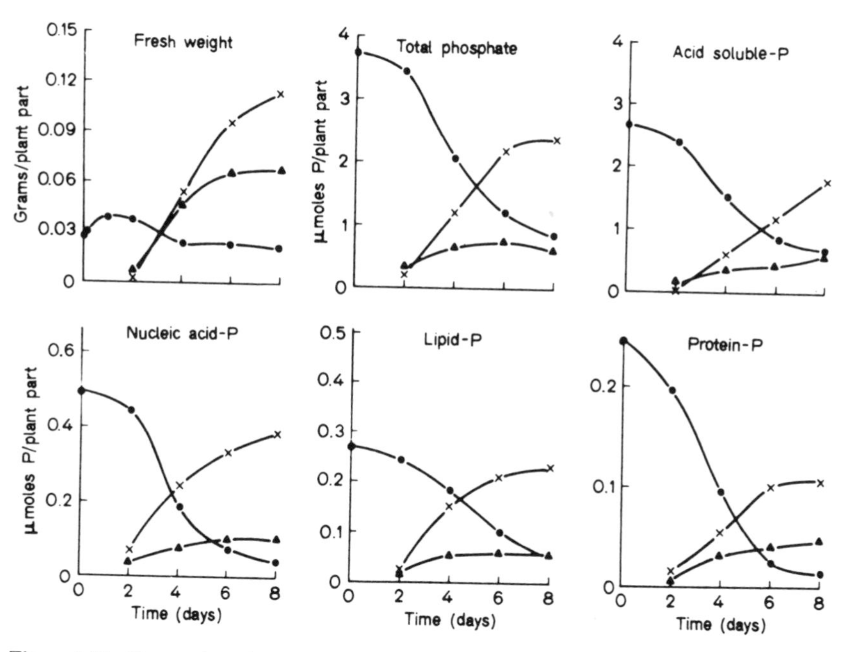

Figure 7.23. Changes in various phosphate components in the roots (▲), shoots (×), and endosperm (●) of germinated oat grains. The phosphate-containing region of the endosperm is the aleurone layer. Acid-soluble phosphate in the endosperm is mostly phytic acid, but in the shoot and root it is inorganic phosphate and some acid-soluble organic phosphate, but not phytic acid. After Hall and Hodges (1966).

is low in the dry seed of dicots and increases with time following imbibition (Fig. 7.24), particularly during the postgerminative mobilization of phytin and protein. After the catabolites of the phosphate-containing compounds are transported to the axis, they are used in the synthesis of cell components such as phospholipids and nucleotides. There is little accumulation of free inorganic phosphate within the cotyledons during the degradation of phosphate-containing compounds, but this released phosphate can account for up to 50% of that present in the growing axis. Interestingly, the germinated embryo may retain its capacity to synthesize phytin, for when isolated embryos from mature castor bean seeds are incubated in phosphate solutions there is an increase in phytin, especially within the cotyledons. The site of phytin deposition is not known, however, but it is not within the protein bodies. Temporary storage of phosphorus in the form of phytin might be a way of conserving this important metabolite during early seedling establishment.

7.8. MOBILIZATION OF NUCLEIC ACIDS FROM THE STORAGE REGIONS

Changes in the nucleic acid phosphate fraction in the endosperm and growing regions of the germinated oat grain are shown in Fig. 7.23. Here, as in other cereals, nucleic acids account for less than 10% of the stored phosphate: most is present in the aleurone layer, with little in the starchy endosperm. Neither the endosperm (including the aleurone layer) nor scutellum can supply enough of the nucleotide precursors required for RNA and DNA synthesis within the

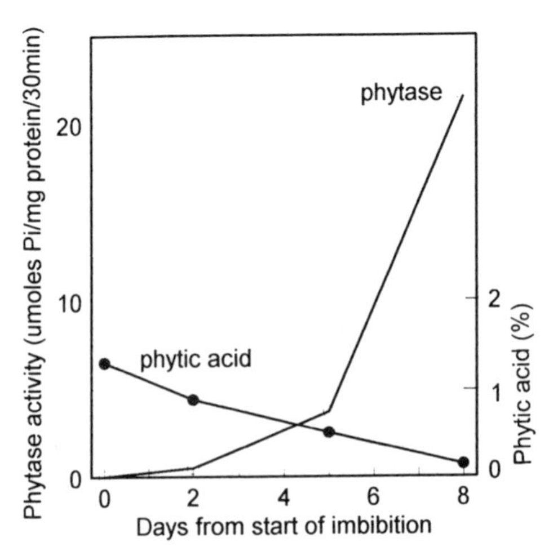

Figure 7.24. Changes in phytic acid content (●) and phytase activity (○) in germinated rapeseed (*Brassica napus* cv. Regent). After Lu *et al.* (1987).

axis. These are provided by *de novo* synthesis, the nitrogen source being the amino acids present in, or transported to, the embryo.

In legume seeds also there is a depletion of RNA and DNA in the cotyledons, particularly during the final disintegration of this tissue. Ribonucleases (and presumably deoxyribonucleases) which increase from a low activity in the dry seed, as mentioned in Section 7.6.3, may be deposited in the protein bodies as they assume their hydrolytic function. The nucleotide products of ribonuclease activity do not accumulate in the cotyledons but are translocated to the growing axis where they may support nucleic acid synthesis. But, as with cereals, the increase in axis RNA is greater than the decline in cotyledonary RNA, indicating net nucleotide synthesis using an alternative nitrogen source.

USEFUL LITERATURE REFERENCES

SECTIONS 7.1–7.4

Ap Rees, T., 1974, in: *Plant Biochemistry. Biochemistry, Series One,* Volume 11 (H. L. Kornberg and D. C. Phillips, eds.), Butterworths, London, pp. 89–127 (carbohydrate catabolism pathways).

Ashford, A. E., and Gubler, F., 1984, in: *Seed Physiology, Volume 2. Germination and Reserve Mobilization* (D. R. Murray, ed.), Academic Press, New York, pp. 117–162 (polysaccharide mobilization in endosperm).

Beck, E., and Ziegler, P., 1989, *Annu. Rev. Plant Physiol. Plant Mol. Biol.* **40:**95–117 (starch metabolism).

Bewley, J. D., Leung, D. W. M., MacIsaac, S., Reid, J. S. G., and Xu, N., 1993, *Plant Physiol. Biochem.* **31:**483–490 (transient starch accumulation in fenugreek cotyledons).

Brown, H. T., and Morris, G. H., 1890, *J. Chem. Soc.* **57:**458–528 (classic studies on cereal reserve mobilization).

Chandra Sekhar, K. N., and DeMason, D. A., 1990, *Planta* **181:**53–61 (mobilization of reserves from date palm endosperm).

Duffus, C. M., 1984, in: *Storage Carbohydrates in Vascular Plants* (D. H. Lewis, ed.), Cambridge University Press, Cambridge, pp. 321–352 (metabolism of reserve starch).

Edelman, J., Shibko, S. I., and Keys, A. J., 1959, *J. Exp. Bot.* **10:**178–189 (sucrose transport and scutellum).

Edwards, M., Dean, I. C. M., Bulpin, P. V., and Reid, J. S. G., 1985, *Planta* **163:**133–140 (xyloglucan mobilization from nasturtium cotyledons).

Fincher, G. B., 1989, *Annu. Rev. Plant Physiol. Plant Mol. Biol.* **40:**95–117 (endosperm mobilization in cereals).

Fincher, G. B., and Stone, B. A., 1974, *Aust. J. Plant Physiol.* **1:**297–311 (hydrolysis in wheat endosperm).

Gibbons, G. C., 1979, *Carlsberg Res. Commun.* **44:**353–356 (amylase in barley).

Halmer, P., 1985, *Physiol. Veg.* **23:**107–125 (review of starch mobilization).

Jones, R. L., and Jacobsen, J. V., 1991, *Int. Rev. Cytol.* **126:**49–87 (mobilizing enzymes in cereals).

Juliano, B. O., and Varner, J. E., 1969, *Plant Physiol.* **44:**886–892 (starch breakdown in pea).

Lauriere, C., Doyen, C., Thevenot, C., and Daussant, J., 1992, *Plant Physiol.* **100:**887–893 (β-amylases in cereals).

Longstaff, M. A., and Bryce, J. H., 1993, *Plant Physiol.* **101:**281–289 (limit dextrinases in barley).
Manners, D., 1985, in: *Biochemistry of Storage Carbohydrates* (P. M. Dey and R. A. Dixon, eds.), Academic Press, New York, pp. 149–204 (review of starch and its metabolism).
Palmer, G. H., 1982, *J. Inst. Brew.* **88:**145–153 (scutellar α-amylase).
Reid, J. S. G., 1971, *Planta* **100:**131–142 (hydrolysis of the fenugreek endosperm).
Reid, J. S. G., and Bewley, J. D., 1979, *Planta* **147:**145–150 (dual role of the fenugreek endosperm).
Zambou, K., Spyropoulos, C. G., Chinou, I., and Kontos, F., 1993, *Planta* **189:**207–212 (saponin-like inhibitors in fenugreek endosperm).

SECTION 7.5

Behrends, W., Thieringer, R., Engeland, K., Kanau, W.-H., and Kindl, H., 1988, *Arch. Biochem. Biophys.* **263:**170–177 (β-oxidation of unsaturated FFA).
Breidenbach, R. W., and Beevers, H., 1967, *Biochem. Biophys. Res. Commun.* **27:**462–469 (discovery of glyoxysome).
Chapman, K. D., and Trelease, R. N., 1991, *J. Cell Biol.* **115:**995–1007 (glyoxysome enlargement in cotyledons).
Donaldson, R. P., and Fang, T. K., 1987, *Plant Physiol.* **85:**792–795 (recycling of NAD/NADH in glyoxysomes).
Gerhardt, B., 1991, in: *Molecular Approaches to Compartmentation and Metabolic Regulation* (A. H. C. Huang and L. Taiz, eds.), ASPP, Rockville, Md., pp. 121–128 (pathways of lipid catabolism).
Gerhardt, B., 1992, *Prog. Lipid Res.* **31:**417–446 (biochemistry of fatty acid degradation).
Gietl, C., 1991, in: *Molecular Approaches to Compartmentation and Metabolic Regulation* (A. H. C. Huang and L. Taiz, eds.), ASPP, Rockville, Md., pp. 138–150 (targeting of MDH to glyoxysomes).
Gonzalez, E., 1986, *Plant Physiol.* **80:**950–955 (formation of glyoxysomes in castor bean endosperm).
Halpin, C., Conder, M. J., and Lord, J. M., 1989, *Planta* **179:**331–339 (import of matrix proteins into castor bean seed glyoxysomes).
Hills, M. J., and Beevers, H., 1987, *Plant Physiol.* **84:**272–276 (lipases and oil mobilization in castor bean endosperm).
Huang, A. H. C., 1975, *Plant Physiol.* **55:**555–558 (glycerol metabolism).
Muto, S., and Beevers, H., 1974, *Plant Physiol.* **54:**23–28 (triacylglycerol hydrolysis in castor bean endosperm).
Olsen, L. J., and Harada, J. J., 1991, in: *Molecular Approaches to Compartmentation and Metabolic Regulation* (A. H. C. Huang and L. Taiz, eds.), ASPP, Rockville, Md., pp. 129–137 (mRNA and glyoxysome enzyme synthesis in cotyledons).
Opute, F. I., 1975, *Ann. Bot.* **39:**1057–1061 (oil palm triacylglycerol catabolism).
Trelease, R. N., and Doman, D. C., 1984, in: *Seed Physiology. Volume 2. Germination and Reserve Metabolism* (D.R. Murray, ed.), Academic Press, New York, pp. 201–245 (overview of triacylglycerol catabolism in seeds).

SECTION 7.6

Baumgartner, B., and Chrispeels, M. J., 1976, *Plant Physiol.* **58:**1–6 (mung bean proteinase inhibitors).
Chrispeels, M. J., and Jones, R. L., 1980/81, *Isr. J. Bot.* **29:**225–245 (endoplasmic reticulum and reserve hydrolysis).
Dilworth, M. F., and Dure, L., III, 1978, *Plant Physiol.* **61:**698–702 (asparagine and glutamine synthesis).

Elpidina, E. N., Voskoboynikova, N. E., Belozersky, M. A., and Dunaevsky, Y. E., 1991, *Planta* **185:**46–52 (proteinase inhibitor in buckwheat).

Ericson, M. C., and Chrispeels, M. J., 1973. *Plant Physiol.* **52:**98–104 (mung bean storage proteins).

Faye, L., Greenwood, J. S., Herman, E. M., Sturm, A., and Chrispeels, M. J., 1988, *Planta* **174:**271–282 (α-mannosidase synthesis).

Hara, I., and Matsubara, H., 1980, *Plant Cell Physiol.* **21:**219–232 (pumpkin seed globulins).

Herman, E. M., Baumgartner, B., and Chrispeels, M. J., 1981, *Eur. J. Cell Biol.* **24:**226–235 (autophagic vacuoles).

Kern, R., and Chrispeels, M. J., 1978, *Plant Physiol.* **62:**815–819 (amides in mung bean).

Kobrehel, K., Wong, J. H., Balogh, A., Kiss, F., Yee, B.C., and Buchanan, B. B., 1992, *Plant Physiol.* **99:**919–924 (thioredoxin and reduction of wheat storage proteins).

Larson, L. A., and Beevers, H., 1965, *Plant Physiol.* **40:**424–432 (homoserine in peas).

Miflin, B. J., Wallsgrove, R. M., and Lea, P. J., 1981, *Curr. Top. Cell. Regul.*, **20:**1–43 (glutamine metabolism in plants).

Mitsuhashi, W., and Oaks, A., 1994, *Plant Physiol.* **104:**401–407 (endopeptidases in maize endosperm).

Reilly, C. C., O'Kennedy, B. T., Titus, J. S., and Splittstoesser, W. E., 1978, *Plant Cell Physiol.* **19:**1235–1246 (pumpkin globulin solubilization).

Richardson, M., 1991, *Methods Plant Biochem.* **5:**259–305 (proteinase inhibitors in seeds).

Salmenkallio, M., and Sopanen, T., 1989, *Plant Physiol.* **89:**1285–1291 (amino acid and peptide uptake by cereal scutella).

Shutov, A. D., and Vaintraub, I. A., 1987, *Phytochemistry* **26:**1557–1566 (review of proteinase A and B activities).

Stewart, C. R., and Beevers, H., 1967, *Plant Physiol.* **42:**1587–1595 (castor bean amino acids).

Van Der Wilden, W., Herman, E. M., and Chrispeels, M. J., 1980, *Proc. Natl. Acad. Sci. USA* **77:**428–432 (autophagic vacuoles).

Walker-Smith, D. J., and Payne, J. W., 1985, *Planta* **164:**550–556 (synthesis of carriers in barley scutellum).

Wilson, K. A., 1986, in: *Plant Proteolytic Enzymes*, Volume 2, (M. J. Dalling, ed.), CRC Press, Boca Raton, Fla., pp. 19–47 (review of protein mobilization in dicot seeds).

Wrobel, R., and Jones, B. L., 1992, *Plant Physiol.* **100:**1508–1516 (proteolytic enzymes in germinated barley grains).

SECTIONS 7.7 AND 7.8

Hall, J. R., and Hodges, T. K., 1966, *Plant Physiol.* **41:**1459–1464 (P metabolism in oats).

Lu, S.-Y., Kim, H., Eskin, N. A. M., Latta, M., and Johnson, S., 1987, *J. Food Sci.* **52:**173–175 (phytases in *Brassica* cultivars).

Maiti, I. B., and Loewus, F. A., 1978, *Planta* **174:**513–517 (*myo*-inositol metabolism).

Organ, M. G., Greenwood, J. S., and Bewley, J. D., 1988, *Planta* **174:**513–517 (phytin synthesis in germinated embryos).

Walker, K. A., 1974, *Planta* **116:**91–98 (phytin in development and germination).

Chapter 8

Control of the Mobilization of Stored Reserves

Most of the work on the control of mobilization of stored reserves has been carried out on cereal grains, and much less is known about the control mechanisms in dicot seeds. For these reasons the following discussion rests heavily on the cereal grain but a consideration of the situation in dicot seeds is included.

8.1. CONTROL OF RESERVE MOBILIZATION IN CEREALS

Modification of barley endosperm by mobilization is the basis of an important industrial process—malting. For this reason, the control of mobilization in grains of this species has long been a topic for investigation. That degradation of the barley endosperm is markedly slowed down or even prevented by removal of the embryo has been known since the end of the last century. But only about 1960 was it appreciated that the role of the embryo is to release a diffusible promotive factor that stimulates hydrolytic events within the endosperm. The circumstantial evidence strongly suggests that this diffusible factor is the regulator, or hormone, gibberellin. Certainly, embryoless grains of many cereal cultivars can be stimulated to commence hydrolytic events by application of gibberellins, with liquefaction of the endosperm occurring virtually as in the intact kernel. And, moreover, it is well established that germinated cereal embryos produce and secrete gibberellin from the scutellum. Notwithstanding this favorable evidence, there is still no absolute, unequivocal demonstration that gibberellin is the diffusible factor, and some researchers have claimed that it is not. The central, regulatory role of gibberellin is not denied, but suggested alternatives include the possibility that another factor induces gibberellin production by the aleurone cells or that these cells become more sensitive to residual gibberellin within them. Regulation of the production of several mobilizing enzymes by the aleurone layer cells of certain cultivars of wheat, barley, maize, and wild oats is, nevertheless, ultimately exerted by gibberellin. A cautionary note, however, is that grains of several cultivars and harvests do not exhibit a clear-cut response to gibberellin and some may synthesize α-amylase without any apparent requirement for this hormone.

8.1.1. Gibberellin and the Induction of α-Amylase Synthesis

Starch is the major stored reserve found within the cereal endosperm. Not unexpectedly, the enzyme that has been most studied in relation to the control of reserve mobilization is the one largely responsible for its hydrolysis, i.e., α-amylase. Much of our understanding of the mode of action of gibberellin has come from work using deembryonated grains or isolated aleurone layers, i.e., those that have been dissected out of the imbibed grain and then incubated in a sterile medium, to which appropriate additions (e.g., gibberellic acid, GA_3) have been made. Isolated aleurone layers of barley incubated in GA_3 solution begin to secrete α-amylase about 8 h after being introduced to the hormone; this release proceeds linearly over the next 16 h (Fig. 8.1). The aleurone layer itself accumulates only a little enzyme and this might be external to the protoplast, within the cell walls. ABA, an antagonist of gibberellin action in many plant systems, is inhibitory to α-amylase production (Fig. 8.1).

Protoplasts isolated from the aleurone cells of barley, wild oat, and wheat respond similarly to GA_3 which is exploited in studies of the molecular basis of regulation by the hormone.

There is now overwhelming evidence that the increase in α-amylase activity is a consequence of *de novo* synthesis of this enzyme. However, there is no accumulation of an inactive precursor or zymogen of α-amylase during its synthesis—the mature, active enzyme is synthesized directly.

Because of the considerable interest in the action of gibberellins at the molecular level, many studies have been conducted to determine how GA_3 induces the synthesis of α-amylase. Consequently, the events of the lag period

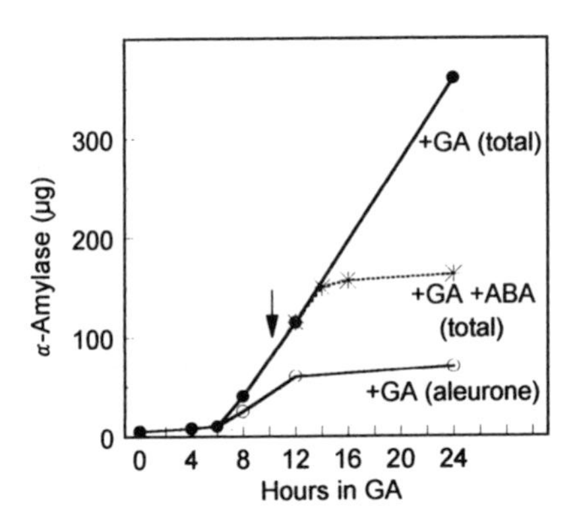

Figure 8.1. Time course of α-amylase synthesis and release by aleurone layers of barley cv. Himalaya isolated from 3-day-old water-imbibed embryoless barley grains and incubated with 1 μM GA_3 and 5 μM ABA. Enzyme activity measured in the medium surrounding the aleurone layers and in the supernatant of a 0.2 M NaCl extract of the aleurone layer (aleurone). Total (for GA, and GA + ABA) refers to the sum of these two activities. Arrow points to the time of ABA application to the GA_3-treated aleurone layers. After Chrispeels and Varner (1967a,b).

between GA_3 addition and the start of α-amylase production (about 8 h in barley; Fig. 8.1) have received much attention. Changes occur in the internal organization of the aleurone layer cells during this period; e.g., at least 2 h before the lag period is completed, the protein-storing aleurone grains lose their spherical appearance and undergo distinct volume and shape changes. This is probably because of the hydrolysis of the proteins within these grains to provide amino acids for the *de novo* synthesis of α-amylase and other enzymes.

An attractive hypothesis, which was popular for some time, was that GA_3 induces an increase in the overall protein-synthesizing capacity of the aleurone layer cells, and that α-amylase is made on newly formed membrane-bound polysomes. Evidence seemed to suggest that GA_3 induced an increase in polysomes during the lag phase, as well as a proliferation of membranes—the polysomes then associated with the membranes, resulting in an increase in rough endoplasmic reticulum. It is now evident, however, that GA_3 does not enhance total lipid, total lipid phosphorus, or membrane phospholipid in aleurone layer cells up to and beyond the time of α-amylase synthesis, nor does it induce an increase in polysomes. The original claims unfortunately were based on results obtained using inadequate techniques.

The addition of GA_3 to isolated aleurone layers causes a qualitative rather than quantitative shift in the pattern of protein synthesis. After 10 h the major protein that is synthesized and secreted (Fig. 8.2) is α-amylase, and later this enzyme accounts for up to 70% of total protein synthesized *de novo* by the aleurone layer. An important question in relation to gibberellin action is, therefore, whether this hormone acts at the nuclear level to induce transcription of the gene for α-amylase. Is the lag period, in fact, the time required for the synthesis of the mRNA for this enzyme? That this indeed is the case was first shown by translating the mRNA extracted at different times from barley aleurone tissue incubated with GA_3 (Fig. 8.3). The preparation of cDNA clones of α-amylase and other gene products provided probes for detecting and quantifying the mRNA induced by GA, and it became clear that a number of different messages are produced by the cells in response to the hormone. Moreover, the promotive effect of GA in some cases can be suppressed by ABA, in accordance with physiological findings (Figs. 8.1, 8.2). The action of GA_3 and ABA on accumulation of α-amylase mRNA is shown in Fig. 8.4. Nuclei isolated from aleurone cell protoplasts of barley and wild oat produce α-amylase mRNA, provided the protoplasts have previously been treated with GA. These nuclear "runoff" experiments show that the GA-induced accumulation of α-amylase mRNA within the cells is indeed the result of enhanced transcription and not because of the prevention by GA of degradation of messages that are independently produced. These observations suggest that, in barley aleurone cells, α-amylase messages start to be transcribed within just a few hours of the addition of GA_3; these are translated over a subsequent 10-h period. In two other well-documented

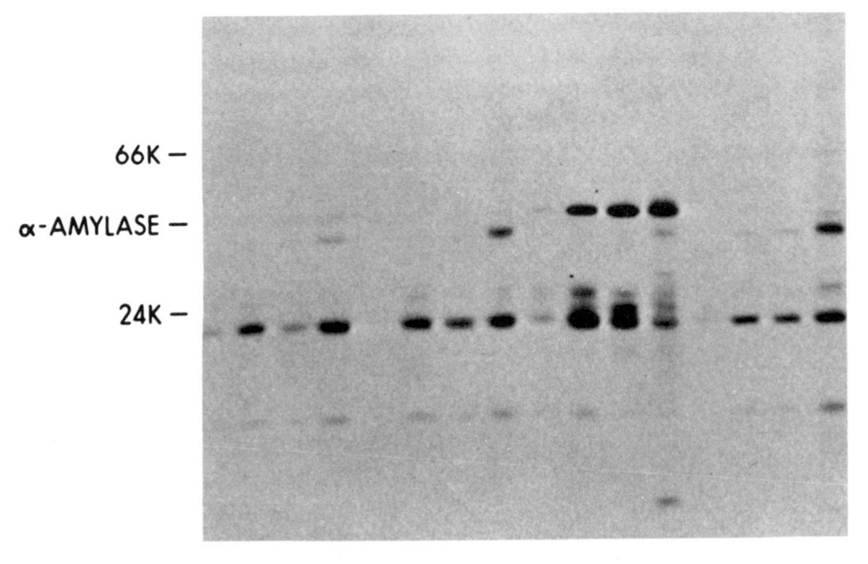

Figure 8.2. Fluorogram of a polyacrylamide gel of *in vivo* ^{35}S-methionine-labeled proteins (i.e., *de novo* synthesized proteins) secreted into the medium of barley aleurone layers incubated for the indicated number of hours with no hormones, with ABA, with GA_3, or with both. The locations of two protein standards of 66 and 24 kDa and of purified α-amylase are indicated on the left. Note that α-amylase is produced and secreted only in GA_3-treated aleurone layers. After Mozer (1980).

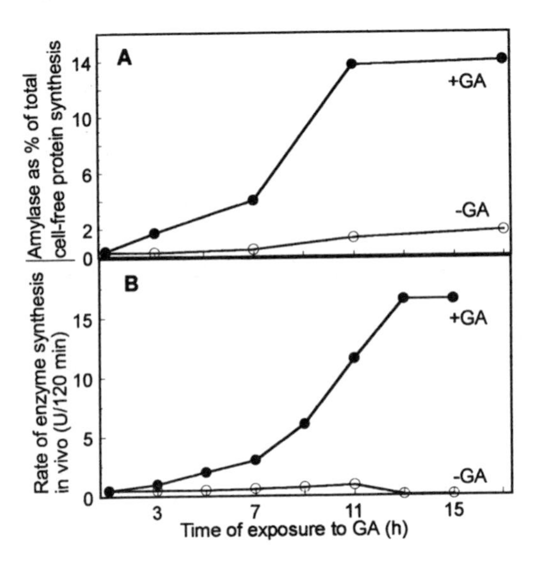

Figure 8.3. The increase with time of (A) the amount of translatable messenger RNA for barley α-amylase and (B) the rate of synthesis of the enzyme *in vivo* in response to GA_3. U, units of α-amylase activity. For (A) the poly(A) RNA was extracted from aleurone layers treated with GA_3 for different time periods and used to support α-amylase synthesis *in vitro*. After Higgins *et al.* (1976).

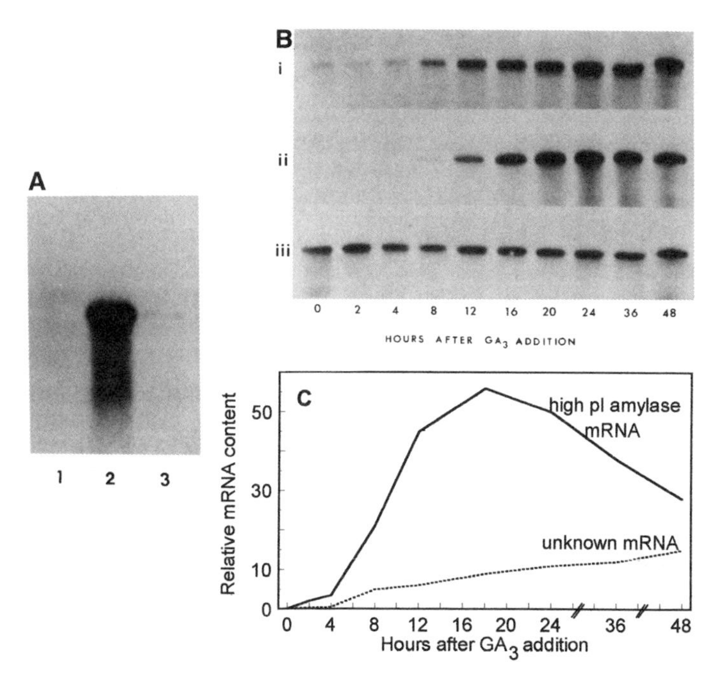

Figure 8.4. GA-stimulated accumulation of mRNA by barley aleurone layers. (A) Aleurone layers were treated with (lane 2) or without (lane 1) 1 μM GA_3, and with 1 μM GA_3 plus 25 μM ABA (lane 3) for 24 h, when the mRNA was extracted. Hybridization was with a cDNA probe for a high-pI amylase. No mRNA accumulated without treatment with GA_3, and GA_3-stimulated accumulation was blocked by ABA. (B) The time course of mRNA accumulation in barley aleurone layers treated with 1 μM GA_3. Line i, an unknown GA-stimulated mRNA; line ii, mRNA for a high-pI α-amylase; line iii, ribosomal RNA (not affected by GA_3). (C) Quasi-quantification of the mRNA content at different times. Adapted from Chandler *et al.* (1984).

cases of GA-induced α-amylase synthesis, wheat and wild oat, the lag period between application of the hormone and appearance of the enzyme is some hours longer. The newly synthesized α-amylase in barley (and wheat) has a slightly higher molecular weight than the secreted enzyme. The larger α-amylase molecule is not a zymogen, but a precursor, which is present transiently until the signal peptide (see Section 2.3.4.1) is cleaved from it.

At this point it should be noted that α-amylase in fact consists of several isoforms separable from each other by methods such as isoelectric focusing (Fig. 8.5). In wheat and barley, most of the isoenzymes fall into two groups, the low pI (AMY I) and high pI (AMY 2), having pIs of approximately 4.5–5.5 and 5.9–6.9, respectively. A third group, AMY 3, is found in wheat and barley and

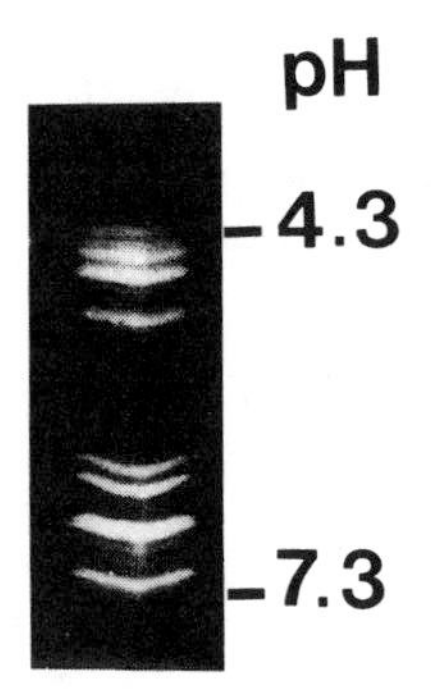

Figure 8.5. α-Amylase isoenzymes from wheat embryos. Isoenzymes in extracts of wheat embryos were separated by isoelectric focusing in a polyacrylamide gel containing β-limit dextrin. After staining with iodine the presence of α-amylase activity is revealed as clear bands. From Garcia-Maya *et al.* (1990).

represents complexes of some of the high-pI isoforms with an α-amylase inhibitor. Several of the isoforms are produced by posttranslational modification of others. Fewer isoforms are found in maize, sorghum, rice, and oat, and they are mostly of the low-pI group. In barley, about eight α-amylase genes have been identified, about three for the low-pI group and five for the high-pI forms. Eight different mRNAs are found, corresponding with these genes. The locations of the low- and high-pI genes are known for both wheat and barley, and in both cases two chromosomes (or groups of chromosomes in the hexaploid wheat) are involved.

The production of the α-amylase isoenzymes is differentially affected by GA. In response to GA_3 the low-pI forms generally appear first in isolated aleurone tissue or aleurone layers of barley; in fact, low amounts of the low-pI group can be produced even in the absence of GA. In addition, the dose–response curves for the two groups are usually different, and rather higher concentrations of GA are required to induce the appearance of the high-pI isoenzymes. Overall, the activity attained by the high-pI group in response to GA is greater than that of the low-pI isoenzymes. In one study, GA induced about a 100-fold increase in the high-pI activity and approximately 20-fold in the low-pI. These differences are matched by the mRNAs, the messages for the high-pI forms appearing later, to a higher final value, and requiring higher concentrations of GA for induction. It must be said, however, that this pattern is not consistent and in the case of isolated protoplasts and the intact, germinated grains, differences in timing and sensitivity can be found which have not yet been explained. These considerations should enable the reader to recognize some of the complexities inherent in the regulation of α-amylase production of GA.

Synthesis of α-amylase by aleurone tissue and scutella (of rice) is also dependent on calcium ions; hence, in all experiments with isolated aleurone layers or protoplasts, Ca^{2+} must be supplied at millimolar concentrations. Because α-amylase is a Ca^{2+}-containing metalloprotein (one Ca^{2+} per molecule),

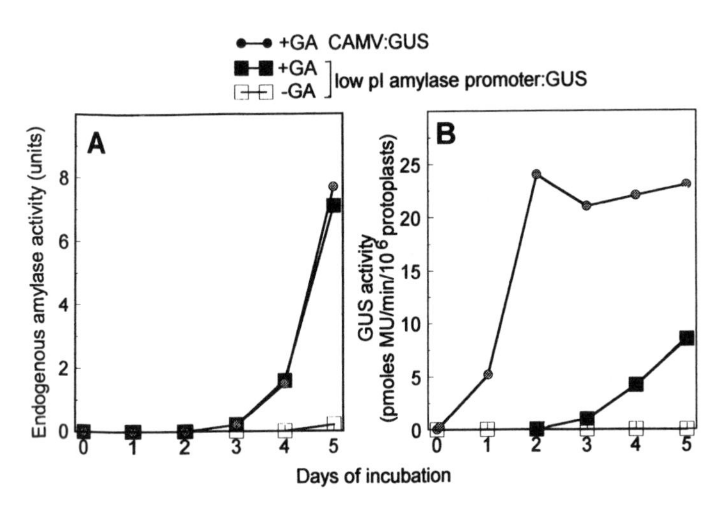

Figure 8.6. Direction of gene expression by a wheat low-pI α-amylase 5′ upstream promoter in response to GA. Protoplasts of wild oat (*Avena fatua*) were transformed with two constructs: the low-pI amylase promoter linked with GUS (■,□), or the 35 S promoter of cauliflower mosaic virus (CaMV) linked with GUS (●). The low-pI amylase promoter is regulated by GA while the 35 S CAMV promoter is not. (A) The time course of α-amylase production in response to GA_3 showing that the protoplasts respond normally and enzyme production is not affected by the presence of the constructs. (B) GA-regulated production of GUS in protoplasts transformed with the low-pI promoter construct. Note the coincidence between the times of GA-regulated α-amylase production and GA-regulated GUS production by the latter construct. Note that when the non-GA-responsive promoter (35 S CAMV) is linked to GUS, expression is not regulated in the correct temporal manner. Adapted from Huttly and Baulcombe (1989).

course very close to that for the formation of native α-amylase. Promoter deletions (see Section 2.3.5.1) reveal that a region extending to -300 bp upstream of the transcription start site is sufficient to drive GA-regulated expression which can be prevented by the inclusion of ABA in the incubation medium. Using a similar approach, it has been shown that a major GA-response element (GARE) in the 5′ upstream promoter of the high-pI barley α-amylase gene is located between −148 and −128 bp from the transcription start site; a construct consisting of six copies of the region fused to CAT is also sensitive to ABA, which suppresses GA action. In another high-pI barley α-amylase, the GARE is in the region −174 to −108, and this again has sensitivity to ABA (Fig. 8.7).

Sequence analysis of the barley 5′ upstream promoter of different α-amylase genes (both low and high pI) reveals three conserved regions, the sequences of which are shown for one of the high-pI genes in Fig. 8.7. The results of deletion mutant studies, together with site-directed mutagenesis, now strongly suggest that in all of the α-amylase genes that have been investigated in detail in both

relatively high concentrations of Ca^{2+} are required for the mai
stability and activity. In barley aleurone tissue, the effect of Ca
entirely at the protein level, transcription being unaffected; but i
Ca^{2+} might also affect α-amylase mRNA accumulation, though p
acting on transcription itself. Since Ca^{2+} is a constituent of the enzy
that intracellular Ca^{2+} is continually utilized and lost from cells th
synthesizing and secreting α-amylase. This Ca^{2+} is replenished
flux of ions into the endoplasmic reticulum, where addition to the en
occurs. This flux, at least from the exterior of the cell, is itself prom
via a calcium transporter located in the plasma membrane. As far
grain is concerned it is not clear how much of the Ca^{2+} is derived f
in the aleurone layer protein bodies and how much is from the
dosperm.

We saw in Fig. 8.1 how the addition of ABA prevents the i
GA_3-induced α-amylase and in Fig. 8.2 that there is virtually no enzym
(nor is it synthesized) in the presence of this inhibitor. Production
especially for the high-pI α-amylase, is suppressed by ABA (Fig. 8.4
addition, there is evidence that ABA can exert an inhibitory acti
translational level. Aleurone cells also respond positively to ABA in t
proteins are induced by the regulator. One particularly interesting exam
inhibitor of α-amylase activity. Hence, ABA can repress α-amylase at t
tion and at the enzyme itself. The regulator also inhibits Ca^{2+} uptake by
cells.

8.1.2. Regulation of the α-Amylase Genes

Much progress has been made in our understanding at the gene level
regulation of α-amylase production by GA. These studies have relied on th
that α-amylase genes (e.g., from wheat and barley) can be isolated, cloned
the 5′ upstream promoter region used to make various gene constructs
Section 2.3.5.1 for a summary of this technique). The constructs are introd
into transient expression systems, such as isolated aleurone protoplasts of ba
and wild oat by osmotic perturbation of the protoplasts, or into intact aleur
layers by means of a particle gun. To distinguish construct expression from
normal, background expression of the native α-amylase genes, the constru
include a reporter gene, e.g., β-glucuronidase (GUS) or chloramphenicol acet
transferase (CAT).

An illustration of the behavior of such a transient expression system
given in Fig. 8.6. Here, the 5′ upstream promoter of wheat low-pI α-amylas
gene, fused to the GUS coding region, was introduced into isolated wild oa
protoplasts. When treated with GA, the protoplasts produced GUS, over a tim

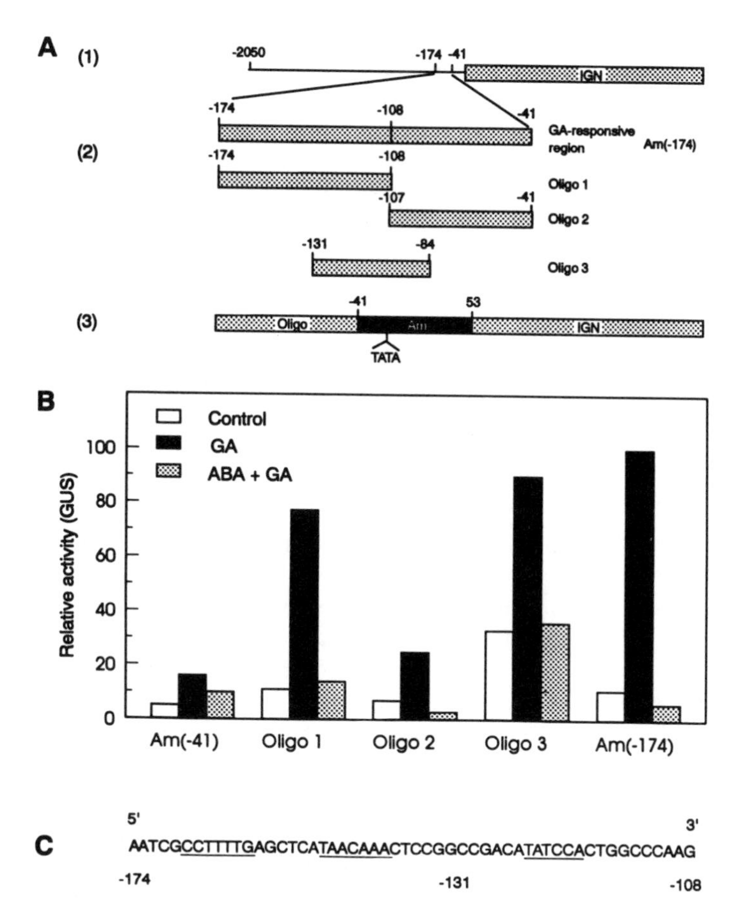

Figure 8.7. Activity of the 5′ upstream promoter of a barley high-pI α-amylase gene. (A, 1) A construct which gives GA-regulated expression of GUS is shown. The 5′ upstream promoter extends to −2050 bp; this is fused to sequences (IGN) which contain the GUS coding region. (A, 2) The GA-responsive region lies between −174 and −41 bp[Am(−174)]. Synthetic oligonucleotides equivalent to various regions are shown (oligo 1,2,3). (A, 3) All of these sequences are fused separately to a minimal α-amylase promoter, Am(−41) linked to IGN (GUS containing). (B) The constructs when introduced into barley aleurone protoplasts express GUS to various extents in response to GA and GA plus ABA. Note: (a) The minimal promoter, Am(−41), exhibits a weak response to GA, (b) Oligo 2 has little GA-responsiveness, (c) Oligo 1 and the full-length GA-response region, Am (−174), have strong GA and ABA responsiveness, (d) oligo 3 (this overlaps oligo 1) confers high degrees of expression, but the effects of GA and ABA are less than for oligo 1 and Am (−174) (compare GA and control columns). These findings indicate that a sequence (−107 to −131) confers high expression but with less GA responsiveness, and a region lying between −131 and −174 has major sensitivity to GA and ABA. (C) The base sequence of the 5′ upstream promoter (−108 to −174) is shown. The regions underlined are highly conserved in all of the barley α-amylase genes, and in some cases, in the wheat low-pI gene. Adapted from Gubler and Jacobsen (1992).

barley and wheat, the sequence TAACAA(G)A at around -135 bp upstream is implicated in GA-regulated gene expression, possibly with the cooperation of the TATCCAC sequence. The first of these two also appears to be the site of sensitivity to ABA. There is also evidence in barley that a sequence corresponding to binding sites for the barley homologue of the maize *Opaque-2* transcriptional regulator (see Section 2.3.5.6) participates in regulation. This sequence, together with an element CCTTT forms part of an "endosperm box" which may be important in modulating the expression of the promoter in response to GA.

Although an understanding of these *cis*-acting factors is now emerging, relatively little information is available on *trans*-acting factors, i.e., DNA-binding proteins, or transcriptional regulators. On theoretical grounds it is very likely that such factors are involved in the action of GA on gene expression but in only two cases have they been reported. A nuclear protein from GA-treated rice aleurone tissue was found to bind to specific regions of the 5′ upstream promoter of a rice α-amylase gene. The DNA-binding protein itself was produced in response to GA, and was absent from nontreated tissue, and from other parts of the plant. There is no information available to encourage us to accept this as a universal feature of GA action on aleurone cells but the finding strengthens the likelihood that such *trans*-acting factors are involved. DNA-binding nuclear proteins have, in fact, been found in wild oat aleurone protoplasts and they bind to different regions of introduced wheat high- and low-pI promoters. Unlike the finding in rice, these *trans* factors are present in the absence of GA. One of them binds to the conserved promoter sequence TAACAGA, a suspected GARE (see above).

It is clear, then, that GA activates the α-amylase genes through *cis*-acting factors (including the GAREs) which probably interact with *trans*-acting, DNA-binding protein(s). This can be regarded as the final set of events initiated by the perception of GA by the aleurone cell. But how is GA perceived? And how is this perception transduced into effects at the gene level? It is not clear if specific GA-binding proteins exist in the aleurone cells; none has yet been isolated, although there are indications that binding can occur. On the other hand, some imaginative experimentation has led to the proposal that GA acts on the plasma membrane. Aleurone protoplasts of wild oat can be stimulated to produce α-amylase mRNA and protein by exposure to GA immobilized on inert beads. Since the GA is tightly bound it cannot enter the protoplasts and hence it seems likely that the hormone has acted on the cell surface, possibly on a receptor protein located within the plasma membrane. This stimulus must then be propagated by a transduction pathway, about which nothing is known, to the *trans*-acting protein and the level of the gene. An interesting aside is that this mechanism is unlike the steroid action mechanism in animal cells which involves binding of the hormone to an intracellular, soluble receptor protein. Elucidation of the perception of the hormonal signal and the transduction pathway remains as the most challenging problem to be approached with respect to GA action on the aleurone cells.

8.1.3. Enzyme Secretion

After synthesis, the α-amylase is secreted from the aleurone layer or, in the appropriate cases, the scutellum. This occurs via the rough endoplasmic reticulum (RER) (Fig. 8.8) and Golgi apparatus. Synthesis of the enzyme takes place on the RER followed by targeting to the lumen. During this process, the signal or leader peptides are excised within the RER, in the case of rice scutellar α-amylase, before translation is completed. Immunocytochemical studies have located the enzyme in both the RER lumen and the Golgi body, and enzyme-containing vesicles derived from both of these are directed to the plasma membrane. Rice scutellar α-amylase is glycosylated, the addition of oligosaccharide taking place in the Golgi apparatus; in fact, glycosylation is inhibited by treating the rice scutella with monensin, which perturbs Golgi activity. In this situation, the addition of galactose and fucose to the α-amylase protein is suppressed and intracellular transport and extracellular secretion are to a large extent blocked.

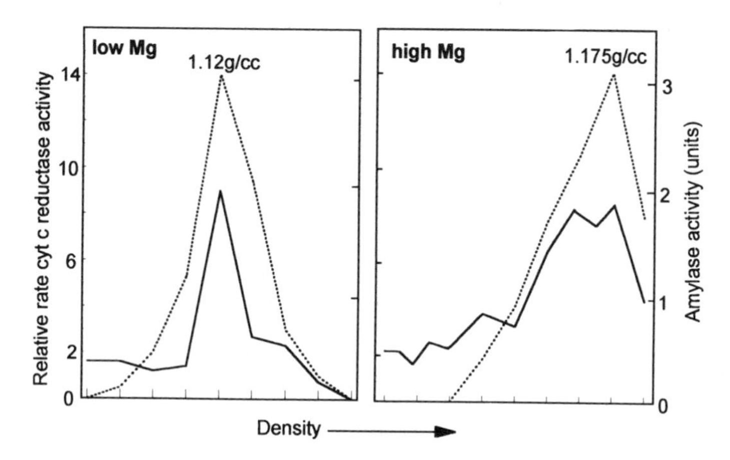

Figure 8.8. Distribution of NADH-cytochrome *c* reductase (- - - -) (an ER-associated enzyme) and α-amylase activity (——) on isopycnic sucrose gradients in an organelle fraction from GA_3-treated aleurone layers of barley. Aleurone layers were incubated for 18.5 h in GA_3, a time when synthesis and secretion of α-amylase are high. Note that when the density of the ER-rich membrane fraction is changed via the presence of different Mg^{2+} concentrations in the gradient (peak density given in grams per cubic centimeter), then so is the density at which α-amylase sediments. This shows that the enzyme is closely associated with the membranes and may be sequestered within ER-derived vesicles. After Chrispeels and Jones (1980/81).

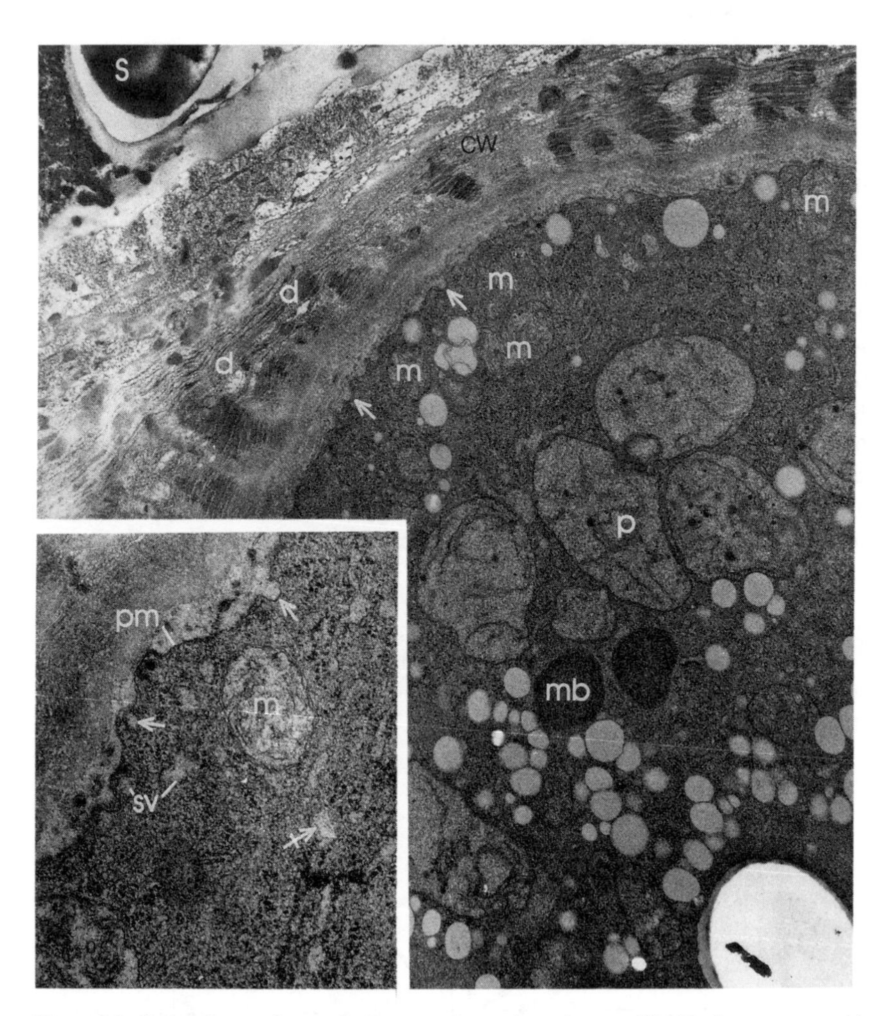

Figure 8.9. Barley aleurone layer cell adjacent to the starchy endosperm (S) 16 h after treatment with GA_3. The cell wall (CW) shows sign of hydrolysis in dissolution zones (d). The plasmalemma (pm) appears to be undulated (arrows), possibly owing to the fusion with small secretory vesicles (V), some of which lie just below the membrane (see inset). Numerous mitochondria (m) are close to the plasmalemma, and plastids (p) and microbodies (mb) are present in the cytoplasm. The rough endoplasmic reticulum has areas where it is devoid of ribosomes (crossed arrow in inset). After Vigil and Ruddat (1973).

Although wheat and barley α-amylases bear no carbohydrate residues, treatment with monensin nevertheless causes the enzyme to accumulate within the Golgi body, indicating the importance of this organelle in secretion. To effect secretion, the transport vesicles from the endoplasmic reticulum (Fig. 8.9) and the Golgi fuse with the plasmalemma whereby the α-amylase and other vesicular contents (e.g., various hydrolases) are expelled from the protoplast. Calcium ions may be required for fusion to take place of the secretory vesicles to the plasma membrane.

Once secretion from the protoplast has been effected, the α-amylase has to traverse the cell walls of the aleurone cell or scutellar epithelium. The aleurone cell wall is rich in arabinoxylan (see Section 7.2.2) in the outer part and (1→3, 1→4)-β-glucan in the inner portion, which together form a gel-like matrix. This offers a barrier to free diffusion of enzyme molecules, but degradation of the wall takes place leaving channels through which the secreted protein can pass into the starchy endosperm. These changes presumably result from the activity of endoxylanases, especially in the outer regions of the wall. The inner wall appears to stay intact until enzyme secretion is almost completed, but the (1→3, 1→4)-β-glucans may be removed before this occurs, leaving what is thought to be a largely proteinaceous skeleton through which secreted enzymes can move (see Section 8.1.5). The inner wall is continuous with structures around the plasmodesmata, which provide a passage for various wall-degrading enzymes en route to the outer wall. Similar degradation takes place in the outer walls of the scutellar epithelial cells, but much less is known about the enzymes involved or their distribution with the cell wall.

8.1.4. Regulation of α-Amylase Production within the Intact Grain

In intact barley grains subjected to malting conditions (steeped in excess water under partially anaerobic conditions at 14°C) the content of gibberellin-like material (i.e., material giving a positive response in a bioassay for gibberellins) rises to a peak by the second day after imbibition starts, a time when α-amylase synthesis commences (Fig. 8.10A). Production of the enzyme continues even as the content of gibberellin-like materials declines; the rate of enzyme synthesis itself declines about a day later. In barley germinated on moist filter paper at 25°C the major increase in gibberellin-like compounds somewhat follows the increase in α-amylase synthesis, which starts to decline when the content of gibberellin-like material is still high (Fig. 8.10B). It could be argued that the low amount of gibberellin present in the germinated grains over the first 1–2 days after imbibition starts is sufficient to trigger α-amylase synthesis, although this does not explain the subsequent and massive synthesis of gibberellins that follows enzyme production. Perhaps this latter event is concerned with embryo growth rather than with reserve mobilization. As we saw earlier, α-amylase consists of two groups of isoenzymes. In the intact, germinated barley grain the high-pI group begins to accumulate in the endosperm before the low-pI isoforms, but the former declines in activity before the latter (Fig. 8.10C). These differences in timing are matched by the accumulation within the aleurone cells of the appropriate mRNA species.

It should be noted that the α-amylase within the endosperm of germinated cereal grains comes from two sources, the scutellum and the aleurone tissue. We must be careful, therefore, not to ascribe the changes in α-amylase content to activity of the aleurone cells alone. It is clear, incidentally, that as far as the

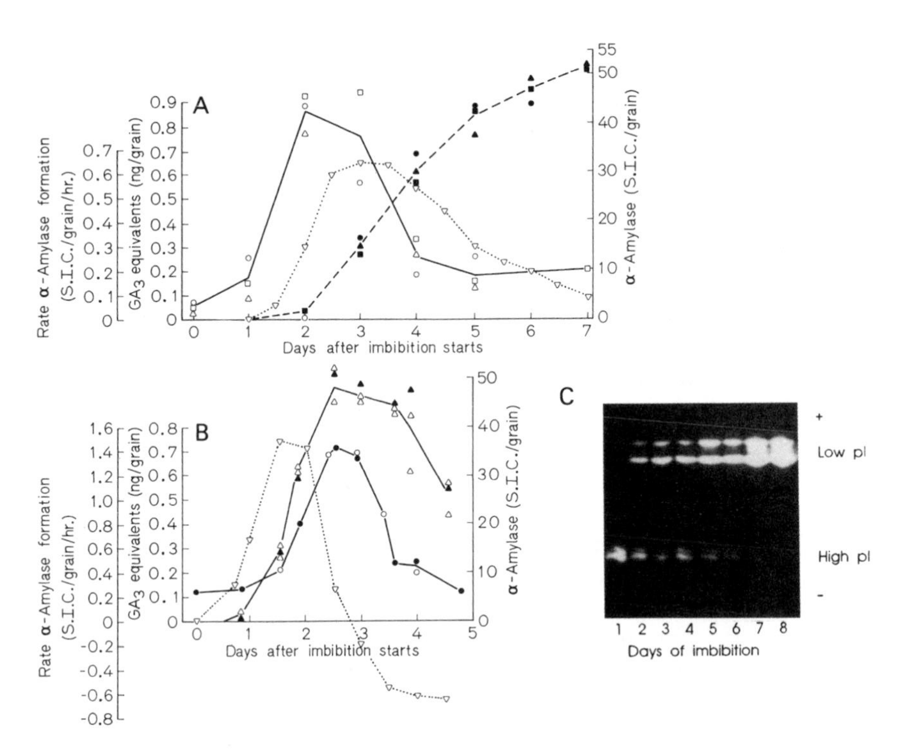

Figure 8.10. Changes in the levels of α-amylase and gibberellin-like materials in barley. In A, barley (*Hordeum distichon* cv. Proctor) was malted at 14°C, in B it was germinated on moistened filter paper at 25°C. Symbols in A (represent values from three separate experiments): □,○,△, gibberellin-like material; ■,●,▲, α-amylase; ▽, rate of formation of α-amylase. Symbols in B (represent values from two experiments): ○, ●, gibberellin-like material; △,▲, α-amylase; ▽, rate of formation of α-amylase. After Groat and Briggs (1969). In C, barley (cv. Himalaya) was germinated on moist filter paper; the α-amylase was extracted at daily intervals and equal units of enzyme activity were separated by isoelectric focusing. Clear bands indicate the isoenzymes. From Chandler and Jacobsen (1991).

kinetics of α-amylase production are concerned, there are some important differences between the isolated aleurone layer and the intact grain, especially regarding the appearance of the high- and low-pI groups. The reasons for this are not clear, but it could be because hormonal regulation within the intact grain is different from that in the *in vitro* experiments with GA_3—possibly attributable to the fact that in the germinated grain GA_1 appears to be the hormone involved.

The terminal decline of α-amylase (Fig. 8.10) occurs partly because the reserve protein in the aleurone cells, which provides the amino acids for enzyme synthesis, is eventually depleted. This cannot be the complete story, however, if only because of the different kinetics of the increase and decline of the high- and low-pI groups (Fig. 8.10C). Factors other than amino acid availability obviously are operating, but we do not yet understand what these are.

The production of α-amylase and its activity in the endosperm might be modified by various factors. There is evidence from *in vitro* experiments with isolated aleurone layers that enzyme production is inhibited by the products of starch hydrolysis (glucose and maltose) at relatively high concentrations (Fig. 8.11A). By about the third day after the start of imbibition, the sugars in the endosperm of a germinated barley grain can reach concentrations as high as 400–500 mOsm so the same effect as that shown in Fig. 8.11A might well operate. The action of the sugars appears to be osmotic since osmotically active agents that are not products of starch hydrolysis, e.g., polyethylene glycol and mannitol, have the same inhibitory action. In wheat embryos, this appears to be through the inhibition of the high-pI group of isoenzymes (Fig. 8.11B) the mRNA for which is also suppressed. It should be mentioned, however, that in

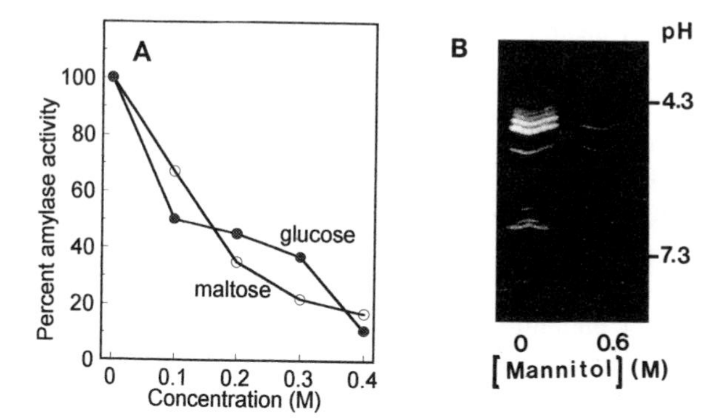

Figure 8.11. (A) Reduction of α-amylase activity achieved by incubating isolated GA_3-treated aleurone layers of barley in glucose and maltose solutions of different concentrations for 24 h. After Jones and Armstrong (1971). (B) Effect of osmoticum (0.6 mOsm mannitol, approximately −1.7 MPa) on high- and low-pI α-amylase activity in wheat embryos. Note the complete inhibition of the high-pI group. From Garcia-Maya *et al.* (1990).

the rice scutellum, the sugars appear to be acting in their own right and not only osmotically. Whatever the cause, the inhibition may play an important part in regulating α-amylase production in the germinated grain, linking it with the utilization of the products of starch hydrolysis by the growing seedling. If for some reason this is slowed down, perhaps by poor growth of the germinated embryo, the feedback mechanism would prevent the continued hydrolysis of the starch reserves, which would resume only when the constraint has been relieved.

Another factor that may be important in determining α-amylase activity in the endosperm is the pH. The pH optimum for α-amylase activity is 5.5–6, and many of the other hydrolases released into the endosperm by the aleurone tissues have even more acidic optima, about 4.5. For several days, the pH of the endosperm of germinated barley is maintained at 4.5–5.1 by the secretion of malic acid from the aleurone layer; similar acidification of the endosperm by the aleurone cells also occurs in wheat. This phenomenon might play a part in the regulation of α-amylase activity.

A question that arises from our previous discussion of the inhibitory effect of applied ABA on α-amylase gene expression is whether this hormone plays any part in the germinated cereal grain. There is a precipitous decline in the ABA content of developing grains during maturation so that the mature grain contains hardly any. Under normal circumstances, the germinated grain does not produce ABA so interference with GA-stimulated α-amylase formation will not occur. Inhibition could become important, nevertheless, under conditions of drought stress when the young seedling, still attached to the endosperm, would respond by producing ABA. Nothing is known about the movement of ABA into the endosperm, but if it does take place the constraint on growth imposed by drought could lead to the arrest of starch mobilization, through the intermediary of the hormone. An experimental investigation of this might be rewarding and it would help to put the role of ABA in α-amylase production into a comprehensible context.

It has been mentioned that considerable variation is encountered in the responses of different cereal species and cultivars to GA. In some cases, factors in the starchy endosperm seem to be involved since the behavior of the aleurone tissue in the intact grain may differ from when it is isolated in *in vitro* experiments. For example, in one cultivar of barley there is evidence that the starchy endosperm has an inhibitory effect on the aleurone tissue which cannot be attributed to ABA and, therefore, it has been suggested that another inhibitor is present.

Thus, the control mechanism for α-amylase synthesis in the aleurone layer of intact barley (and probably wheat) grain might operate as follows (details can be followed in Fig. 8.12). Following imbibition, a factor, probably GA, is synthesized within the embryo (A); it then diffuses across the starchy endosperm and into the aleurone layer (B). In the meantime, the aleurone layer cells undergo some metabolic changes making them receptive to stimulation by GA which then promotes the synthesis of various hydrolases, but particularly α-amylase. This is secreted into the starchy endosperm (C), and there hydrolysis of the reserve starch commences (D). Some of the maltose and glucose released by amylolysis may be converted to sucrose by the aleurone cells and transported as such to the growing embryo, but most is absorbed directly through the scutellum where sucrose is formed (F). If the production of these monosaccharide sugars exceeds their rate of transport to, and utilization by, the growing embryo, they accumulate in the endosperm. There they act as an effective switch to stop production of further α-amylase (E), an enzyme whose activity in the endosperm already exceeds the requirement for the products of its hydrolytic activity (i.e., sugars).

In closing this section, it is worth noting that the initial amylolytic activity released from the scutellum, prior to or during α-amylase production by the aleurone layer (Section 7.2.2), is not known to be under the control of gibberellins and the mechanism for regulating scutellar α-amylase synthesis remains to be determined.

8.1.5. Regulation of Other Hydrolases in the Cereals

α-Amylase is not the only hydrolase whose synthesis in, or release from, the aleurone layer is controlled by gibberellin (Table 8.1). The GA_3-induced *de novo* synthesis of a cysteine proteinase in barley exhibits the same time course of synthesis and release from the aleurone layer as does α-amylase; moreover,

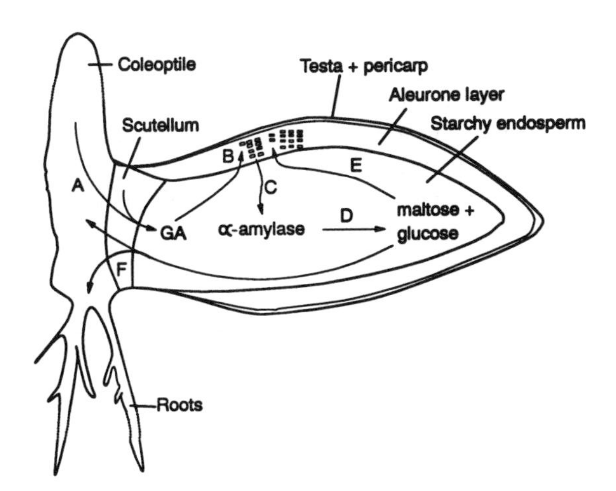

Figure 8.12. Summary diagram of the control mechanism for α-amylase production in intact barley grains. See text for details. After Jones and Armstrong (1971).

Table 8.1. Some GA-Regulated Enzymes Produced by Cereal Aleurone Layer Cells[a]

Enzyme	Secreted or not
α-Amylase	Yes
Cysteine proteinase	Yes
Carboxypeptidase	Yes
Ribonuclease	Yes
DNase	Yes
(1→3)-β-Glucanase	Yes
(1→3)(1→4)-β-Glucanase	Yes
Acid phosphatase	Yes
Endoxylanase	Yes
Arabinofuranosidase	Yes
α-Glucosidase	Yes
(1→6)-α-Glucanase	Yes
Isocitrate lyase	No
Polyphenol oxidase	No

[a]Adapted from Jones and Jacobsen (1991).

the GA_3 dose–response curves for the two enzymes are very similar. The role of this proteinase in reserve mobilization has not been fully elucidated, but it is secreted into the starchy endosperm and is able to hydrolyze the reserve protein hordein there; it may also be responsible for the activation of the bound enzyme, β-amylase (Section 7.2.2).

Several other proteinases occur in the endosperm, some secreted by the aleurone tissue and some by the scutellum. Especially well known are the five carboxypeptidases, one of which is specific to the scutellum. Two carboxypeptidases are synthesized *de novo*, which in wheat (but apparently not in barley) are regulated by GA, while two others increase from a very low value during endosperm mobilization. The endopeptidases (e.g., cysteine proteinase) and exopeptidases (e.g., carboxypeptidase) probably act in concert to degrade the storage protein into amino acids and small peptides. Proteinases also hydrolyze the reserve proteins in the aleurone layer cells (in the aleurone grains) to provide amino acids for the synthesis of other enzymes.

As mentioned in Section 8.1.3, release of α-amylase and other enzymes from the aleurone layer is facilitated by digestion of the aleurone-layer cell wall in the basal area, toward the starchy endosperm (Fig. 8.9). The cell wall has a high content of arabinoxylan (85%), which has a linear (1→4)-β-xylan backbone, and only a little cellulose (8%). Three pentosanases capable of degrading this polymer, viz., (1→4)-β-endoxylanase, β-xylanopyranosidase, and α-arabinofuranosidase, increase in activity in germinated barley grains. They are also stimulated by GA_3 in isolated aleurone layers, resulting in an increase in released pentose residues (60% xylose, 40% arabinose) into the incubation medium (Fig. 8.13). The pentosanases may also play a role in degrading the walls of the starchy endosperm, thus rendering the starch and other reserves more accessible for enzymatic attack. (1→3)-β-Glucanase is synthesized by the aleurone layers in the absence of GA_3, but it is only released in the presence of this hormone. Since there is relatively little (1→3)-β-glucan in the endosperm (e.g., of barley) the enzyme is not thought to be important in endosperm cell wall breakdown. Current thinking is that it may have a defensive role against fungal pathogens whose walls consist of a high proportion of (1→3,1→6)-β-glucan. The enzymes which are responsible for breaking down the walls of the starchy endosperm are the (1→3,1→4)-β-glucanases. For example, two enzymes are known in barley, one which emanates from the scutellum and one from the aleurone cells, the latter being induced by GA. A similar situation is also found in wheat, although here low amounts of synthesis of the aleurone layer enzyme occur in the absence of GA, but the hormone enhances production by two- to three-fold. Approximately 70% of the starchy endosperm cell wall consists of (1→3,1→4)-β-glucan, and the walls must be degraded to allow easy access of the other hydrolytic enzymes to the starch and protein reserves. The (1→3,1→4)-β-glucanases succeed in doing this, and, together with β-glucosidases, hydrolyze the glucan completely to glucose. It has been estimated that up to 18.5% of all of the glucose made available in the endosperm of the germinated barley grain is from this cell wall source.

Both α-amylase and acid phosphatase increase in amount in the walls of the aleurone layer prior to their release. Isolated aleurone layers incubated in the absence of GA_3 accumulate acid phosphatase (Fig. 8.14) in the inner region of the wall, indicating that the synthesis and release of this enzyme from the cytoplasm are not gibberellin dependent. But release from the cell walls (and

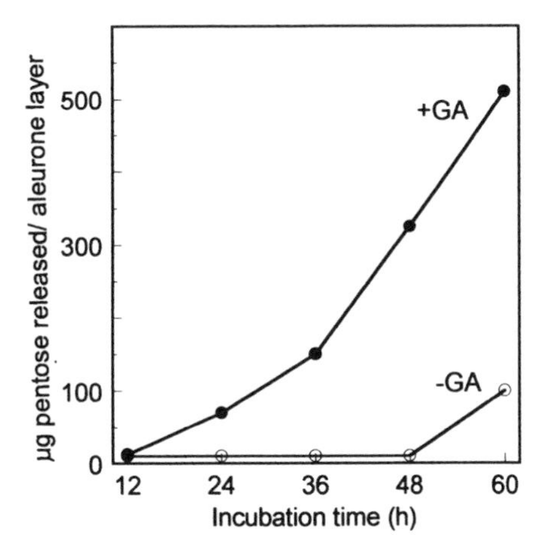

Figure 8.13. Release of pentose residues by isolated aleurone layers of barley into the incubation medium in the presence (●) or absence (○) of 1 μM GA_3. The extent of pentose release is proportional to the amount of pentosanases induced by the hormone. After Dashek and Chrispeels (1977).

into the surrounding medium, in the case of isolated layers) is GA_3 dependent (Fig. 8.14). Release of acid phosphatase commences 6–12 h after the introduction of the hormone and occurs via areas of the cell wall that under the microscope show obvious signs of digestion. It is likely, then, that the release of this enzyme [and also of (1→3)-β-glucanase] is only indirectly attributable to the action of GA_3: first, synthesis of the cell-wall-hydrolyzing enzymes, the pentosanases, is induced and these digest channels in the aleurone-layer cell wall to liberate enzymes that are trapped [e.g., acid phosphatase, (1→3)-β-glucanase)]or enzymes whose release is impeded by the cell wall (e.g., α-amylase). It is likely that the cell wall channels eventually provide a common path for the release of all enzymes from the aleurone layer.

Two other enzymes that are intimately involved in starch hydrolysis are synthesized *de novo* in aleurone layer cells in response to GA_3; these are limit dextrinase and α-glucosidase. β-Amylase is not synthesized *de novo* in the presence or absence of GA_3 but is carried over in an inactive form from the developing grain. As mentioned above, it may be activated indirectly by GA through induction of a proteinase that releases it from its latent, bound form.

Besides stimulating the production and release of hydrolases, GA_3 can also retard enzyme synthesis. Consistent with the observation that GA_3 induces the degradation of aleurone-layer cell walls is the observation that it inhibits the synthesis of pentosan (xylose and arabinose) components of the wall, in part owing to the reduced activity of a membrane-bound arabinosyl transferase and perhaps also of xylosyl transferase. Changes in such membrane-bound enzymes could reflect a shift in the role of membranes from the production of cell wall components to the synthesis and packaging of hydrolases.

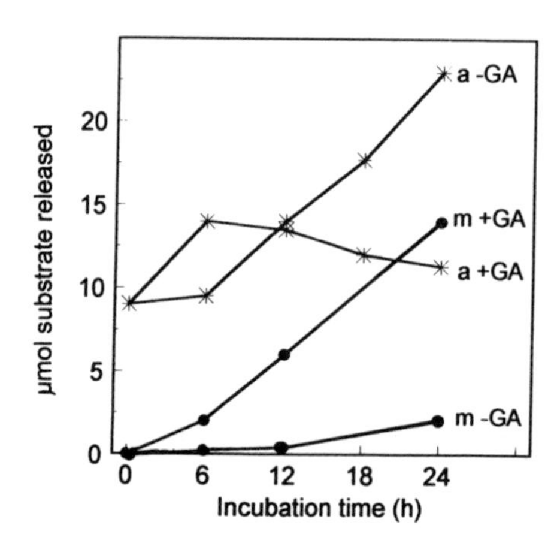

Figure 8.14. Changes in acid phosphatase activity in the incubation medium (m, ●) and in isolated aleurone layers (a, *) of barley incubated in the presence or absence of 1 μM GA_3. After Ashford and Jacobsen (1974).

In summary, then, gibberellin released from the embryo induces the *de novo* synthesis of α-amylase, limit dextrinase, and α-glucosidase in the aleurone layer, which are released into the starchy endosperm to effect starch hydrolysis. This release is aided by the action of GA-induced pentosanases, produced by the aleurone-layer cells and secreted into their walls which are thus hydrolyzed. The cell walls of the starchy endosperm are broken down by (1→3,1→4)-β-glucanases, a process which renders the starch grains readily accessible to α-amylase. The synthesis of α-amylase (and perhaps other carbohydrases) is reduced by the low-molecular-weight products of starch hydrolysis. The "feedback" signal presumably signifies that more sugars are available to the growing seedling than can be transported there, and that sufficient enzymes have been produced to complete hydrolysis of the stored starch.

8.2. CONTROL PROCESSES IN OTHER SEEDS

In cereals and some other grasses, almost all of the mobilizing enzymes are produced in a tissue separate from the location of the reserves. Some, possibly the minor quantity, of these enzymes are formed in the embryo, but most come from a digestive tissue, the aleurone layer, which is activated by hormonal factors—probably gibberellins—from the germinated embryo. No strictly comparable system occurs in other seeds, however. In most species, the mobilizing enzymes are generally produced at the same tissue site as the reserves, i.e., in the endosperm or cotyledons. And even in the relatively few instances where there is a discrete enzyme-secreting tissue, such as the aleurone layer of fenugreek (which makes enzymes to attack the endosperm galactomannans), there is no evidence that factors from the embryo participate in a regulatory fashion. But mobilization of storage reserves must be associated with germination and embryo growth just as it is in cereals. If this were not the case, some, possibly substantial, breakdown of reserves could occur in an imbibed, ungerminated seed (e.g., a dormant seed), with deleterious consequences. What are the controls that operate to secure production and activity of mobilizing enzymes at the right time to link these with the embryo and subsequent seedling formation? These are the questions we will attempt to answer in the following sections.

8.2.1. When Are the Mobilizing Enzymes Produced?

Little or no activity of the major hydrolytic enzymes—proteinases, lipases, and amylases—can be detected in dry or newly imbibed seeds, but in most cases activity increases gradually as the seeds hydrate (Fig. 8.15). It seems likely that enzyme synthesis generally accounts for the rise in extractable activity, a conclusion that arises largely from the observation in some species that inhibitors of protein and RNA synthesis prevent the increase in certain mobilizing enzymes. Examples of enzymes that can be inhibited in this way are isocitrate lyase in squash cotyledons, lipase in castor bean endosperm, and α-amylase in French bean cotyledons. Better evidence that the enzymes are newly synthesized is rare; however, when hydrated cotyledons of watermelon and mung bean are supplied with D_2O the isocitrate lyase of the former and the endopeptidase of the latter become density labeled. Similar labeling of α-amylase occurs during its production in pea cotyledons. A few apparent exceptions have been reported; for example, a lipase which is present in newly hydrated castor bean endosperm is likely to have been in the dry seed (Fig. 7.11). And in lettuce, α-galactosidase and β-mannosidase, which participate in the final stages of hydrolysis of the galactomannans of the endosperm cell walls, are found in the cotyledons of the dry seeds.

It is not clear why activity of the reserve-mobilizing enzymes diminishes after a time (see Fig. 8.15). The fall does not always accompany the completion of mobilization of the stored reserves but often precedes it. One suggestion is that products of reserve hydrolysis (e.g., amino acids from protein) act by feedback mechanisms to arrest the synthesis of further enzyme molecules. Although the application of such products does reduce enzyme formation in some cases (e.g., proteinases in pea and cucumber cotyledons), it fails to do so in others (e.g., endopeptidase in mung bean cotyledons), and so wide experimental proof for the concept of feedback inhibition is lacking. There is some evidence

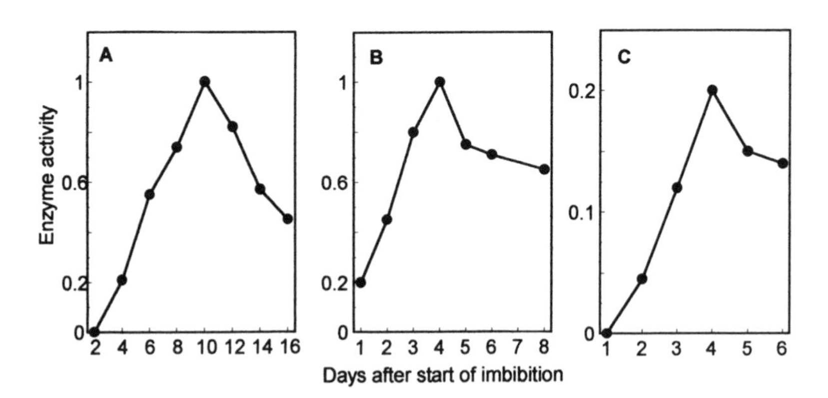

Figure 8.15. Changes in activity of some reserve-mobilizing enzymes in dicot seeds. All seeds were allowed to complete germination in darkness. (A) α-Amylase in French bean cotyledons. After Van Onckelen *et al.* (1977). (See also Fig. 7.5.) (B) Lipase in cucumber cotyledons. After Davies *et al.* (1981). (See also Fig. 7.11.) (C) Proteinase (carboxypeptidase) in castor bean endosperm. After Yamamoto *et al.* (1982). (See also Figs. 7.20 and 8.16B.)

that synthesis of an enzyme closely associated with protein mobilization in mung beans—asparagine synthetase—is arrested when an enzyme inhibitor develops, but it is not known if such a mechanism is widespread.

8.2.2. Regulation of Mobilization

What regulates the rise in mobilizing activity that follows the germination of a seed? This question really inquires about two processes, enzyme formation and enzyme activity, since regulation can be achieved by controlling either or both of these. Unfortunately, it is not always possible to determine which is responsible simply by measuring the overall rates of mobilization of the reserves. Further, measurements of the activity of extracted enzymes do not necessarily reflect their *in vivo* activity, and storage tissues that have the same extractable enzyme activity may be mobilizing reserves to quite different extents. Therefore, an understanding of the regulation of mobilization should include knowledge about rates of reserve utilization, rates of enzyme formation, and *in vitro* and *in vivo* enzyme activity.

It is clear in many cases that mobilization is controlled by the embryonic axis, i.e., the radicle/hypocotyl and plumule, or, where the endosperm is the storage tissue, by the embryo as a whole. This point has been established by means of surgical experiments in which the axis (or embryo) is excised at different times during germination and seedling growth. An illustration of the role of the axis is provided by the example shown in Fig. 8.16—protein mobilization in the cotyledons of mung bean. In the 6 days following the start of imbibition, the seed germinates and seedling growth ensues. During this time, the amount of storage protein in the cotyledons falls by about 75%, a change that is accompanied by a rise in activity of extractable proteinase (vicilin peptidohydrolase) (Fig. 8.16A,B). The enzymatic breakdown of stored protein leads at first to an accumulation of free amino acids in the cotyledons, but these decrease after about 3 days as they, or their products, are transported into the growing axis (Fig. 8.16C). The pattern is quite different, however, in isolated, hydrated cotyledons, i.e., when the axis is previously carefully removed from the dry seeds. Here, the rates of protein hydrolysis and the increase in peptidohydrolase activity are reduced by about 75%, and the amino acids that arise from the limited protein breakdown accumulate in the cotyledons (Fig. 8.16A–C). Hence, the mung bean axis appears to regulate the breakdown of the proteins stored in the cotyledons, at least partially, by controlling the formation of peptidohydrolase.

A similar effect of the axis is found in relation to the food reserves of several species, and in many mobilization and the formation and activity of hydrolytic enzymes depend on its presence (Table 8.2). But there are also some cases where removal of the axis has little or no effect on enzyme activity, and

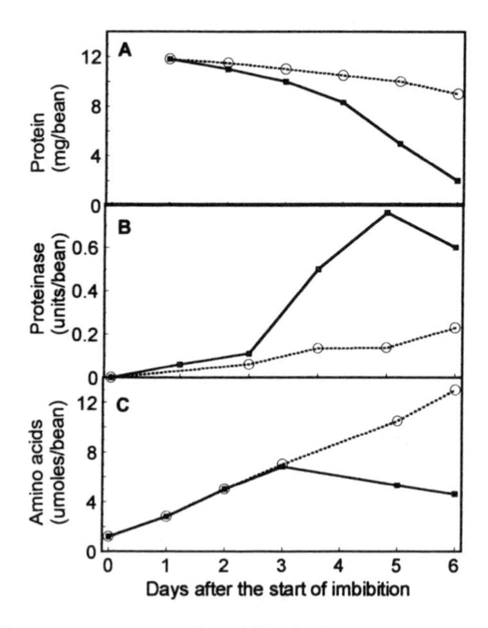

Figure 8.16. The effect of the axis on protein mobilization in mung bean cotyledons. The embryonic axis was carefully removed from dry seeds. The isolated cotyledons (- - - -) were placed on moist sand and the intact seeds (——) on wet vermiculite in darkness (where radicle emergence occurred). Storage protein (A), proteinase (vicilin peptidohydrolase) (B), and free amino acids (C) were measured at daily intervals over 6 days. After Kern and Chrispeels (1978).

so the tissues containing the enzyme seem to be autonomous. To add to the complexity, even with the same species (e.g., pea) some researchers find effects of the axis while others do not! We should be clear, however, that the increase in *extractable* enzyme activity in the absence of the axis does not necessarily reflect the change in activity within the storage tissue itself. Some enzymes are sensitive to product inhibition, and their activity ceases when the products of even limited reserve hydrolysis accumulate, a situation that would occur if exit of materials was impeded by the absence of the growing axis. Thus, the enzyme would be inactive *in vivo* but active when extracted and assayed *in vitro*.

An interesting case is the cucumber seed, which illustrates the important regulatory role of the tissues enclosing its storage organs, the cotyledons. After germination of an intact seed, the triacylglycerols stored in the cotyledons rapidly become depleted, but if the axis is removed from a newly imbibed seed, the amount of these stored reserves hardly falls (Fig. 8.17, curve A). Removal of the testa and the inner membrane around the axis-free cotyledons allows

Table 8.2. Effect of the Attached Axis on Extractable Enzyme Activity

Enzyme	Species	Effect of axis[a]
Peptidohydrolase	Mung bean	+
Proteinase	Pea	+
	Squash	+
	Cucumber	–
	Castor bean	+
α-Amylase	Pea	+
	Mung bean	+
	Chick-pea	+
Amylolytic activity	French bean	+
	Groundnut	–
Lipase	Cotton	+
	Cucumber	–
Isocitrate lyase	Cucumber	–
	Squash	+
	Groundnut	+
	Castor bean	–
Glutamine synthetase	Mung bean	–
Asparagine synthetase	Mung bean	–
β-Mannanase	Lettuce	+
α-Galactosidase	Lettuce	+

[a]+, extractable enzyme activity promoted by the attached axis; –, no promotive effect of the axis.

substantial triacylglycerol breakdown to proceed, however, indicating that these enclosing tissues normally inhibit the process (Fig. 8.17, curve B). The inhibition is probably caused by a limitation of oxygen entry, thereby affecting both enzyme synthesis and the oxidation of fatty acids, for which molecular oxygen is needed. The coat is apparently important in seeds which have had no "surgical" treatments and which have germinated normally. Triacylglycerol breakdown in the cotyledons of these seeds does not begin when the axis elongates but only when the testa is pushed off as a result of its being wedged against a peg of tissue on the elongating hypocotyl (Fig. 8.17, curve D). Moreover, the start of triacylglycerol mobilization is advanced when the testa is experimentally removed from newly germinated seeds (Fig. 8.17, curve C). In cucumber, therefore, both the axis and the testa have important regulatory influences in the mobilization of triacylglycerols by the cotyledons. The testa also inhibits α-amylase formation and starch mobilization in pea and mung bean cotyledons. In the case of the latter it has been suggested that the testa acts by preventing the loss of a diffusible inhibitor which, in some way, is also nullified when the axis is present.

In conclusion, some of the processes taking place in reserve mobilization are regulated by the axis, which is needed for the formation of enzymes in some species (e.g., vicilin peptidohydrolase in mung bean) and/or for the maintenance of enzyme activity in others. The testa plays an important, additional role in cucumber, probably by affecting entry of oxygen, and it seems likely that this kind of regulation also occurs in other species.

8.2.3. Mode of Regulation by the Axis

Two possibilities have been considered to explain regulation by the axis: (1) Specific regulatory substances move from the axis to the storage organs or tissues where enzyme formation occurs, i.e., a hormonal mechanism. This explanation invokes a system similar to the one in certain cereal grains, where the embryo regulates enzyme production in the aleurone layer through the action of gibberellins that it secretes. (2) The axis is a sink, drawing off the products of reserve mobilization in the cotyledons or endosperm, which would otherwise arrest continued enzymatic activity by feedback inhibition mechanisms. What is the evidence for these two possibilities?

8.2.3.1. Hormonal Control by the Axis

Most of the support for the possibility that hormones control the development of activity of the mobilizing enzymes comes from testing the effects of adding growth regulators to isolated storage tissue. This approach attempts to answer the question, can these chemicals replace the influence of the axis? In many cases, applied growth regulators induce greater breakdown of reserves in isolated tissue and/or increase the activity of enzymes concerned with mobilization. Cytokinins, for example, cause increases in amylolytic and proteolytic

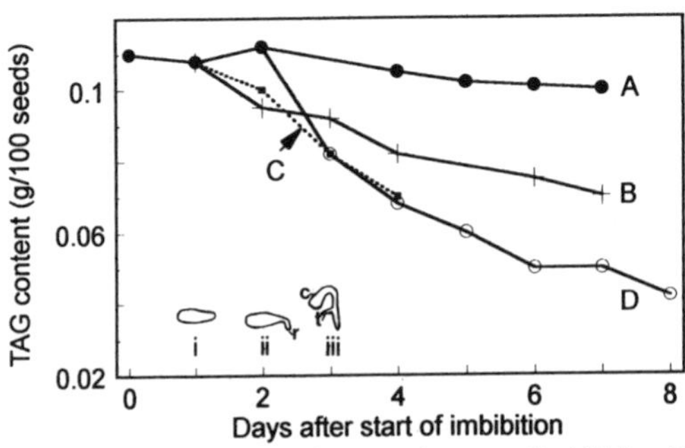

Figure 8.17. Triacylglycerol (TAG) mobilization in cucumber cotyledons. Total TAG was extracted and measured at daily intervals from cotyledons of seeds treated in the following ways: (A) Radicle/hypocotyl axis removed from the dry seed; testa still around the cotyledons. (B) As A but testa and membrane also removed. (C) Testa removed from newly imbibed seed; radicle/hypocotyl axis intact. (D) Fully intact seed; axis present and testa intact at the start of imbibition. The appearance of the intact seed/seedling over the first $2\frac{1}{2}$ days is shown at the bottom left of the diagram: Stage i, dry seed; stage ii, radicle emergence; stage iii, testa displacement. The testa becomes displaced at day 2; TAG breakdown in the cotyledons then begins. c, cotyledons; t, testa; r, radicle. Adapted from Slack *et al.* (1977).

activity of isolated chick pea cotyledons as well as mobilization of carbohydrate and protein (Fig. 8.18), and in the activity of certain proteolytic enzymes and protein hydrolysis in excised squash cotyledons. They also enhance the activities of isocitrate lyase (in the glyoxylate cycle, Section 7.5.1) in watermelon and sunflower cotyledons. Activities of enzymes of the glyoxylate cycle in hazel cotyledons, for β-oxidation of fatty acids and for hydrolysis of stored protein reserves in castor bean endosperm, are increased by gibberellin, which also promotes α-amylase activity in excised pea cotyledons. Auxin has also been found to be effective in the latter. It is important to note, however, that the effects of applied growth regulators are generally relatively small, even though sometimes comparable to the action of the axis (see Fig. 8.18), and nowhere match the dramatic control of enzyme production that is achieved by gibberellin in cereal aleurone cells.

Direct evidence for the involvement of hormones is tenuous. There are some instances, such as fructose-1,6-bisphosphatase in castor bean endosperm and isocitrate lyase in megagametophytic tissue of Ponderosa pine, where diffusates or extracts of embryos stimulate enzyme activity in isolated storage

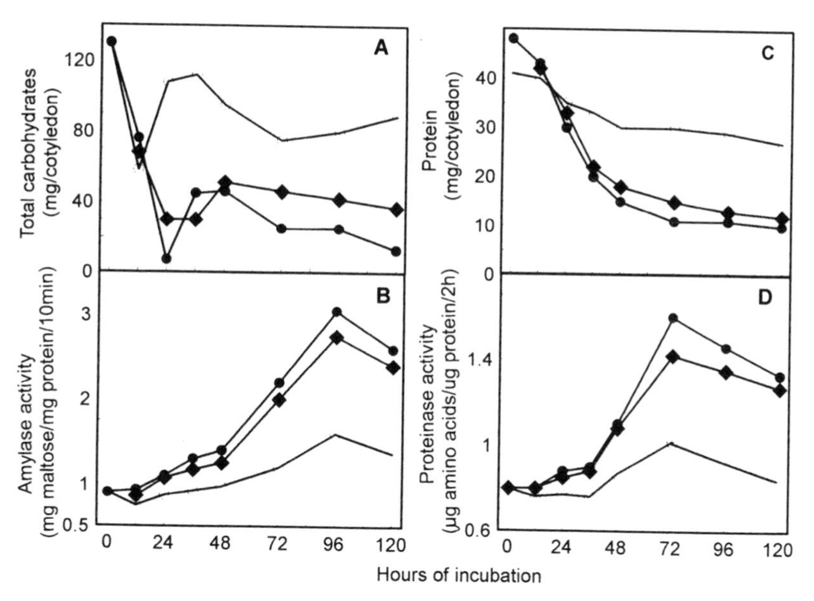

Figure 8.18. Effect of cytokinin on amylolytic and proteolytic activity of cotyledons of chick pea (*Cicer arietinum*). Activities are shown of cotyledons in the intact seed (●), of excised cotyledons on water (O), and of excised cotyledons which were treated so as to restore the normal concentration of endogenous, native cytokinins (♦), A, total carbohydrate; B, amylase; C, protein; D, proteolytic activity. Adapted from Munoz *et al.* (1990).

tissue. But in no case has it been shown convincingly that a known hormone, e.g., a cytokinin or gibberellin, moves from the embryo or axis to the storage tissue and thereby regulates mobilizing activity. Evidence from detailed investigations on lettuce seeds does, nevertheless, support the concept of hormonal control. Although the endosperm of lettuce makes a qualitatively minor contribution to the overall reserve economy, it does appear to furnish usable food material in the earliest stages of seedling growth, before utilization of the cotyledonary reserves begins. The mannans (polymers of mannose, with galactose side groups) of the endosperm cell walls are hydrolyzed by endo-β-mannanase, produced by the endosperm itself, and the products—oligomannans—move into the cotyledons, to be attacked by β-mannosidase and α-galactosidase. The β-mannosidase is present in the dry seed, and its formation does not depend on axial influences after germination. But endo-β-mannanase forms in the endosperm of the intact seed only when the axis has been stimulated to germinate, e.g., by light. Also, the amount of α-galactosidase activity in the cotyledons increases in response to factors coming from the axis, within 2–3 h after illumination. As far as these two enzymes are concerned, the effect of the axis can be replaced by a combination of added gibberellin and cytokinin. The presence of axes, stimulated to germinate by light, in the same incubation medium as isolated cotyledons, promotes an increase in α-galactosidase in these organs. There is good evidence that an inhibitor in the endosperm, probably ABA, normally stops the formation of endo-β-mannanase, and that the promotive factor(s) from the axis are needed to overcome the inhibition. The lettuce seed is therefore a case where mobilizing enzymes present at two sites—the endosperm and the cotyledons—are controlled by promotive influences coming from the axis, which may be cytokinins and gibberellins (Fig. 8.19A).

8.2.3.2. The Axis as a Sink

Many enzymes are inhibited by the products of the reactions they catalyze. The effect can involve repression of enzyme synthesis and inhibition of the activity of already existing enzyme molecules. Such feedback inhibition may be important in the regulation of activity of the enzymes for reserve mobilization. Growth of the axis uses the products of reserve breakdown, which therefore do not accumulate in the storage tissues. The continual withdrawal of these products by the axis could account for the rise and maintenance of activity of the mobilizing enzymes; i.e., the axis regulates enzyme activity simply by virtue of its action as a sink.

This mechanism seems, in several cases, to account for the beneficial effect of the axis on mobilization. In cucumber, for example, although removal of the axis does not hinder development of several lipolytic enzymes in the cotyledons, triacylglycerol breakdown itself is much reduced; hence, the activity of the

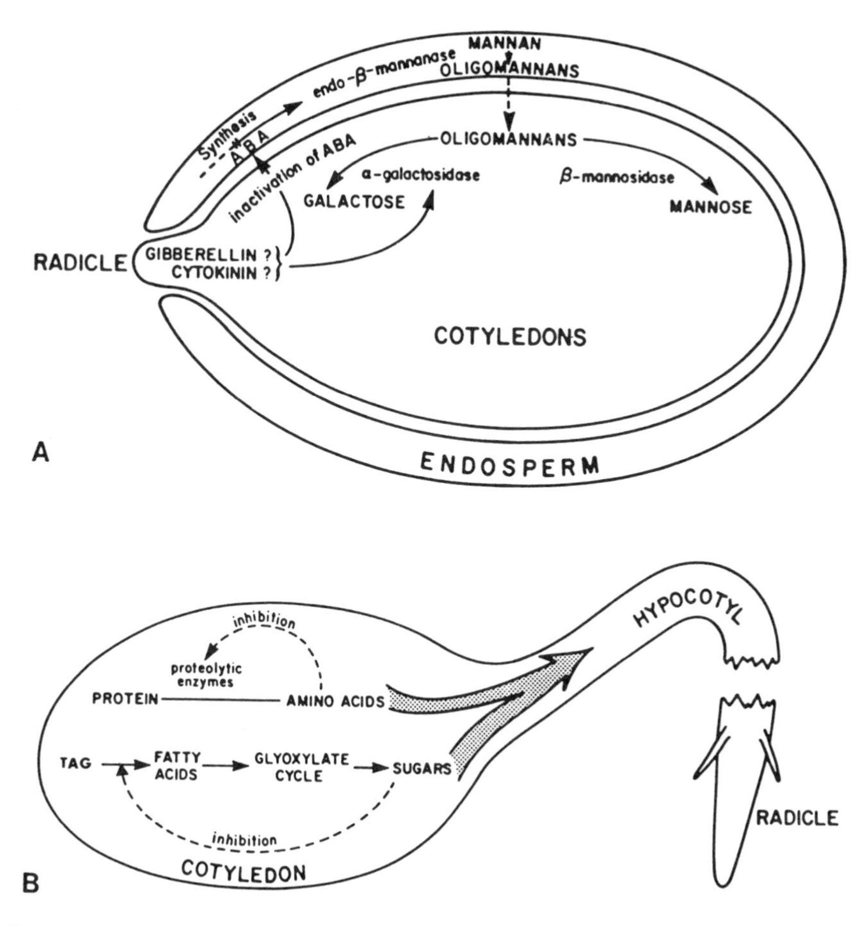

Figure 8.19. Hormonal and feedback regulation of mobilization. (A) Galactomannan reserves in the cell walls of the lettuce endosperm. After Bewley and Halmer (1980/81) and from data of Bewley *et al.* (1983). (B) Reserves in the cotyledons of cucumber. From data of Davies and Chapman (1980) and Slack *et al.* (1977).

enzymes apparently stops in the absence of the axis. Reducing sugars and sucrose accumulate in the excised cotyledons as lipolytic activity slows down. Moreover, addition of sucrose to isolated cotyledons leads to an even greater inhibition of lipolysis. The fact that triacylglycerol breakdown proceeds in isolated cotyledons when the testa is removed might be thought to argue against regulation by a sink, since the normal sink—the growing axis—is missing. However, such isolated cotyledons enlarge, making additional cell wall material (e.g., cellulose) as they do so, presumably from the sugars coming from triacylglycerol utilization; and, in addition to that of cellulose, synthesis of starch also occurs. So even though the axial sink is absent, two other sinks—cellulose and starch synthesis—serve to drain off the products of triacylglycerol mobilization and, by preventing their accumulation, permit the activity of the lipolytic enzymes to continue. The activity of extracted proteolytic enzymes from the cotyledons is similarly unaffected by removal of the axis, but within the cotyledons protein hydrolysis itself is minimal. Accumulated amino acids, especially leucine and phenylalanine, as well as the dipeptide tryptophylphenylalanine, inhibit the activity of aminopeptidase in the cotyledons thus preventing protein degradation.

USEFUL LITERATURE REFERENCES

SECTION 8.1

Ashford, A. E., and Jacobsen, J. V., 1974, *Planta* **120:**81–105 (release of acid phosphatase).

Azakawa, T., Mitsui, T., and Hayashi, M., 1988, in: *The Biochemistry of Plants* (J. Preiss, ed.) Academic Press, New York, pp. 465–492 (α-amylase biosynthesis, including in rice).

Chandler, P. M., and Jacobsen, J. V., 1991, *Plant Mol. Biol.* **16:**637–645 (GA regulation of α-amylase transcription).

Chandler, P. M., Zwar, J. A., Jacobsen, J. V., Higgins, T. J. V., and Inglis, A. S., 1984, *Plant Mol. Biol.* **3:**407–418 (GA control of α-amylase gene expression).

Chrispeels, M. J., and Jones, R.L., 1980/81, *Isr. J. Bot.* **29:**222–245 (ER and α-amylase secretion).

Chrispeels, M. J., and Varner, J. E., 1967a, *Plant Physiol.* **42:**398–406 (α-amylase synthesis and release).

Chrispeels, M. J., and Varner, J. E., 1967b, *Plant Physiol.* **42:**1008–1016 (GA_3 and ABA action on α-amylase).

Dashek, W. V., and Chrispeels, M. J., 1977, *Planta* **134:**251–256 (cell-wall-hydrolyzing enzymes).

Fernandez, D. F., and Staehelin, L. A., 1985, *Planta* **165:**455–468 (secretion via Golgi bodies; GA action).

Fincher, G. B., 1989, *Annu. Rev. Plant Physiol. Plant Mol. Biol.* **40:**305–346 (endosperm mobilization in cereals).

Garcia-Maya, M., Chapman, J., and Black, M., 1990, *Planta* **181:**296–303 (regulation of α-amylase in wheat embryos).

Groat, J. I., and Briggs, D. E., 1969, *Phytochemistry* **8:**1615–1627 (GA and α-amylase in germinated barley).

Gubler, F., and Jacobsen, J. V., 1992, *Plant Cell* **4:**1435–1441 (GA-responsive elements in the barley high-pI amylase gene).

Harvey, B. M. R., and Oaks, A., 1974, *Planta* **121:**67–74 (GA and enzyme induction in maize).

Higgins, T. J. V., Zwar, J. A., and Jacobsen, J. V., 1976, *Nature* **260:**166–169 (GA_3 and mRNA synthesis).

Hooley, R., Beale, M. H., and Smith, S. J., 1991, *Planta* **183:**274–280 (GA perception at the plasma membrane).

Huttly, A. K., and Baulcombe, D. C., 1989, *EMBO J.* **8:**1907–1913 (wheat α-amylase promoter).

Jones, R. L., and Armstrong, J. E., 1971, *Plant Physiol.* **48:**137–142 (osmotic regulation of α-amylase).

Jones, R. L., and Jacobsen, J. V., 1991, *Int. Rev. Cytology* **126:**49–88 (review of synthesis and secretion of proteins in cereal aleurone cells).

Koehler, S. M., and Ho, T.-H. D., 1990, *Plant Physiol.* **94:**251–258 (GA-induced proteinase in barley).

Lai, D. M. L., Slade, A. M., and Fincher, G. B., 1993, *Seed Sci. Res.* **3:**65–74 (glucanases in wheat).

Mozer, T. J., 1980, *Cell* **20:**479–485 (transcriptional and translational control by GA and ABA).

Naylor, J. M., 1966, *Can. J. Bot.* **44:**19–32 (α-amylase induction in wild oat).

Ou-Lee, T.-M., Turgeon, R., and Wu, R., 1988, *Proc. Natl. Acad. Sci. USA* **85:**6366–6369 (GA induces a DNA-binding protein).

Rogers, J. C., and Rogers, S. W., 1992, *Plant Cell* **4:**1443–1451 (GA and ABA response complexes in barley).

Rushton, P. J., Hooley, R., and Lazarus, C. M., 1992, *Plant Mol. Biol.* **19:**891–901 (nuclear proteins bind to wheat α-amylase promoter).

Simpson, G. M., and Naylor, J. M., 1962, *Can. J. Bot.* **40:**1659–1673 (GA, maltase, α-amylase in wild oat).

Skriver, K., Olsen, F. L., Rogers, J. C., and Mundy, J., 1991, *Proc. Natl. Acad. Sci. USA* **88:**7266–7270 (*cis*-acting elements responsive to GA).

Stuart, I. M., Loi, L., and Fincher, G. B., 1986, *Plant Physiol.* **80:**310–314 (glucanases in barley).

Vigil, E. L., and Ruddat, M., 1973, *Plant Physiol.* **51:**549–558 (EM study of secretory vesicles).

Zwar, J. A., and Hooley, R., 1986, *Plant Physiol.* **80:**459–463 (nuclear runoff in wild oat, response to GA).

SECTION 8.2

Bewley, J. D., and Halmer, P., 1980/81, *Isr. J. Bot.* **29:**118–132 (endosperm/embryo interactions).

Bewley, J. D., Leung, D. W. M., and Ouellette, F. B., 1983, in: *Mobilization of Reserves in Germination* (C. Nozzolillo, P. J. Lea, and F. A. Loewus, eds.), Recent Advances in Phytochemistry, Volume 13, Plenum Press, New York, pp. 137–152 (regulation of enzymes for endosperm and oligomannan utilization in lettuce).

Davies, H. V., and Chapman, J. M., 1980, *Planta* **149:**288–291 (protein mobilization, feedback inhibition in cucumber).

Davies, H. V., and Slack, P. T., 1981, *New Phytol.* **88:**41–51 (review on regulation of reserve mobilization).

Davies, H. V., Gaba, V., Black, M., and Chapman, J. M., 1981, *Planta* **152:**70–73 (lipid mobilization in cucumber).

Dunaevsky, Y. E., and Belozersky, M. A., 1993, *Physiol. Plant.* **88:**60–64 (embryonic axis and phytohormone effects in buckwheat).

Gifford, D. J., Thakore, E., and Bewley, J. D., 1984, *J. Exp. Bot.* **35:**669–677 (GA and protein mobilization in castor bean).

Kern, R., and Chrispeels, M. J., 1978, *Plant Physiol.* **62:**815–819 (protein mobilization in mung bean).

Munoz, J. L., Martin, L., Nicolas, G., and Villalobos, N., 1990, *Plant Physiol.* **93:**1011–1016 (cytokinin effects on mobilization in chick pea).

Slack, P. T., Black, M., and Chapman, J. M., 1977, *J. Exp. Bot.* **28:**569–577 (testa and lipid mobilization in cucumber).

Van Onckelen, H. A., Caubergs, R., and DeGreef, J., 1977, *Plant Cell Physiol.* **18:**1029–1040 (hormonal control of α-amylase in bean).

Yamamoto, T., Shimoda, T., and Funatsu, G., 1982, *Sci. Bull. Fac. Agric. Kyushu Univ.* **36:**71–78 (proteinase in castor bean)

Chapter 9

Seeds and Germination
Some Agricultural and Industrial Aspects

9.1. INTRODUCTION

The fundamental importance of germination physiology to agriculture and horticulture is so obvious that it need hardly be stated, for almost all of our reliance on plants depends ultimately on the germinability of their seeds. The most straightforward dependence is when seeds are the starting materials for crops; in this case, we require that they have high viability, that their germination capacity is high (and therefore that they have no dormancy, at least under the conditions experienced during cultivation), and that germination is completed uniformly so as to produce vigorous plants closely similar in their stage of growth. All of these requirements concern aspects of seed physiology and biochemistry that have been covered in previous chapters. Also critical, especially when seeds are used directly as human or animal food, are the events occurring during seed development and maturation when the seeds' storage reserves are deposited, for the processes taking place then govern the quality and amount of materials that are nutritionally important. Another extremely important consideration is the place of seeds in the conservation of biodiversity, and as sources of material for plant breeding and plant improvement. Seeds are gene repositories which in most cases can be conveniently stored and preserved. It is extremely important that the efficiency with which this is done is maximized to secure high levels of seed longevity and quality.

For this final chapter, we have chosen some aspects of industry, agriculture and conservation in which seed physiology is important. These examples place events such as viability, the processes of germination, dormancy, and facets of reserve mobilization and seed production in an applied context. It is hoped that the reader will obtain from them some further appreciation of how the physiology and biochemistry of seeds are of fundamental relevance to humans.

9.2. MALTING

This section is not intended as a comprehensive account of malting (see the reference list for other sources) but instead it highlights the germination physiology on which the process rests.

Malt is cereal grain that has been allowed to germinate under controlled conditions, to make a very limited amount of seedling growth, and then been dried and lightly cooked. Wheat, oat, rye, millet, sorghum, and triticale can be malted, but by far the most commonly used cereal is barley. In some countries, barley is grown almost exclusively for malting, and in others, such as the United Kingdom, the malting industry consumes a substantial proportion (ca. 25%) of the barley crop. There are many uses for malt—malt extracts, syrups, malt flour—but most of it is used to provide the fermentable materials during the manufacture of beers, spirits such as whisky, gin, and vodka, and malt vinegar. It is estimated that the annual world production of barley exceeds 180 million metric tons, of which about 35% is destined for malting. This industry, then, depends on germination under controlled conditions on a massive scale.

The function of malting is to achieve the production of enzymes which, at a later stage, hydrolyze the starch reserves of the endosperm to make sugars available for fermentation. Amylolytic enzymes are of prime importance, but other enzymes also play important roles. Other essential processes occur, for example the production of flavor compounds, all of which contribute to the quality of the malt. The enzymes are derived principally from the aleurone layer and to a lesser extent from the embryo. The physiological and biochemical basis of malting therefore centers on the control of reserve mobilization in cereals, which was described in Chapter 8. It is necessary, however, that mobilization does not proceed to completion, when the reserves are utilized by the growing seedling; thus, an important feature of malting is the minimization of reserve mobilization and growth so as to avoid loss of potentially fermentable material brought about by the seedling. To this extent, then, malting involves the imposition of certain constraints on some of the normal germinative and postgerminative events.

Malting begins with imbibition by the barley grains during the process of steeping (Fig. 9.1), when large quantities of grain are submerged in water in steep tanks. The aim of steeping is to hydrate the grain to a water content sufficient to support germination, enzyme formation, and the movement of the mobilizing enzymes into the starchy endosperm, but not to encourage excessive growth. Steeping may continue for 50–70 h, but unless certain measures are adopted the quality of the grain can be severely impaired. For example, prolonged submergence in water restricts germination because of the so-called water sensitivity of the grain. This results from the anaerobic conditions imposed by excess water and leads to considerable delays in the germination of individual grains. Germination is then uneven, with a consequent reduction in extract yield of the malt. Nonuniform germination is avoided by interpolating periods of "air resting," when the water is drained off and the grain is exposed to air, say for 16 h. However, this treatment can itself have some undesirable effects, because the rootlets of the grains begin to grow ("chitting"), a process that utilizes some of the endosperm reserves and thus lowers the eventual yield of malt. Chitting and continued rootlet elongation may be checked by the application, for short time periods, of various chemicals, such as calcium hydroxide, sodium carbonate, or potassium bromate, a chemical that also decreases proteolysis. In some malting procedures, steeped grain is heated to 45°C in order to inhibit rootlet growth.

The overall water content of the grain is generally allowed to reach about 43% by the end of steeping, though slower-malting or high-nitrogen cultivars may be permitted higher moisture levels (ca. 48%). The early phase of steeping takes the grain to about 35% water content, the minimum needed for subsequent embryo growth; there then follows the air rest to encourage even germination; and, finally, overall moisture contents of 35–45% hydrate the endosperm sufficiently to permit enzyme production, migration, and some activity. Several steps may be taken to accelerate water uptake, including warm water or abrasion of the grain. It is essential, however, that the water content should not rise above the quoted levels, otherwise seedling growth will be excessive, at the expense of the storage reserves of the endosperm and with a reduction in malt yield.

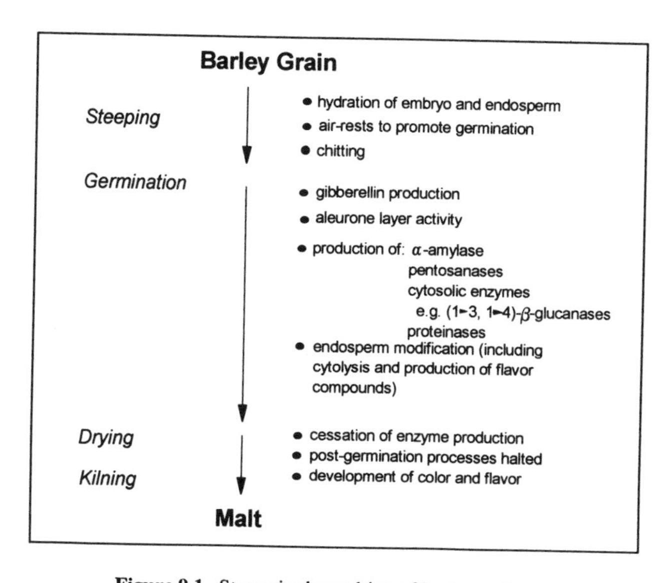

Figure 9.1. Stages in the malting of barley grains.

The conditions during steeping can also differentially affect enzymes that are important in malting. Higher temperatures (e.g., 20°C) depress the formation of (1→3,1→4)-β-glucanase, a cytolytic enzyme that hydrolyzes components of the endosperm cell walls [(1→3,1→4)-β-glucan], to expose the starch grains and make them more accessible for amylolytic attack. Steep temperatures of 13–15°C are consequently preferred, to encourage the production of this enzyme.

By careful practice, therefore, grains emerge from steeping with a water content set within narrow limits, with a small, even amount of germinative growth, with some cytolytic enzymes produced, but still without the substantial quantities of the amylolytic enzymes necessary for modification of the endosperm. Production of these enzymes, especially α-amylase, occurs largely after the steeped grain has been transferred to so-called germination beds, when growth of the germinated embryo continues to some extent, but more importantly, when the production of enzymes proceeds apace (Fig. 9.1). This stage generally takes place in huge beds accommodating as much as 50 metric tons of grain, which is turned and aerated periodically. The time spent in the germination bed varies with the malting practice but is generally from 4 to 8 days. During this time, the water content and the temperature are carefully regulated so as to encourage efficient enzyme production and satisfactory endosperm modification, but not utilization of the products of hydrolysis by the seedling which would lead to losses of malt yield (see Fig. 9.2). Addition of chemicals to the malting barley accelerates endosperm modification (gibberellic acid) and reduces rootlet growth (potassium bromate).

The production of gibberellins is considered to be the crucial role of the germinated embryo in malting. These hormones diffuse from the scutellum to the aleurone layer, to promote enzyme synthesis and secretion into the starchy

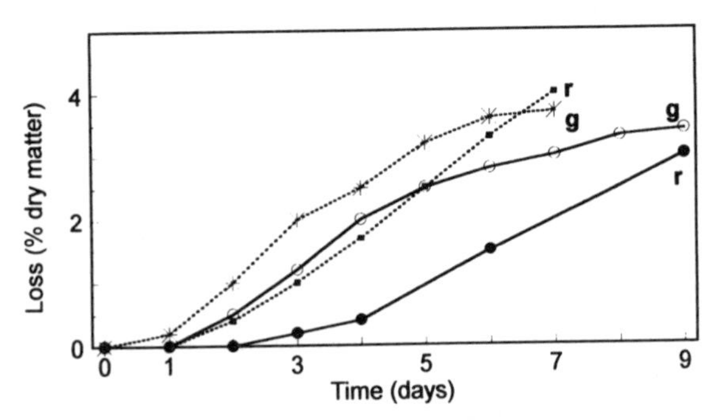

Figure 9.2. Effect of temperature on losses of seed dry matter occurring during malting. Temperatures during germination and growth: - - - -, 19–22°C; ——, 13–17°C. r, Losses related to respiration; g, losses related to growth of rootlets (chitting). After Briggs (1978).

endosperm (Section 8.1). As a consequence of enzyme activity, modification of the endosperm occurs with the pattern shown in Fig. 9.3. There is some debate, however, concerning the role of embryo-derived α-amylase in this process, and when the embryonic contribution is significant, the modification pattern may be altered accordingly. Addition of gibberellic acid to malting grains is now a fairly common, even a routine procedure in those countries where it is permissible. This growth regulator accelerates malting (thus cutting costs) and makes endosperm modification more uniform. Even greater rates and uniformity of malting are achieved by abrading grains before the addition of GA_3. The entry of the growth regulator over the whole of the grain surface, to reach the underlying aleurone layer, is hence greatly enhanced. α-Amylase is the gibberellin-induced enzyme whose activity is central to endosperm modification during malting (Section 8.1.1). But the activities of other enzymes such as pentosanases, β-glucanases, and proteinases are also of great importance. The pentosanases (which are also gibberellin induced) act on the aleurone cell walls and thus assist in the outward passage of other enzymes, especially α-amylase, into the starchy endosperm. As we noted previously, (1→3,1→4)-β-glucanase is a cytolytic enzyme responsible for the breakdown of the cell walls of the starchy endosperm [which consist of about 70% (1→3,1→4)-β-glucan] thus facilitating the accessibility of the enclosed starch grains for α-amylase, especially when the malt is later mixed with water ("mashing") to make fermentable wort. The synthesis and secretion of this cytolytic enzyme are promoted by gibberellin. Proteolytic attack

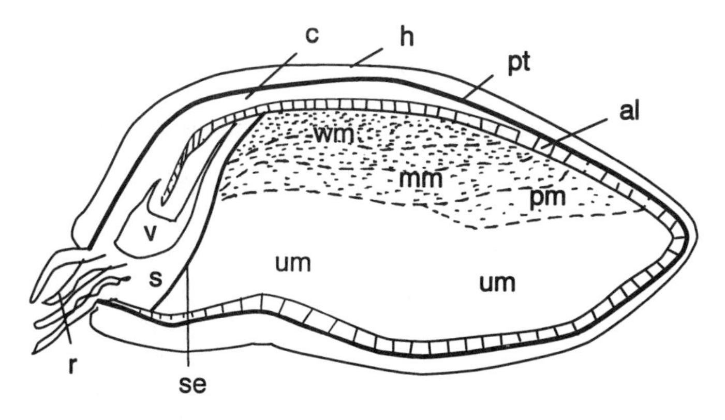

Figure 9.3. Endosperm modification in a malted barley grain. wm, Well-modified endosperm; mm, moderately modified endosperm; pm, partly modified endosperm; um, undermodified endosperm; al, aleurone layer; c, coleoptile; h, husk; pt, pericarp/testa; r, rootlets; s, scutellum; se, scutellar epithelium; v, vascular tissue. For a comparison with the pattern of α-amylase production under nonmalting conditions, see Fig. 7.3. After Palmer (1980).

by the gibberellin-induced proteinases also assists in rendering starch grains more available for amylolysis by removing the surrounding protein bodies.

By judicious regulation of the conditions, as outlined previously, satisfactory malt is thus obtained. When the malting process is terminated, the malted grain consists of friable endosperm containing high amount of enzymes, including α-amylase (which will later participate in the hydrolysis of the starch to produce sugars for fermentation), altered protein, and various other chemicals that are important for the quality of the final product.

Since germination is clearly central to this industrial process, nongerminable grains cannot be malted! Hence, the viability and dormancy of barley selected for malting are of critical importance. Viabilities lower than 98% are unacceptable, as is dormancy of more than 3–4% of the grain. Dormancy can be a serious problem in freshly harvested grain, especially in some seasons and geographical locations. Such grain has to be stored, sometimes at elevated temperatures so as to accelerate afterripening, which increases the costs incurred in malting. On the other hand, if the grain does not have some dormancy, especially while still on the ear, it may be subject to preharvest sprouting in wet, cool weather (see Section 9.3).

In conclusion, we can see how the important industrial process of malt production involves so many different aspects of seed physiology and biochemistry, including water uptake, embryo growth, hormone production, enzyme production, some of the processes of reserve mobilization, dormancy, and viability. Successful, economically efficient malting requires an understanding of the basic seed biology in order to control and manipulate all of these events.

9.3. PREHARVEST SPROUTING IN CEREALS

Cereal grains, especially wheat, maize, rice, and barley, are prone to germinate while still on the ear of the parent plant. This is the situation in many "wild type" cultivars, i.e., it is not a genetically altered viviparous state. Germination can occur just after harvest when the stooked sheaves are wetted by rain, but in parts of the world where mechanical harvesting and threshing are practiced, germination at this stage is avoided. In all cases, however, sprouting can take place on plants still standing in the field; this is called preharvest sprouting ("preharvest" refers to the time before the grain is gathered in and threshed) and is a phenomenon that in some years is responsible for large losses to the agricultural industry. In this section, we will concentrate on its occurrence in wheat, an advanced state of which is shown in Fig. 9.4.

Preharvest sprouting occurs in wet or humid conditions in many regions of the world, including northwest Europe, North and South America, Australia, and New Zealand. The extent of the problem is illustrated by a few examples. In

Figure 9.4. Sprouting in the wheat ear. The ear (B) shows a very advanced case of sprouting, induced in the laboratory. This ear would also be prone to sprouting in the field but germination would not be as extensive. The ear on the far right (C) shows sprouting as it is often found in the field. Two nonsprouted ears (A) are shown for comparison. Photograph courtesy of M. Gale.

Australia, where white-grained cultivars predominate, 1.8 million metric tons of grain was spoiled in northern New South Wales in 1969 as a result of germination in the ear. Though most sprouting damage in United State occurs in the white-grained wheat-producing areas of the northwestern seaboard and Idaho (as a result of which Japan ceased importing these wheats for a period in 1968), preharvest sprouting sometimes also appears in the plains states from Texas to North Dakota. In Nebraska in 1977, for example, about 12% of the red-grained wheat and 19% of the durum wheat were affected. Preharvest sprouting is considered to be the most serious problem facing wheat growers in the northern region of Brazil, and it causes great losses to the producer. In Europe, the United Kingdom experienced particularly bad years in 1985 and 1987 especially in southeast England, where average sprouting levels of about 20% were recorded. Consequently, very little of the wheat crop could be used for milling, and farmers suffered cash losses of 50–60 million pounds sterling in both years. To redress the loss of homegrown material, extra wheat had to be imported, leading to substantial balance of payments deficits. The economy of Poland suffered considerable losses when sprouting of wheat occurred in 3 years in the period 1984–1988, amounting to a value of about U.S. $92 million per year.

The extent of sprouting varies according to cultivar, provenance, and growing season. Many countries routinely screen wheat cultivars and grade them for sprouting susceptibility under local conditions. Wheat in regions that are warm and dry during grain development and maturation is not likely to suffer from preharvest sprouting, but that in cooler, wet localities is at risk. Obviously, since meteorological conditions vary from year to year, crops of some seasons may completely escape sprouting damage, whereas those in other years are badly afflicted.

The period over which sprouting can occur is wide, extending from 2–3 weeks after fertilization until harvest; hence, sprouting is not confined to grains that have been dried but also occurs when moisture contents are still quite high. This was seen in the 1977 United Kingdom crop (Fig. 9.5), when high sprouting levels were observed in so-called dough-ripe grains—of about 45% (FW) water content. In this figure, two modes of sprouting are expressed: visible and nonvisible. The former category includes grains in which some rootlet and coleoptile growth occurs, i.e., a typical, germinated grain. In nonvisible sprouting, however, there is no radicle or rootlet growth, and though the coleoptile elongates, it fails to emerge from beneath the testa/pericarp, extending, instead, underneath these enclosing tissues. Hence, it is not immediately apparent that the grains have germinated, and only close inspection reveals that they have done so.

Sprouted grains are obviously of imperfect appearance and will not be purchased for milling, though it can be diverted into animal feed. The major objection to sprouted grain is not aesthetic, however, but relates to of its relatively high α-amylase content. Enzymatic hydrolysis of starch by this enzyme proceeds in the sprouted grain, but since the enzyme persists in the flour made from the grain, it also continues during the baking process, producing

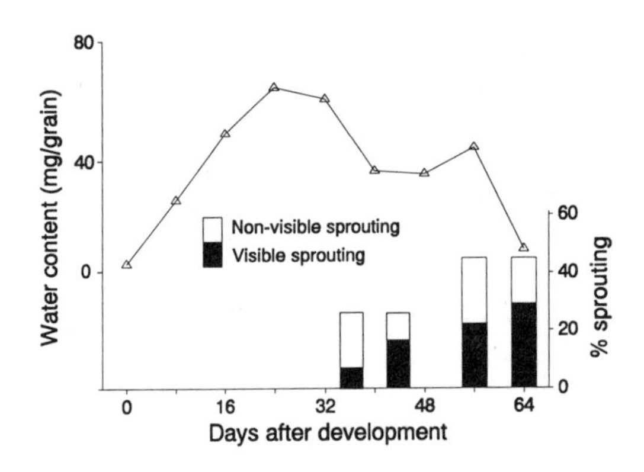

Figure 9.5. Time course of preharvest sprouting. Sprouting on the ear of wheat grains of different ages, southeast England, 1977. Water content is also shown (△). See text for explanation of visible and nonvisible sprouting. After Mitchell *et al.* (1980).

undesirable quantities of sugars. The bread-making quality of the flour is thus seriously impaired, for the loaf has sticky crumb and a darkly pigmented crust caused by caramelization of the sugars. The production line for mass-produced, sliced loaves is halted when the slicing blades cease to cope with the gummy texture of the bread. It can be appreciated, therefore, why millers reject even slightly sprouted grain. Certain cultivars, however, have a high α-amylase content even in the absence of germinative sprouting, a feature that can be detected in the routine testing of incoming batches of grain being offered for purchase. To be selected as a good milling and baking wheat, a cultivar must produce grains having less than a certain critical content of α-amylase.

9.3.1. The Physiology of Preharvest Sprouting in Wheat

Preharvest sprouting occurs because grain is too germinable! During the life of a seed, a period of rest (quiescence, sometimes also dormancy) normally intervenes between embryo development and germination itself. The controlling factors enforcing rest are poorly understood, but they involve constraints imposed by parental tissues, and possibly effects of inhibitors such as ABA. These controls, which should operate in the early phases of seed development, obviously are ineffectual in those cases where sprouting occurs in very young grain. It is not known why, but the absence of the inhibitor ABA does not seem to be the explanation, although sensitivity to the hormone might be implicated. In older grain, embryo development has long been completed and drying has begun. Such grain will sprout when wetted as long as there is no dormancy; hence, preharvest sprouting is intimately dependent on the presence or absence of dormancy. Dormancy in wheat is strongly expressed at germination temperatures above about 13°C (see Fig. 5.11), a condition referred to as relative dormancy. Sprouting, therefore, will take place even in grains with relative dormancy when the temperature is low. This is one reason why preharvest sprouting is common when cool, wet weather prevails at certain times during maturation.

For a further appreciation of preharvest sprouting, and of how this occurs at different times on the ear, we clearly need to know something about the time course of dormancy in the developing and maturing grain. The degree of dormancy is dependent on the environmental factors operating during seed development, particularly the temperature. Grains of wheat, barley, and wild oat are more dormant at maturity when development has taken place at relatively low temperatures (e.g., 10–20°C) than at higher temperatures (e.g., 20–28°C). Especially important seems to be the temperature occurring later in grain maturation (Fig. 9.6). This explains why sprouting susceptibility of *mature* grain (i.e., at or close to harvest dryness) is less when temperatures during maturation have been low.

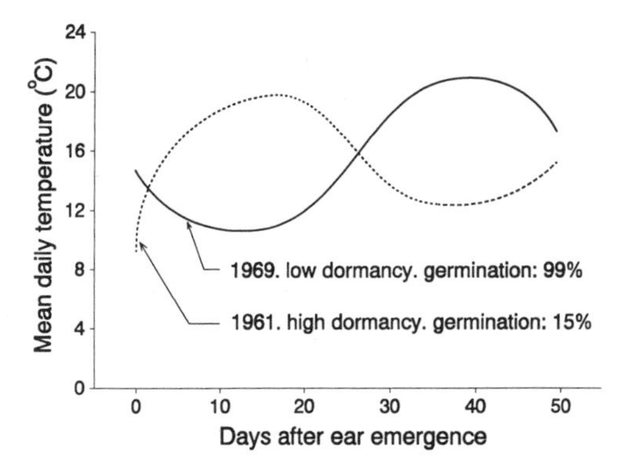

Figure 9.6. Effect of temperature during grain development on the dormancy of mature barley grain. Mean daily temperatures experienced over the course of grain development and maturation are shown for two crops of barley. Germinability was measured 3 weeks after harvest. After Reiner and Loch (1975).

When wheat grains are examined at intervals during their development, it can be seen how the pattern of dormancy varies with the temperature (Fig. 9.7). The first point to note is how the low dormancy of mature grains is achieved if higher temperatures (20°C) are obtained during development. Such grains actually enter a period of dormancy (at 30–40 days, when the germination tested at 20°C is low) but they rapidly pass out of it as they afterripen on the ear. Afterripening occurs because the water content of 40-day-old grain is low (ca. 30–35%), and it is also enhanced by the higher temperature. On the other hand,

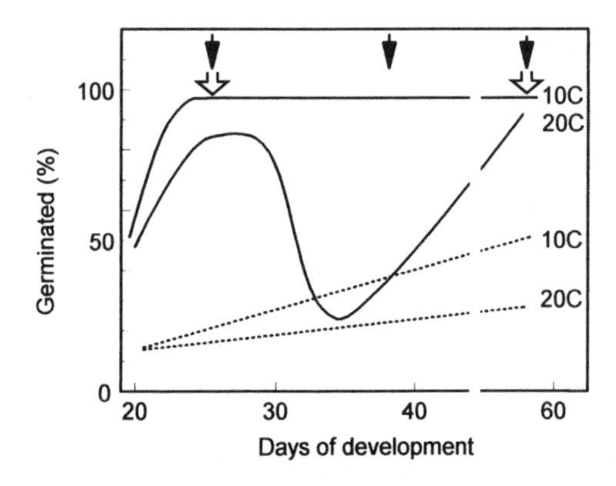

Figure 9.7. Diagram showing patterns of dormancy in developing wheat grains (cv. Sappo). Plants bearing developing grains were transferred to two different temperature regimens: ——, 16 h day at 25°C; - - - -, 16 h day at 10°C, 8 h night at 10°C. At intervals throughout development and maturation grains were removed from the ears and tested for germinability (i.e., dormancy) at two temperatures, 10 and 20°C. Arrows indicate sprouting susceptibility: black arrows, at a germination temperature of 10°C; white arrows, at a germination temperature of 20°C. From data in Black *et al.* (1987) and unpublished observations.

grains developing and ripening at 10°C have a low germinability when young (ca. 25 days old) and show only marginal afterripening; so even at 60 days old they are still substantially dormant and do not respond expeditiously even to a germination temperature of 10°C.

Let us consider briefly what might be the fate of these two batches of grain—developing at 10°C and 25°C/20°C, respectively—if subjected to rain at different times. Grain developing and maturing at 25°C/20°C has a fairly low dormancy at ages up to about 25 days and is therefore highly susceptible to sprouting when wetted, whether the temperature remains at about 20°C or falls to 10°C for a few days. Thereafter—up to about 40 days—grains would have a propensity for sprouting only if cool, wet conditions arise, because at a germination temperature of 20°C they express dormancy. After about 40 days, cool, wet weather encourages sprouting, but since the grains are afterripening, heavy rainfall even at warmer temperatures increasingly induces grain germination. At maturity, when afterripening has removed dormancy, grains are highly susceptible to sprouting in both cool and warm wet weather. Grains that are in the trough of dormancy (30–40 days old) are sensitive to a day or two of low temperatures (e.g., 10°C), which promotes subsequent germination, probably because of the low-temperature breaking of dormancy. But when subjected to cool conditions throughout their entire developmental and maturation period, grains remain dormant and resist sprouting even when severely wetted. In nature, such extreme temperature differences during development as those quoted here will not occur; notwithstanding this, temperature-induced variation in dormancy and germination will arise to some extent or other.

We can see, therefore, that preharvest sprouting in wheat depends substantially on the dormancy characteristics of the grain, on the temperature dependence of dormancy expression, and on factors involved in the relief of dormancy (low temperatures and afterripening) which are effective even on the ear.

Some steps can be taken to reduce the incidence of preharvest sprouting. Taking the weather conditions into account, sprouting susceptibility can sometimes be predicted, and late sprouting can thus be circumvented by harvesting the crop earlier than usual. Further, breeding programs can produce wheat cultivars with deeper grain dormancy. Though protecting against sprouting, this can have undesirable consequences, for it can interfere with seed testing after harvest or delay germination in the field, which is particularly disadvantageous for winter wheats sown in the fall. It could be argued that real progress in dealing with the sprouting problem may have to await the time when we have a fuller knowledge of the biochemical basis of dormancy, since only when we understand the mechanisms of dormancy can we begin to control it. There is evidence in wheat that sprouting-susceptible cultivars (i.e., low dormancy types) have a much reduced sensitivity to ABA, and so, although their ABA content is close to that of the sprouting-resistant cultivars, they may fail to respond to it. Thus, it follows that if dormancy is induced by ABA the less-sensitive cultivars will be less dormant, possibly because expression of the dormancy genes does not occur, or is much reduced (Section 5.3.2). If we can begin to identify sprouting susceptibility in terms of molecular genetics, an important step will have been taken in the systematic, soundly based development of sprouting control.

9.4. VIABILITY AND LONGEVITY OF SEEDS

While much of the seed that is harvested is used immediately for consumption, that which is replanted must be stored until the time it is sown. Additionally, long-term storage of the plant gene pool is required for conservation, breeding, and improvement purposes, and this is best achieved by maintaining dry seeds under suitable conditions. In this section we consider some of the research directions being pursued in order to understand and improve seed quality and longevity in storage.

9.4.1. Ancient Seeds

Considerable controversy has surrounded the claims that seeds remain viable for many hundreds or even thousands of years. A list of some of the more spectacular claims is presented in Table 9.1, along with reasons for accepting or rejecting them. Perhaps the most persistent myth concerning seed longevity is that viable grains of wheat and barley were uncovered during archaeological excavations of ancient Egyptian buildings. Reports that "mummy" grains could germinate and produce seedlings were given considerable publicity and credence during the 19th and early 20th centuries. But recent, scientifically rigorous studies show unequivocally that most ancient grains (particularly the embryos) have undergone severe morphological and physiological degradation (including carbonization) with accompanying total loss of viability. Some stored grains retain their original shape and even much of their cell fine structure, although on hydration considerable disintegration occurs. It is worth reemphasizing, then, that *there is no scientific proof for the retention of viability by ancient cereal grains*.

The most extreme claim for longevity is for the arctic lupin (*Lupinus arcticus*), seeds of which were discovered frozen and buried in the Yukon Territory in Canada. These seeds were removed from ancient rodent burrows containing remnants of a nest, fecal matter, and the skull of a lemming species. Dating of nests and remains of arctic ground squirrels found buried under similar conditions in central Alaska showed them to be over 10,000 years old, and hence it was concluded that the seeds in the Yukon must be of this age too. The highly

Table 9.1. An Examination of the Claims for Extended Longevity in Seeds

Species	Location	Age	Status of seed	Comments
Stored in dry conditions				
Barley	Tomb of King Tutankhamun	ca. 3350 years	Nonviable	Extensively carbonized
Wheat	Various ancient Egyptian tombs	3000 years or more	Not known	Claims for their viability have been made, but age and source of grains never authenticated
Wheat	Thebes	4000–5000 years	Nonviable	Some cell fine structure conserved, although degraded on rehydration
	Feyum	6400 years	Nonviable	
Canna compacta	Santa Rosa de Tastil, Argentina	ca. 600 years	Viable	Enclosed in a nutshell forming part of rattle. Shell dated at 600 years and seed probably of same age
Albizzia julibrissin	China to British Museum	200 years	Viable in 1940	Germination started accidentally. Viable seedlings produced
Cassia multijuga	Museum of Natural History, Paris	158 years	Viable	Wholly authenticated history from collection to sowing
Buried in soil or water				
Arctic lupin (*Lupinus articus*)	Miller Creek, Yukon	>10,000 years	Viable	Doubtful. Age was derived indirectly from geological data which could be wrong. No direct dating evidence
Indian, Asiatic, Oriental or sacred lotus (*Nelumbo nucifera*)	Pulantien basin, Southern Manchuria	150–several thousand years	Viable	Evidence of extreme age from indirect geological and lake-draining data. Radiocarbon dating suggests there are viable seeds 430–705 (± 165) years old.
Indian lotus	Kemigawa, near Tokyo	3000 years	Viable	Seeds found on submerged boat radio-dated at 3000 years. No direct measurements of age of seed, which could have settled in sediments after shedding from modern plants.
Chenopodium album *Spergula arvenis*	Denmark and Sweden archaeological digs	>1700 years	Viable	No direct dating of seeds. Could be modern seeds dispersed into archaeological digs.

circumstantial nature of the "evidence" militates against its acceptability, and the putative longevity of the lupin seeds still requires confirmation using direct dating techniques. In the absence of reliable evidence from radiocarbon or alternative dating methods, other claims for extended longevity, e.g., *Nelumbo nucifera* in Japan, *Chenopodium album* and *Spergula arvensis* (~1700 years), must be regarded with considerable skepticism (Table 9.1).

The longevity of a *Canna compacta* seed has been put at about 600 years. The seed was collected from a tomb in Argentina, enclosed in a *Juglans australis* nutshell forming part of a rattle necklace. Radiocarbon analysis of the nutshell and surrounding charcoal remnants was used to date the seed. Insuffcient seed material was available for analysis, but the only way in which the *C. compacta* seed could have arrived inside the *J. australis* nutshell is for it to have been inserted there through the still-developing nutshell, while it was soft. Then the nutshell hardened and dried with the seed inside, forming a rattle. Hence, the seed must have been at least the same age as the shell, and possibly older. In his book on "Seed Aging" Priestley gives an excellent critical and historical account of the many claims for seed longevity. From his account, seeds from Pulantien basin in China appear to have retained viability for over 700 years, with the possibility of there being some 1000-year-old viable seeds.

Some seeds stored in museums and herbaria are known to have survived for more than 100 years. Specimens of *Albizzia julibrissin* collected in China in 1793 and deposited in the British Museum, London, germinated after attempts to quench a fire started by an incendiary bomb that hit the museum in 1940. Three of the resulting seedlings were planted nearby, but two eventually met their end during a bombing raid the following year. Tests on old collections of seeds from the Museum of Natural History in Paris in 1906 and 1934 showed that their longevity ranged from 55 to 158 years (Table 9.2); most of the long life span of these and other specimens was attained even though the storage conditions were arbitrary (generally, warm temperatures and low relative humidity). Longevity might have been greater had different conditions for storage been used.

A number of projects have been carried out, or are in progress, to determine longevity of seeds buried in soil. A study of buried weed seed, scheduled to last for 50 years, was initiated in 1972. After only 2.5 years of burial as few as 4 of the 20 species tested retained more than 50% viability. The longest controlled burial experiment to date is one initiated by W. J. Beal in 1879. He selected seeds of 21 different species of plants common in the vicinity of Michigan Agricultural College in East Lansing. Fifty seed lots of each species were mixed with moist sand in unstoppered bottles and buried in a sandy knoll. At regular intervals since, bottles have been unearthed and the germinabilty of the seeds tested. At the 100th anniversary of the experiment, of the 21 species originally buried, only *Verbascum blattaria* seeds had retained their viability throughout, although many species were capable of germinating after 35–50 years (Table 9.3).

Table 9.2. Viability Record of Some Old Seeds from the Museum of Natural History, Paris[a]

Species	Date collected	% Germinated in		Longevity (years)
		1906	1934	
Mimosa glomerata	1853	50	50	81
Melilotus lutea	1851	30	0	55
Cytisus austriacus	1843	10	0	63
Dioclea pauciflora	1841	10	0	93
Trifolium arvense	1838	20	0	68
Stachys nepetifolia	1829	10	0	77
Cassia bicapsularis	1819	30	40	115
Cassia multijuga	1776	—	100	158

[a]After Becquerel (1934).

Many different factors determine the longevity of seeds buried in soil, even in controlled experiments. Soil characteristics such as moisture content and pH are important, as are seed attributes such as depth of dormancy and hardness or impermeability of the seed coat, which vary within any population of seeds of the same species.

9.4.2. Viability of Seeds in Storage

The majority of seed species retain their viability when dried; in fact, drying is the normal, final phase of maturation for most seeds growing in temperate climates. Hence, it is common for seeds to be stored in a "dry" state, or more correctly, with a low moisture content. There are, however, some seed species that must retain a relatively high moisture content during storage in order to maintain maximum viability. These are the so-called "recalcitrant" seeds, which were discussed in Chapter 3.

9.4.2.1. The Relationship between Temperature and Moisture during Storage

Seeds that can be stored in a state of low moisture content are called "orthodox," and their viability under certain storage conditions conforms to some general rules, as follows:

1. For each 1% decrease in seed moisture content the storage life of the seed is doubled.
2. For each 10°F (5.6°C) decrease in seed storage temperature the storage life of a seed is doubled.

Table 9.3. Longevity of Some of the Seed Species from W. J. Beal's Buried-Seed Experiment Started in 1879[a]

Species	Longevity (years)
Agrostemma githago	<5
Amaranthus retroflexus	40
Brassica nigra	50
Capsella bursa-pastoris	35
Euphorbia maculata	<5
Lepidium virginicum	40
Polygonum hydropiper	50
Trifolium repens	5
Oenothera biennis	80 (10%)[b]
Verbascum blattaria	100 (42%)

[a]After Kivilaan and Bandurski (1981).
[b]Figures in parentheses are percent germinated after the time indicated.

3. The arithmetic sum of the storage temperature in degrees F and the percent relative humidity (RH) should not exceed 100, with no more than half the sum contributed by the temperature.

These "rules of thumb" clearly indicate that temperature and moisture content of the seed are major factors in determining viability in storage. Let us now consider these factors, along with others that might play a role, as follows:

1. *Moisture content.* This is defined by International Seed Testing Association (ISTA) practices as:

$$\%\ \text{Moisture content} = \frac{\text{Fresh weight of seed} - \text{dry weight of seed}}{\text{Fresh weight of seed}} \times 100$$

When the moisture content is high (> 30%), nondormant seeds may germinate, and from 18 to 30% moisture content rapid deterioration by microorganisms can occur (see Section 9.4.2.3). Seeds stored at moisture contents > 18–20% will respire, and in poor ventilation the generated heat will kill them. Below 8–9% moisture content there is little or no insect activity, and below 4–5% moisture content seeds are immune from attack by insects and storage fungi, but they may deteriorate faster than those maintained at a slightly higher moisture content. The activities of seed storage fungi are ultimately more influenced by the RH of the interseed atmosphere than by the moisture content of the seeds themselves. This is because the moisture content of some seeds (e.g., oil seeds) may be different from that of others (e.g., starchy seeds) even though both are in equilibrium with the same atmospheric RH. For example, all cereals and many of the legumes that are high in starch and low in oil have a moisture content of about 11% at 45% RH, whereas oil-containing seeds (e.g., rape) have a moisture content of only 4–6% at this RH (Fig. 9.8).

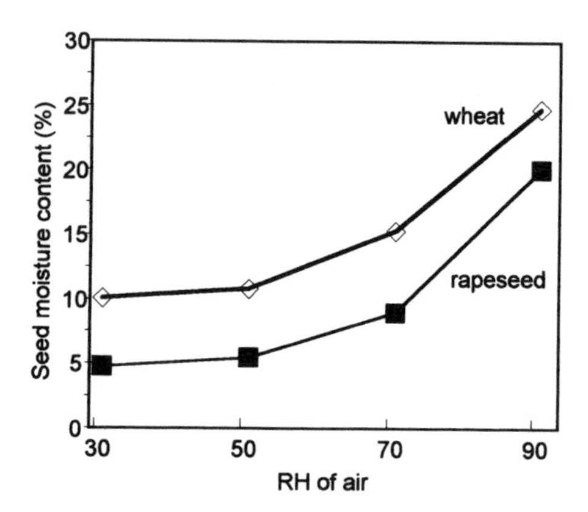

Figure 9.8. Changes in equilibrium moisture content of wheat and rapeseed with relative humidity of the ambient air. Based on work of Kreyger (1972) cited in Thomson (1979).

In Chapter 3 we considered the types of water that are found in dry seeds. Many agronomically important seeds stored for short periods prior to sowing probably contain at least Type 3 water, or are more hydrated. Those stored for longer times, under more carefully controlled conditions, contain at least Type 2 water. The viability of seeds obviously will be affected by the types of water they contain.

2. *Temperature.* Cold storage of seeds at 0–5°C is generally desirable, although this may be inadvisable unless they are sealed in moisture-proof containers or stored in a dehumidified atmosphere. Otherwise the RH of storage could be high, causing the seeds to gain moisture; if they are then brought out to a higher temperature (e.g., for transport), they might deteriorate because of their high moisture content. At moisture contents below 14%, no ice crystals form within cells on freezing; thus, storage of dry seeds at subzero temperatures after freezing in a dry atmosphere should improve longevity. Freeze-drying of certain seeds improves their longevity in storage, but others may be killed by this treatment. A method that is increasing in popularity, particularly for small batches of seeds for gene banks (Section 9.4.3.2), is to keep them immersed in liquid nitrogen. Under this condition seeds should survive indefinitely.

3. *Interrelationships.* Several mathematical models have been proposed to relate the viability of seeds to their storage environment. Discussion of these models is beyond the scope of this text, but the reader may obtain details by reading the publications and reviews of E. H. Roberts and his colleagues (Roberts, 1972, 1973; Ellis and Roberts, 1980), some of which are referred to at the end of this chapter. Their studies of the survival curves of seven species (wheat, rice, barley, pea, broad bean, onion, and tomato) have shown that under a given set of constant conditions a sample of seeds has a particular mean

viability period, and that there is a random distribution of the viability periods in a population around this mean value. This is illustrated in Fig. 9.9, from which it can be seen that although the spread of distribution (of which σ, the standard deviation, is a measure) increases if the mean viability period (p) of the seed lot is enhanced by improving the storage conditions, the coefficient of variation (which is the standard deviation expressed as a percentage of the mean value of distribution—$\sigma/p \times 100$) remains the same. The resulting survival curves are sigmoid curves (negative cumulative normal distributions, or ogives). When percentage viability is plotted on a probability scale, survival curves are transformed to straight lines. This implies that under any constant storage conditions

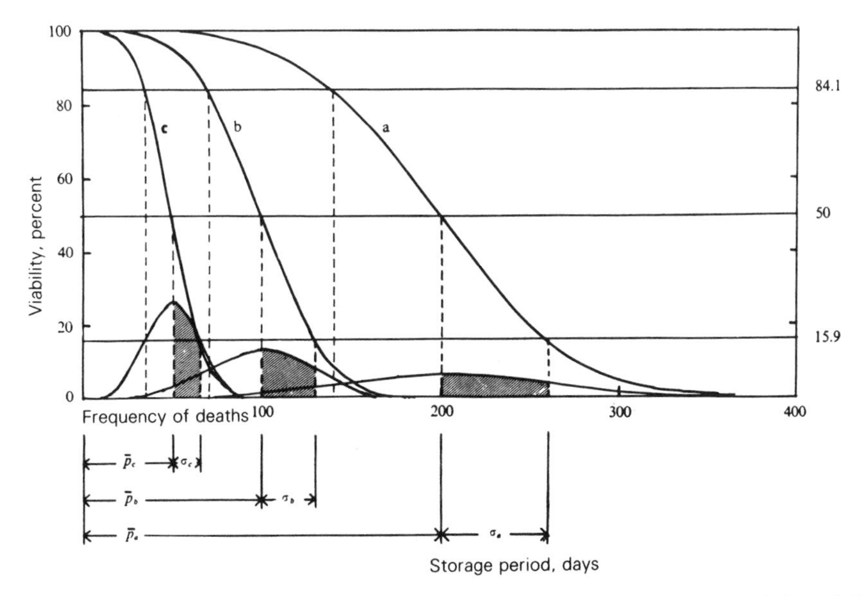

Figure 9.9. The relationship between the mean viability period and standard deviation of the distribution of the viability periods of individual seeds of a seed lot. The figure shows that an estimate of mean viability period (p) may be made by noting the point on the time scale at which the survival curve intersects the 50% level of germination. The standard deviation (σ) may be estimated by noting the point at which the survival curve intersects the 15.9% (or 84.1%) level and measuring the distance from this point on the time scale to the point representing the mean viability period. This is based on the fact that the area under the normal curve between the mean and 1 S.D. contains 34.1% of the area under the whole curve (shaded areas). It is assumed that three seed samples from the same lot have been stored under different conditions such that the mean viability period of sample a ($\bar{p}_a$) is twice that of sample b ($\bar{p}_b$) and four times that of sample c($\bar{p}_c$). It can be seen that the ratio $\bar{p}:\sigma$ is the same for all samples. The value p can be estimated quite accurately graphically when the data for percentage viability are transformed to probit values or when they are plotted on a probability scale, since survival curves then become straight lines. After Roberts (1973).

the frequency distribution of seed deaths in time is normal, even in adverse storage environments where death is rapid.

The relationship between mean storage temperature, moisture content, and mean viability periods is relatively simple, and the data for *Vicia faba* are plotted in Fig. 9.10.

9.4.2.2. Other Factors That Affect Seed Viability during Storage

Besides the interrelationships between temperature, moisture content, and time, other factors must be considered when attempting to determine the optimum storage conditions for a particular species:

1. Cultivar and harvest variability. Different cultivars and harvests of a particular species may show different viability characteristics under the same

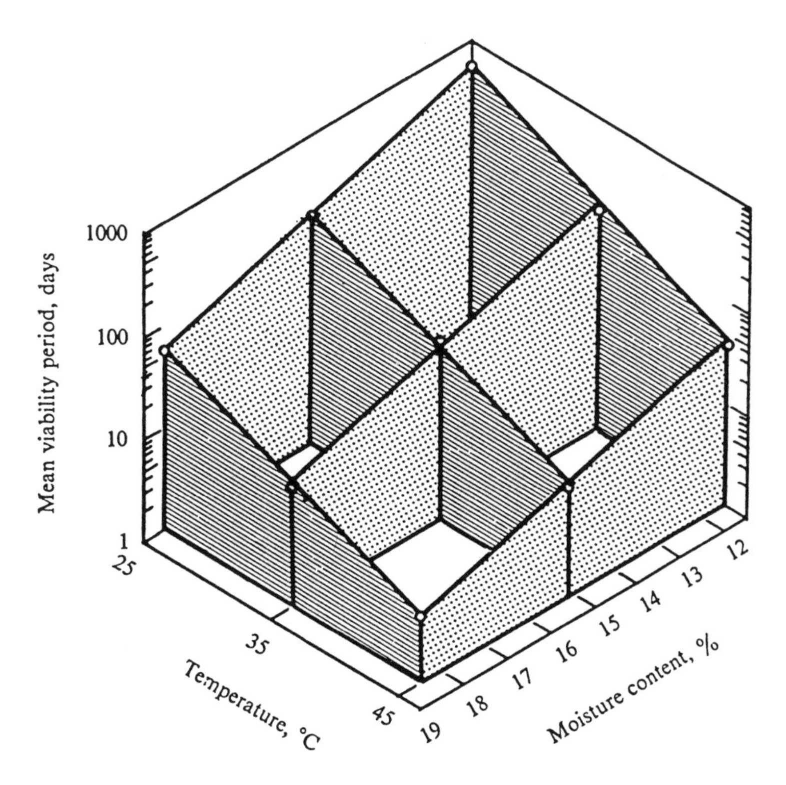

Figure 9.10. Isometric three-dimensional graph showing the relationship between moisture content, temperature in storage, and mean viability period for broad bean (*Vicia faba*). Time (days) is plotted on a log scale. The open circles represent the experimental points for the observed times for germination to decline to 50%. After Roberts (1973).

storage conditions. These differences are relatively small under good storage conditions, but under adverse conditions (elevated temperature and RH) they can be quite large. For example, the relatively hard flint and dent varieties of maize remain viable longer than starchy or sweet ones in open (unsealed conditions) storage. But in closed storage, at fairly constant moisture contents, few differences are evident.

Until recently, it was thought that the maximum potential longevity of seeds in storage occurred if seeds were harvested at physiological maturity (end of the seed-filling period). However, studies on the seeds of several species (e.g., barley, wheat, pearl millet, and tomato) have shown that, in terms of seed storage physiology, they continue to mature after the completion of seed filling, and this can result in increased longevity in subsequent storage. In spring wheat grains, for example, maximum potential longevity was achieved 10 days after physiological maturity, and 12–13 days after in grains of spring barley.

2. Pre- and postharvest conditions. Environmental variation during seed development usually has little effect on the viability of seeds, unless the ripening process is interrupted by premature harvesting. Weathering of maturing seeds in the field, particularly in conditions of excess moisture or freezing temperatures, results in a product with inferior storage potential. Mechanical damage inflicted during harvesting can severely reduce the viability of some seeds, e.g., certain large-seeded legumes. Cereals are largely immune from mechanical injury, presumably because of the protective outer structures, the palea and lemma. Small seeds tend to escape injury during harvest, and seeds that are spherical tend to suffer less damage than do elongated or irregularly shaped ones. During storage, injured or deeply bruised areas may serve as centers for infection and result in accelerated deterioration. Injuries close to vital parts of the embryonic axis or near the point of attachment of cotyledons to the axis usually bring about the most rapid losses of viability. High temperatures during drying, or drying too quickly or excessively, can dramatically reduce viability.

3. Oxygen pressure during storage. If seeds are not maintained in hermetic (closed or airtight) storage at low moisture contents, then even under conditions of constant temperature and moisture the gaseous environment may change as a result of respiratory activity of the seeds and associated microflora. For example, it takes pea seeds about 11 weeks, when stored at 18.4% moisture content at 25°C (rather extreme conditions) in a closed atmosphere, to decrease the oxygen from 21% to 1.4% and increase the carbon dioxide from 0.03% to 12%. During this time, there is a 50% loss of viability. In open storage, seeds maintained in an atmosphere of nitrogen may retain their viability considerably longer than those placed in replenished air or oxygen, although there is probably no advantage over storing seeds hermetically sealed in air. In fact, for storage at relatively low

temperatures and moisture there is probably little benefit in using controlled atmospheres, i.e., reduced oxygen pressures. There may be advantages for short-term storage in conditions of high temperature or moisture content, but it is probably simpler to reduce either the temperature or moisture content. Hermetic sealing provides a simple and convenient method of controlling seed moisture content after the seeds have been dried adequately.

4. Fluctuating storage conditions. Onion and dandelion seeds stored under conditions of alternating high and low RH lose viability proportionately to the length of time that the seeds are subjected to the high RH. During lengthy (e.g., 8-week) cycles of high and low RH, however, seeds deteriorate as rapidly as when kept only at the higher RH. In contrast, increases in temperature *per se* at intervals during cold storage do not necessarily have deleterious effects on viability. For example, batches of red clover (*Trifolium pratense*) seeds stored at −5 to −15°C for 13 years and thawed for 24 h at annual, semiannual, monthly, or weekly (in all, 670 times) intervals do not show any appreciable loss of viability compared with controls that were not subjected to periodic thawing. Thus, fluctuating storage conditions may be harmful to some species but not to others; this must be determined empirically.

9.4.2.3. Microflora and Seed Deterioration

Bacteria probably do not play a significant role in seed deterioration, for germination is rarely reduced unless infection has progressed beyond the point of obvious decay. Since bacterial populations require free water to grow, they are unlikely to increase in stored seeds because the latter are usually too dry. If conditions were moist enough, this would encourage growth of fungi which would suppress bacterial growth.

Two types of fungi invade seeds: field fungi and storage fungi. The former invade seeds during their development on plants in the field or following harvesting while the plants are standing in the field. Field fungi need a high moisture content for growth (as high as 33% for cereals) and hence are infective only under conditions where seeds fail to follow their normal pattern of maturation drying. A period of high rainfall at harvest time, therefore, can result in extensive grain deterioration. The main fungal species found associated with wheat or barley in the field are *Alternaria, Fusarium,* and *Helminthosporium* spp., although several others have been recorded. Seeds that are sheltered from airborne pathogens by pods, fleshy fruits, or other surrounding structures (e.g., pea, tomato, melon, maize) are generally less susceptible to field fungi than seeds that are more exposed (e.g., wheat, oat, barley).

Storage fungi, almost exclusively of the genera *Aspergillus* and *Penicillium,* infest seeds only under storage conditions and are never present before, even in seeds of plants left standing in the field after harvesting. Each species of storage fungus has a sharply defined minimum of seed moisture content below which it will not grow, although other factors also determine virulence, e.g., ability to penetrate the seed, condition of seed, nutrient availability, and temperature. The major deleterious effects of storage fungi are to (1) decrease viability, (2) cause discoloration, (3) produce mycotoxins, (4) cause heat production, and (5) develop mustiness and caking. Fungi will not grow at seed moisture contents that are in equilibrium with an ambient RH below 68%; hence, they are not responsible for deterioration that occurs at moisture contents below about 13% in starchy seeds and below 7–8% in oily seeds (Fig. 9.8).

Deterioration of seeds by insects and mites is a serious problem, particularly in warm and humid climates. Weevils, flour beetles, or borers are rarely active below 8% moisture content and 18–20°C, but are increasingly destructive as the moisture content rises to 15% and the temperature to 30–35°C. Mites do not thrive below 60% RH, although they have temperature tolerance that extends close to freezing.

9.4.3. Seed Storage Facilities

9.4.3.1. Short-Term Storage

It is obvious from the previous account that seed stores should incorporate various features into their design to minimize the chances for deterioration of their contents. Ideally, an establishment should contain rooms in which temperature and RH can be closely controlled, within which batches of seeds should be stored after hermetic sealing in containers. But such establishments are impractical on logistic and economic grounds for storage of large quantities of one particular species of seed, especially if it has to be stored only from harvest until the next planting season. Storage of seeds under ambient conditions is possible if certain precautions are taken. The storage structure should be:

1. Protected from water. The roof and sidewalls must be free of holes and cracks that permit entry of rain and snow. Structures should have a waterproof floor. A wooden floor should be elevated and a concrete floor should have a moisture barrier beneath it.
2. Protected from cross contamination. Storage facilities should be constructed to provide maximum protection from chance contamination. For bulk storage, a separate bin should be provided for each cultivar. For bag storage, seeds of each cultivar should be stacked separately.
3. Ventilated and aerated. Fans or blowers are a useful addition to provide ventilation, although their covers should be tightly closed when not in use. Where it is impractical to provide electricity for forced-air ventilation, an arrangement of covered (insect-proof) ducts to allow flow-

through of air is desirable. In the tropics, double roofs and heat-insulating materials are helpful. Seed or seed containers should not be piled against the walls since this reduces airflow.

4. Protected from rodents. In some countries, e.g., India, seed losses related to rodent attack are enormous. Metal and concrete buildings normally provide good protection from rodents. Small isolated stores should have a floor raised about a meter above the ground, a rat-proof door, and a removable entrance ramp. Metal seed storage bins with tight covers also protect against rodents.
5. Protected against insects. A storage facility should be fumigated each time it is emptied, as should the bins and boxes in which seeds are stored. Areas where bags and boxes are placed should be kept free of loose seeds and garbage at all times. An entrance constructed of a door leading into a small annex, from which there is a door entering into the main storage area, also cuts down direct access of insects to the seeds, as well as minimizing interior temperature and humidity fluctuations.
6. Protected against fungi. Since most fungi grow best under warm, humid conditions, storage structures should provide cool, dry conditions. Damage from storage fungi can be minimized by drying seeds to a safe moisture content and holding them under dry conditions. This is aided by ventilation to prevent the accumulation of translocated moisture. Treatment of seeds with fungicide may control some storage fungi (although many fungicides are most effective on soil fungi), and spraying the storage area when empty may be advantageous.

9.4.3.2. Long-Term Genetic Conservation—Seed Gene Banks

To meet the demands of increasing population and industrialization, society has developed an array of plants for consumption by people themselves or by their domesticated animals. Yet only about 15 species actually feed the world; these include five cereals (rice, wheat, maize, barley, and sorghum), two sugar plants (sugarcane and sugar beet), three subterranean crops (potato, sweet potato, and cassava), three legumes (bean, soybean, and peanut), and two tree crops (coconut and banana). Of these, just three species, wheat, rice, and maize, produce nearly 70% of the world's seed crop. Thus, the fate of hundreds of millions of lives hangs on the precarious balance of the genetic systems of these three crops, their diseases and pests, and their interactions with their environment. Unfortunately, the current trend to develop a few cultivars of high-yielding crops is also increasing genetic uniformity, at the expense of genetic diversity or variability, setting up the potential for global disaster should such crops become susceptible to specific diseases or to a changing environment. For example, losses on a relatively major scale occurred in the United States in 1970, when overuse of a single genetic strain of maize allowed southern corn leaf blight to become rampant. Losses reached 50% in some states and 15% nationally. Genetic resources are being lost also by cultivation of undisturbed lands where wild progenitors might be found, by abandonment of old farming systems, and by the destruction or deterioration of cultivars no longer in use. This inestimable loss of potentially valuable genes necessary for future plant improvement, particularly in relation to pest, disease, and stress resistance, has reached alarming proportions. The gradual loss of genetic variability in the form of germplasm is known as genetic erosion. Slowly, necessary steps are being taken to assemble germplasm resources of our cultivated plants and their relatives in order to preserve them in germplasm banks—gene banks.

Ideally, we should conserve as complete as possible a range of genetic diversity of all species that have actual or potential economic importance. Clearly this is impossible, for even ignoring the problem of deciding the potential value of species not yet investigated we would need to sample a known species right through its range of genetic diversity. This requires at least $10\text{–}20 \times 10^3$ samples per single crop species, and probably more for the major world crop cereals. Moreover, each sample should contain about 3000 seeds to encompass the normal genetic variation within a population. Thus, there are definite limits to the number of samples that can be handled effectively in programs for the conservation and utilization of crop genetic resources. These limits are imposed by financial and personnel considerations, and since governments and world bodies have been generally reluctant to commit themselves to genetic conservation programs, gene banks are few and, on a global basis, distressingly inadequate.

Several stages involved in the establishment of a gene bank are outlined in the following flow chart.

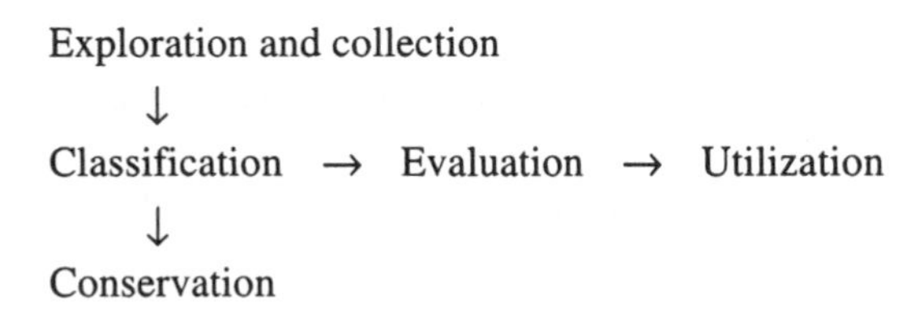

For most species there is one major limiting factor in these processes which determines what the sample size will be. For example, in the case of the primitive land races of the temperate cereals in the Mediterranean basin, which are rapidly facing extinction, adequate personnel resources are not available to collect the material before it is lost forever. On the other hand, in the case of the hexaploid weed relatives of cultivated oat there is an abundance of material present in the field, and the major limiting factor is the breeders' capacity to evaluate and utilize the collected material.

It is not appropriate here to discuss the economic, political, and logistical difficulties associated with exploring for and collecting populations of seeds, nor with the evaluation by breeders of the potential usefulness, in relation to certain desirable genetic traits. Instead, let us turn briefly to the critical features germane to the establishment of the seed gene bank itself. But we should also recognize that for some species, particularly those whose seeds have low storability, the formation of seed gene banks is inappropriate, and other methods, e.g., meristem banks, seedling banks, might have to be found.

There is little point in collecting the maximum amount of genetic diversity in a species unless the material can be adequately conserved to prevent subsequent loss of this diversity. To this end, the International Board for Plant Genetic Resources (IBPGR) has been involved, for the last decade, in the establishment and development of a world network of crop genetic conservation activities. It is on their guidelines and recommendations for conservation of material in gene banks that some of the following section is based.

After collection, seeds must be dried under carefully controlled conditions to avoid damage, then cleaned and sorted for storage in suitable containers, e.g., hermetically sealed cans or laminated foil packages. Prior to storage, germination tests must be carried out to determine initial viability. Since germination is essentially a destructive process as far as storage is concerned, tests must use the minimum number of seeds to yield statistically significant results and, at the same time, not deplete the stored stocks. A general rule of thumb is that for heterogeneous materials 12,000 seeds are required for storage, and for homogeneous material 4000 seeds. These should make up a "base" collection, i.e., one that is undisturbed and specifically laid down for long-term conservation. A "working" collection of the same seeds is also desirable, and these seeds can be used for medium-term storage, regeneration, evaluation, and distribution to other users. If a seed collection is initially unsuitable for storage, or its viability deteriorates to an unacceptably low level during storage, then facilities must be available for the propagation and regeneration of the species with the same genetic composition of the original seed.

Gene banks vary in their size and design. In the National Seed Storage Laboratory in Fort Collins, Colorado, for example, different storage rooms are maintained at between 4 and 12°C and up to 35% relative humidity. The accessions are arranged in numbered steel trays, placed in numbered steel racks (Fig. 9.11); each room has a capacity of about 180,000 pint cans. Liquid nitrogen storage is a newly developing technique and potentially can confer infinite longevity on seeds. An advantage of this method is that the need for germination tests is removed. Certainly, the lower the storage temperature, even below −12°C, the less frequently will monitoring of viability be necessary.

9.4.4. Metabolic Consequences and Causes of Viability Loss

Storage of seeds under adverse conditions results in the production of "aged" seeds exhibiting a variety of symptoms ranging from reduced viability or germinability (sometimes to zero germination) to more or less full viability (i.e., no obvious decline in germinability) but with abnormal development of the seedling (i.e., poor vigor). It can be appreciated, therefore, that it is difficult to make comparisons among deteriorated seeds that show such diversity in their final response, particularly since the metabolic lesions affecting viability might be manifest in the seedling stage in one case but prior to the completion of germination in another. In an attempt to simplify the situation, we will distinguish between the metabolic changes wrought in completely nonviable seeds and those occurring in populations of deteriorated seed showing all manner of viabilities and degrees of vigor. We do not distinguish here, however, between seed aging occurring under "natural" conditions, i.e., over time in storage, and "accelerated" aging brought about relatively quickly (in days, as opposed to weeks or years) by subjecting seeds to elevated temperatures (e.g., 40–45°C) at high relative humidities (up to 100%). While accelerated aging is useful for

Figure 9.11. Many thousands of seed containers, each labeled as to serial number, kind of seed, and storage location, are housed in the National Seed Storage Laboratory in Fort Collins, Colorado. Photograph courtesy of P. C. Stanwood (pictured).

lowering viability quickly, for experimental purposes, it is likely that the deteriorative changes brought about by this treatment will differ from those resulting from longer-term storage, especially since the extent of hydration of the seeds following accelerated and natural aging will be quite different.

As was discussed in Chapter 4, early and important events during germination include the establishment of respiration and an ATP energy-producing system and the commencement of RNA and protein synthesis. Not surprisingly, these are perturbed in seeds with reduced viability, although not always in a consistent and predictive (diagnostic) manner.

In nonviable embryos of rye, rice, and maize, the mitochondria appear to be swollen, and their internal membrane structure is distorted. Whereas in viable embryos mitochondrial organization becomes more ordered as germination progresses, in nonviable embryos these organelles become increasingly disorganized after imbibition, eventually leading to their complete lysis. Rye embryos that fail to germinate after storage exhibit little respiratory activity. In some species there appears to be some correlation between the decline in gaseous exchange and reduced activity of respiratory enzymes. In nonviable rice grains, for example, activities of cytochrome oxidase, succinic, glutamic, malic, and alcohol dehydrogenases, catalase, and peroxidase are all lower than in viable grains. It is not known, however, to what extent this decline in enzyme activity contributes to respiratory failure or if it is simply a manifestation of deterioration of mitochondria and loss of cell compartmentation. Not surprisingly, the ATP content of nonviable seeds is considerably lower than that of viable controls (as little as 1% in *Trifolium incarnatum*), and presumably it is insufficient to support metabolic processes essential for germination.

Respiratory patterns of deteriorated but still viable (at least partially) populations of seeds are both complex and variable, and often they are difficult to interpret. One of the more clear-cut responses of seeds to deterioration is found in isolated mitochondria of aged soybean seeds. Mitochondria from axes of aged seeds take up 10–40% more oxygen than those from seedling axes of fresh seeds, but the amount of ATP produced per volume of oxygen consumed by mitochondria from fresh seeds is over twice that of mitochondria from old seeds. Perhaps, then, mitochondria extracted from deteriorated seeds are at least partially uncoupled. Some attempts have been made to correlate ATP content of imbibed seed with seed vigor (i.e., seed weight, seedling weight, and hypocotyl length). In nine lettuce cultivars, including Great Lakes, highly significant correlations have been found, as they have also for rapeseed, ryegrass, and crimson clover seeds. Unfortunately, however, ATP measurements cannot be used universally as a test for seed viability and seedling vigor, for in several other species aged under a variety of conditions the correlation between seedling growth and ATP content of the imbibed seed is poor. Whether this is because of differences in seed species or differences in the aging conditions used is not known.

Nonviable embryos and embryonic axes fail to conduct protein synthesis on imbibition. Even aged but viable embryos exhibit signs of reduced protein synthesis. For example, rye embryos from a population possessing 86% viability (which have reduced root growth after germination, i.e., loss of vigor) have only about 20% of the protein-synthetic capacity (as measured by incorporation of radioactive amino acids) of those with 95% viability (and which show no sign of vigor loss). Since protein synthesis involves several diverse components (e.g., ribosomes, messenger RNA, aminoacylated tRNA, initiation and elongation factors) within the cell, a decline in activity or content of any one will result in a decline in translational capacity. In nonviable cereal embryos and dicot embryonic axes the inability to synthesize proteins is accompanied by a marked loss of capacity for RNA synthesis. Moreover, in several seeds, including rye embryos, a gradual loss of viability is associated with a decline in synthesis of all major classes of RNA, and in the processing of ribosomal RNA (rRNA) into the subunits of the ribosome. In aged but still viable seeds of tobacco there appears to be some correlation between the integrity of rRNA and their rate of germination (Table 9.4).

The synthesis of mRNAs during germination is an important event, for the synthesis of new messages and for replacing those messages present in the dry seed which are gradually degraded as the seed imbibes. This normal turnover of mRNAs is disrupted in seeds with low viability, or vigor, as illustrated in Fig. 9.12 for germinating wheat embryos. In the embryos with different viabilities or vigor the degradation of messages occurs during the early hours of germination, but in low viability/vigor embryos new synthesis to replace them is impaired. It remains to be determined whether this impediment is due to a block to transcription, or results from degradation by ribonucleases, for example, of the newly transcribed messages after their release from the nucleus.

9.4.5. Damage to Chromosomes and DNA

Almost any combination of time, temperature, and moisture content that leads to loss of viability of seeds in storage will lead to some genetic damage in

Table 9.4. Correlation between the Integrity of Ribsomal RNA in Dry Seeds of *Nicotiana tabacum* (Tobacco) and Germinability and Vigor[a]

Year of harvest	% Germinated	Time to complete germination (h)	% rRNA integrity
1978	82	5.1	62
1973	72	17.1	2

[a]After Brocklehurst and Fraser (1980).

the survivors. In several species, including garden pea, broad bean, and barley, this damage is manifest as chromosomal breakages that appear during the anaphase stage of the first mitotic divisions of the root tips. The relationship between viability and the mean frequency of aberrant cells for broad bean seeds is shown in Fig. 9.13A. There is an increase in chromosome damage with increase in the period of storage, as perceived by the increase in the number of aberrant cells (Fig. 9.13B) and concomitant loss of viability (Fig. 9.13C). Under the most severe storage conditions (45°C at 18% moisture content), where the mean viability period is less than 1 week, the frequency of aberrant cells is considerably less than for all other treatments (Fig. 9.13A). Perhaps under these unusually harsh conditions of aging some nonnuclear (cytoplasmic) lesions occur which are in themselves lethal and which lead to loss of viability at a faster rate than the breakage of chromosomes. With this exception, though, there is generally an excellent correlation between loss of viability and accumulation of chromosome damage in survivors.

Most aberrant cells in surviving seeds do not persist beyond the first cell division, and they are lost from the apical region before the roots grow extensively. Minor genetic damage such as recessive gene mutations may persist, but being masked by their dominant allele, they may have little obvious effect. Nevertheless, some recessive genes may be lethal in haploid cells, and their frequency can be related to the frequency of pollen abortion in mature plants grown from aged seeds. Accumulation of genetic damage in stored seed lots that are to be planted for food or feed production (i.e., whose progeny will not be replanted) may be of little consequence, unless the seeds germinate slowly or produce stunted seedlings. In these cases, the progeny may be more sus-

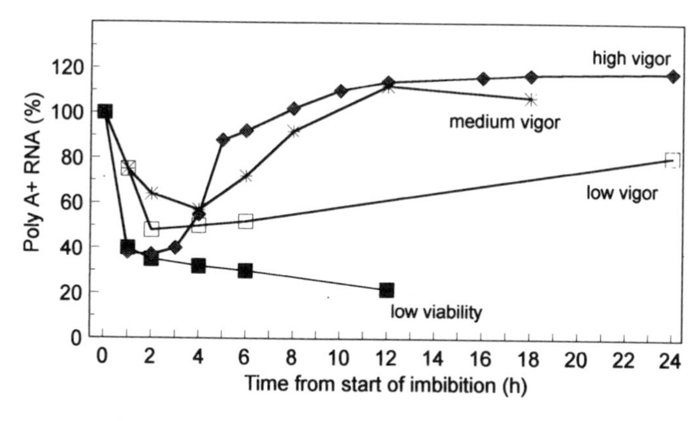

Figure 9.12. Poly A^+ RNA (mRNA) content of germinating wheat embryos of high, medium, and low vigor, and of low viability. After Rushton and Bray (1987).

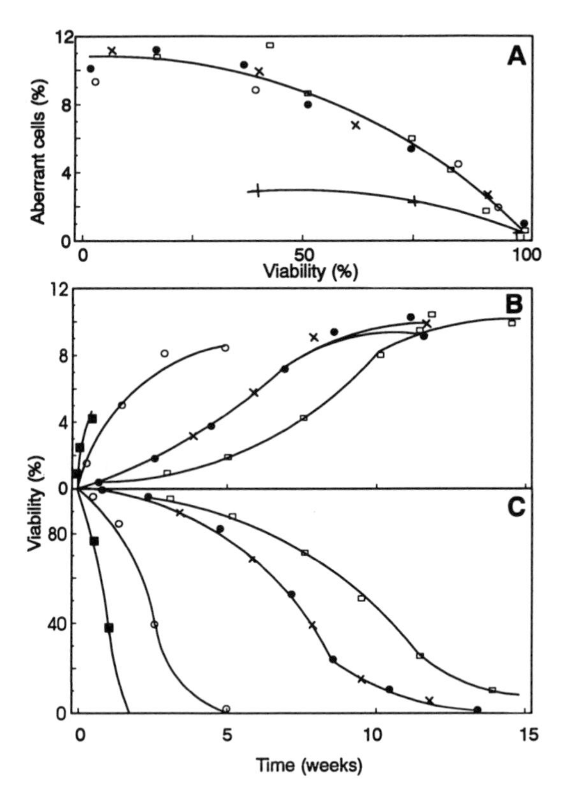

Figure 9.13. (A) The relationship between percentage viability and the mean frequency of aberrant cells, and (B) the increase in mean frequency of aberrant cells in surviving populations of seeds of *Vicia faba* stored under various conditions of temperature and moisture content (m.c.). (C) Seed survival curves for the same treatments. Conditions: 45°C, 18% m.c. (■); 45°C, 11% m.c. (●); 35°C, 18% m.c. (○); 35°C, 15% m.c. (×); 25°C, 18% m.c. (□). After Roberts (1972).

ceptible to pathogens, to adverse environmental changes, or to competition from other plants, e.g., weeds. Genetic changes in seeds of crops grown for germplasm stock may have more serious long-term implications. For example, a stored seed lot with little reduction in viability still could harbor a considerable number of mutations, which would not be expressed immediately in the crop grown from that seed but which would begin to segregate in the subsequent generation, and continue to do so in all generations thereafter. Thus, poor storage of a "pure line" would soon result in a loss in desired purity.

Why aging seeds accumulate chromosome aberrations is not understood. One suggestion is that there is a buildup of chemical mutagens within seeds during aging, but the published evidence in favor of this is not convincing. Another possibility is that nuclear DNases are activated during aging, resulting in partially degraded DNA molecules. DNase activity in dry rye embryos is higher in nonviable than in viable material. How this increase in DNase occurs in the dry seed and how it is active therein are matters for conjecture. There could be activation of a preexisting latent enzyme or the removal of a DNase inhibitor during aging. If the now active DNase is present in the nucleus and is closely associated with nuclear chromatin, then even at low moisture contents slow enzymatic fragmentation of DNA might occur. The total amount of DNA in dry nonviable and viable embryos of rye is the same, although in the former the total amount of intact high-molecular-weight DNA is less (Table 9.5). This and other evidence is indicative of at least partial DNA fragmentation into lower-molecular-weight components during aging in dry storage. Minor repair to DNA might occur when viable seeds are imbibed; loss of viability could be accounted for if the fragmentation of DNA in the dry seed is too extensive for repair to be effected on subsequent hydration, or if the DNA repair system itself loses its integrity during aging.

When lettuce (*Lactuca sativa*) and ash (*Fraxinus americana*) seed are stored in a fully imbibed but dormant state, they maintain their full capacity for germination for long periods (at least 3 years) and sustain very little chromosome damage. Loss of viability from seeds stored at a low moisture content (9.7%) might result from the inactivity of enzyme systems capable of repairing storage-induced damage to DNA and also to other essential macromolecules (and organelles) in the cytoplasm. At even lower water contents (5–7%) there is not enough damage to affect germination, for degradative enzymes cannot operate at such low water contents. Thus, at any given temperature, the lower the moisture content, the slower is the increase in accumulation of chromosomal aberrations. Assuming that repair can only occur successfully in the imbibed state, then any damage suffered by the wet-stored seeds will be continuously repaired and will not accumulate. Although in the dry-stored seeds the repair

Table 9.5. Nuclear DNA Content and Extractable High-Molecular-Weight DNA in Rye Embryos of Different Viability[a]

Viability	Nuclear DNA (relative OD_{254})	High-molecular-weight DNA content ($mg \cdot g^{-1}$ dry wt)
95	8.4	10.2
64	—	9.7
15	—	7.8
0	8.1	3.1

[a]After Osborne *et al.* (1980/81).

mechanisms will become activated on imbibition after storage, by this time the damage might be so extensive and beyond restitution that viability will be lost. This, for the moment, must remain only a hypothesis, for biochemical evidence in its favor has still to be obtained. Nevertheless, the concept of repair mechanisms is worthy of consideration, for many seeds may remain in soil in the imbibed state for long periods (e.g., when dormant) before conditions for germination become favorable.

9.4.6. Aging and the Deterioration of Membranes

Of the many factors involved in maintaining the control of metabolism within cells, spatial separation of metabolic components and the correct alignment of synthetic complexes are very important. Not only are cooperative enzymes of a metabolic pathway linked together within organelles (e.g., respiratory enzymes within mitochondria) but often they are intimately associated with or integrated into membrane structures. Thus, disruption of membranes because of aging could lead to diverse metabolic changes, all of which contribute to different extents to seed deterioration and loss of viability and vigor. Imbibition by viable seeds is accompanied by a rapid but transient efflux of inorganic and organic compounds into the surrounding solution because membrane integrity is incomplete for at least several minutes after water uptake. But the situation is reversed with time, the membranes either physically reverting to their most stable configuration or else being repaired by some still unidentified enzymatic mechanism. In low- or nonviable seeds such repair mechanisms might be absent or inefficient, or the membranes might be so badly damaged that repair is impossible.

Loss of membrane integrity in deteriorated seeds is suggested by the observation that on imbibition more substances leak into the medium from such seeds than from viable ones. Excess leakage of sugars may represent loss of respirable substrate from some seed species, whereas others leak more amino acids than sugars. To what extent this leakage affects germinability is not known, and it may be only one manifestation of the severe perturbations that occur to membrane systems within the cells. Increased leakage of organic metabolites from deteriorated seeds might indirectly add to their demise by encouraging the growth of contaminating microorganisms.

How membranes and macromolecules become destabilized during storage under adverse conditions has been a subject of research for several decades. The unsaturated fatty acids in membrane phospholipids are highly susceptible to peroxidative degradation. This results both in the destruction of the fatty acid and the generation of a number of potentially toxic products. Peroxidation in dry-stored seeds is likely to be a consequence of autoxidation, but at moisture contents sufficient to permit enzyme activity, lipoxygenase may contribute to the degradative process.

Peroxidation of an unsaturated fatty acid occurs as illustrated in Fig. 9.14. Initially, a hydrogen atom (H·) is removed from a methylene ($-CH_2-$) group adjacent to a double bond to yield an organic free radical. This reacts with molecular oxygen and results in a rearrangement in the fatty acid chain, and the formation of a peroxy radical. In turn, this reacts with a neighboring unsaturated fatty acid (RH) to form a lipid hydroperoxide, which is unstable and degrades to generate new free-radical species, thus perpetuating the process and yielding olefins, alcohols, alkanes, and carbonyl compounds as decomposition products. The accumulation of one such product, malondialdehyde (MDA), has been used as a measure of lipid peroxidation within stored seeds. Fatty acids with several double bonds are more susceptible to degradation than those with fewer, and the presence of free-radical scavengers (antioxidants) within a seed, e.g., tocopherol, ascorbic acid, and glutathione, can limit the extent of peroxidation damage. Abstractions of hydrogen from fatty acids causes a decline in lipid fluidity, and the changes in structure resulting from peroxidation inactivate membrane-bound proteins (transport proteins and receptors), thus changing membrane permeability. These changes, along with the production of the cytotoxic decomposition products are symptoms of membrane deterioration.

There is still a considerable debate in the research literature as to whether lipid peroxidation is a major cause of loss of viability within seeds, or whether it occurs only after the demise of the seed, as part of the general deterioration of the cells. In many instances, the evidence supporting a causative role for lipid peroxidation is only circumstantial. The variations in the reported data are quite remarkable; e.g., some researchers have found that during aging there are marked decreases in seed phospholipid content, a decline in saturated fatty acids, an increase in degradation products, a reduction in tocopherol content, and the

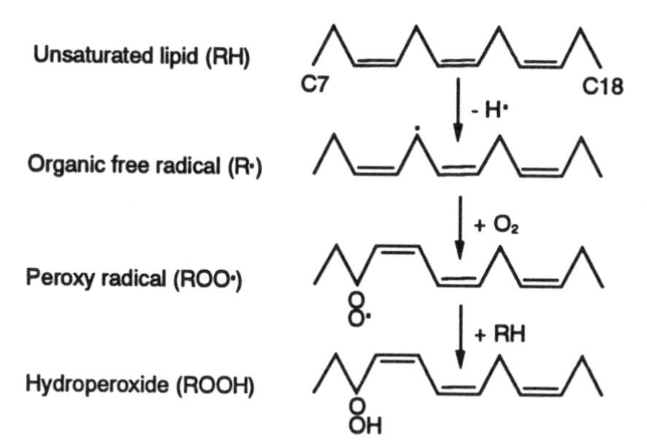

Figure 9.14. Peroxidation of a polyunsaturated fatty acid, e.g., linolenic acid.

presence of free radicals. Other researchers using seeds of the same or different species have reported no substantial changes. Sometimes the contradictions can be attributed to differences in the way in which the seeds were aged (especially accelerated versus long-term); or to the tissues that were studied (e.g., cotyledons versus axis); or to the time of free-radical measurement in relation to viability loss (e.g., during deterioration versus post mortem). At other times, there are no apparent reasons for the lack of consistency in the observations. Thus, we are led to draw the unsatisfactory conclusion that aging in some seeds under some conditions could be mediated by destabilization of membranes by free-radical-initiated lipid peroxidation. For an interesting and reasoned discussion on free-radical processes and seed longevity, the reader is directed to the review by Hendry (1993).

A summary of the variety of metabolic lesions that can occur during storage and which result in a loss of viability is found in Fig. 9.15. The reader can appreciate that although any single lesion can result in the loss of viability or reduction of vigor, it is likely that aging elicits a number of changes. Which one occurs first or is the most important is impossible to tell, but it makes a continuing topic for debate!

Attempts have been made to restore the viability and vigor of aged seeds after storage. One method which has achieved some success is osmopriming, which involves placing dry seeds in solutions of osmotica, thus limiting the rate of imbibition and initially slowing the germination processes. It may be that under conditions of limited water availability, which can be achieved also by placing dry seeds in a water-saturated atmosphere, repair processes are effected and that age-induced damage is sufficiently diminished before the germinative events commence. If the time for, or extent of, repair to cellular damage is inadequate, e.g., when the aged seeds are imbibed in water, then the integrity or metabolism of the cell remains disrupted, and germination not only cannot proceed, but further deterioration occurs following imbibition.

9.5. SOMATIC EMBRYOGENESIS

Somatic embryogenesis is the development of embryos from cells of somatic tissues, i.e., those which are not a direct product of gametic fusion. They include diploid vegetative tissues (e.g., petioles, cotyledons, and roots), haploid cells (e.g., microspores), and even the triploid endosperm. During development, these embryos undergo similar ontogenetic changes to their zygotic counterparts, and contain a similar range of reserve materials at maturity. The initiation and development of embryos from somatic tissues was first achieved in the late 1950s, independently by Steward and Reinert, in cultures of tissues derived from the taproot of wild carrot. Since then, there have been reports of similar successes

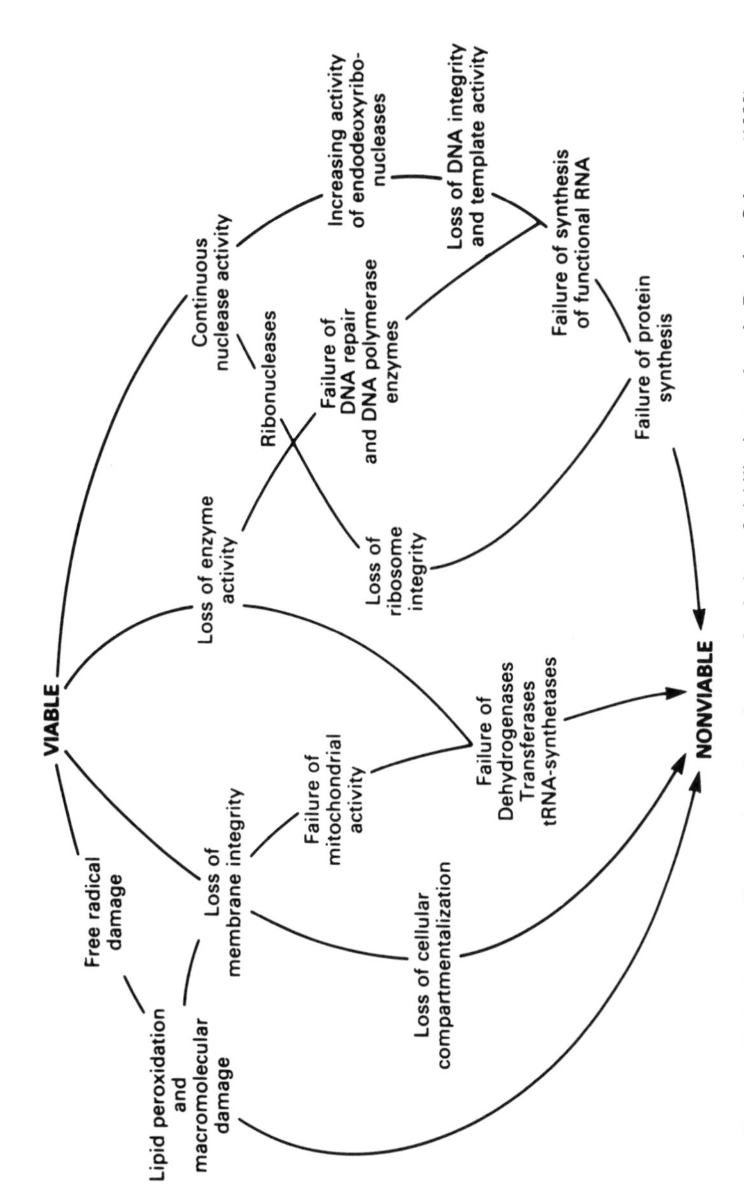

Figure 9.15. A scheme to illustrate the variety of causes for the loss of viability in stored seeds. Based on Osborne (1980).

with a large number of other species, in many families and genera, especially among the angiosperms, but also in the gymnosperms.

Some vegetative tissues are already embryogenic, e.g., hypocotyls and cotyledons, and can form embryos directly (direct embryogenesis), without an intervening callus phase, when placed on an appropriate culture medium. Indirect embryogenesis from predominantly differentiated or vegetative tissues requires a treatment to initiate mitosis, which leads to the formation of an undifferentiated cell mass, the callus. This is then subjected to further manipulations to stimulate redifferentiation of some of the cells, i.e., those with embryogenic potential, to form somatic embryos. Growth regulators are required for these events to occur, for reentry into mitosis, for determination of the embryogenic state, and for maturity of the embryo. There are countless variations in the techniques used to achieve somatic embryogenesis, and frequently the conditions necessary for acquiring embryos which will germinate and grow successfully into seedlings (undergo "conversion") are very stringent, and are species (even cultivar) specific.

As an example of the type of manipulations that are involved in achieving somatic embryogenesis, we will illustrate what is required to obtain embryos from petiole tissues of alfalfa (*Medicago sativa*). In this case, the callus is derived from explanted petioles, transferred to liquid culture, and then plated out on a solid embryo-development medium (Fig. 9.16). The various steps are presented in Table 9.6, omitting many details. It is important to note here that even within the species *Medicago sativa*, only very few of the many available cultivars have been found to possess embryogenic potential (i.e., it is determined genotypically), and of these, only vigorously growing plants can be used as a source of explants; any stress on the vegetative plant will drastically reduce both the quantity and quality of the embryos produced.

The embryogenic cells from which somatic embryos are derived have features that are characteristic of rapidly proliferating meristematic cells. They are small, with dense cytoplasm, small vacuoles, large nuclei with enlarged nucleoli, and often contain many starch grains. In alfalfa, dense masses of potentially embryogenic cells (Fig. 9.17A) form in the suspension cell mass, plated onto hormone-free medium (Table 9.6, step 3). While auxin is required for the first phase of embryogenesis (induction), the formation of embryogenic cell clusters, it is inhibitory from this stage onwards. Further development of somatic embryos occurs, with the formation of globular, heart, torpedo, and cotyledon-stage embryos (Fig. 9.17). During somatic embryogenesis of alfalfa, no obvious suspensor is formed; in other species this structure is evident during development (e.g., embryos derived from *Brassica* microspores and from shoots of young black spruce seedlings). The mature embryos of alfalfa, like those of many other species, have poorly developed cotyledons, perhaps resulting from incomplete differentiation of the embryo following the induction

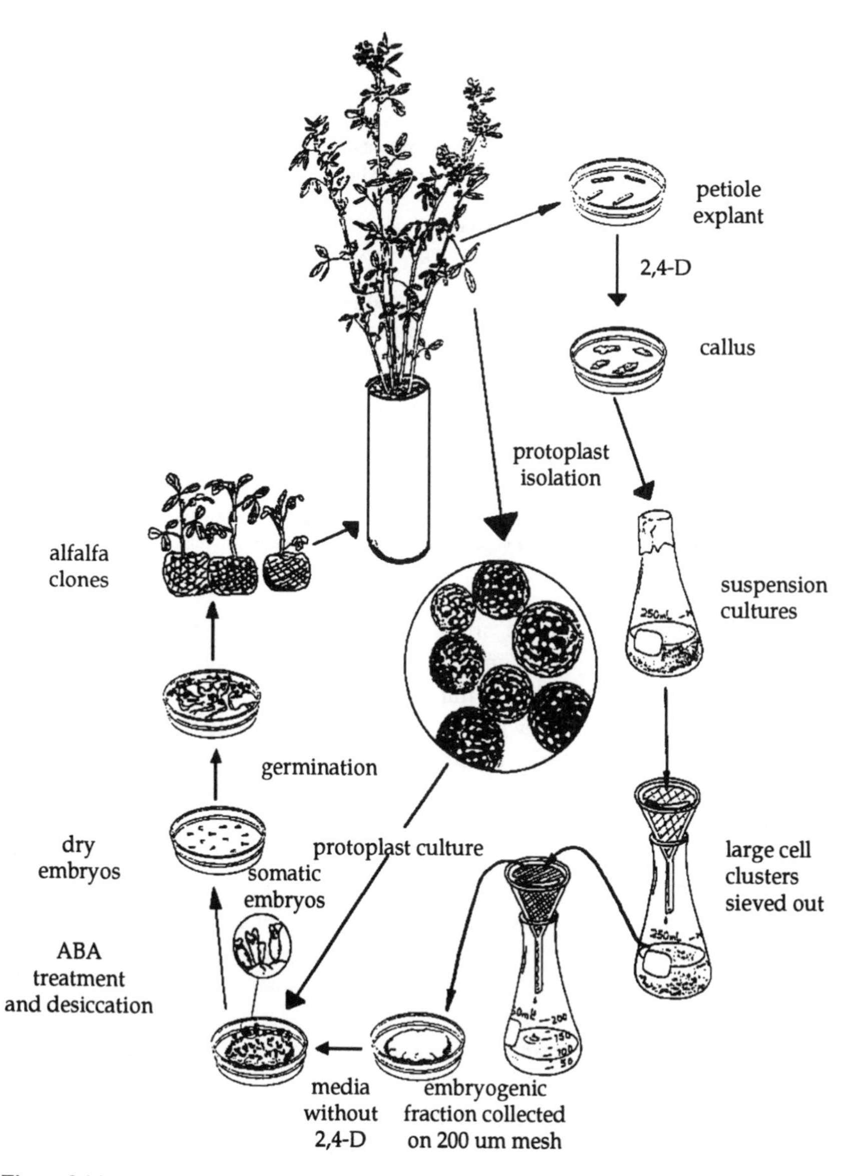

Figure 9.16. An illustration of the major steps required to achieve somatic embryogenesis in alfalfa (*Medicago sativa*), either from explants or through the production of protoplasts. To obtain plantlets, somatic embryos are first transferred to a nutrient-containing sterile medium, and then to peat pots. Courtesy of K. P. Pauls.

Table 9.6. An Outline of the Steps Necessary to Achieve Somatic Embryogenesis from Petiole Explants of Alfalfa[a]

Step	Treatment	Results
1	Cut petioles and place on solid B5H medium containing 1 mg/l 2,4D and 0.2 mg/l kinetin, 25°C, low light/dark: 16h/8h, 2-4 weeks	Soft, yellowish-green callus
2	Transfer callus to liquid suspension culture medium containing B5G, 1 mg/l 2,4D and 0.1 mg/l NAA. Shake at 125 rpm at 25°C, low light/dark 16h/8h, 7-10 days.	Thick cell mass
3	Sieve cell mass through nylon mesh screens (500 and 200 μm), and place cell clumps on smaller-sized screen onto hormone-free solidified Boi2Y medium at 25°C, low light/dark: 16h/8h, up to 20 days.	Green somatic embryos
4	Transfer to medium containing 20 μM ABA, as step 3, for 10 days, and dry slowly.	Yellow, desiccation-tolerant somatic embryos

[a]Method based on Atanassov and Brown (1984) and Senaratna *et al.* (1989).
For details on the culture media, the reader is directed toward an appropriate handbook on tissue culture techniques, e.g., Bhojwani and Razdan (1983). For a consideration of the explant and culture conditions that are particularly important in initiating and controlling somatic embryo development, see Ammirato (1983).

phase. Some species produce similarly defective cotyledons (e.g., carrot) whereas the mature, somatic embryos of others have normal-looking ones. Auxin polar transport may play an important role in cotyledon formation during embryo development, as shown for globular- to heart-stage zygotic embryos of *Brassica juncea*. Interference with this transport results in the formation of embryos with fused cotyledons; likewise, in culture conditions, the development of somatic embryos could be affected by the concentration and distribution of available auxin.

The morphology of mature somatic embryos, therefore, is variable between species, and is influenced in any one species by culture conditions; this profoundly influences germinability and subsequent formation of plantlets. Improved cotyledon morphology *per se* may not enhance conversion or establishment of plantlets, but decreased storage reserve deposition, which may be a consequence of poor cotyledon development, can influence the vigor with which seedlings become established.

Because the initial stages of zygotic embryogenesis occur within the confines of the ovule, it is difficult to obtain sufficient quantities of the microscopic embryos to study and understand gene expression during the early developmental stages. Somatic embryogenesis has been proposed as a model system with which to circumvent this problem, since large quantities of reasonably well-synchronized embryos can be produced. To date, the expression of a few genes which are exclusive to early embryogenesis has been identified, particularly in carrot somatic embryos. The importance and function of their

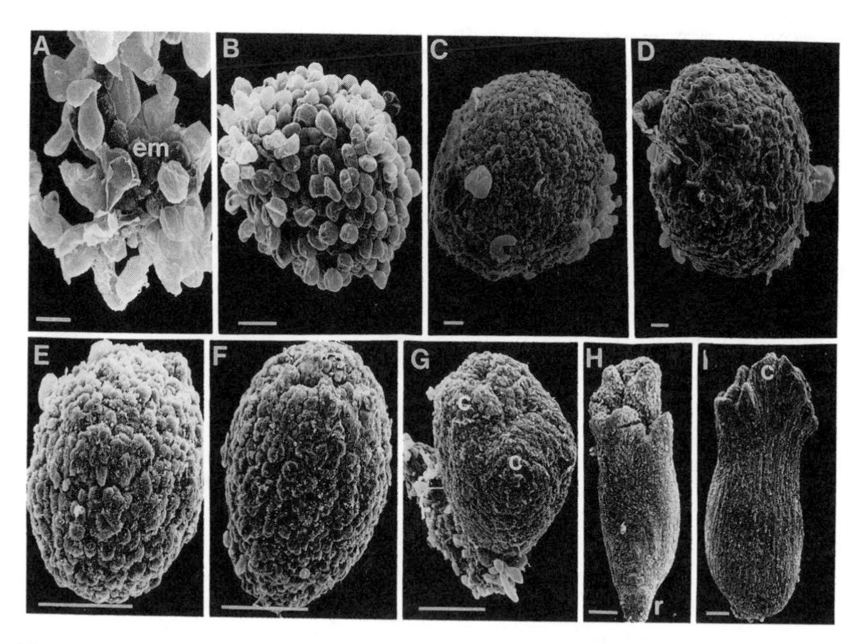

Figure 9.17. The origin and development of alfalfa somatic embryos in culture as shown using scanning electron microscopy. (A,B) Embryo induction in suspension culture; (C–I) embryo growth and maturation on solid agar medium. (A) Dense, small relatively round embryonic cells (em) surrounded by large white, vacuolated callus cells; (B) somatic proembryo with vacuolated cells on the surface; (C) early globular stage, the embryo is round; (D) middle globular stage, the embryo is oval shaped; (E) late globular stage, note the large size and the elongated oval shape; (F) early heart stage, showing cotyledon initiation; (G) late heart stage, showing the beginning of cotyledon elongation (c); (H) torpedo stage, note the formation of radicle (r); (I) cotyledon (c) elongation stage. A–D, bars = 30 μm; E–I, bars = 200 μm. See Fig. 2.2 for equivalent stages of zygotic embryogenesis. After Xu and Bewley (1992).

products in embryogenesis are not known, for those which have been identified are LEA-type proteins (Section 3.3.2) which are normally associated with the late, maturation stages of zygotic embryogenesis.

9.5.1. Somatic Embryos and Synthetic Seed Production

Studies of somatic embryo development provide us with opportunities to study embryogenesis. In those instances where such embryos are produced from transformed plants, they also are a useful regenerative system in which to study the expression of introduced genes. But, from a practical point of view, the use of *in vitro* somatic embryogenesis to produce "synthetic" or "artificial" seeds offers the potential for efficient vegetative propagation. Some of the ways in which somatic embryo production could provide a viable alternative to zygotic seed production are outlined below:

1. Synthetic cultivars are genetically heterozygous (e.g., alfalfa and orchard grass) and in a production field the plants differ in both genotype and phenotype. Cross-pollination and genetic recombination between relatives leads to inbreeding depression and loss in plant vigor, in economic yield, and dilution of desirable traits (e.g., stress tolerance, disease resistance). Use of uniformly produced synthetic seeds would allow propagation of hybrids with superior traits, and the need for fertilization to increase seed stocks will be circumvented.

2. Planting efficiency of crops that are vegetatively propagated because of self-incompatibility or long breeding cycles, e.g., some hardwood, nut, and fruit trees, could be increased by the use of synthetic seed instead of cuttings, although existing methods are cost-effective. Use of synthetic seed for germplasm conservation could be the most advantageous application for somatic embryos, however, since the germplasm of grape, for example, is currently maintained as living plants in the vineyard.

3. Softwood species which make up the temperate forests are planted as seed or containerized seedlings. Improvement by conventional breeding is very time-consuming because of the long life cycle of conifers. Also the populations of conifers are very heterozygous so that seeds from high-quality individuals may not result in improved progeny. Synthetic seeds provide the possibility of cloning elite trees at reasonable costs. The problems encountered to date, however, include poor maturation of somatic embryos, variable germination, and poor soil-establishment. The potential economic benefits to the forest industry are huge if such difficulties can be overcome.

4. It is difficult to produce commercial quantities of hybrid seed of some seed-propagated crops (e.g., soybean and cotton) because the flowers are closed (cleistogamous), or they abscind prematurely. Seeds of most existing cultivars are derived from self-pollination. New hybrid lines could be produced by hand-pollination, and somatic embryos mass-produced from these, thus exploiting hybrid vigor.

5. For crops such as tomato, asparagus, and seedless watermelon, the high cost of producing hybrid seed is offset by the value of the crop. However, the production of synthetic seed could be less expensive than conventionally produced hybrid seed (Table 9.7).

6. The corn industry relies on inbred parentals to produce uniform hybrid seed, and mass hybridization is possible because male-sterile lines are used as females. The higher production costs which are incurred are offset by the superior yield and quality of the product, but the development and maintenance of parental lines is time-consuming, and incorporation of new germplasm is slow. Synthetic seed could be used to propagate new hybrids and eliminate the need for male steriles and parental inbreds.

7. Species which produce recalcitrant seeds might yield somatic embryos which have better storage properties.

The production of synthetic seed using somatic embryos still remains to be achieved on a large scale. Some of the requirements, and limitations, are included in the following list:

a. Large-scale production of somatic embryos is essential. Embryogenic cultures will need to be scaled up in bioreactors, in liquid culture. But embryos frequently do not develop well or uniformly under such conditions, and yields and quality have generally been disappointing.

Table 9.7. Potential Applications of Synthetic Seed Technology for Selected Crop Species

	Somatic embryo quality[a]	Relative seed cost[b]	Application[c]	Relative need for crop[d]
Alfalfa	h	l	s	m
Corn	p	m	i	m
Cotton	p	m	h	m
Grape	h	na	s,g	m
Loblolly pine	p	h	c	h
Norway spruce	h	—	c	h
Orchard grass	h	l	s	m
Ornamentals	p	na	d,s	h
Soybean	p	m	h	m
Hybrid tomato	n	v	d	h
Seedless watermelon	n	v	d	h

[a]Relative somatic embryo quality: h = highly developed embryos; p = poorly developed embryos; n = somatic embryos not obtained.
[b]Relative cost of seed: v = seed cost limits planting; h = seed is costly; m = moderate; l = relatively inexpensive; na = seed is not used.
[c]Application for synthetic seed: c = circumvent long breeding cycles; d = decrease seed and/or plant cost; g = germplasm conservation; h = mass production of hybrids; i = eliminate need for inbreds; s = circumvent self-incompatibility.
[d]Relative need: h = highly useful if implemented; m = existing methods are effective but implementation should yield improvements.
After Gray and Purohit (1991).

b. Embryos will need to be sorted into appropriate size and shape ranges to separate those of high quality. This is essentially an engineering problem, but the embryos will have to be handled gently and maintained under reasonably sterile conditions.

c. Since naked embryos cannot be planted in soil, they will have to be encapsulated. Ideally, embryos should be desiccated before encapsulation; this increases their robustness and ease of handling, and allows for prolonged storage. While desiccation-tolerant embryos have been produced, the frequency of their subsequent germination, and especially their conversion to seedlings, has been unpredictable. Somatic embryos are both unprotected and generally lacking in storage reserves compared with their coat-enclosed zygotic counterparts. Encapsulation provides physical protection, and within the capsules, nutrients, growth regulators, fungicides, etc., can be included. Wet encapsulation, e.g., in a hydrogel, is possible for somatic embryos that will be planted very soon after their production, e.g., some greenhouse-grown vegetable and flower crops. The rate of establishment of synthetic seed in soil has been sufficiently low, so far, to suggest that the appropriate technologies associated with encapsulation have far to go.

d. Finally, the cost of producing artificial seed may be prohibitive for many species, with the resulting increase in yield, quality, etc., being too low to offset the required investment. Table 9.7 provides a rough cost–benefit analysis for several of the types of crops mentioned in this section.

Research on somatic embryogenesis is proceeding rapidly, with reports monthly on successes with more species and cultivars, and improved technologies. Many challenges lie ahead before synthetic seed production becomes a widespread and commercially viable enterprise.

USEFUL LITERATURE REFERENCES

SECTION 9.2

Briggs, D. E., 1978, *Barley,* Chapman and Hall, London (barley utilization, including malting).

Macleod, A. M., 1979, in: *Brewing Science,* Volume 1 (J. R. A. Pollock, ed.), Academic Press, New York, pp. 145–232 (physiology of malting).

Palmer, G. H., 1980, in: *Cereals for Food and Beverages* (G. E. Inglett and L. Munck, eds.), Academic Press, New York, pp. 301–338 (morphology and physiology of malting barleys).

Palmer, G. H., and Bathgate, G. N., 1976, in: *Advances in Cereal Science and Technology* (Y. Pomeranz, ed.), American Association of Cereal Chemists, St. Paul, Minn., pp. 237–324 (malting and brewing).

SECTION 9.3

Black, M., Butler, J., and Hughes, M., 1987, in: *Fourth International Symposium on Pre-Harvest Sprouting in Cereals* (D. J. Mares, ed.), Westview Press, Boulder, Colo., pp. 379–392 (dormancy and preharvest sprouting).

Mitchell, B., Armstrong, C., Black, M., and Chapman, J., 1980, in: *Seed Production* (P. Hebblethwaite, ed.), Butterworths, London, pp. 339–356 (physiology of preharvest sprouting).

Reiner, L., and Loch, U., 1975, *Cereal Res. Commun.* **4:**107–110 (temperature and barley dormancy).

Ringlund, K., Mosleth, E., and Mares, D. J. (eds.), 1990, *Fifth International Symposium on Pre-Harvest Sprouting in Cereals,* Westview Press, Boulder, Colo. (articles on pre-harvest sprouting).

Sawhney, R., and Naylor, J. M., 1979, *Can. J. Bot.* **57:**59–63 (temperature and dormancy in wild oats).

SECTION 9.4

Becquerel, M. P., 1934, *C. R. Acad. Sci.* **199:**1662–1664 (viability records).

Brocklehurst, P. A., and Fraser, R. S. S., 1980, *Planta* **148:**417–421 (rRNA integrity and vigor).

Buchvarov, P., and Grantcheff, T., 1984, *Physiol. Plant.* **73:**85–91 (free radical production in aging soybean axes).

Ellis, R. H., and Pieta Filho, C., 1992, *Seed Sci. Res.* **2:**9–15 (potential longevity of seeds in storage, in relation to maturity).

Ellis, R. H., and Roberts, E. H., 1980, *Ann. Bot.* **45:**13–30 (seed longevity equations).

Hendry, G. A. F., 1993, *Seed Sci. Res.* **3:**141–153 (review of oxygen, free radical formation, and seed longevity).

Kivilaan, O., and Bandurski, R. S., 1981, *Am. J. Bot.* **68:**1290–1292 (Beal's 100-year burial experiment).

Leprince, O., Hendry, G. A. F., and McKersie, B. D., 1993, *Seed Sci. Res.* **3:**275–290 (membranes, protection, desiccation, and aging).

Osborne, D. J., 1980, in: *Senescence in Plants* (K. V. Thimann, ed.), CRC Press, Boca Raton, Fla, pp. 13–37 (review of seed aging).

Osborne, D. J., Sharon, R., and Ben-Ishai, R., 1980/81, *Isr. J. Bot.* **29:**259–272 (DNA integrity and repair).

Pandy, D. K., 1989, *Seed Sci. Technol.* **17:**391–397 (recovery of aged seeds by osmopriming).

Plucknett, D. L., Smith, N. J. H., Williams, J. T., and Anishetty, N. M., 1987, *Gene Banks and the World's Food,* Princeton University Press, Princeton, N.J. (germplasm collection and storage).

Priestley, D. A., 1986, *Seed Aging,* Cornell University Press (Comstock), Ithaca, N.Y. (critical coverage of seed storage and aging).

Rao, N. K., Roberts, E. H., and Ellis, R. H., 1987, *Ann. Bot.* **60:**85–96 (chromosome aberrations and viability of lettuce seeds).

Roberts, E. H., 1972, in: *Viability of Seeds* (E. H. Roberts, ed.), Chapman and Hall, London, pp. 253–306 (cellular and genetic changes during viability loss).

Roberts, E. H., 1973, *Seed Sci. Technol.* **1:**499–514 (predicting longevity in storage).

Roberts, E. H., 1975, in: *Crop Genetic Resources for Today and Tomorrow,* Volume 2, International Biological Program, Cambridge University Press, Cambridge, pp. 269–296 (storage and genetic changes).

Rushton, P. J., and Bray, C. M., 1987, *Plant Sci.* **51:**51–59 (mRNA and loss of vigor and viability in wheat grains).

Sen, S., and Osborne, D. J., 1977, *Biochem. J.* **166:**33–38 (RNA and protein synthesis in nonviable rye).

Senaratna, T., Gusse, J. F., and McKersie, B. D., 1988, *Physiol. Plant.* **73:**85–91 (membrane changes in aging soybean axes).

Styer, R. C., Cantliffe, D. J., and Hall, C. B., 1980, *J. Am. Soc. Hortic. Sci.* **105**:298–303 (noncorrelation between ATP and vigor).
Thomson, J. R., 1979, *An Introduction to Seed Technology*, Wiley, New York (seed storage methods).
Villiers, T. A., 1974, *Plant Physiol.* **53**:875–878 (storage in hydrated state).
Wilson, D. O., Jr., and McDonald, M. B., Jr., 1986, *Seed Sci. Technol.* **14**:269–300 (review of lipid peroxidation and aging).

SECTION 9.5

Altree, S. M., and Fowke, L. C., 1991, in: *Biotechnology in Agriculture and Forestry*, Volume 14, Springer, Berlin, pp. 53–70 (somatic embryogenesis in conifers).
Ammirato, P. V., 1983, in: *Handbook of Plant Cell Culture*, Volume 1, Macmillan Co., New York, pp. 82-123 (important variables in somatic embryo culture).
Atanassov, A., and Brown, D. C. W., 1984, *Plant Cell Tissue Org. Culture* **3**:149–162 (somatic embryo production from alfalfa).
Bhojwani, S. S., and Razdan, M. K., 1983, *Plant Tissue Culture: Theory and Practice*. Elsevier, Amsterdam (tissue culture techniques).
Gray, D. J., 1989, in: *Recent Advances in the Development and Germination of Seeds* (R. B. Taylorson, ed.), Plenum Press, New York, pp. 29–45 (overview of somatic embryogenesis).
Gray, D. J., and Purohit, A., 1991, *Crit. Rev. Plant Sci.* **10**:33–61 (practical aspects of somatic embryo production).
Liu, C., Xu, Z., and Chua, N.-H., 1993, *Plant Cell* **5**:621–630 (auxin polar transport and cotyledon formation).
Raghavan, V., 1986, *Embryogenesis in Angiosperms*, Cambridge University Press, Cambridge (zygotic and somatic embryogenesis).
Redenbaugh, K. (ed.), 1993, *Synseeds. Application of Synthetic Seeds to Crop Improvement*, CRC Press, Boca Raton, Fla (somatic embryos and synthetic seed technology).
Senaratna, T., McKersie, B. D., and Bowley, S. R., 1989, *Plant Sci.* **65**:253–258 (inducing desiccation-tolerance in somatic embryos).
Wurtele, E. S., Wang, H., Durgerian, S., Nikolau, B. J., and Ulrich, T. H., 1993, *Plant Physiol.* **102**:303-312 (genes expressed during early somatic embryogenesis).
Xu, N., and Bewley, J. D., 1992, *Plant Cell Rep.* **11**:279–284 (morphology of alfalfa somatic embryo development).

Index

The number in boldface is the first of several in which the subject receives frequent or extensive reference.